SEDR 300
VOLUME I

NASA
PROJECT GEMINI
familiarization manual

LONG RANGE and MODIFIED CONFIGURATIONS

MANNED SATELLITE SPACECRAFT

MCDONNELL

©2011 Periscope Film LLC All Rights Reserved ISBN #978-1-935700-69-2

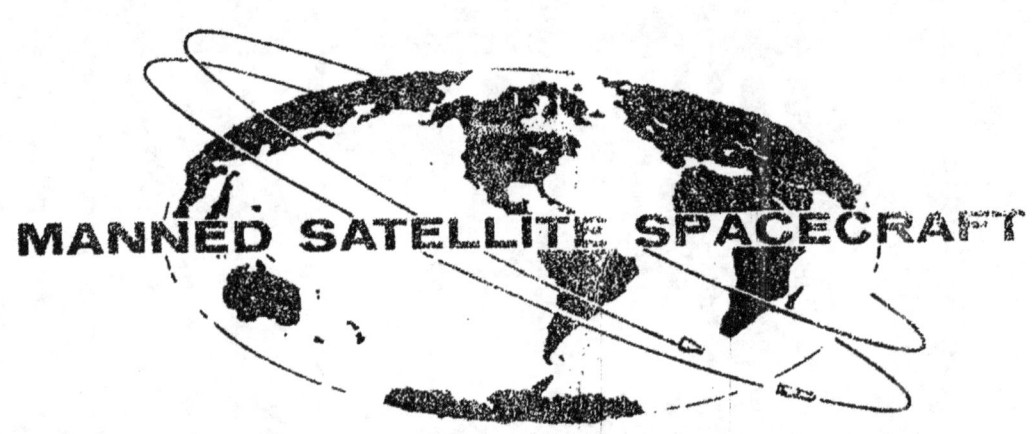

This text has been digitally watermarked to prevent illegal duplication.

NASA PROJECT GEMINI

familiarization manual

SEDR 300 **COPY NO. ________**

LONG RANGE and MODIFIED CONFIGURATIONS

THIS DOCUMENT SUPERSEDES DOCUMENT DATED 15 MARCH 1964 AND INCLUDES CHANGE DATED 31 DECEMBER 1964.

SECTION 8 IS CONTAINED IN A CONFIDENTIAL SUPPLEMENT TO THIS MANUAL

MCDONNELL

Hooven-Dayton Co., November 1965, 1000

30 SEPTEMBER 1965

SEDR 300

 **PROJECT GEMINI**

LIST OF EFFECTIVE PAGES

INSERT LATEST CHANGED PAGES. DESTROY SUPERSEDED PAGES.

NOTE: The portion of the text affected by the changes is indicated by a vertical line in the outer margins of the page.

TOTAL NUMBER OF PAGES IN THIS PUBLICATION IS 773, CONSISTING OF THE FOLLOWING:

Page No.	Issue
Title	Original
A thru D	Original
1-1 thru 1-5	Original
1-6 blank	Original
2-1 thru 2-25	Original
2-26 blank	Original
3-1 thru 3-31	Original
3-32 blank	Original
4-1 thru 4-65	Original
4-66 blank	Original
5-1 thru 5-35	Original
5-36 blank	Original
6-1 thru 6-54	Original
7-1 thru 7-22	Original
8-1	Original
8-2 blank	Original
8-3 thru 8-292	Original
9-1 thru 9-83	Original
9-84 blank	Original
10-1 thru 10-63	Original
10-64 blank	Original
11-1 thru 11-75	Original
11-76 blank	Original
12-1 thru 12-19	Original
12-20 blank	Original

*The asterisk indicates pages changed, added, or deleted by the current change.

A

 SEDR 300

PROJECT GEMINI

FOREWORD

The purpose of this manual is to present, clearly and concisely, the description and operation of the Gemini spacecraft systems and major components. The primary usages of the manual are as a familiarization-indoctrination aid, and as a ready reference for detailed information on a specific system or component.

The manual is sectionalized by spacecraft systems or major assemblies. Each section is as complete as is practical to minimize the necessity for cross referencing.

The information contained in this manual (SEDR 300, Vol. I) is applicable to Long Range and Modified missions only, and is accurate as of 30 September 1965. For information pertaining to Rendezvous Mission Spacecraft refer to SEDR 300, Vol. II.

PREPARED BY MAINTENANCE ENGINEERING PROJECT GEMINI

Reviewed by *R. E. Moon*
Maintenance Engineer

Reviewed by *J. H. Brick*
Supervisor - Maintenance Engineering

Reviewed by *C. Gray*
NASA Resident Manager

 SEDR 300
PROJECT GEMINI

INTRODUCTION

Initiated by the NASA and implemented by McDonnell Aircraft Corporation, Project Gemini is the next logical step in the field of manned space exploration. Closely allied to Project Mercury in concept, and utilizing the knowledge gained from the Mercury flights, Project Gemini will orbit a two-man spacecraft considerably more sophisticated than any employed so far.

The Gemini spacecraft is maneuverable within its orbit and will rendezvous with and connect to a second orbiting vehicle. Depending upon the specific mission objective, it can stay in orbit up to fourteen days. Finally, upon re-entry, the re-entry portion of the spacecraft can be controlled in a relatively conventional landing.

The modified and long range configurations of the spacecraft, however, with which this manual is specifically concerned, perform a variety of missions.

SEDR 300

 # PROJECT GEMINI

SECTION INDEX

SECTION I
    SPACECRAFT MISSION -- 1-1

SECTION II
    MAJOR STRUCTURAL ASSEMBLIES --- 2-1

SECTION III
    CABIN INTERIOR ARRANGEMENT --- 3-1

SECTION IV
    SEQUENCE SYSTEM -- 4-1

SECTION V
    ELECTRICAL POWER SYSTEM -- 5-1

SECTION VI
    ENVIRONMENTAL CONTROL SYSTEM --- 6-1

SECTION VI
    COOLING SYSTEM --- 7-1

SECTION VIII
    GUIDANCE AND CONTROL SYSTEMS --- 8-1

SECTION IX
    COMMUNICATIONS SYSTEM -- 9-1

SECTION X
    INSTRUMENTATION AND RECORDING SYSTEM --- 10-1

SECTION XI
    PYROTECHNICS AND RETROGRADE ROCKET --- 11-1

SECTION XII
    LANDING SYSTEM --- 12-1

D

SPACECRAFT MISSION

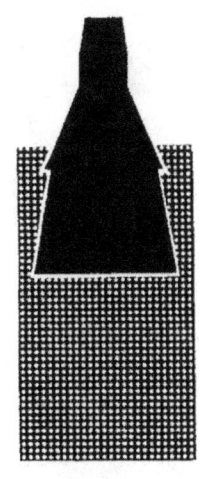

Section 1

TABLE OF CONTENTS

TITLE	PAGE
MISSION DESCRIPTION	1-3
MISSION OBJECTIVES	1-3
SPACECRAFT DESCRIPTION	1-4
LAUNCH VEHICLE DESCRIPTION	1-5
CREW REQUIREMENTS	1-5
SPACECRAFT RECOVERY	1-5

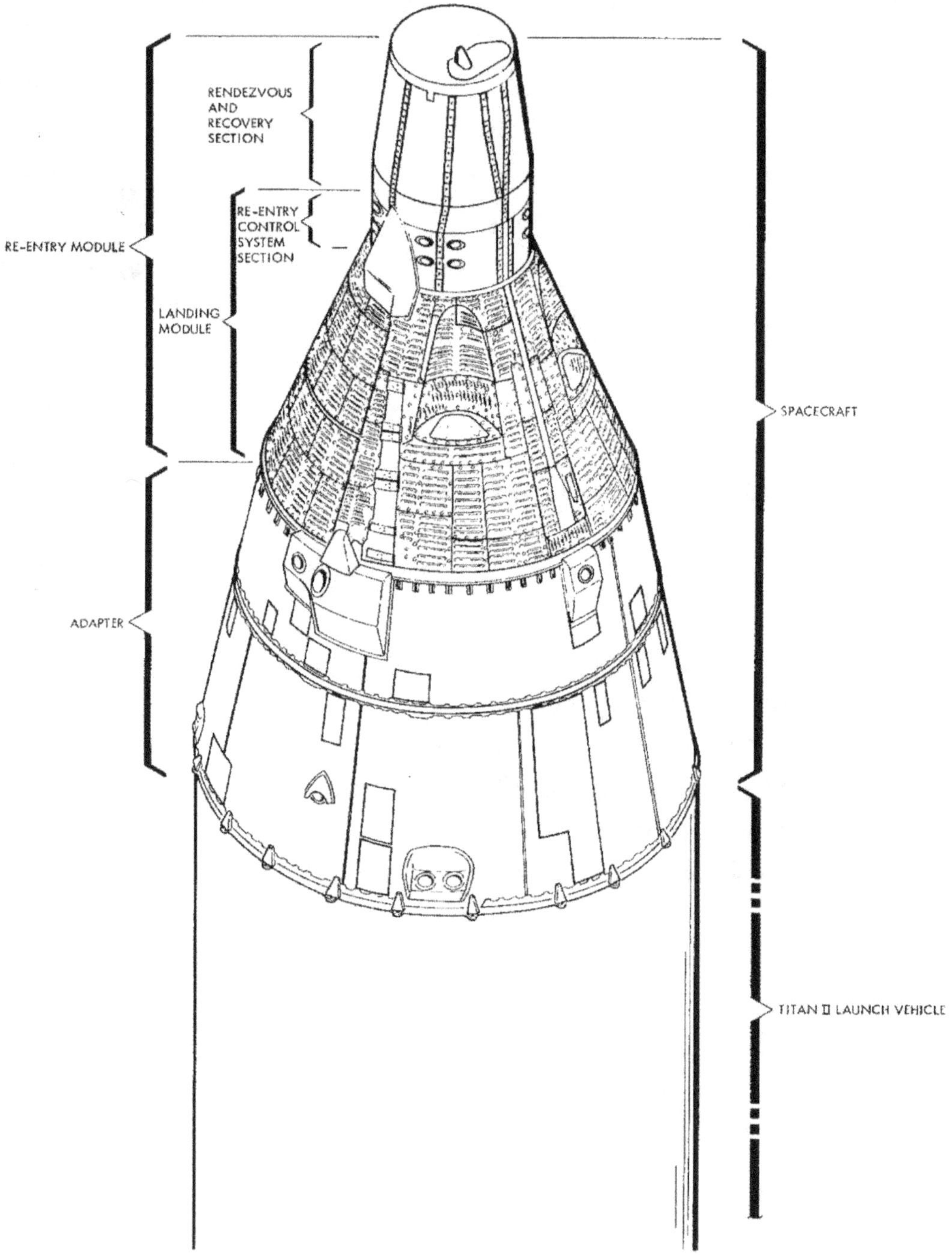

SEDR 300

PROJECT GEMINI

Figure 1-1 Spacecraft Pre-Launch Configuration

 **SEDR 300** **PROJECT GEMINI**

SECTION I SPACECRAFT MISSION

MISSION DESCRIPTION

Fundamentally, the mission of Project Gemini is the insertion of a two-man spacecraft into a semi-permanent orbit about the earth, the study of human capabilities during extended missions in space, and the subsequent safe return of the vehicle and its occupants to the earth's surface. Early missions included an unmanned orbital flight, an unmanned sub-orbital flight, a manned 3 orbit flight and a manned 62 orbit flight. Subsequent missions include rendezvousing and docking with an orbiting Agena spacecraft.

MISSION OBJECTIVES

Specifically, the project will seek to:

1. Demonstrate the ability of the pilots and spacecraft to perform in space in manual and/or automatic modes of operation.

2. Perform a simulated rendezvous for system qualification; assessment of the general problems to be encountered; and rendezvous and dock with an orbiting Agena.

3. Evaluate the adequacy of the spacecraft major systems, such as environmental control system, the electrical power system, communications system, etc.

4. Verify the functional relationships of the major systems and their integration into the spacecraft.

5. Determine man's requirements, necessities, and performance capabilities in a space environment for future extended missions.

6. Determine man's interface problems, and develop operational techniques for the most efficient use of on-board capabilities.

SEDR 300

PROJECT GEMINI

7. Develop controlled re-entry techniques required for landing in a predicted touchdown area.

8. Develop operational recovery techniques of both spacecraft and pilots.

SPACECRAFT DESCRIPTION

GENERAL

The Gemini Spacecraft (Figure 1-1) is a conical structure consisting basically of a re-entry module and an adapter.

RE-ENTRY MODULE

The re-entry module consists of the heat shield, the crew and equipment section, re-entry control section and the rendezvous and recovery section. The crew and equipment section contains a pressurized area suitable for human occupation, and a number of non-pressurized compartments for housing equipment. External access doors are provided for equipment compartments. The re-entry control section contains the major re-entry control system components. The rendezvous and recovery section contains the rendezvous radar equipment, the drogue and pilot chute assembly and the parachute assembly. The rendezvous and recovery section is jettisoned after re-entry along with the drogue chute.

ADAPTER

The adapter consists of the launch vehicle mating section, the equipment section and the retrograde section. The launch vehicle mating section is bolted to the launch vehicle. A portion of this section remains with the launch vehicle at spacecraft-launch vehicle separation. The equipment section contains major components of electrical power system, the maneuvering propulsion system, the equipment cooling system, and the primary oxygen supply for the environmental control system. The retrograde section contains the retrograde rockets and some components of the equipment cooling system.

LAUNCH VEHICLE DESCRIPTION

The vehicle used to launch the Gemini spacecraft is the Gemini - Titan II, built by the Martin Company, which is a Titan II modified structurally and functionally to accept the Gemini adapter and to provide for the interchange of electrical signals.

CREW REQUIREMENTS

The Gemini spacecraft utilizes a two-man crew seated side by side. The man on the left is referred to as the "Command Pilot" and functions as spacecraft commander. The man on the right is referred to as the "Pilot." Crew members are selected from the NASA astronaut group.

SPACECRAFT RECOVERY

The Gemini landing module will make a water landing in a pre-determined area. A task force of ships, planes, and personnel will be standing by for locating and retrieving the spacecraft and crew. In the event an abort or other abnormal occurrence results in the spacecraft landing in a remote location, electronic and visual recovery aids and survival kits are provided in the spacecraft to facilitate spacecraft retrieval and crew survival, respectively.

MAJOR STRUCTURAL ASSEMBLIES

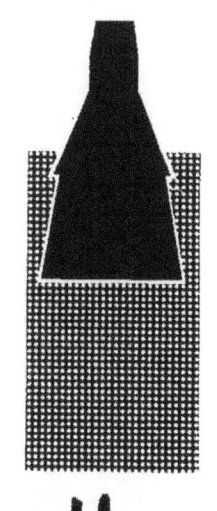

Section II

TABLE OF CONTENTS

TITLE	PAGE
GENERAL INFORMATION	2-3
RE-ENTRY MODULE	2-3
RENDEZVOUS AND RECOVERY SECTION	2-3
RE-ENTRY CONTROL SYSTEM SECTION	2-8
CABIN	2-8
ADAPTER	2-21
RETROGRADE SECTION	2-23
ADAPTER EQUIPMENT SECTION	2-23

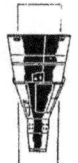

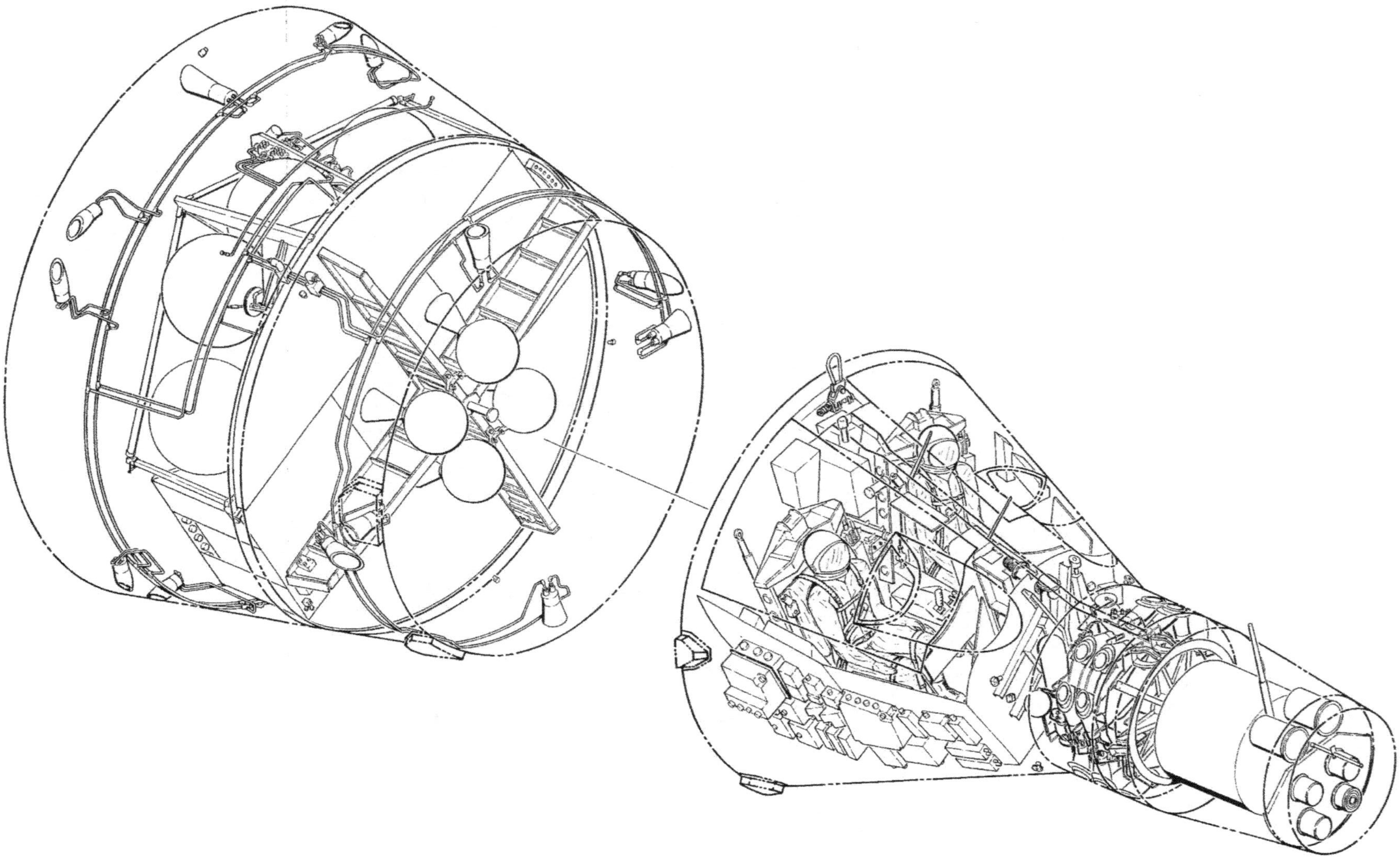

SEDR 300

PROJECT GEMINI

Figure 2-1 Interior Arrangement (Typical)

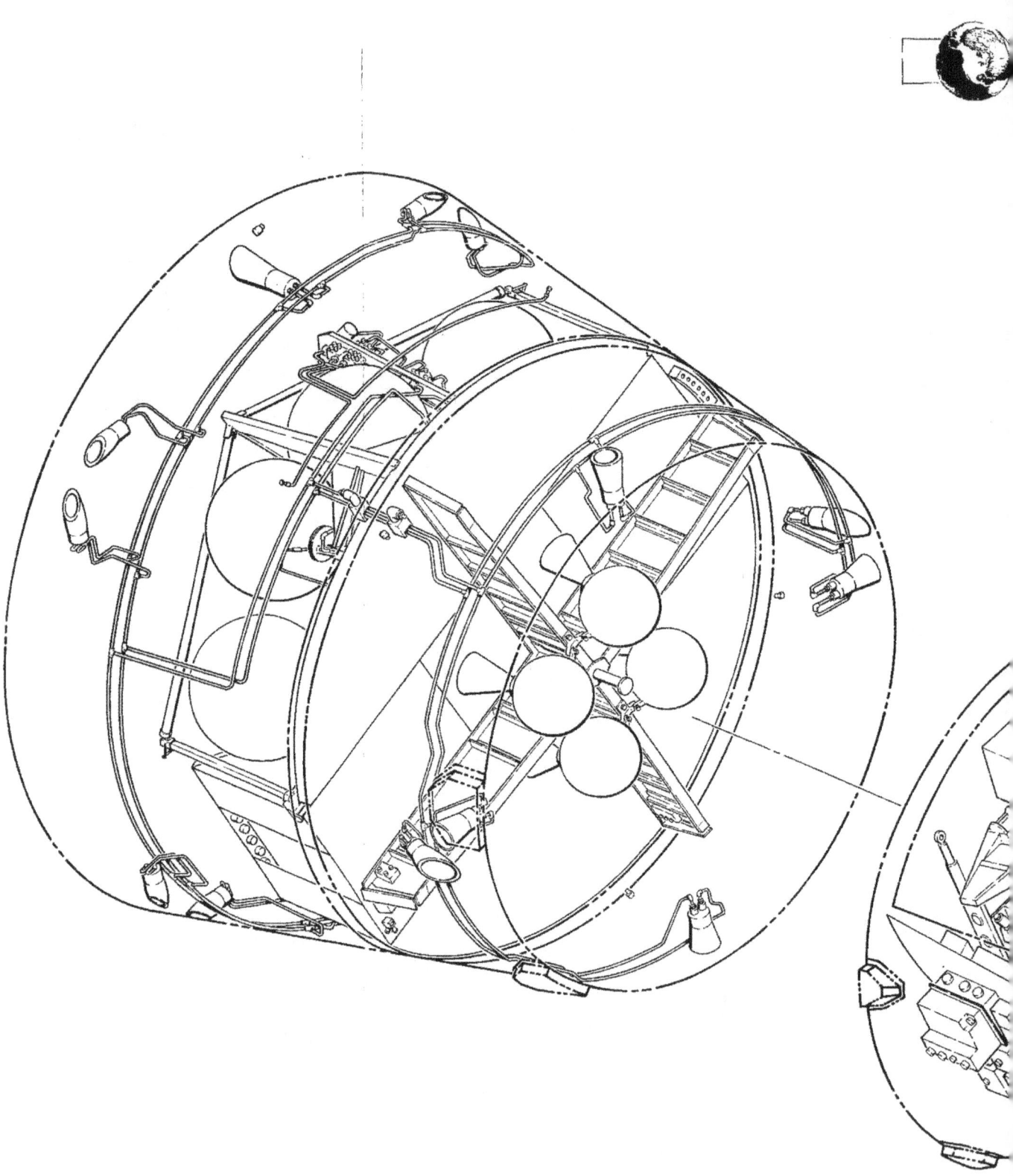

 SEDR 300
PROJECT GEMINI

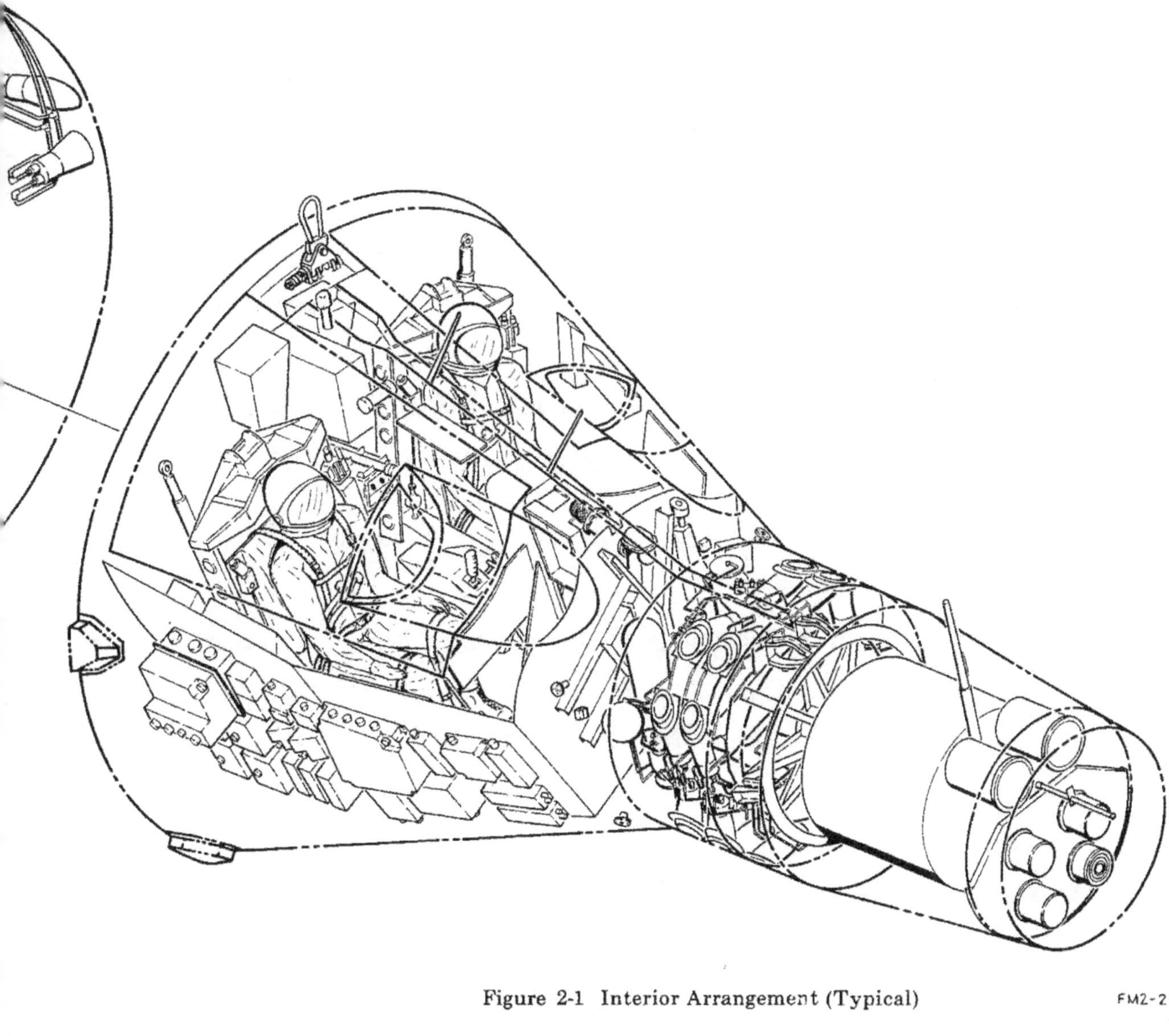

Figure 2-1 Interior Arrangement (Typical)

 SEDR 300
PROJECT GEMINI
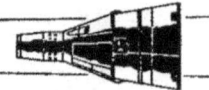

SECTION II MAJOR STRUCTURAL ASSEMBLIES

GENERAL INFORMATION

The Gemini Spacecraft is basically of a conical configuration (See Figure 2-1) consisting of a re-entry module and an adapter as the two major assemblies. Spacecraft construction is semimonocoque utilizing titanium for the primary structure. It is designed to shield the cabin pressure vessel from excessive temperature variations, noise and meteorite penetration (See Figure 2-2). During spacecraft flight, the spacecraft adapter is aft with respect to flight path. See Figures 2-3 and 2-4 for spacecraft orientation.

RE-ENTRY MODULE

The re-entry module (Figure 2-5) is separated into three primary sections which include the rendezvous and recovery section (R & R), re-entry control system section (RCS) and the cabin section. Also incorporated in the re-entry module is the heat shield which is attached to the cabin, and a nose fairing which is attached to the forward end of the rendezvous and recovery section. The nose fairing is ejected during launch.

RENDEZVOUS AND RECOVERY SECTION

The rendezvous and recovery section (R & R) (See Figure 2-5), the forward section of the spacecraft, is semi-conical in shape and is attached to the re-entry control system section with twenty-four bolts. Incorporated in this joint is a pyrotechnic device which severs all bolts causing the rendezvous section to separate from the re-entry control system section on signal for parachute deployment. A drogue parachute will assist in the removal of this section. The R & R

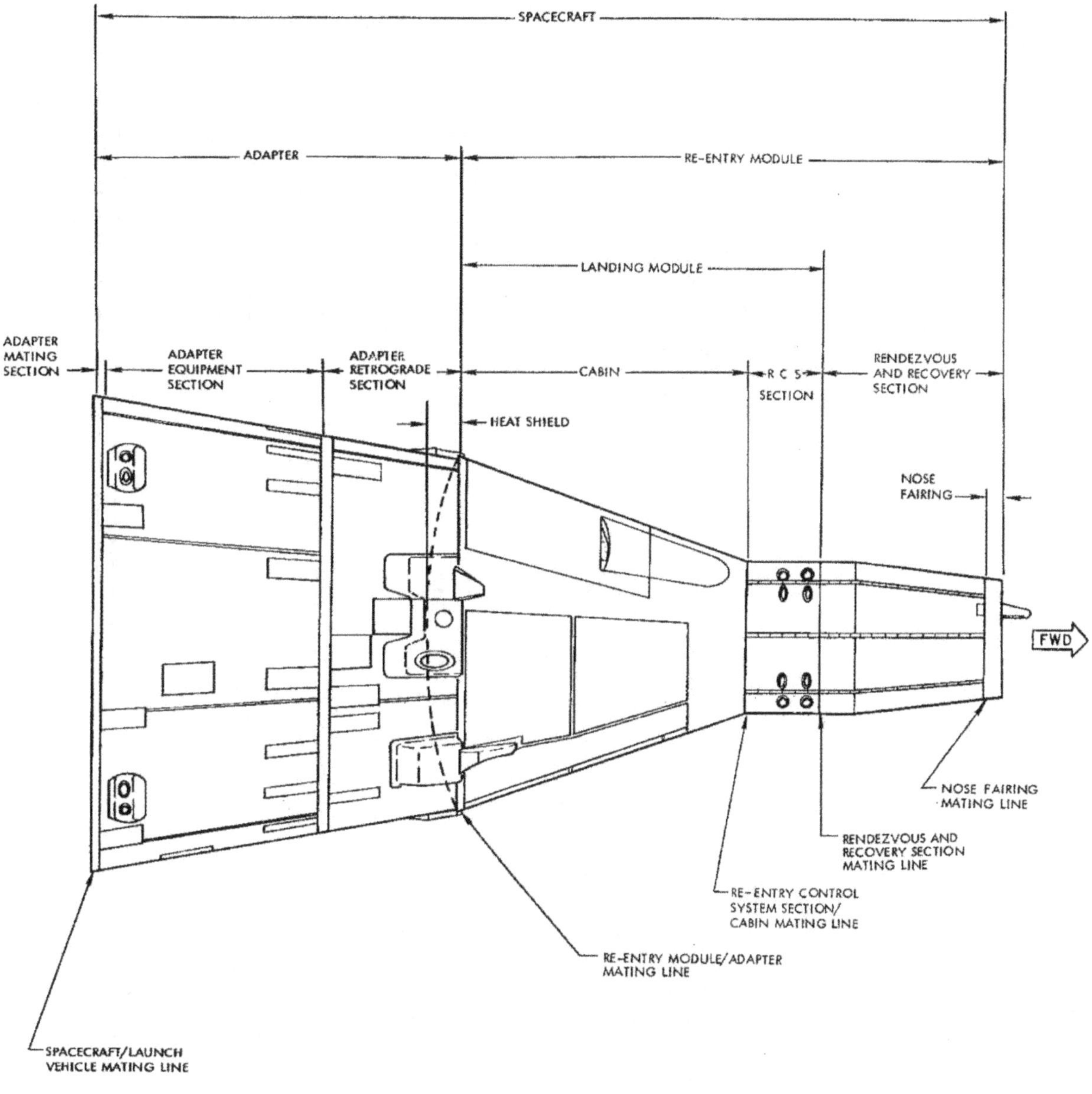

SEDR 300

PROJECT GEMINI

Figure 2-2 Spacecraft General Nomenclature

SEDR 300

PROJECT GEMINI

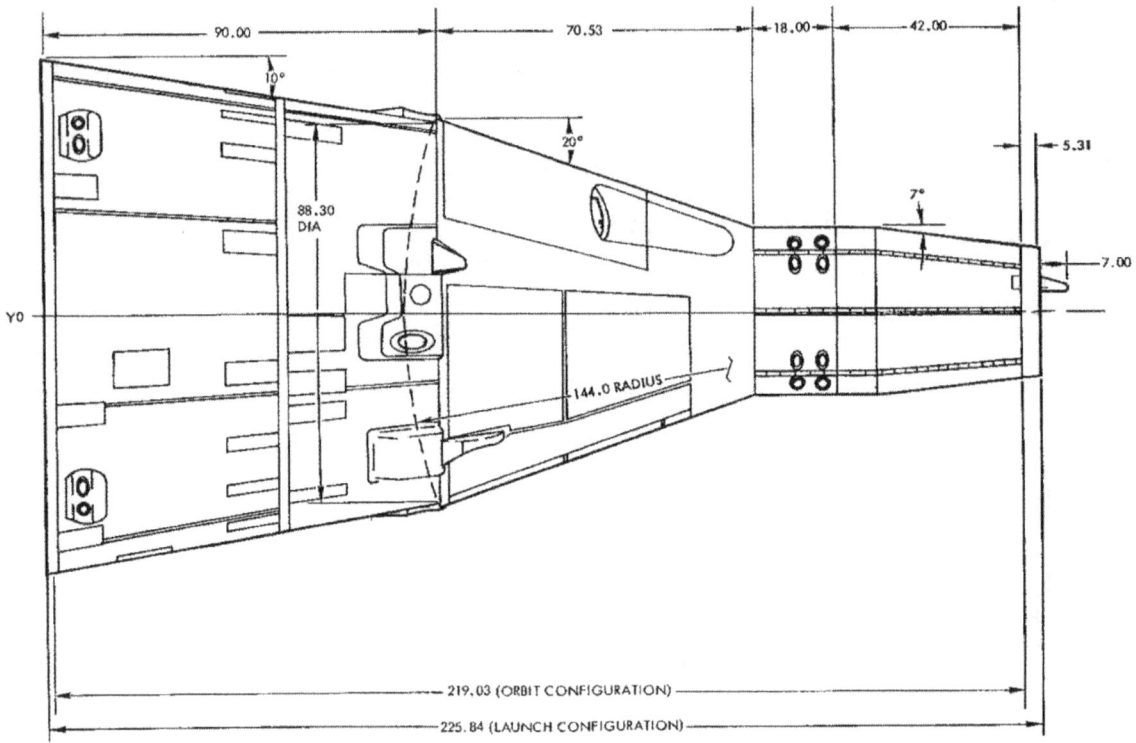

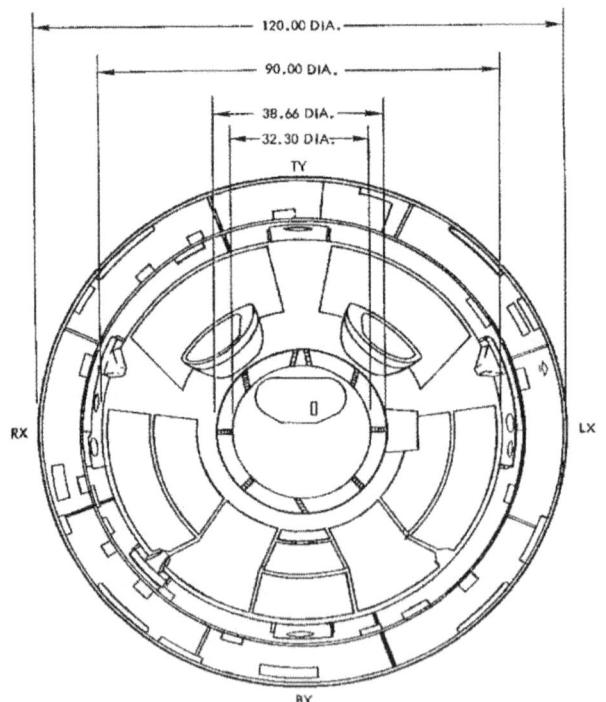

Figure 2-3 Spacecraft Dimensions

 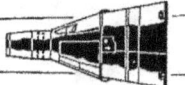

SEDR 300
PROJECT GEMINI

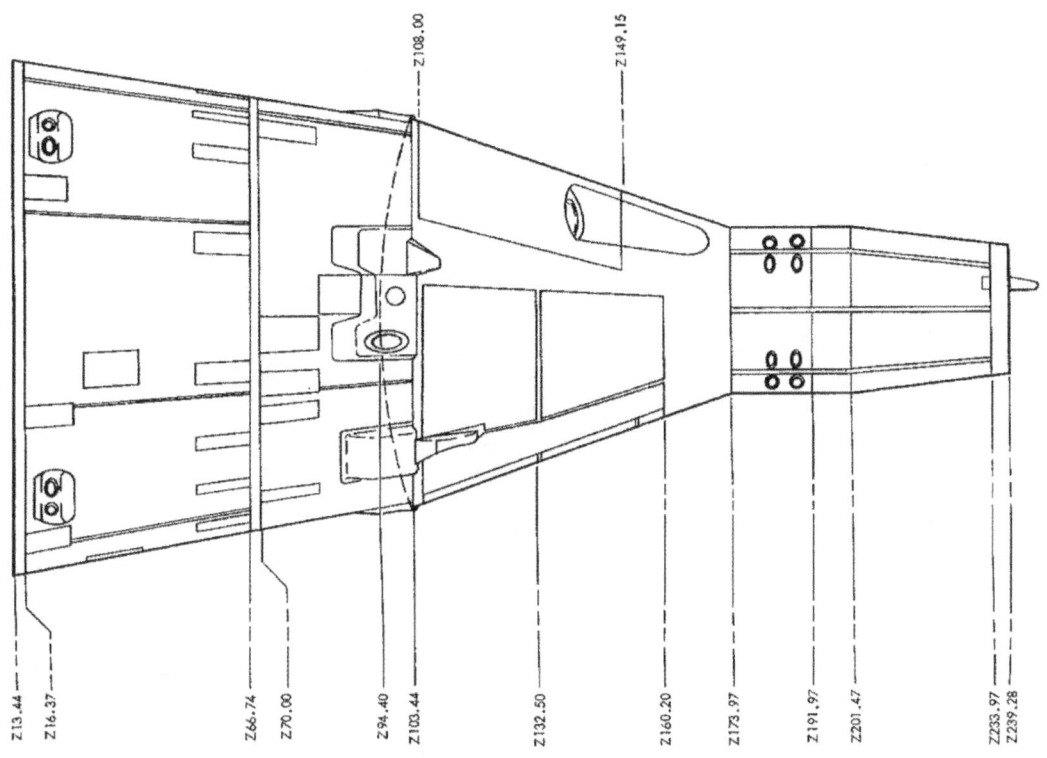

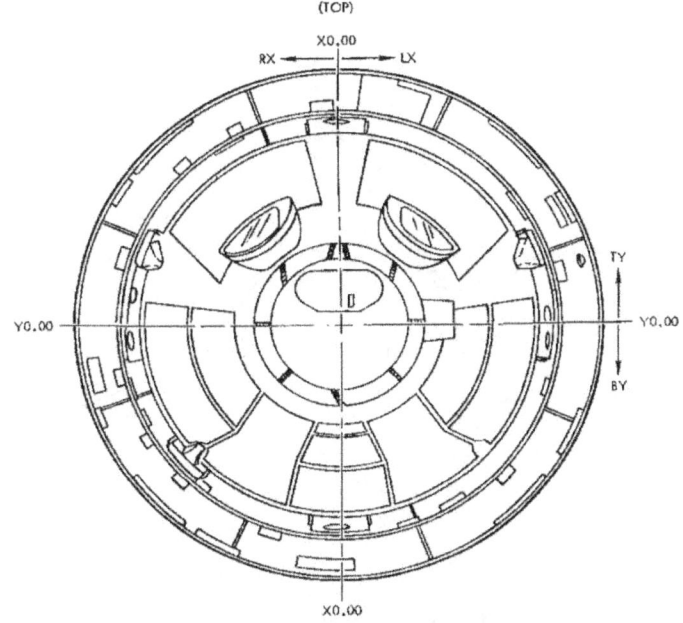

Figure 2-4 Stations Diagram

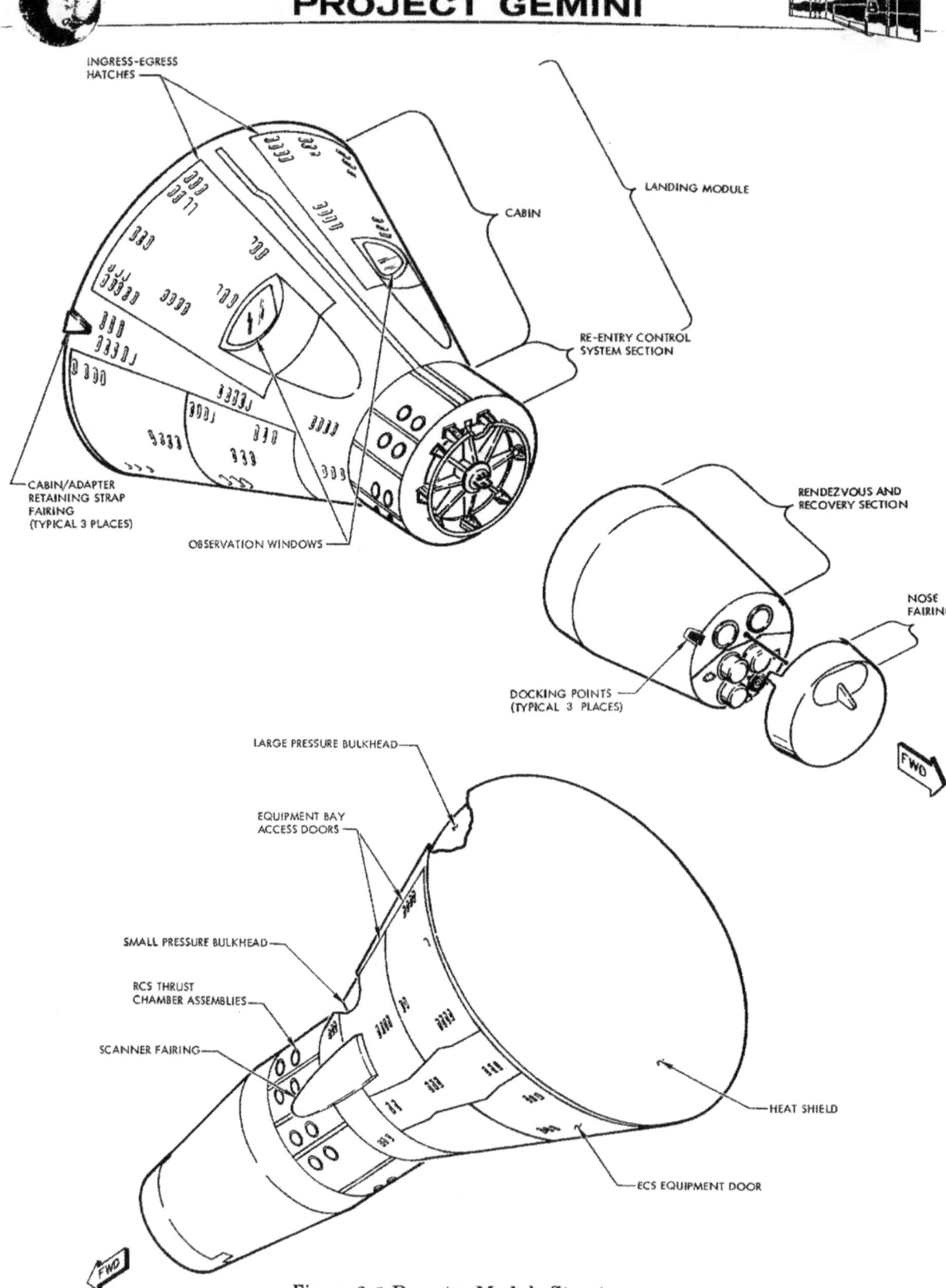

Figure 2-5 Re-entry Module Structure

section utilizes rings, stringers and bulkheads of titanium for its primary structure. The external surface is composed of beryllium shingles, except for the nose fairing. The nose fairing is composed of fiberglass reinforced plastic laminate.

RE-ENTRY CONTROL SYSTEM SECTION

The re-entry control system (RCS) section is located between, and mated to, the rendezvous and recovery and cabin sections of the spacecraft (See Figure 2-5). This section is cylindrical in shape and is constructed of an inner titanium alloy cylinder, eight stringers, two rings and eight beryllium shingles for its outer skin. The RCS section is designed to house the fuel and oxidizer tanks, valves, tube assemblies, and thrust chamber assemblies (TCA) for the re-entry control system.

A parachute adapter assembly is installed on the forward face of the RCS section for attachment of the main parachute.

CABIN

The cabin (Figure 2-5), similar in shape to a truncated cone, is mated to the re-entry control system section and the adapter. The cabin has an internal pressure vessel (Figure 2-6) shaped to provide an adequate crew station with a proper water flotation attitude. The shape of the pressure vessel also allows space between it and the outer conical shell for the installation of equipment.

The basic cabin structure consists of a fusion welded titanium frame assembly to which the side panels, small and large pressure bulkheads and hatch sill are seam welded. The side panels, small and large pressure bulkheads are of double skin construction and reinforced by stiffeners spotwelded in place. Two hatches

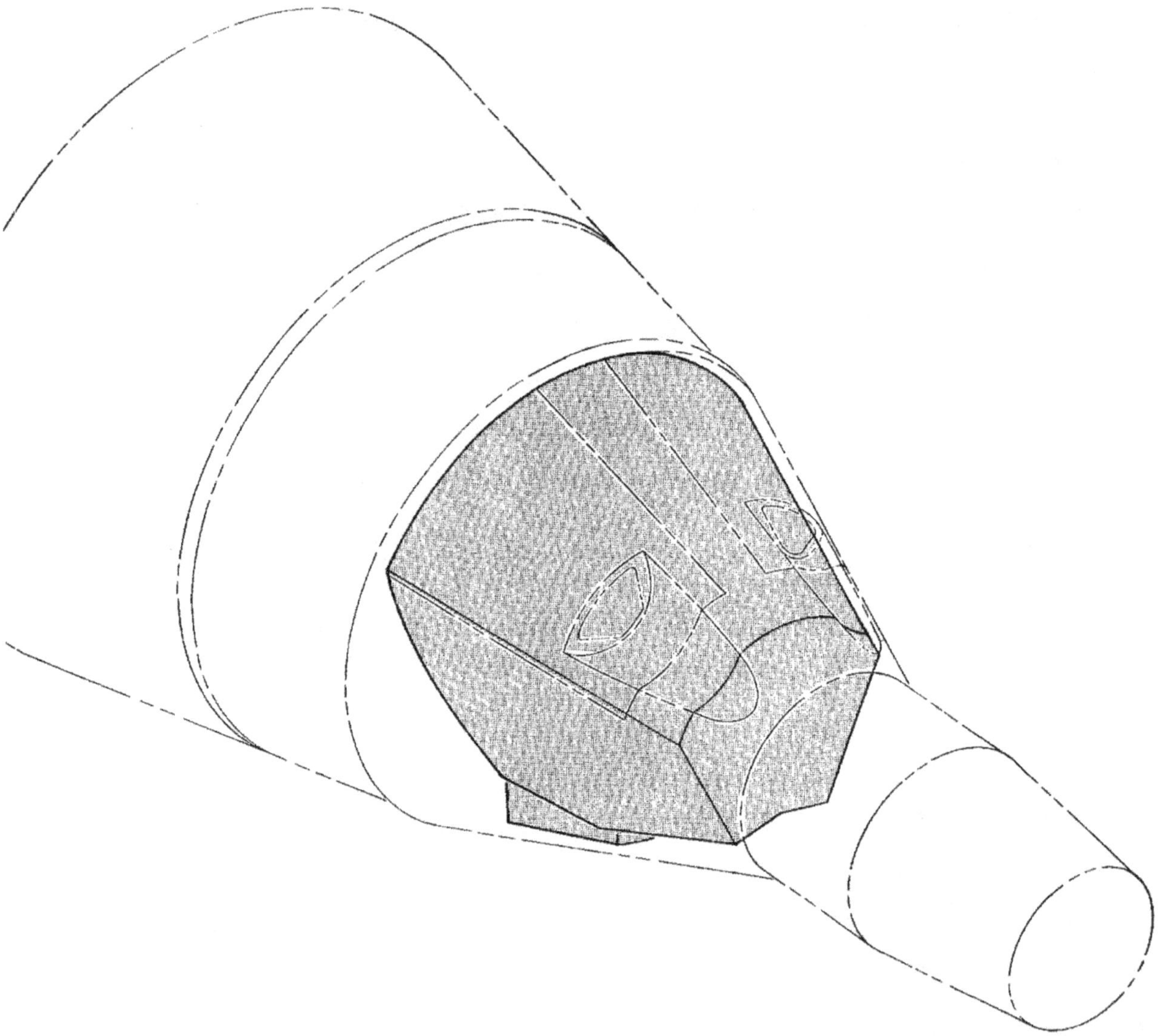

SEDR 300

 **PROJECT GEMINI**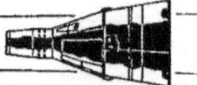

Figure 2-6 Cabin Pressure Vessel

are hinged to the hatch sill for pilot ingress and egress. For heat protection, the outer conical surface is covered with Rene' 41 shingles and an ablative heat shield is attached to the large end of the cabin section.

A spring loaded hoist loop, located near the heat shield between the hatch openings, is errected after landing to facilitate engagement of a hoisting hook for spacecraft retrieval.

Equipment Bays

The equipment bays are located outside the cabin pressure vessel (Figure 2-7). Two bays are located outboard of the side panels and one bay beneath the pressure vessel floor. The bays are structurally designed for mounting of the equipment.

Doors

To enclose the side equipment bays, two structural doors are provided on each side of the cabin (Figure 2-7). These doors provide access to the components installed in the equipment bays. The main landing gear bays, located below the left and right equipment bays, are each enclosed by one door. The landing gears are not installed but fittings are provided for the attachment of the gears for future spacecraft. On the bottom of the cabin, between the landing gear doors, two additional doors are installed. The forward door allows access to the lower equipment compartment and the aft door provides access to the ECS compartment which is a portion of the pressure vessel.

Hatches

Two large structural hatches (Figure 2-8) are incorporated for sealing the cabin ingress or egress openings. The hatches are symmetrically spaced on the top

SEDR 300

PROJECT GEMINI

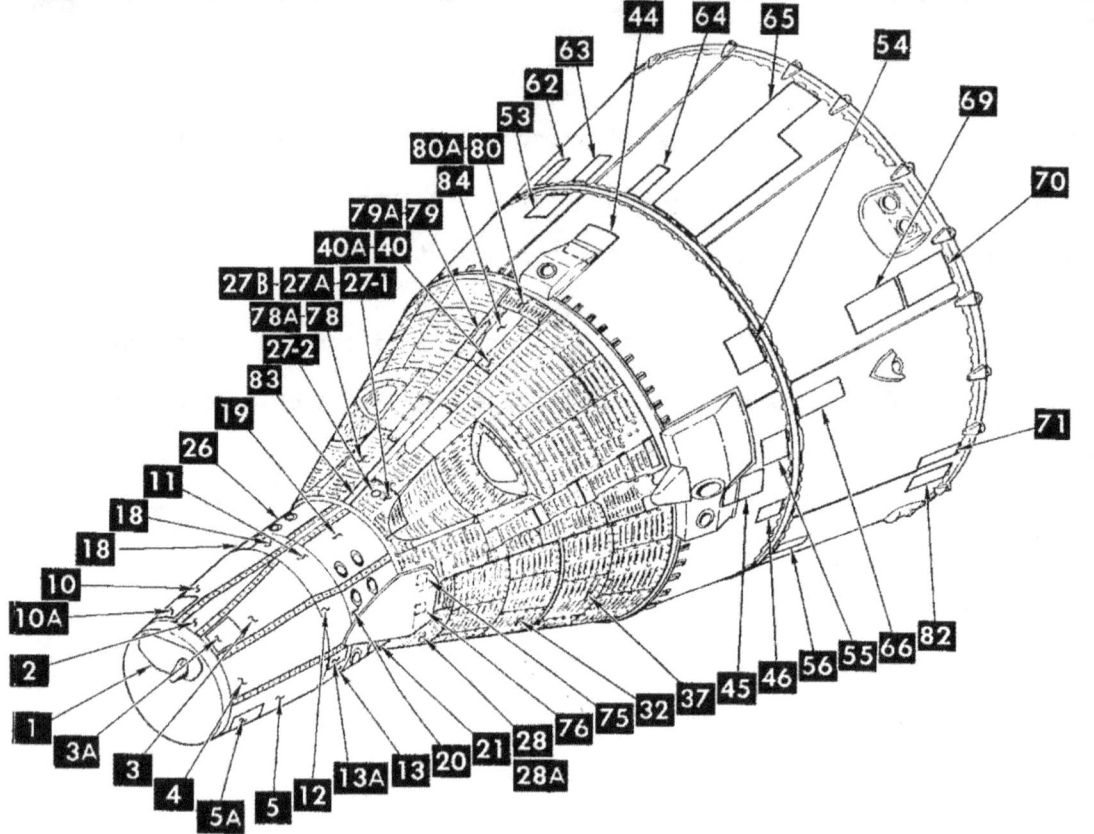

LEGEND					
NO.	DESCRIPTION	NO.	DESCRIPTION	NO.	DESCRIPTION
1	DROGUE CHUTE DOOR	26	RE-ENTRY CONTROL SYSTEM ACCESS	64	OAMS OXIDIZER PURGE ACCESS
2	DOCKING BAR CARTRIDGE ACCESS	27-1	SHINGLE 27-2 FRESH AIR DOOR	65	OAMS MODULE SERVICE ACCESS
3	SHINGLE	27A 27B Z160.20 EQUIPMENT ACCESS		66	ECS SERVICE ACCESS
3A	DROGUE MORTAR CARTRIDGE ACCESS	28	SHINGLE	69	ECS PUMP MODULE SERVICE ACCESS
4	EMERGENCY DOCKING RELEASE CARTRIDGE AND GUILLOTINE CARTRIDGE ACCESS	28A	Z160.20 EQUIPMENT ACCESS	70	ECS PUMP MODULE SERVICE ACCESS
5	SHINGLE	32	FORWARD EQUIPMENT BAY DOOR-LEFT	71	SEPERATION SENSING SWITCH ACCESS
5A	RADAR ACCESS	37	AFT EQUIPMENT BAY DOOR-LEFT	75	ELECTRICAL DISCONNECT ACCESS
10	SHINGLE	40	SHINGLE	76	ELECTRICAL DISCONNECT ACCESS
10A	NOSE FAIRING RELEASE CARTRIDGE ACCESS	40A	RECOVERY LIGHT AND HOIST LOOP RIGGING AND CARTRIDGE ACCESS	78	SHINGLE
11	INTERFACE ACCESS	44	VERTICAL MANEUVERING ENGINE ACCESS	78A	Z160.20 EQUIPMENT ACCESS
12	INTERFACE ACCESS	45	LATERAL MANEUVERING ENGINE ACCESS	79	RECOVERY LIGHT DOOR
13	INTERFACE ACCESS	46	SEPARATION SENSING SWITCH ACCESS	79	RECOVERY LIGHT DOOR RELEASE MECHANISM
13A	GUILLOTINE CARTRIDGE ACCESS	53	OAMS LINE GUILLOTINE ACCESS	80	HOIST LOOP DOOR
18	INTERFACE ACCESS	54	F.L.S.C. TUBING CUTTER ACCESS	80A	HOIST LOOP DOOR RELEASE MECHANISM
18A	PYROTECHNIC SWITCH CARTRIDGE AND BRIDLE DISCONNECT CARTRIDGE ACCESS	55	FORWARD MANEUVERING ENGINE ACCESS	82	ACCESS
19	RE-ENTRY CONTROL SYSTEM ACCESS	56	FUEL CELL SERVICE ACCESS	83	COVER ASS'Y.-PARAGLIDER OR PARACHUTE CONTROL CABLES.
20	RE-ENTRY CONTROL SYSTEM ACCESS	62	OAMS OXIDIZER PURGE ACCESS	84	COVER ASS'Y.-PARAGLIDER OR PARACHUTE CONTROL CABLES.
21	RE-ENTRY CONTROL SYSTEM ACCESS	63	OAMS LINE GUILLOTINE ACCESS		

Figure 2-7 Access Doors (Sheet 1 of 6) (S/C 3)

2-11

SEDR 300

PROJECT GEMINI

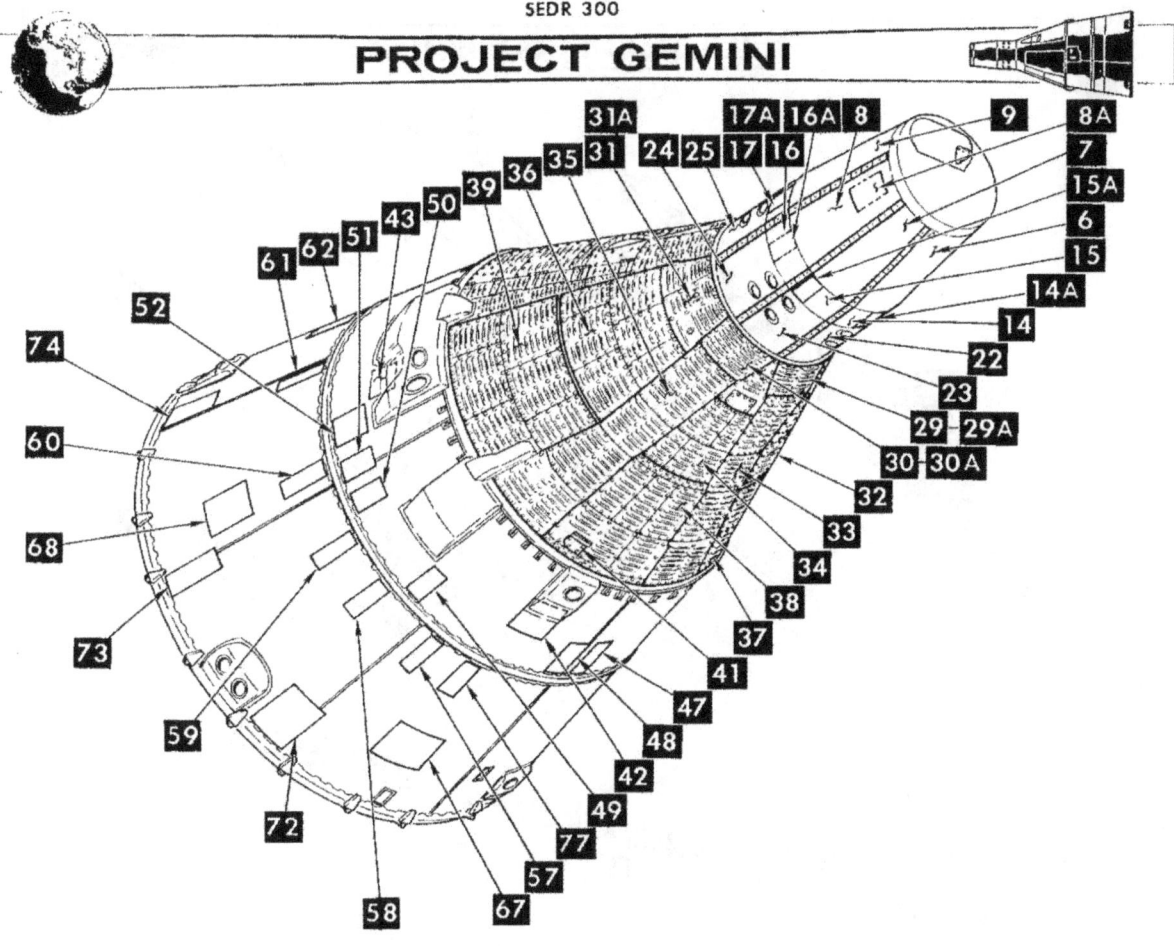

LEGEND					
NO.	DESCRIPTION	NO.	DESCRIPTION	NO.	DESCRIPTION
6	EMERGENCY DOCKING RELEASE CARTRIDGE AND GUILLOTINE CARTRIDGE ACCESS	29	SHINGLE	47	RELAY PANEL ACCESS
7	SHINGLE	29A	Z160.20 EQUIPMENT ACCESS	48	RELAY PANEL ACCESS
8	SHINGLE	30	SHINGLE	49	SEPARATION SENSING SWITCH ACCESS
8A	RADAR ACCESS	30A	Z160.20 EQUIPMENT ACCESS	50	GUILLOTINE CARTRIDGE ACCESS
9	EMERGENCY DOCKING RELEASE CARTRIDGE AND GUILLOTINE CARTRIDGE ACCESS	31	SHINGLE	51	B.I.A. RELAY PANEL ACCESS
14	INTERFACE ACCESS	31A	Z160.20 EQUIPMENT ACCESS	52	FORWARD MANEUVERING ENGINE ACCESS
14A	GUILLOTINE ANVIL ACCESS	32	FORWARD EQUIPMENT BAY DOOR - LEFT	57	SHAPED CHARGE DETONATOR ACCESS
15	INTERFACE ACCESS	33	MAIN LANDING GEAR DOOR - LEFT	58	FUEL CELL SERVICE ACCESS
15A	GUILLOTINE ANVIL ACCESS	34	CENTER EQUIPMENT BAY DOOR - FORWARD	59	GUILLOTINE CARTRIDGE ACCESS
16	INTERFACE ACCESS	35	MAIN LANDING GEAR DOOR - RIGHT	60	GUILLOTINE CARTRIDGE ACCESS
16A	GUILLOTINE CARTRIDGE ACCESS	36	FORWARD EQUIPMENT BAY DOOR - RIGHT	61	OAMS FUEL PURGE ACCESS
17	INTERFACE ACCESS	37	AFT. EQUIPMENT BAY DOOR - LEFT	67	ENGINE TO SCUPPER INTERFACE ACCESS
17A	PARAGLIDER ELECT. CONTROL BOX ACCESS	38	E.C.S. BAY DOOR	68	ELECTRONIC MODULE TEST ACCESS
22	RE-ENTRY CONTROL SYSTEM ACCESS	39	AFT. EQUIPMENT BAY DOOR - RIGHT	72	GUILLOTINE CARTRIDGE AND LAUNCH VEHICLE ELEC. CONN. ACCESS
23	RE-ENTRY CONTROL SYSTEM ACCESS	41	PURGE FITTING ACCESS	73	SEPARATION SENSING SWITCH ACCESS
24	RE-ENTRY CONTROL SYSTEM ACCESS	42	VERTICAL MANEUVERING ENGINE ACCESS	74	SHAPED CHARGE DETONATOR ACCESS
25	RE-ENTRY CONTROL SYSTEM ACCESS	43	LATERAL MANEUVERING ENGINE ACCESS	77	FUEL CELL PURGE ACCESS

Figure 2-7 Access Doors (Sheet 2 of 6) (S/C 3)

 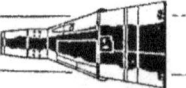

PROJECT GEMINI

SEDR 300

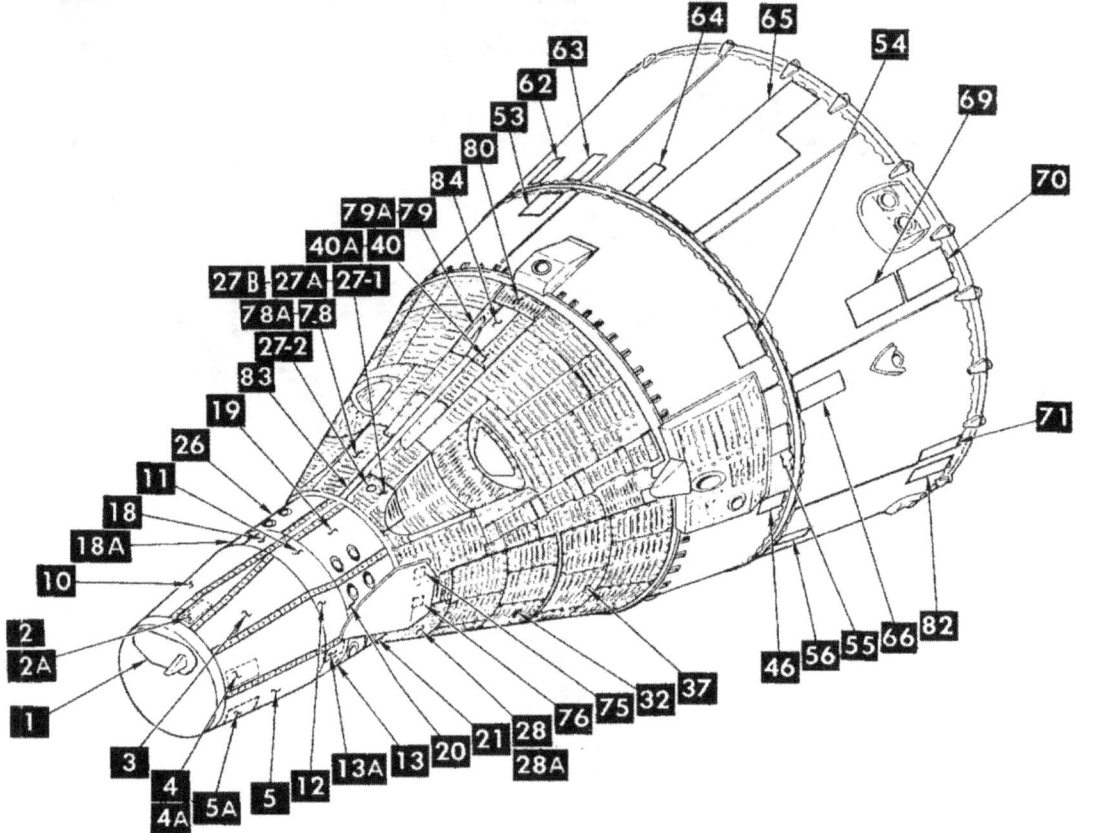

LEGEND					
NO.	DESCRIPTION	NO.	DESCRIPTION	NO.	DESCRIPTION
1	DROGUE CHUTE DOOR	26	RE-ENTRY CONTROL SYSTEM ACCESS	64	OAMS OXIDIZER PURGE ACCESS
2	DOCKING BAR CARTRIDGE ACCESS	27-1	SHINGLE	65	OAMS MODULE SERVICE ACCESS
2A	PYRO ELECTRICAL DISCONNECT ACCESS	27-2	FRESH AIR DOOR	66	ECS SERVICE ACCESS
3	SHINGLE	27A	Z160.20 EQUIPMENT ACCESS	69	ECS PUMP MODULE SERVICE ACCESS
4	EMERGENCY DOCKING RELEASE CARTRIDGE AND GUILLOTINE CARTRIDGE ACCESS	27B	Z160.20 EQUIPMENT ACCESS	70	ECS PUMP MODULE SERVICE ACCESS
4A	PILOT CHUTE DEPLOY SENSOR SWITCH ACCESS	28	SHINGLE	71	SEPERATION SENSING SWITCH ACCESS
5	SHINGLE	28A	Z160.20 EQUIPMENT ACCESS	75	ELECTRICAL DISCONNECT ACCESS
5A	RADAR ACCESS	32	FORWARD EQUIPMENT BAY DOOR-LEFT	76	ELECTRICAL DISCONNECT ACCESS
10	SHINGLE	37	AFT EQUIPMENT BAY DOOR-LEFT	78	SHINGLE
11	INTERFACE ACCESS	40	SHINGLE	78A	Z160.20 EQUIPMENT ACCESS
12	INTERFACE ACCESS	40A	RECOVERY LIGHT AND HOIST LOOP RIGGING AND CARTRIDGE ACCESS	79	RECOVERY LIGHT DOOR
13	INTERFACE ACCESS	46	SEPARATION SENSING SWITCH ACCESS	79A	RECOVERY LIGHT DOOR RELEASE MECHANISM
13A	GUILLOTINE CARTRIDGE ACCESS	53	OAMS LINE GUILLOTINE ACCESS	80	HOIST LOOP DOOR
18	INTERFACE ACCESS	54	F.L.S.C. TUBING CUTTER ACCESS	82	SHAPED CHARGE DETONATOR ACCESS
18A	PYROTECHNIC SWITCH CARTRIDGE AND BRIDLE DISCONNECT CARTRIDGE ACCESS	55	FORWARD MANEUVERING ENGINE ACCESS	83	COVER ASS'Y.-PARAGLIDER OR PARACHUTE CONTROL CABLES.
19	RE-ENTRY CONTROL SYSTEM ACCESS	56	FUEL CELL SERVICE ACCESS	84	COVER ASS'Y.-PARAGLIDER OR PARACHUTE CONTROL CABLES.
20	RE-ENTRY CONTROL SYSTEM ACCESS	62	OAMS OXIDIZER PURGE ACCESS		
21	RE-ENTRY CONTROL SYSTEM ACCESS	63	OAMS LINE GUILLOTINE ACCESS		

Figure 2-7 Access Doors (Sheet 3 of 6) (S/C 4)

SEDR 300

PROJECT GEMINI

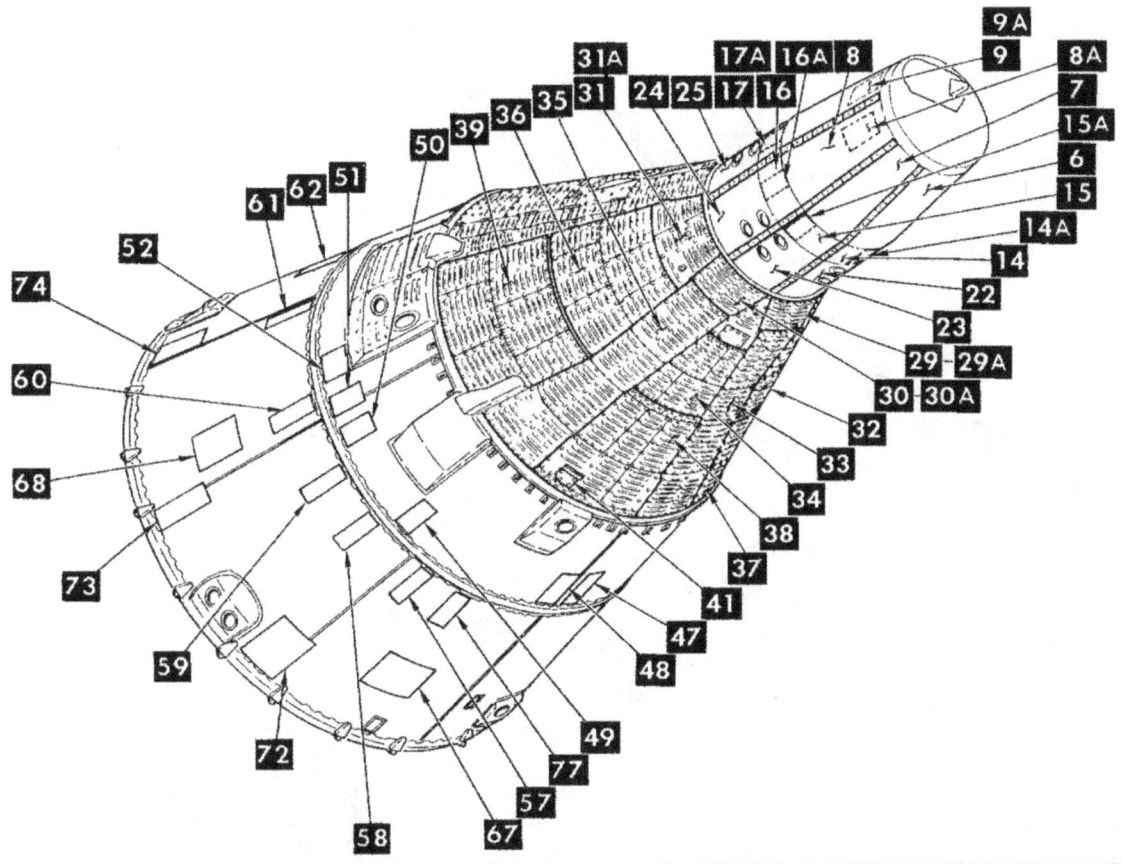

LEGEND					
NO.	DESCRIPTION	NO.	DESCRIPTION	NO.	DESCRIPTION
6	EMERGENCY DOCKING RELEASE CARTRIDGE AND GUILLOTINE CARTRIDGE ACCESS	25	RE-ENTRY CONTROL SYSTEM ACCESS	48	RELAY PANEL ACCESS
7	SHINGLE	29	SHINGLE	49	SEPARATION SENSING SWITCH ACCESS
8	SHINGLE	29A	Z160.20 EQUIPMENT ACCESS	50	GUILLOTINE CARTRIDGE ACCESS
8A	RADAR ACCESS	30	SHINGLE	51	B.I.A. RELAY PANEL ACCESS
9	EMERGENCY DOCKING RELEASE CARTRIDGE AND GUILLOTINE CARTRIDGE ACCESS	30A	Z160.20 EQUIPMENT ACCESS	52	FORWARD MANEUVERING ENGINE ACCESS
9A	DROGUE CHUTE DEPLOY SENSOR SWITCH ACCESS	31	SHINGLE	57	SHAPED CHARGE DETONATOR ACCESS
14	INTERFACE ACCESS	31A	Z160.20 EQUIPMENT ACCESS	58	FUEL CELL SERVICE ACCESS
14A	GUILLOTINE ANVIL ACCESS	32	FORWARD EQUIPMENT BAY DOOR - LEFT	59	GUILLOTINE CARTRIDGE ACCESS
15	INTERFACE ACCESS	33	MAIN LANDING GEAR DOOR - LEFT	60	GUILLOTINE CARTRIDGE ACCESS
15A	GUILLOTINE ANVIL ACCESS	34	CENTER EQUIPMENT BAY DOOR - FORWARD	61	OAMS FUEL PURGE ACCESS
16	INTERFACE ACCESS	35	MAIN LANDING GEAR DOOR - RIGHT	67	ENGINE TO SCUPPER INTERFACE ACCESS
16A	GUILLOTINE CARTRIDGE ACCESS	36	FORWARD EQUIPMENT BAY DOOR - RIGHT	68	ELECTRONIC MODULE TEST ACCESS
17	INTERFACE ACCESS	37	AFT. EQUIPMENT BAY DOOR - LEFT	72	GUILLOTINE CARTRIDGE AND LAUNCH VEHICLE ELEC. CONN. ACCESS
17A	PARAGLIDER ELECT. CONTROL BOX ACCESS	38	E.C.S. BAY DOOR	73	SEPARATION SENSING SWITCH ACCESS
22	RE-ENTRY CONTROL SYSTEM ACCESS	39	AFT. EQUIPMENT BAY DOOR - RIGHT	74	SHAPED CHARGE DETONATOR ACCESS
23	RE-ENTRY CONTROL SYSTEM ACCESS	41	PURGE FITTING ACCESS	77	FUEL CELL PURGE ACCESS
24	RE-ENTRY CONTROL SYSTEM ACCESS	47	RELAY PANEL ACCESS		

Figure 2-7 Access Doors (Sheet 4 of 6) (S/C 4)

SEDR 300
PROJECT GEMINI
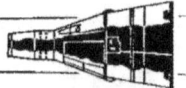

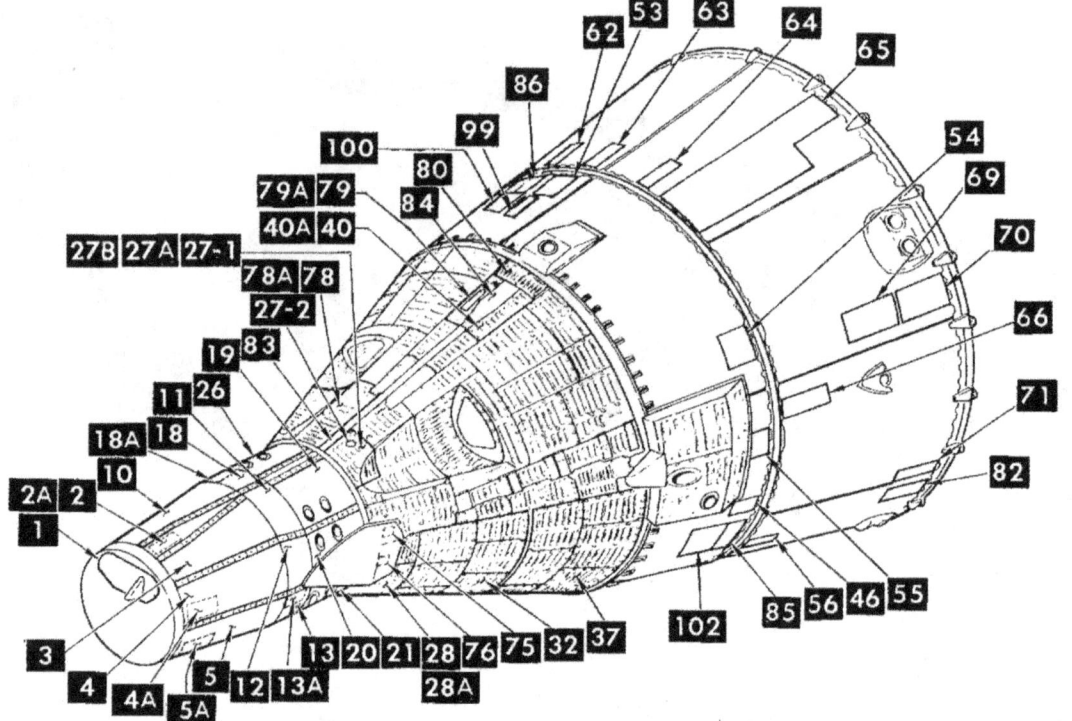

NO.	DESCRIPTION	NO.	DESCRIPTION	NO.	DESCRIPTION
1	DROGUE CHUTE DOOR	28	SHINGLE	79	RECOVERY LIGHT DOOR
2	DOCKING BAR CARTRIDGE ACCESS	28A	Z160.20 EQUIPMENT ACCESS	79A	RECOVERY LIGHT DOOR RELEASE MECHANISM
2A	PYRO ELECTRICAL DISCONNECT ACCESS	32	FORWARD EQUIPMENT BAY DOOR - LEFT	80	HOIST LOOP DOOR
3	SHINGLE	37	AFT EQUIPMENT BAY DOOR - LEFT	82	SHAPED CHARGE DETONATOR ACCESS
4	EMERGENCY DOCKING RELEASE CART. AND GUILLOTINE CARTRIDGE ACCESS	40	SHINGLE	83	COVER ASS'Y - PARACHUTE CONTROL CABLES
4A	PILOT CHUTE DEPLOY SENSOR SWITCH ACCESS	40A	RECOVERY LIGHT AND HOIST LOOP RIGGING AND CARTRIDGE ACCESS	84	COVER ASS'Y. - PARACHUTE CONTROL CABLES
5	SHINGLE	46	SEPARATION SENSING SWITCH ACCESS	85	RADIOMETER
5A	RADAR ACCESS	53	OAMS LINE GUILLOTINE ACCESS	86	CRYO SPECTROMETER/INTERFEROMETER
10	SHINGLE	54	F.L.S.C. TUBING CUTTER ACCESS	99	NUCLEAR EMULSION
11	INTERFACE ACCESS	55	FORWARD MANEUVERING ENGINE ACCESS	100	CYRO SPECTROMETER/INTERFEROMETER ACCESS
12	INTERFACE ACCESS	56	FUEL CELL SERVICE ACCESS	102	RADIOMETER ACCESS
13	INTERFACE ACCESS	62	OAMS OXIDIZER PURGE ACCESS		
13A	GUILLOTINE CARTRIDGE ACCESS	63	OAMS LINE GUILLOTINE ACCESS		
18	INTERFACE ACCESS	64	OAMS OXIDIZER PURGE ACCESS		
18A	PYROTECHNIC SWITCH CARTRIDGE AND BRIDLE DISCONNECT CARTRIDGE ACCESS	65	OAMS MODULE SERVICE ACCESS		
19	RE-ENTRY CONTROL SYSTEM ACCESS	66	ECS SERVICE ACCESS		
20	RE-ENTRY CONTROL SYSTEM ACCESS	69	ECS PUMP MODULE SERVICE ACCESS		
21	RE-ENTRY CONTROL SYSTEM ACCESS	70	ECS PUMP MODULE SERVICE ACCESS		
26	RE-ENTRY CONTROL SYSTEM ACCESS	71	SEPARATION SENSING SWITCH ACCESS		
27-1	SHINGLE	75	ELECTRICAL DISCONNECT ACCESS		
27-2	FRESH AIR DOOR	76	ELECTRICAL DISCONNECT ACCESS		
27A	Z160.20 EQUIPMENT ACCESS	78	SHINGLE		
27B	Z160.20 EQUIPMENT ACCESS	78A	Z160.20 EQUIPMENT ACCESS		

Figure 2-7 Access Doors (Sheet 5 of 6) (S/C 7)

FM2-2-7

SEDR 300

PROJECT GEMINI

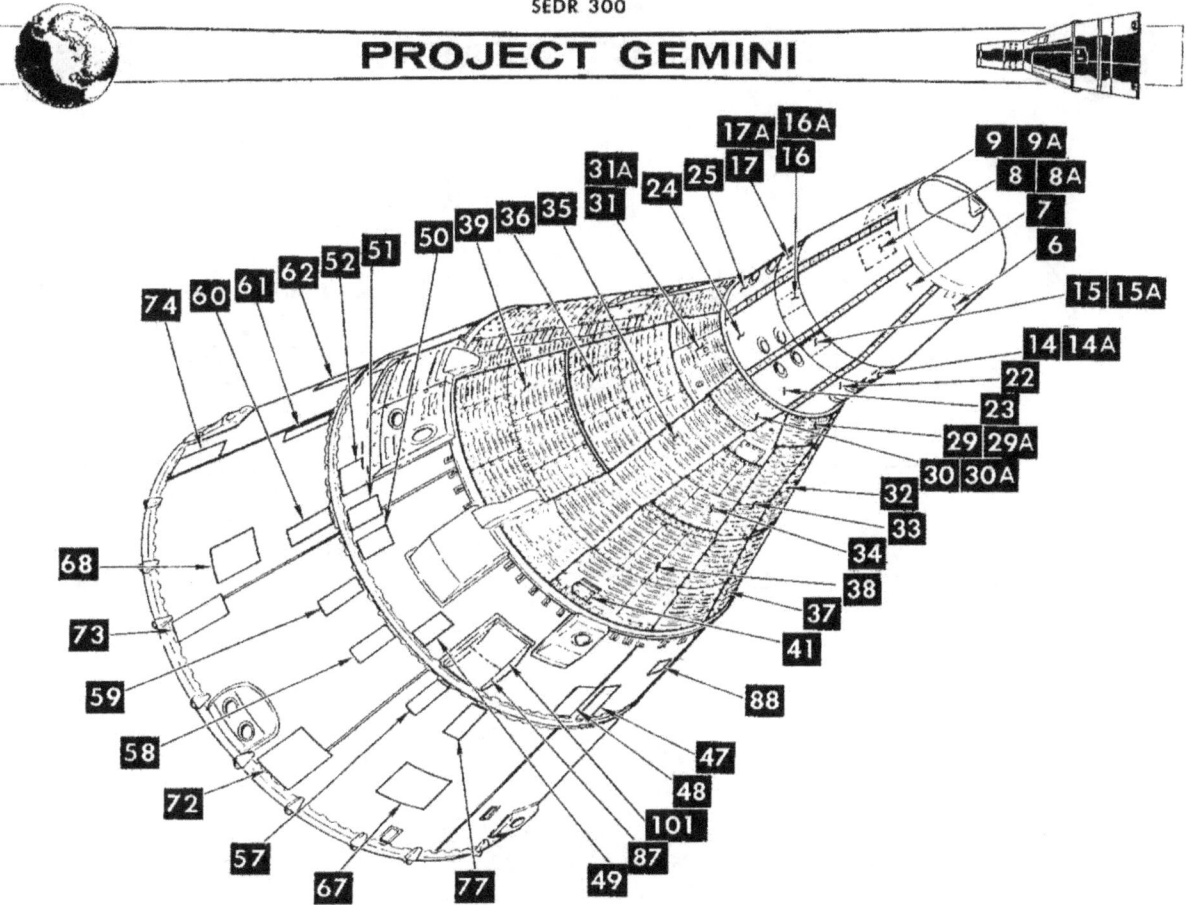

NO.	DESCRIPTION	NO.	DESCRIPTION	NO.	DESCRIPTION
6	EMERGENCY DOCKING RELEASE CARTRIDGE AND GUILLOTINE CARTRIDGE ACCESS	29A	Z160.20 EQUIPMENT ACCESS	51	B.I.A. RELAY PANEL ACCESS
7	SHINGLE	30	SHINGLE	52	FORWARD MANEUVERING ENGINE ACCESS
8	SHINGLE	30A	Z160.20 EQUIPMENT ACCESS	57	SHAPED CHARGE DETONATOR ACCESS
8A	RADAR ACCESS	31	SHINGLE	58	FUEL CELL SERVICE ACCESS
9	EMERGENCY DOCKING RELEASE CARTRIDGE AND GUILLOTINE CARTRIDGE ACCESS	31A	Z160.20 EQUIPMENT ACCESS	59	GUILLOTINE CARTRIDGE ACCESS
9A	DROGUE CHUTE DEPLOY SENSOR SWITCH ACCESS	32	FORWARD EQUIPMENT BAY DOOR - LEFT	60	GUILLOTINE CARTRIDGE ACCESS
14	INTERFACE ACCESS	33	MAIN LANDING GEAR DOOR - LEFT	61	OAMS FUEL PURGE ACCESS
14A	GUILLOTINE ANVIL ACCESS	34	CENTER EQUIPMENT BAY DOOR - FORWARD	67	ENGINE TO SCUPPER INTERFACE ACCESS
15	INTERFACE ACCESS	35	MAIN LANDING GEAR DOOR - RIGHT	68	ELECTRONIC MODULE TEST ACCESS
15A	GUILLOTINE ANVIL ACCESS	36	FORWARD EQUIPMENT BAY DOOR - RIGHT	72	GUILLOTINE CARTRIDGE AND LAUNCH VEHICLE ELEC. CONN. ACCESS
16	INTERFACE ACCESS	37	AFT EQUIPMENT BAY DOOR - LEFT	73	SEPARATION SENSING SWITCH ACCESS
16A	GUILLOTINE CARTRIDGE ACCESS	38	E.C.S. BAY DOOR	74	SHAPED CHARGE DETONATOR ACCESS
17	INTERFACE ACCESS	39	AFT EQUIPMENT BAY DOOR - RIGHT	77	FUEL CELL PURGE ACCESS
17A	ELEC. DISCONNECT ACCESS	41	PURGE FITTING ACCESS	87	SPECTROMETER/INTERFEROMETER
22	RE-ENTRY CONTROL SYSTEM ACCESS	47	RELAY PANEL ACCESS	88	ELECTROSTATIC CHARGE SENSOR
23	RE-ENTRY CONTROL SYSTEM ACCESS	48	RELAY PANEL ACCESS	97	PITCH SENOR SYSTEM
24	RE-ENTRY CONTROL SYSTEM ACCESS	49	SEPARATION SENSING SWITCH ACCESS	101	SPECTROMETER/ INTERFEROMETER ACCESS
25	RE-ENTRY CONTROL SYSTEM ACCESS	50	GUILLOTINE CARTRIDGE ACCESS		
29	SHINGLE				

Figure 2-7 Access Doors (Sheet 6 of 6) (S/C 7)

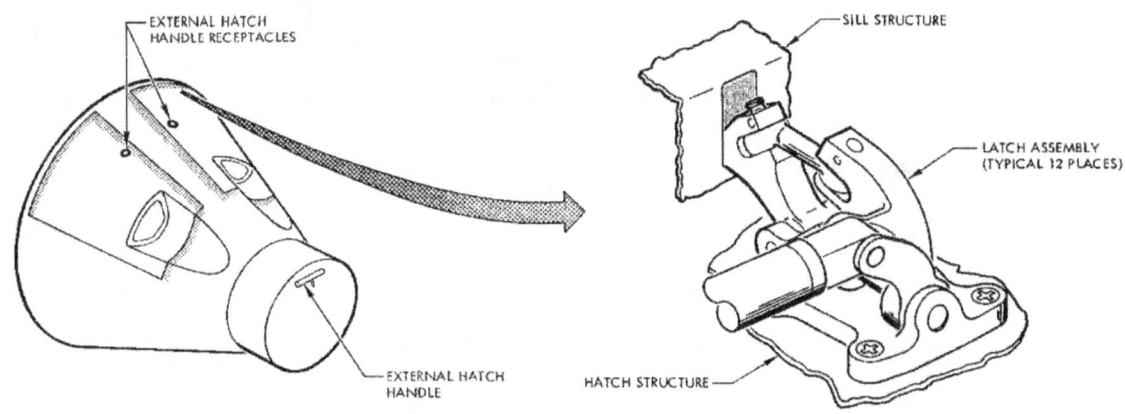
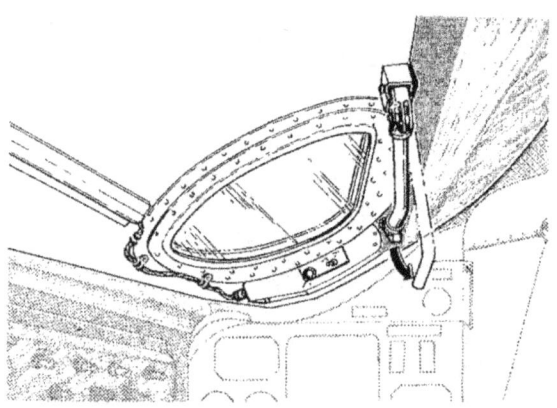
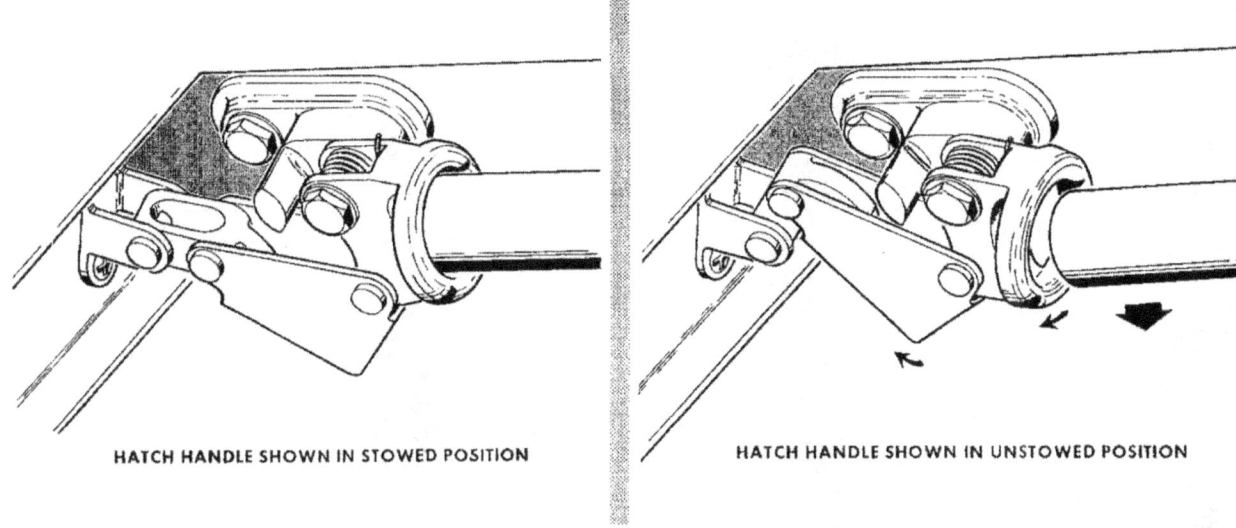

Figure 2-8 Spacecraft Ingress/Egress Hatches

SEDR 300

PROJECT GEMINI

side of the cabin section. Each hatch is manually operated by means of a handle and mechanical latching mechanism. Each is hinged on the outboard side. In an emergency, the hatches are opened in a three sequence operation employing pyrotechnic actuators. When initiated, the actuators simultaneously unlock and open the mechanical latches, open the hatches and supply hot gases to ignite the ejection seat rocket catapults. An external hatch linkage fitting is incorporated to allow a recovery hatch handle to be inserted for opening the hatches from the outside. The recovery hatch handle is stowed on the main parachute adapter assembly located on the forward face of the RCS section. A hatch curtain (Figure 2-9) is stowed along the hinge of each hatch. After water landing, when the hatches are open, the curtains are installed to help prevent water from entering the cabin.

Windows

Each of the ingress/egress hatches incorporates a visual observation window (Figure 2-10). Each window consists of an inner and outer glass assembly. The outer assembly is a single flat pane and the inner panel assembly consists of two flat panes. The panes consist of vycor (96% silica). The panes in the right window are optically ground for better resolution. Each surface of each pane, with the exception of the outer surface of the outer pane, is coated to lessen reflection and glare from cabin lights and to aid in impeding ultraviolet radiation into the cabin compartment.

Heat Shield

The heat shield is a dish-shaped structure composed of silicone elastomer filled phenolic impregnated fiberglass honeycomb. It is an ablative device, 90 inches in diameter with a spherical radius of 144 inches. The shield is designed to

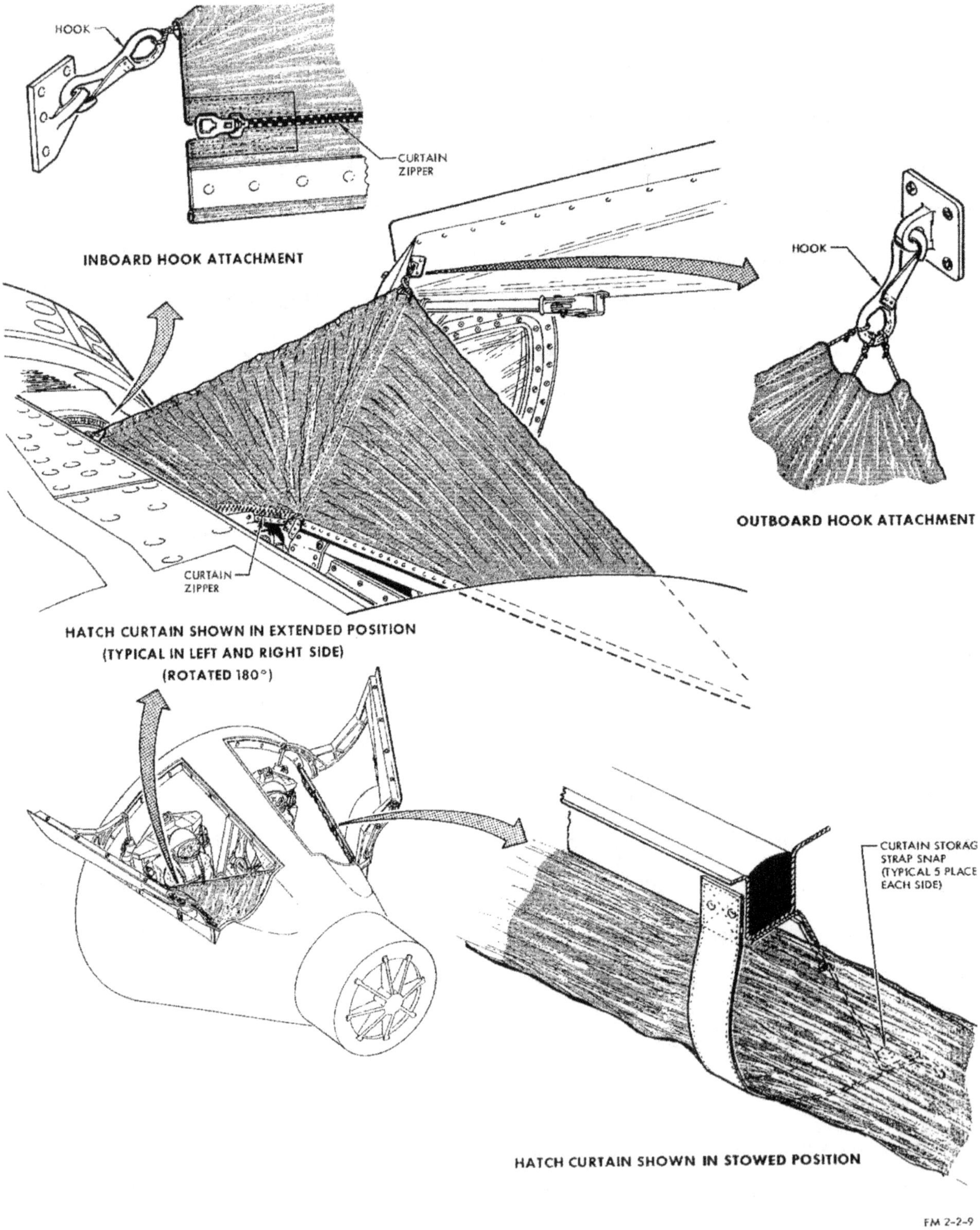

SEDR 300

PROJECT GEMINI

INBOARD HOOK ATTACHMENT

CURTAIN ZIPPER

OUTBOARD HOOK ATTACHMENT

HATCH CURTAIN SHOWN IN EXTENDED POSITION
(TYPICAL IN LEFT AND RIGHT SIDE)
(ROTATED 180°)

CURTAIN STORAGE STRAP SNAP
(TYPICAL 5 PLACE EACH SIDE)

HATCH CURTAIN SHOWN IN STOWED POSITION

FM 2-2-9

Figure 2-9 Hatch Curtain

2-19

 SEDR 300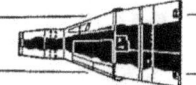

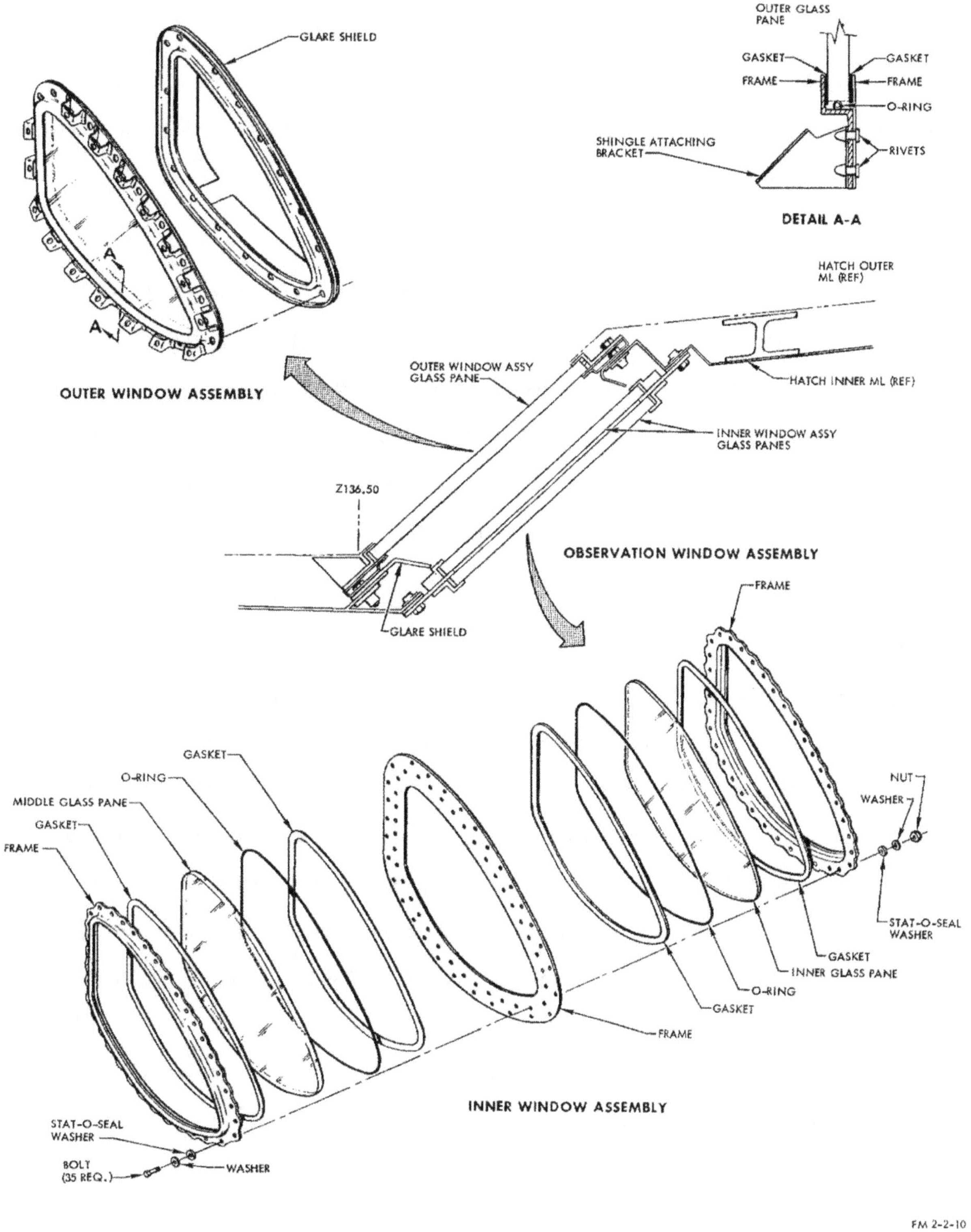

PROJECT GEMINI

Figure 2-10 Observation Window

2-20

protect the re-entry module from extreme thermal conditions during re-entry into the atmosphere. The device is attached to the large diameter end of the cabin structure by 1/4 inch bolts.

Shingles

The external surface of the cabin is made up of beaded shingles of Rene' 41. The R & R and RCS section surfaces are made up of unbeaded shingles of beryllium. The shingles protect the re-entry module structure from excessive heat and provide additional rigidity for the cabin. The shingles are black on the outer surface to control thermal radiation. The inner surface of the beryllium shingles are coated with gold to provide a low emissivity surface.

ADAPTER

In the spacecraft configuration, the adapter functions to mate the spacecraft to the launch vehicle, to provide provisions for mounting equipment, and to serve as a radiator for the spacecraft coolant system. The adapter (See Figure 2-2) is a truncated cone-shape, semimonocoque structure consisting of circumferential aluminum rings, extruded magnesium alloy stringers, and magnesium skin. The extruded stringers are designed in a bulb-tee shape to provide a flow path for the liquid coolant which transfers heat to the adapter skin for radiation to space. The outer surface of the skin is coated with white ceramic type paint and the inner surface is covered with aluminum foil. The forward end of the adapter is coupled to the aft end of the re-entry module by utilizing three titanium tension straps (See Figure 2-11).

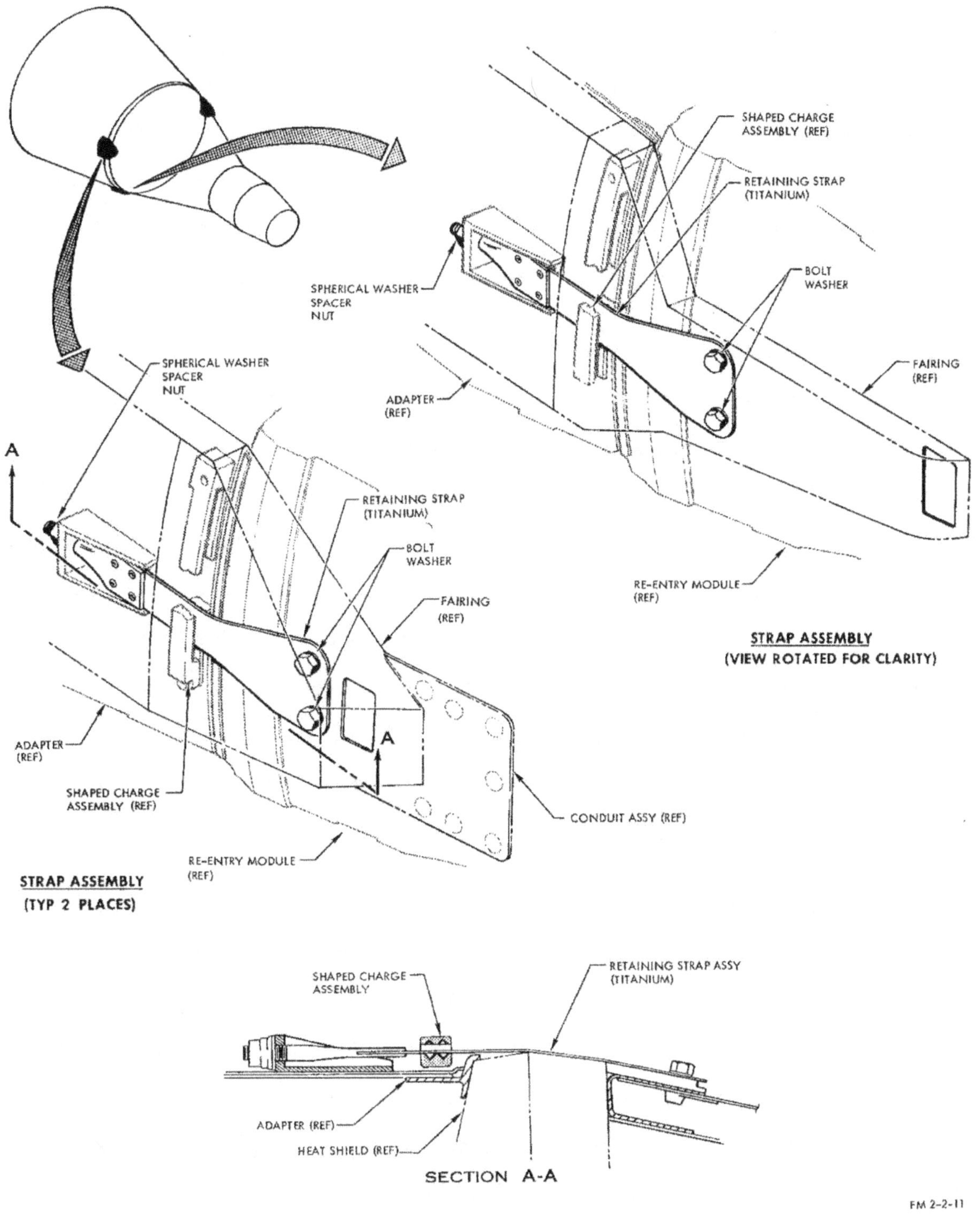

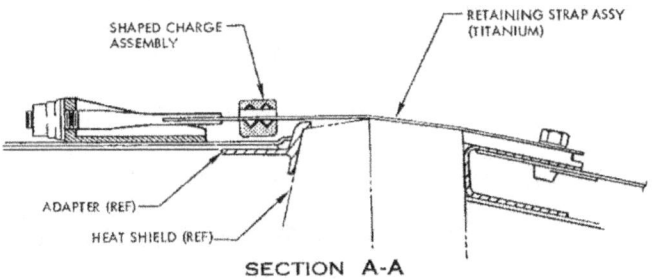

Figure 2-11 Re-Entry Module-Adapter Retaining Straps

PROJECT GEMINI

SEDR 300

RETROGRADE SECTION

The retrograde section, the smaller end of the adapter, provides for installation of four retrograde rockets and six OAMS thrust chamber assemblies. To provide for the installation of the retrograde rockets, the retrograde section employs an aluminum "I" beam support assembly. The "I" beams are assembled in the form of a cruciform with one retrograde rocket mounted in each quadrant.

ADAPTER EQUIPMENT SECTION

The adapter equipment section is the larger diameter end of the adapter. The section provides hard points for the attachment of structural modules for the OAMS tanks, E.C.S. primary oxygen supply, fuel cell (batteries on S/C 3 and 4), coolant, electrical and electronic components. A honeycomb blast shield is provided above the modules to shield the equipment section and booster dome from excessive heat during retro-rocket firing under abort conditions. Ten OAMS thrust chamber assemblies are mounted on the large diameter end of the equipment section. A gold deposited fiberglass temperature control cover protects the equipment from solar radiation thru the open end of the adapter after separation from the launch vehicle.

Spacecraft-Launch Vehicle Mating

The spacecraft is mated to the Titan II launch vehicle with a machined aluminum alloy ring (See Figure 2-12). This ring, 120 inches in diameter, mates with the launch vehicle mating ring. Twenty bolts secure the rings together. To provide for alignment, the launch vehicle incorporates one steel 3/26 inch diameter alignment pin located at "TY" and four index marks. To separate the spacecraft from the launch vehicle, a pyrotechnic charge is fired, severing the adapter

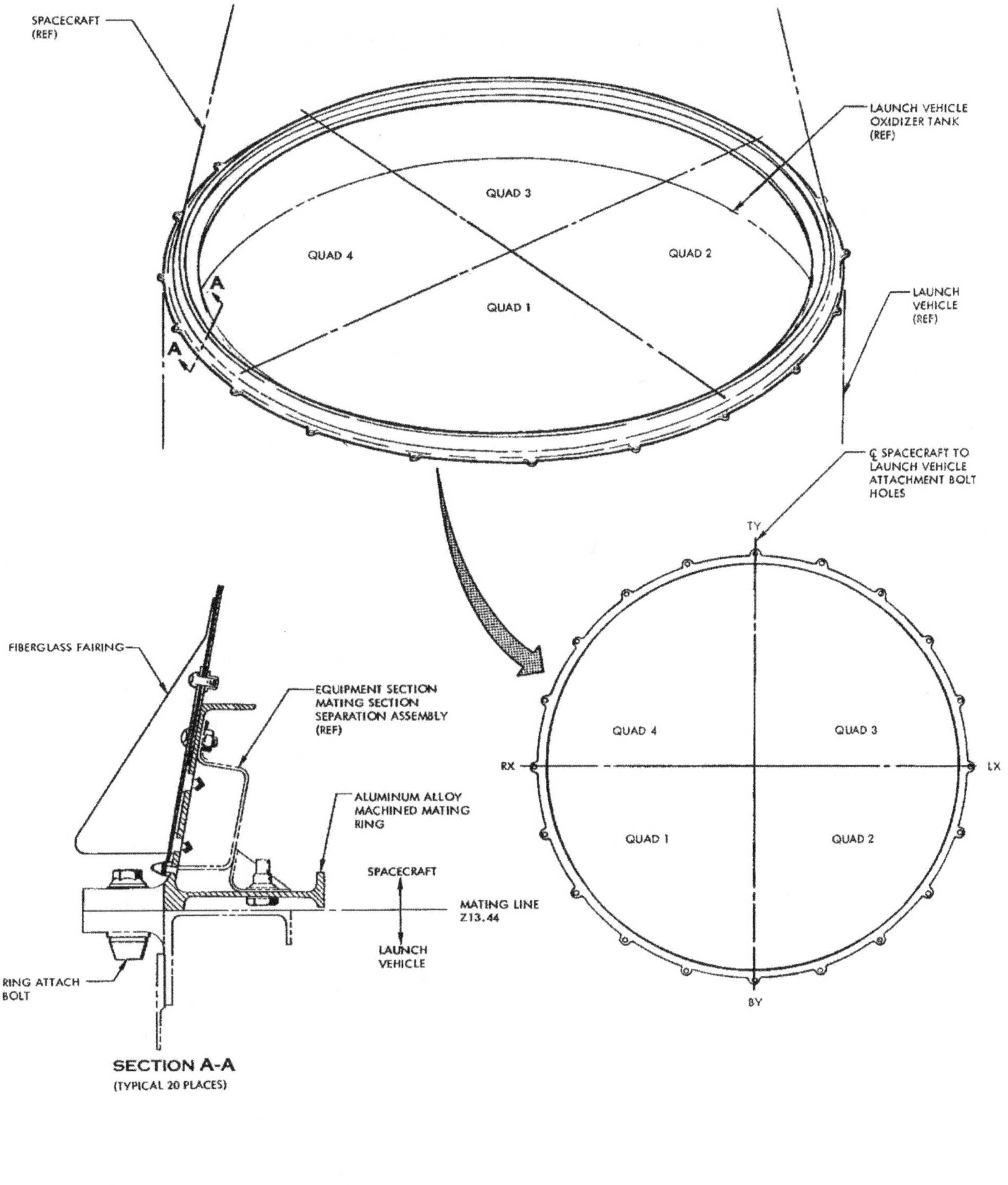

Figure 2-12 Spacecraft/Launch Vehicle Mating Ring

section approximately 1 1/2 inches above the launch vehicle/spacecraft mating point.

CABIN INTERIOR ARRANGEMENT

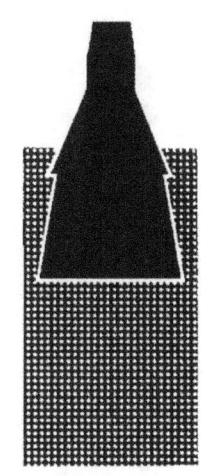

Section III

TABLE OF CONTENTS

TITLE	PAGE
GENERAL	3-3
CREW SEATING	3-3
SEAT DESCRIPTION	3-3
SEAT EJECTION SYSTEM	3-5
RESTRAINT SYSTEM	3-10
EGRESS KIT	3-13
BACKBOARD ASSEMBLY	3-14
PELVIC BLOCK	3-15
BALLUTE SYSTEM	3-15
PERSONNEL PARACHUTE	3-15
PARACHUTE DROGUE MORTAR	3-16
HARNESS ASSEMBLY	3-16
SURVIVAL KIT	3-16
PYROTECHNIC DEVICES	3-19
DEBRIS GUARDS	3-19
INSTRUMENT PANELS	3-20
CABIN INTERIOR LIGHTING	3-20
STATIC SYSTEM	3-26
FOOD, WATER and EQUIPMENT STOWAGE	3-26
WASTE DISPOSAL	3-31
STOWAGE PROVISIONS	3-31

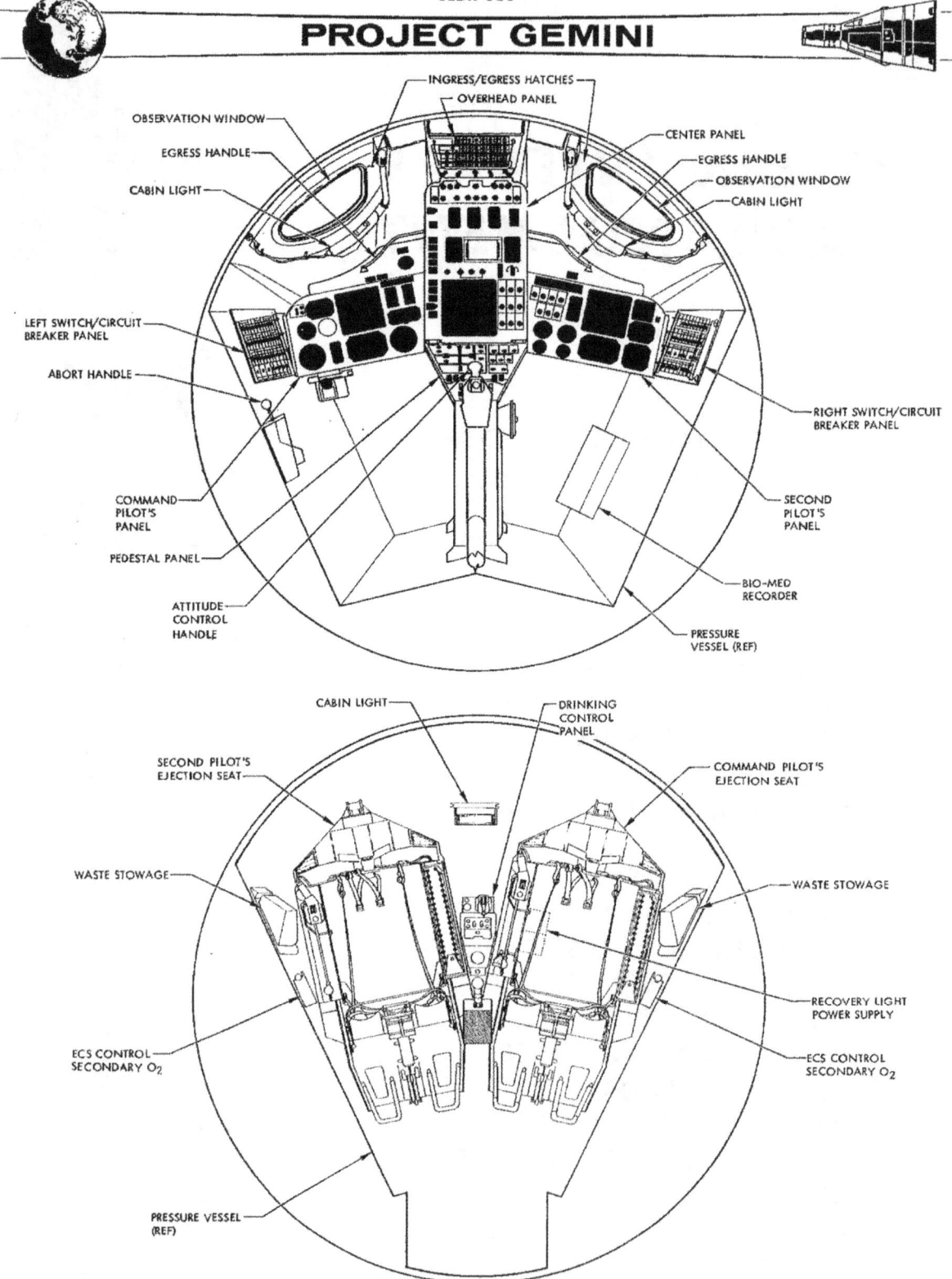

Figure 3-1 Cabin Equipment (Typical)

SEDR 300

PROJECT GEMINI

SECTION III CABIN INTERIOR ARRANGEMENT

GENERAL

The equipment within the cabin is arranged to permit the Command Pilot, seated to the left, and the Pilot, seated to the right, to operate the controls and observe displays and instruments in full pressure suits in the restrained or unrestrained position. The cabin air outflow is regulated during launch to establish and maintain a 5.5 psi differential pressure between the cabin and outside air. The cabin is maintained at a nominal 5.1 psia throughout the flight by a cabin pressure regulator. The cabin equipment (Figure 3-1) basically consists of crew seats, instrument and control panels, controls, lighting, food, water, waste collection, and miscellaneous equipment.

CREW SEATING

The crew members are seated in the typical pilot and co-pilot fashion, faced toward the small end of the re-entry module. The seats are canted 12° outboard and 8° forward to assure separation and to provide required elevation in the event an off the pad ejection is necessitated.

Crew seating provisions include seats, restraint mechanisms, ejection devices, seat man separator, survival gear, and egress kit assembly.

SEAT DESCRIPTION

The crew seats (Figure 3-2) are all metal built-up assemblies consisting of a torque box framed seat bucket, channeled backs and arm rests. The seat has lateral and vertical stiffeners, designed for a single moment of thrust. The

NOTE

COMMAND PILOT EJECTION SEAT ILLUSTRATED. HARNESS RELEASE ACTUATOR IS LOCATED ON OUTBOARD SIDE OF SEAT.

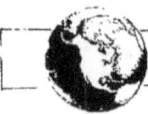

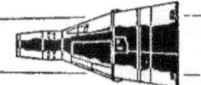

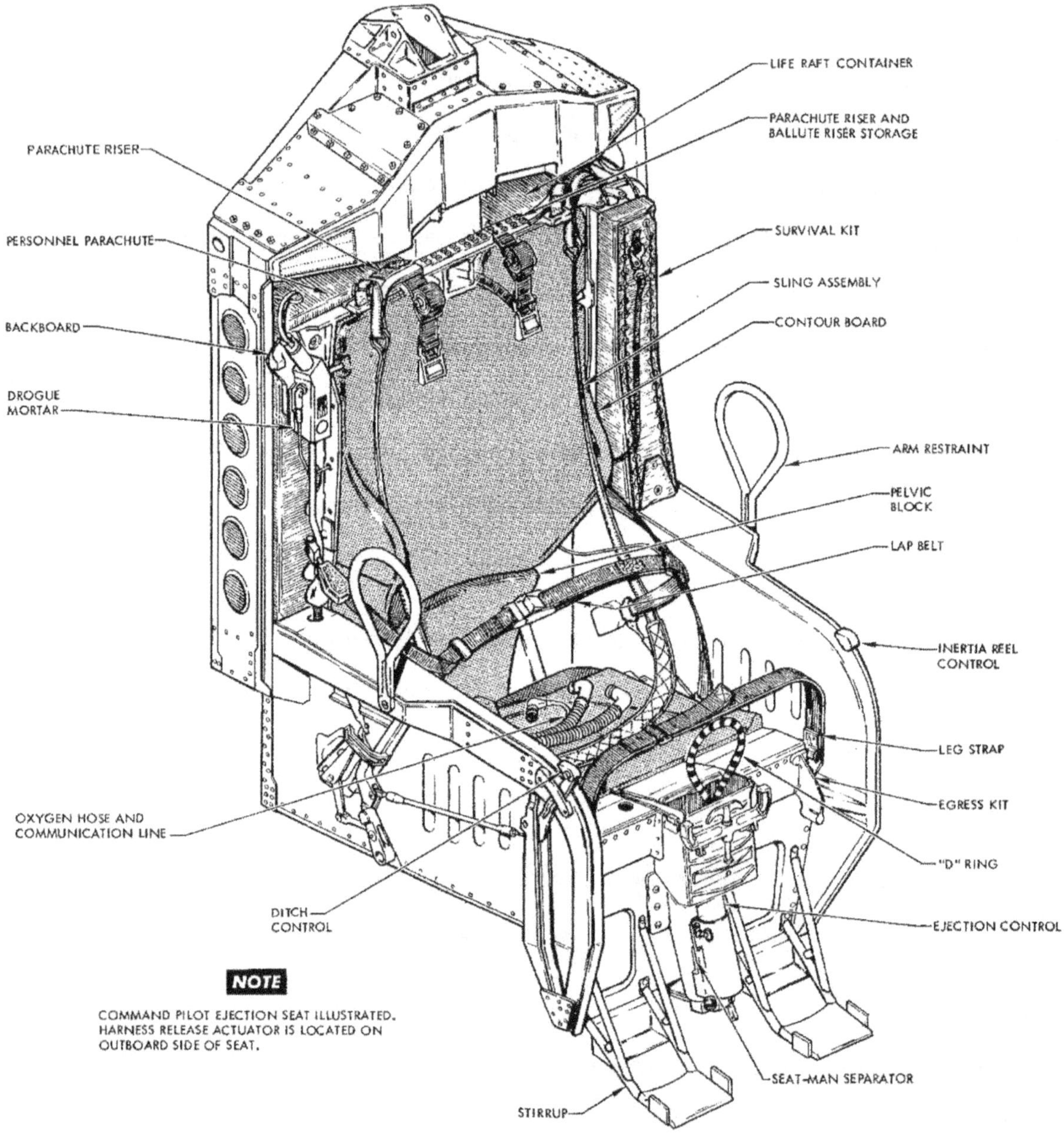

Figure 3-2 Gemini Ejection Seat Assembly

seat is supported at a single point at the top of the seat back. At this point, the seat bolts to the rocket/catapult. Each seat is supported against fore, aft, and side movement by slide blocks mounted on the seats and retained in tee type rail assemblies attached to the large pressure bulkhead. The seats incorporate a padded contoured headrest to support the pilots helmet. Each seat also incorporates a restraint system, harness release system and a seat-man separator.

SEAT EJECTION SYSTEM

The seat ejection system (Figure 3-3) provides the pilot with a means of escaping from the vicinity of the spacecraft in the event of an abort or in an emergency condition during launch or re-entry. Crew member seats are ejected by means of rocket catapults. Hot gas from each of the hatch actuators is routed to the appropriate seat catapult where dual firing pins strike dual percussion primers, thereby igniting the seat rocket catapult main charge and ejecting the seats from the spacecraft. Hot gas from the rocket catapult main charge ignites the sustainer rocket and the rocket provides additional separation from the spacecraft. In the event ejection becomes necessary, after deployment of main landing system parachute and while descending in the two point suspension, it is mandatory that the main landing system parachute be jettisoned before ejecting from the spacecraft.

The ejection sequence is initiated by manually pulling either "D" ring located on front of the seat buckets. During the launch phase of flight, each pilot erects and holds on to the "D" ring. This action aids in stabilizing the pilots' arms and at the same time places them in a position for instant response. The

SEDR 300
PROJECT GEMINI

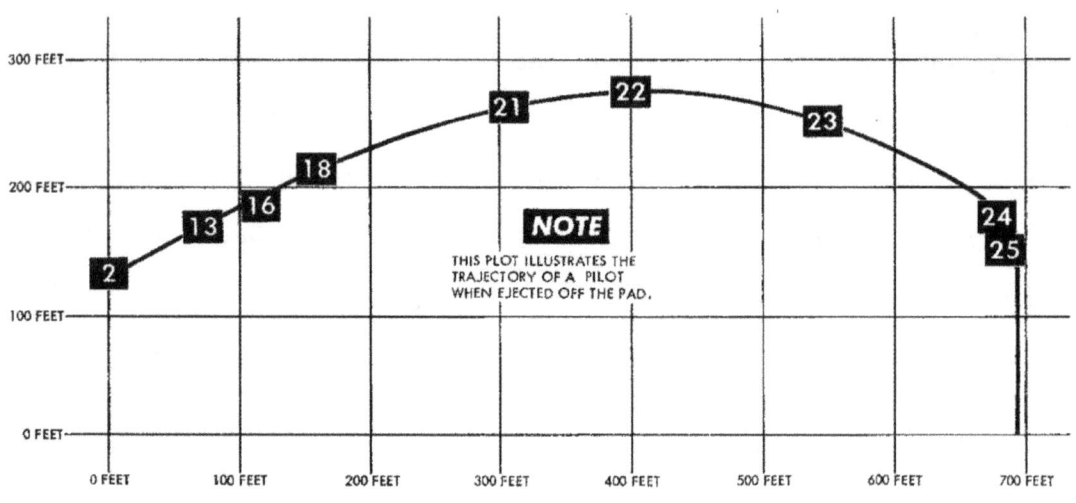

EJECTION SEAT TRAJECTORY PLOT

NOTE
THIS PLOT ILLUSTRATES THE TRAJECTORY OF A PILOT WHEN EJECTED OFF THE PAD.

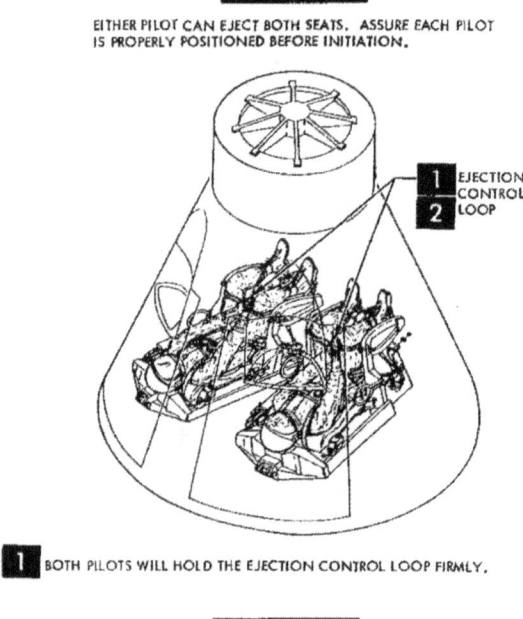

WARNING
EITHER PILOT CAN EJECT BOTH SEATS. ASSURE EACH PILOT IS PROPERLY POSITIONED BEFORE INITIATION.

1 EJECTION CONTROL LOOP
2

1 BOTH PILOTS WILL HOLD THE EJECTION CONTROL LOOP FIRMLY.

WARNING
THE CONTROL LOOP WILL BE HELD FIRMLY TO KEEP ARMS WITHIN EGRESS AREA LIMITS.

2 INITIATE EJECTION

EVENT TIME
0

INDIVIDUAL ACTION	
TIME	TOLERANCE
0	

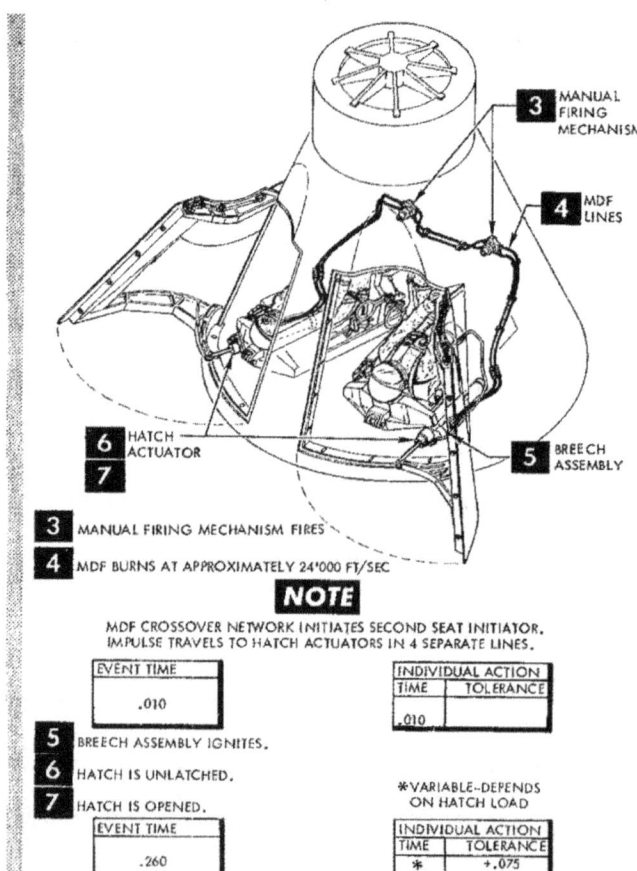

3 MANUAL FIRING MECHANISM
4 MDF LINES
5 BREECH ASSEMBLY
6 HATCH ACTUATOR
7

3 MANUAL FIRING MECHANISM FIRES
4 MDF BURNS AT APPROXIMATELY 24,000 FT/SEC

NOTE
MDF CROSSOVER NETWORK INITIATES SECOND SEAT INITIATOR. IMPULSE TRAVELS TO HATCH ACTUATORS IN 4 SEPARATE LINES.

EVENT TIME
.010

INDIVIDUAL ACTION	
TIME	TOLERANCE
.010	

5 BREECH ASSEMBLY IGNITES.
6 HATCH IS UNLATCHED.
7 HATCH IS OPENED.

EVENT TIME
.260

*VARIABLE--DEPENDS ON HATCH LOAD

INDIVIDUAL ACTION	
TIME	TOLERANCE
*	+.075
.250	-.05

Figure 3-3 Ejection Seat Sequence Of Operation (Sheet 1 of 4)

 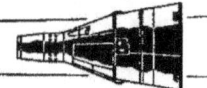

SEDR 300

PROJECT GEMINI

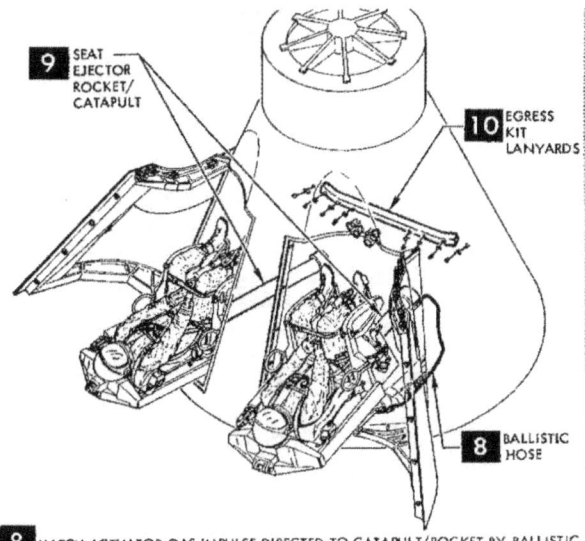

8 HATCH ACTUATOR GAS IMPULSE DIRECTED TO CATAPULT/ROCKET BY BALLISTIC HOSE.

9 CATAPULT FIRES

10 HARNESS RELEASE ACTUATORS INITIATED. EGRESS KIT LANYARDS PULLED. PILOT SUPPLIED WITH EGRESS OXYGEN PRESSURE. COMMUNICATION SEVERED.

EVENT TIME	INDIVIDUAL ACTION	
	TIME	TOLERANCE
.333	.073	+.007 / -.008

11 EJECTION SEAT MOVING UP, IGNITES EJECTION ROCKET APPROXIMATELY 4 INCHES FROM END OF RAIL TRAVEL.

EVENT TIME	INDIVIDUAL ACTION	
	TIME	TOLERANCE
.394	.134	+.009 / -.008

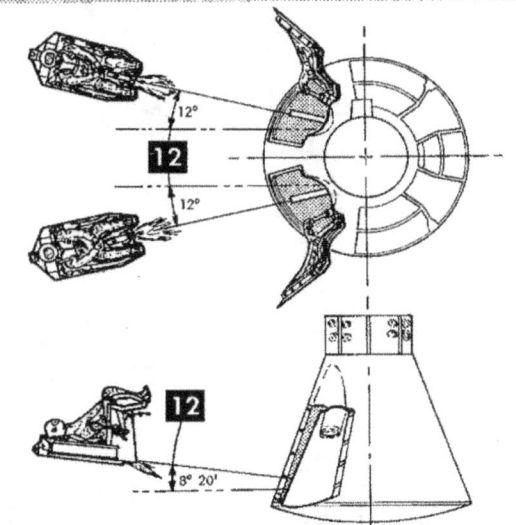

12 EGRESS CONTINUES UNDER ROCKET POWER. PILOTS LEAVE SPACECRAFT AT 12° OUTBOARD OF "X" AXIS AND 8° 20' FORWARD OF "Z" AXIS.

13 ROCKET BURN OUT.

EVENT TIME	INDIVIDUAL ACTION	
	TIME	TOLERANCE
.664	.270	± .02

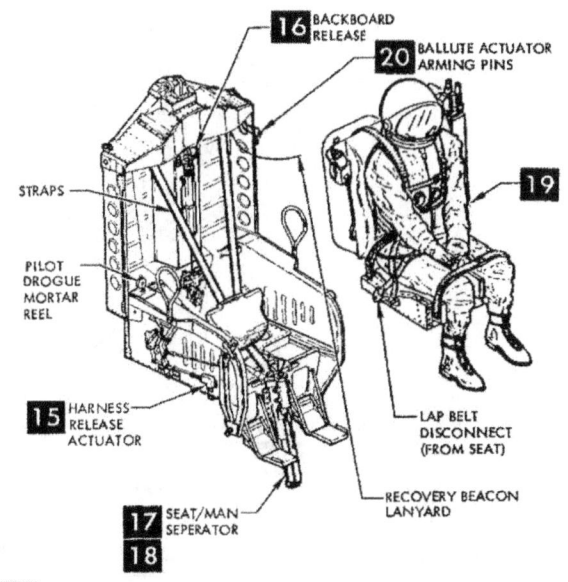

14 EJECTION SEAT CONTINUES ON TRAJECTORY

15 HARNESS RELEASE ACTUATOR FIRES

16 LAP BELT RELEASE ASSEMBLY ACTIVATED. BACKBOARD AND SURVIVAL GEAR ASSEMBLY RELEASED FROM SEAT.

EVENT TIME	INDIVIDUAL ACTION	
	TIME	TOLERANCE
1.418	1.085	± .162

17 HARNESS RELEASE ACTUATOR GAS IMPULSE DELIVERED TO SEAT/MAN SEPARATOR BY BALLISTIC HOSE.

18 SEAT/MAN SEPARATOR SHOE EXTENDS AND REMOVES SLACK FROM STRAP ASSEMBLY.

19 PILOT WITH BACKBOARD AND SURVIVAL GEAR, AND EGRESS KIT SEPARATE FROM SEAT.

20 PILOT DROGUE MORTAR, BALLUTE SYSTEM AND RECOVERY BEACON INITIATED BY LANYARDS CONNECTED TO SEAT STRUCTURE.

EVENT TIME	INDIVIDUAL ACTION	
	TIME	TOLERANCE
1.508	.090	± .015

Figure 3-3 Ejection Seat Sequence Of Operation (Sheet 2 of 4)

SEDR 300
PROJECT GEMINI

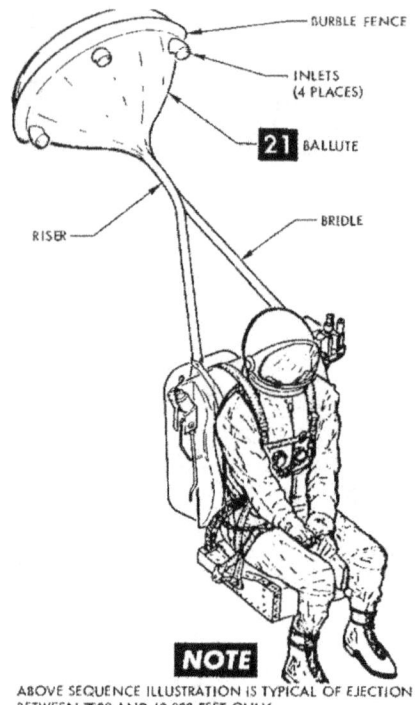

- BUBBLE FENCE
- INLETS (4 PLACES)
- **21** BALLUTE
- RISER
- BRIDLE

NOTE
ABOVE SEQUENCE ILLUSTRATION IS TYPICAL OF EJECTION BETWEEN 7500 AND 40,000 FEET ONLY.

21 BALLUTE DEPLOYS AFTER A 5 SECOND DELAY.

NOTE
1. BALLUTE BAROSTAT HAS BEEN ACTIVATED TO JETTISON THE BALLUTE AT 7500 FEET.
2. TIME CHART APPLICABLE TO EJECTION ABOVE ONLY.

EVENT TIME	INDIVIDUAL ACTION	
	TIME	TOLERANCE
6.508	5.00	

- PILOT CHUTE
- SUSPENSION LINES
- LANYARD
- **23** DROGUE MORTAR SLUG
- **22** PERSONNEL PARACHUTE DROGUE MORTAR

NOTE
DROGUE MORTAR BAROSTAT IS ACTIVATED DURING SEAT/MAN SEPARATION TO DEPLOY THE PARACHUTE AT 5700 FEET OR BELOW.

22 DROGUE MORTAR FIRES.

EVENT TIME	INDIVIDUAL ACTION	
	TIME	TOLERANCE
3.808	2.3	+.575 / -.460

NOTE
EVENT TIMES ARE FOR EJECTION BELOW 5700 FEET ONLY.

23 MORTAR SLUG DEPLOYES AND INFLATES PILOT AND MAIN PARACHUTES.

24 PARACHUTE FULLY INFLATED.   *VARIABLE-DEPENDS ON DYNAMIC PRESSURE.

EVENT TIME	INDIVIDUAL ACTION	
	TIME	TOLERANCE
6.808	3.0	*

25 BACKBOARD AND EGRESS KIT SEPARATED FROM PILOT. (TIME FROM DROUGE MORTAR FIRING)

EVENT TIME	INDIVIDUAL ACTION	
	TIME	TOLERANCE
8.808	5.0	+1.25 / -1.00

Figure 3-3 Ejection Seat Sequence Of Operation (Sheet 3 of 4)

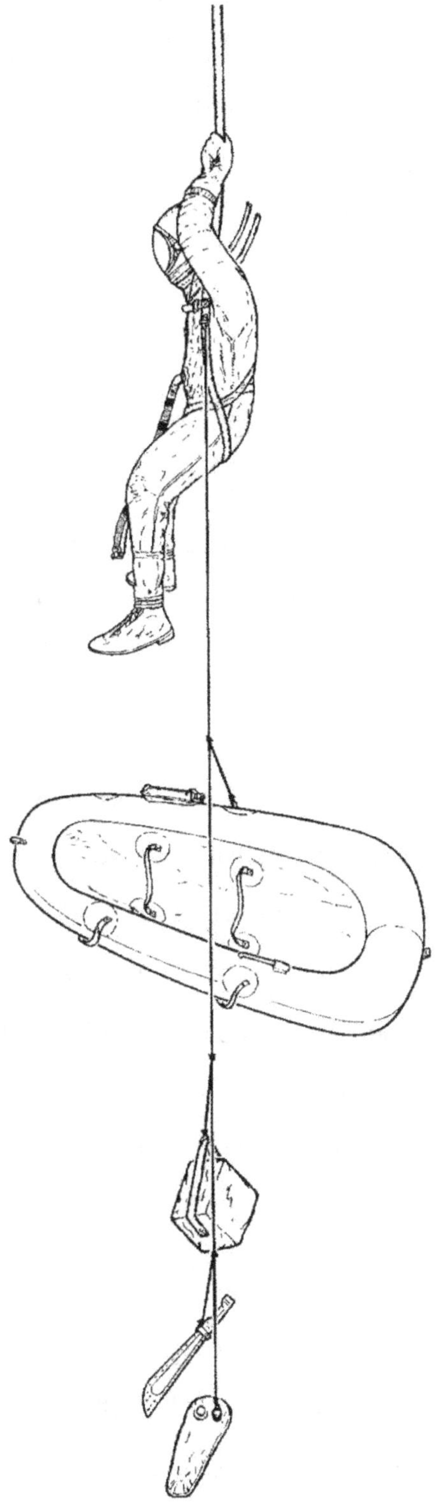

PROJECT GEMINI

NOTE

SURVIVAL EQUIPMENT DEPLOYED AS BACKBOARD FALLS FROM PILOT. THE LIFE RAFT IS INFLATED MANUALLY DURING PARACHUTE DESCENT ALL SURVIVAL EQUIPMENT IS SECURED TO PILOT BY A LANYARD.

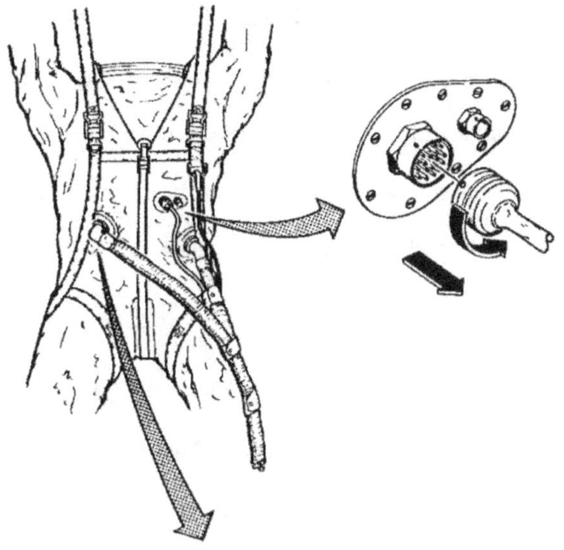

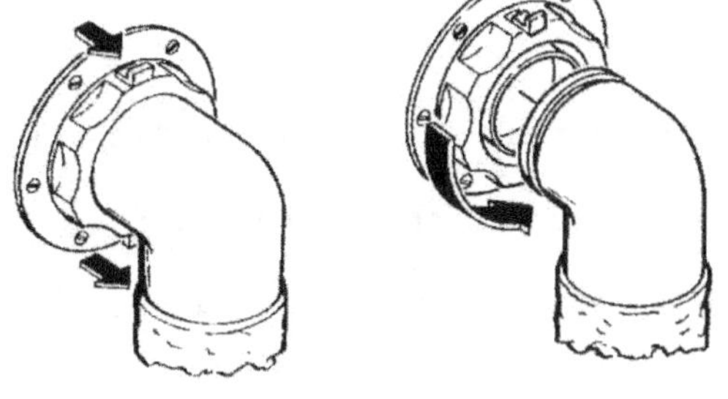

NOTE

PILOT DISCONNECTS OXYGEN INLET AND OUTLET HOSES. OXYGEN CONNECTION IN PRESSURE SUIT IS SEALED CLOSED WHEN OXYGEN HOSE IS REMOVED.

Figure 3-3 Ejection Seat Sequence Of Operation (Sheet 4 of 4)

"D" rings are normally stowed under a sliding door on the front of egress kit and are locked into place via a pip pin on the front of the structure. This pin is removed during launch and replaced for spaceflight. The pip pin is removed for re-entry.

RESTRAINT SYSTEM

Each pilot is restrained in his ejection seat by a restraint system (Figure 3-4) consisting of arm restraint loops, leg restraint straps, foot stirrups, elbow restraint, lap belt, shoulder harness and inertia reel assembly. The restraint system provides adequate support and restraint during conditions of maximum acceleration and deceleration.

INERTIA REEL

The inertia reel (Figure 3-4) is a two position locking device, located on the rear of the backboard. Two straps connect the inertia reel and the pilot's harness to restrain the pilot's forward movement. The inertia reel control handle is located on the front of the left arm rest and has two positions, "manual lock" and "automatic lock." Orbital flight is accomplished with the inertia reel in the "automatic lock" position. Manual lock position is used during launch and re-entry. The manual lock position prevents the pilot's shoulders from moving forward.

To release his shoulders, when the inertia reel is in the manual lock position, the pilot must position the control handle to the automatic position. The "automatic lock" allows the astronaut to move forward slowly but will lock with a shock movement of 3 "G's." When the automatic lock has engaged, the lock will

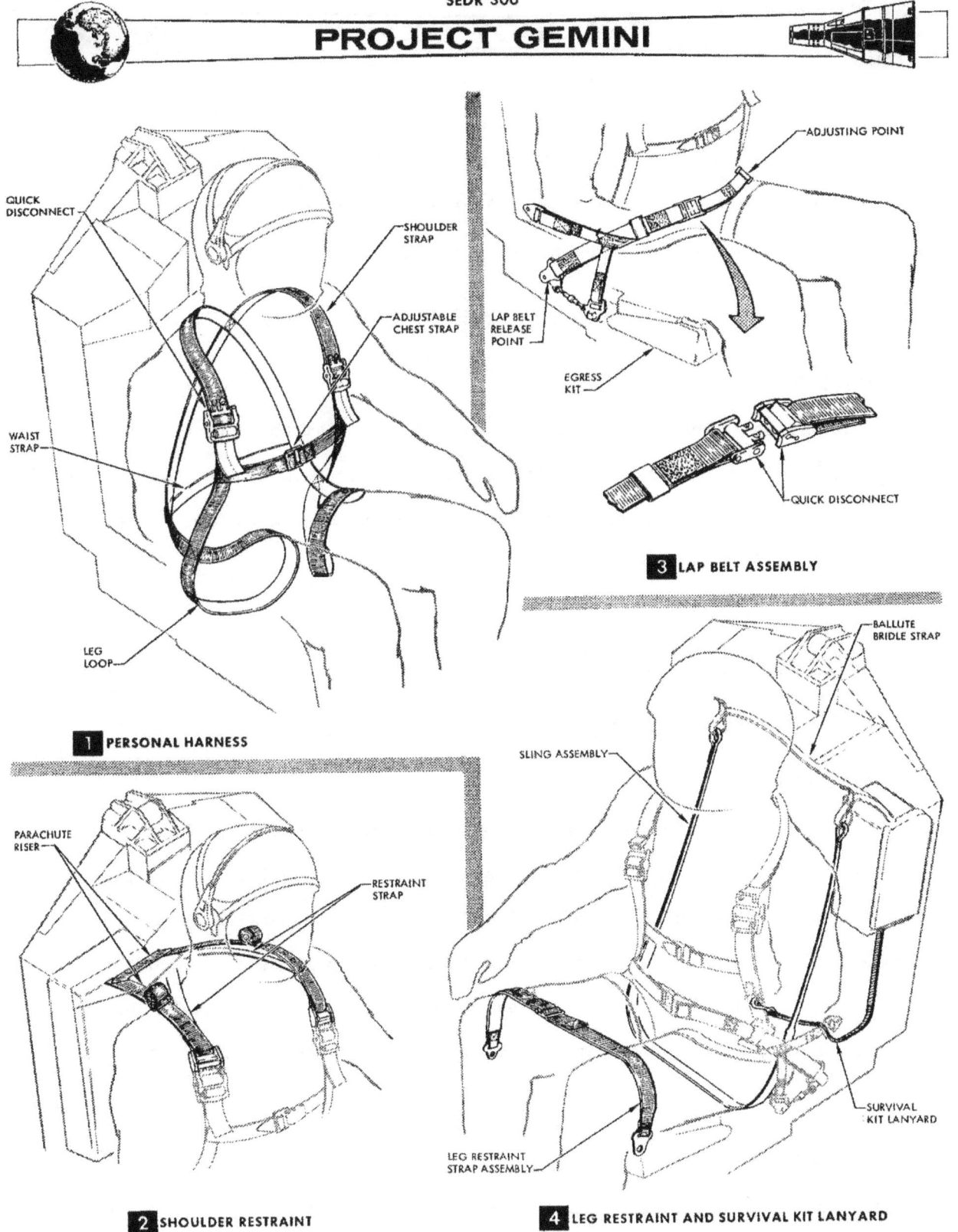

Figure 3-4 Restraint System

SEDR 300

PROJECT GEMINI

ratchet and permit movement back into the seat, but will not permit forward movement. The release of the automatic lock is accomplished by cycling the control handle to manual and back to automatic lock.

Maximum extension of the shoulder strap from the inertia reel is 18 inches.

ARM RESTRAINT

The arm restraint (Figure 3-4) is a welded, 1/2 inch diameter tube assembly made up in the form of a loop. A loop is installed on each arm rest to retain the pilot's arms within the ejection envelope. When the arm restraint loop is not required, it may be rotated back and down.

ELBOW RESTRAINT

An elbow restraint is provided for the command pilot only. It is used to stabilize his forearm during manual re-entry.

LEG RESTRAINT STRAP

The leg restraint (Figure 3-4) consists of two straps of dacron webbing. One end of each strap is secured to the egress kit by round metal eyelets. The left strap of each leg restraint has a metal end assembly that permits the right strap to fold back on itself. Velcro tape on the right strap is used to secure the strap in position when the strap is drawn tight over the pilot's legs. During seat/man separation, the restraint strap eyelets are automatically released from the egress kit.

EJECTION SEAT FOOT STIRRUP

The ejection seat foot stirrups (Figure 3-2) consist of two welded frames attached to the front of the ejection seat. Each stirrup has a short protruding platform

PROJECT GEMINI

with small vertical edges rising along the outboard side. The stirrup is so constructed that the pilot's shoe heel will lock in place and prevent forward movement of the foot while the small vertical edges will prevent side movement. During seat ejection, the pilot's feet must be in place.

LAP BELT

The lap belt (Figure 3-4) is an arrangement of dacron and nylon straps, designed to restrain the pilot in the seat structure. Load carrying straps from the lap belt are fastened to the backboard and egress kit. The lap belt has a manual quick disconnect and a pyrotechnic release fitting near the center of the pilot's lap. The manual quick disconnect can be released with one finger. Lap belt tension is adjusted by sliding excess strap through the pyrotechnic release. During ejection, the lap belt ends attached to the seat structure are released just prior to seat/man separation. During separation, the lap belt remains with the pilot. Five seconds after the backboard drogue mortar fires, the pyrotechnic lap belt release activates and allows the lap belt, backboard and egress kit to fall free.

A second manual release for the lap belt is also available to the pilot. It is located forward on the right arm rest and is referred to as the ditch handle. Releasing the lap belt from the seat structure with the ditching handle allows the pilot to egress from the landing module with the backboard and egress kit.

EGRESS KIT

The egress kit assembly contains the bail out oxygen for an ejected pilot. The egress kit rests in the ejection seat bucket and forms a mounting surface for the egress kit cushion. The egress kit contains an oxygen supply, for breathing and suit pressurization; a composite disconnect, which when separated closes the port and prevents escape of egress oxygen; a relief valve, to prevent pressure build up in the pressure suit; a regulator, to reduce high pressure to a controlled flow of low pressure oxygen; a pressure gage, for visually checking egress oxygen pressure; and connecting lines. Three lanyards are attached between the egress kit and the spacecraft. These lanyards pull release pins to allow the composite disconnect to separate, allow the oxygen to flow through the pressure regulator and allow the relief valve to control the pilot's suit pressure. When the drogue mortar deploys the pilot parachute, a 5 second pyrotechnic time delay is initiated and at burn out the egress kit with the backboard is separated from the pilot.

EGRESS KIT CUSHION

The egress kit cushion (Figure 3-2) has a universal type of contour and is attached to the top of the egress kit. The cushion is positioned forward of the pelvic block and up to the access door to the ejection control "D" ring.

BACKBOARD ASSEMBLY

The backboard assembly (Figure 3-2) is machined aluminum, designed and stressed to retain the inertia reel, ballute, ballute release and deploy mechanism, drogue mortar, personnel parachute and survival kit. A cushion, contoured to the individual pilot's body requirements, is positioned on the forward surface of the backboard. The cushion is provided to supply support and comfort to the pilot's back. The inertia reel straps and lap belt secures the pilot to the backboard. The backboard accompanies the pilot through seat ejection to parachute deployment. Five seconds after parachute deployment, the backboard with the egress kit is separated from the pilot.

PELVIC BLOCK

The pelvic block (Figure 3-2), contoured to the lower torso of each pilot, is positioned between the backboard assembly and the egress kit. The block supports the pilot's lower vertebra and pelvic structure. It remains with the seat structure upon seat man separation.

BALLUTE SYSTEM

The ballute system (Figure 3-2) consists of a barostat controlled pyrotechnic initiator, combined with a pyrotechnic gas generator, cutters and a packaged ballute. The ballute, located on the back and lower left side of the pilot's backboard, is an aluminized nylon fabric enclosed cone. It is inflated by ram air passing through four inlets located symmetrically around the upper periphery. The ballute is connected to the backboard through an 8" riser, a 5 ft. dual bridle, and by a 1.00 inch wide dacron webbing passing through a pyrotechnic actuated cutter. The ballute provides the pilot with a stabilized, feet into the wind, attitude for all ejections over 7500 feet. The system is fully automatic and is actuated at seat man separation. At altitudes below 7,500 feet, the barostat prevents deployment of the ballute.

PERSONNEL PARACHUTE

The personnel parachute (Figure 3-2) is a standard 28 ft. dia. nylon parachute. The parachute is located on the right rear of the pilot's backboard. It is deployed by the drogue mortar slug and pilot chute. The parachute risers are attached to the pilot's personal harness.

PARACHUTE DROGUE MORTAR

The parachute drogue mortar (Figure 3-2) is a pyrotechnic device designed to eject a 10 oz. drogue slug with sufficient energy to deploy the pilot chute of the personnel parachute. The drogue mortar is a barostat operated firing mechanism, but can be fired manually. It will fire and deploy the parachute at or below 5700 feet plus a 2.3 seconds time delay from seat/man separation. An MDF chain is initiated by the drogue mortar to separate the backboard and egress kit from the pilot.

HARNESS ASSEMBLY

The harness assembly (Figure 3-4) provides a light, strong, and comfortable arrangement to attach the personnel parachute to the pilot. The harness is constructed from nylon webbing formed into a double figure "8". The two figure "8's are joined by two cross straps, the waist strap, and the chest strap. Only the chest strap is adjustable. A quick disconnect is placed forward and below each shoulder for connection of the parachute risers and inertia reel straps. Below the left quick disconnect, a small ring is incorporated to attach the survival equipment lanyard.

SURVIVAL KIT

The survival kit (Figure 3-2) is a packaged group of specially designed equipment for the use of a downed pilot. Articles in this kit are intended to aid in preserving life under varying climatic conditions. Deployment of the survival kit is automatic if the pilot ejects but is available to the pilot if he lands with the re-entry vehicle.

 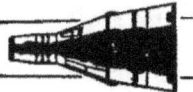

PROJECT GEMINI

Deployment of the survival kit, during the ejection cycle, takes place as the backboard and egress kit falls away from the parachuting pilot. As the backboard falls, the survival equipment lanyard, connected to the pilot's harness, pulls a pin on the life raft container. When the pin is removed, the "daisy chain" loops are disengaged and the life raft and rucksack are extracted from the container. The survival equipment lanyard repeats the extraction process in removing the machete and water bottle from the second container. The machete and water bottle are stowed in a survival equipment container on the left front side of the backboard.

During seat/man separation, a lanyard between the seat structure and the rucksack activates the radio/beacon. As the pilot descends on his parachute, the survival equipment is suspended below and the radio beacon transmits on emergency frequency. Direction finding equipment on aircraft and aboard ship can plot the pilot's position.

Survival equipment is divided into two major stowage containers. The life raft container mounted on the left rear of the backboard has the following items:

    Life raft container

        1 Life Raft

            1 Sea Anchor

            1 4 x 4 Foam Rubber Pad

            1 CO_2 Cylinder

            1 Sea Dye Marker

            1 Sun Bonnet

 SEDR 300
PROJECT GEMINI

Rucksack

    1 Survival light

        1 Strobe light
        1 Flash light

    4 Fish hooks

    Fish line

    2 Sewing Needles

    Thread

    1 Compass

    1 Fire Starter

    4 Fire Fuel

    1 Whistle

    1 Signal Mirror

    14 Water Purification tablets

    1 De-salter kit (less can)

        8 De-salter tablets

        1 Water Bag

        1 Repair kit

    1 Medication kit

        6 Tablet Packets

        1 Small Injector (1 CC)

        1 Large Injector (2 CC)

        1 3 x 3 Compress

        1 12 x 12 Aluminum Foil

        1 Tube Zink Oxide

    1 pr Sun glasses

SEDR 300
PROJECT GEMINI

1 Radio/Beacon

The forward survival kit, mounted on the forward surface of the backboard to the left of the pilot's shoulder, contains the following:

1 Water container with 3 lb of water

1 Machete with sheath

PYROTECHNIC DEVICES

There are 18 pyrotechnic devices incorporated in the cabin all of which pertain to seat ejection, restraint release and parachute deployment. The pyrotechnic devices are: 2 hatch actuators, 2 seat rocket catapults, 2 ballute deployment and release mechanisms, 2 backboard and egress kit jettisons, 2 drogue mortars, 2 harness release actuators, 2 seat/man separator actuators, 2 hatch actuator initiators and 2 hatch MDF harnesses. The pyrotechnic devices, except the drogue mortar, are safetied by stowing the ejection control handle and installing a safety pin through the mechanically actuated initiator and a pip pin through the egress kit. The safety pins will prevent seat ejection if control handle is inadvertently pulled.

DEBRIS GUARDS

Zero gravity in space poses problems with items not attached to the cabin interior. Under normal gravity conditions, objects tend to remain fixed when placed on the floor or any other flat surface. During zero gravity, the movement of the cabin oxygen can displace heavy objects. Because of movement of objects during zero gravity, pilots must exercise a great deal of care to enclose or secure each item or piece of material during flight. As an aid in keeping debris away from areas not accessible to the pilots, debris guards have been installed. Areas requiring

SEDR 300
PROJECT GEMINI

protection against entry of debris are around the ejection seats and under the instrument panel. The instrument panel debris guards are shaped nonmagnetic wire mesh and held in place by metal fasteners. The debris guards around the ejection seats are taylored from neoprene coated nylon fabric and secured to the spacecraft and ejection seats by velcro tape. Fencing off the areas makes it easier for the pilots to capture any floating object when policing the cabin interior.

INSTRUMENT PANELS

Instrument panels, switch and circuit breaker panels and pedestal (Figure 3-5) are arranged to place controls and indicators within reach and convenient view of each crew member while in a full pressure suit. A swizzle stick, stowed by the overhead circuit breaker panel, enables a pilot to position switches and rotate selectors on the opposite side of the cabin. With this arrangement, one pilot can control the complete spacecraft and temporarily free the second pilot of all duties.

CABIN INTERIOR LIGHTING

Basic lighting provisions consists of three incandescent flood lamps. One lamp is located at each side of the crew compartment and one in the center for crew station interior and instrument panels. Full range intensity control is available through a rheostat on each lamp. A three position selector switch on each of the hatch-mounted panel floodlights allows selection of WHITE-OFF-RED. The center floodlight WHITE-OFF-RED, control switch is in the forward row of the overhead panel. The rheostat is located on the lamp fixture. The units contain individual bulbs for red and white light, with permanent filters and lenses to prevent

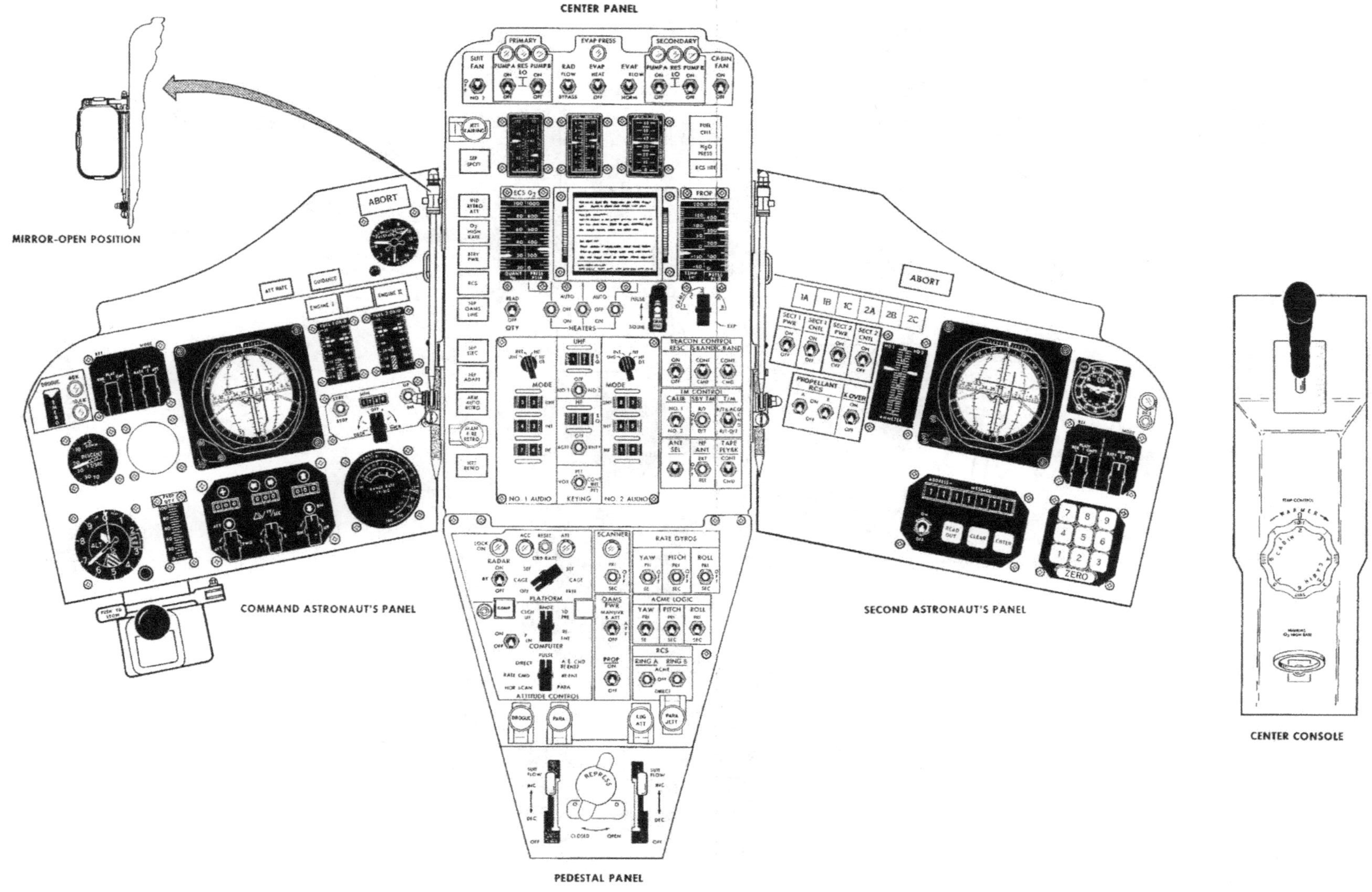

SEDR 300
PROJECT GEMINI

CENTER CONSOLE

SECOND ASTRONAUT'S PANEL

CENTER PANEL

COMMAND ASTRONAUT'S PANEL

PEDESTAL PANEL

MIRROR-OPEN POSITION

Figure 3-5 Instrument Panels and Controls (Sheet 1 of 5) (S/C 3)

3-21

SEDR 300

PROJECT GEMINI

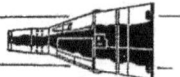

Figure 3-5 Instrument Panels and Controls (Sheet 1 of 5) (S/C 3)

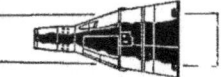
CENTER PANEL

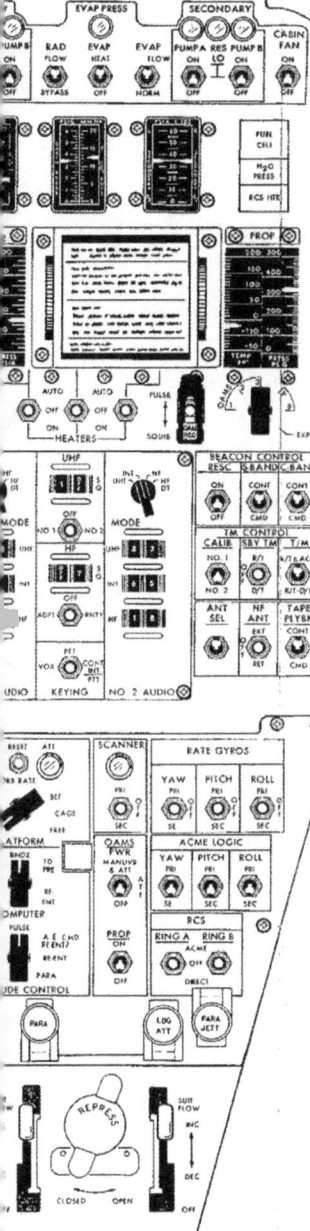
PEDESTAL PANEL

(C 3)

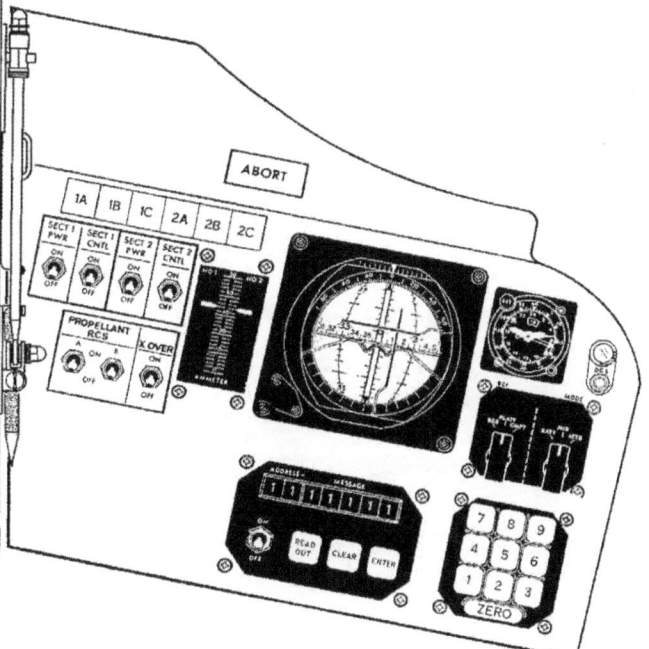
SECOND ASTRONAUT'S PANEL

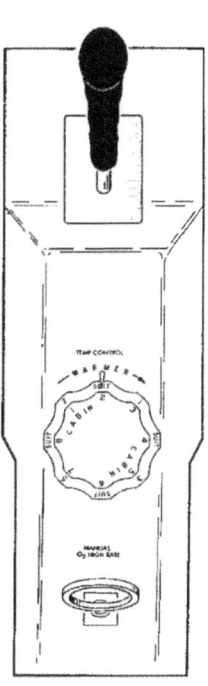
CENTER CONSOLE

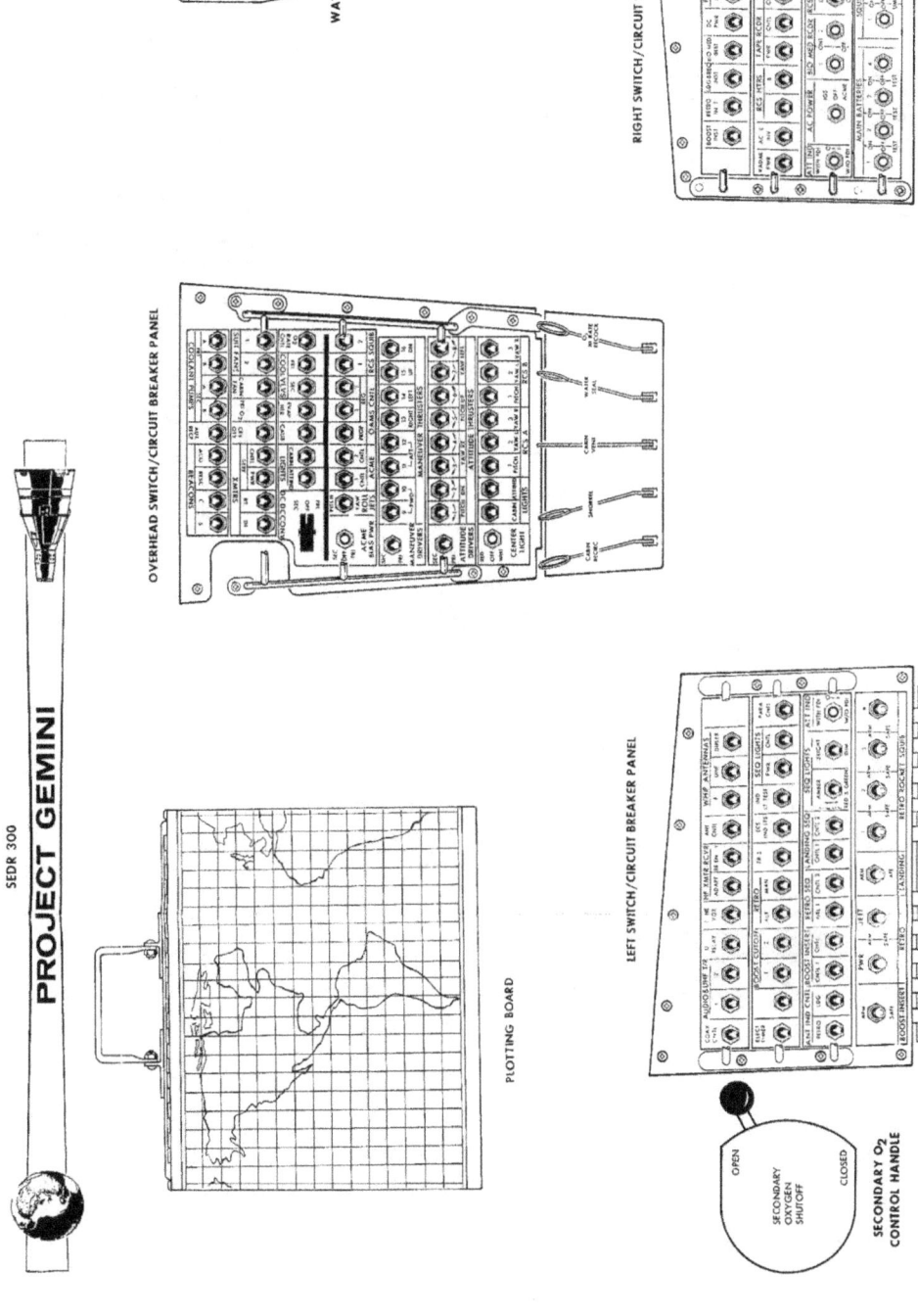

Figure 3-5 Instrument Panels and Controls (Sheet 2 of 5) (S/C 3)

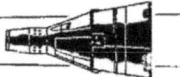

PROJECT GEMINI

SEDR 300

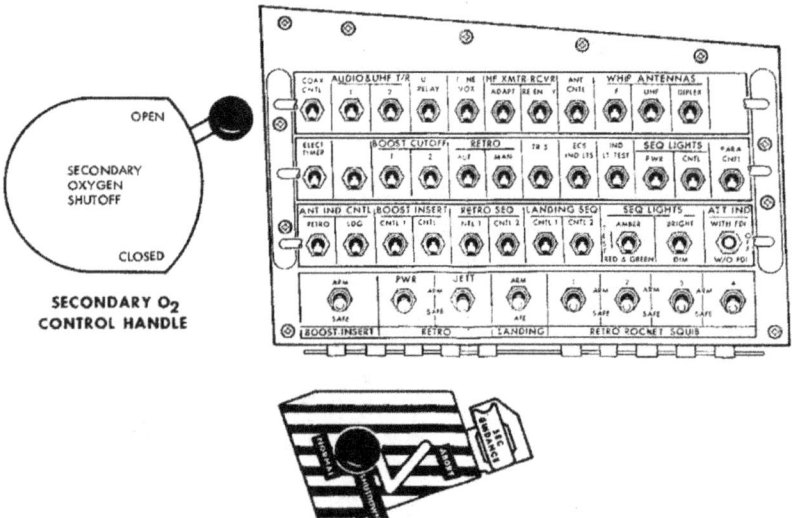

Figure 3-5 Instrument Panels and Controls (Sheet 2 of 5) (S/C 3)

OVERHEAD SWITCH/CIRCUIT BREAKER PANEL

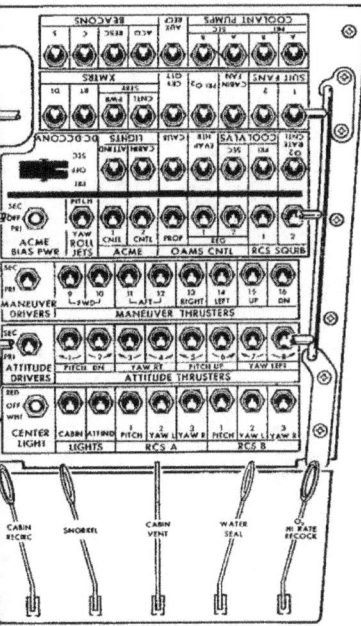

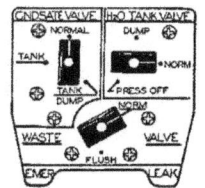

WATER MANAGEMENT PANEL

RIGHT SWITCH/CIRCUIT BREAKER PANEL

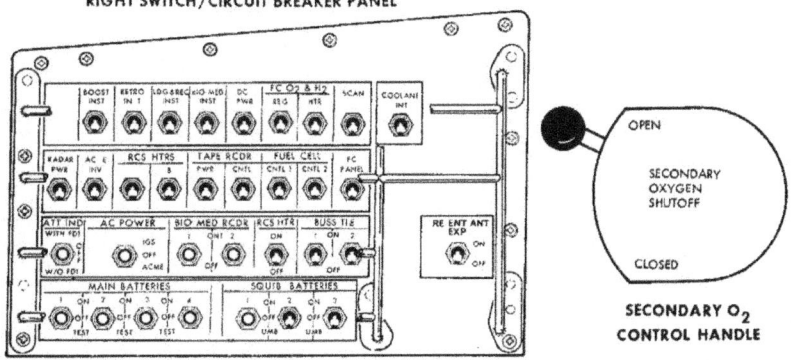

SECONDARY O₂ CONTROL HANDLE

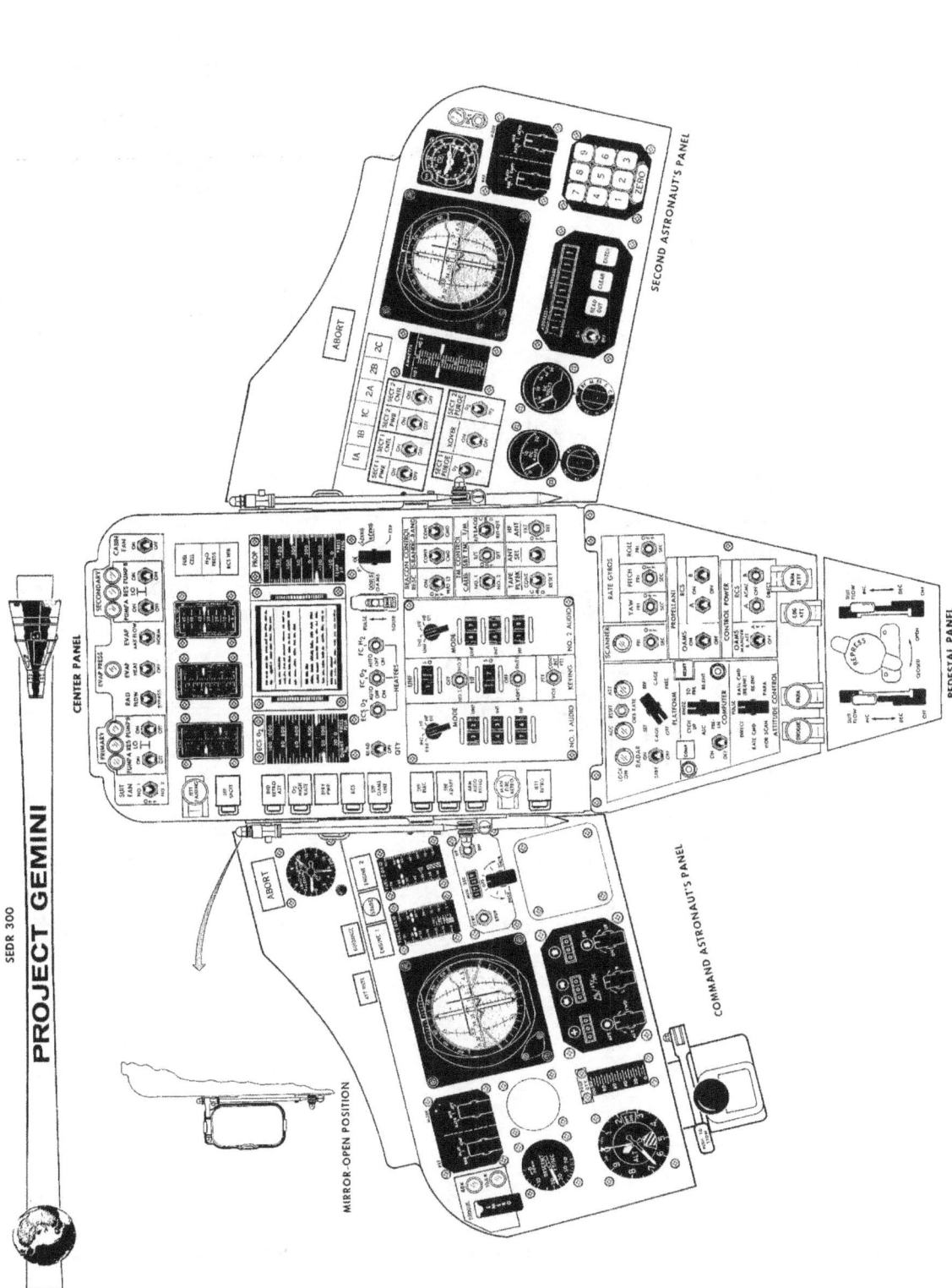

Figure 3-5 Instrument Panels and Controls (Sheet 3 of 5) (S/C 4)

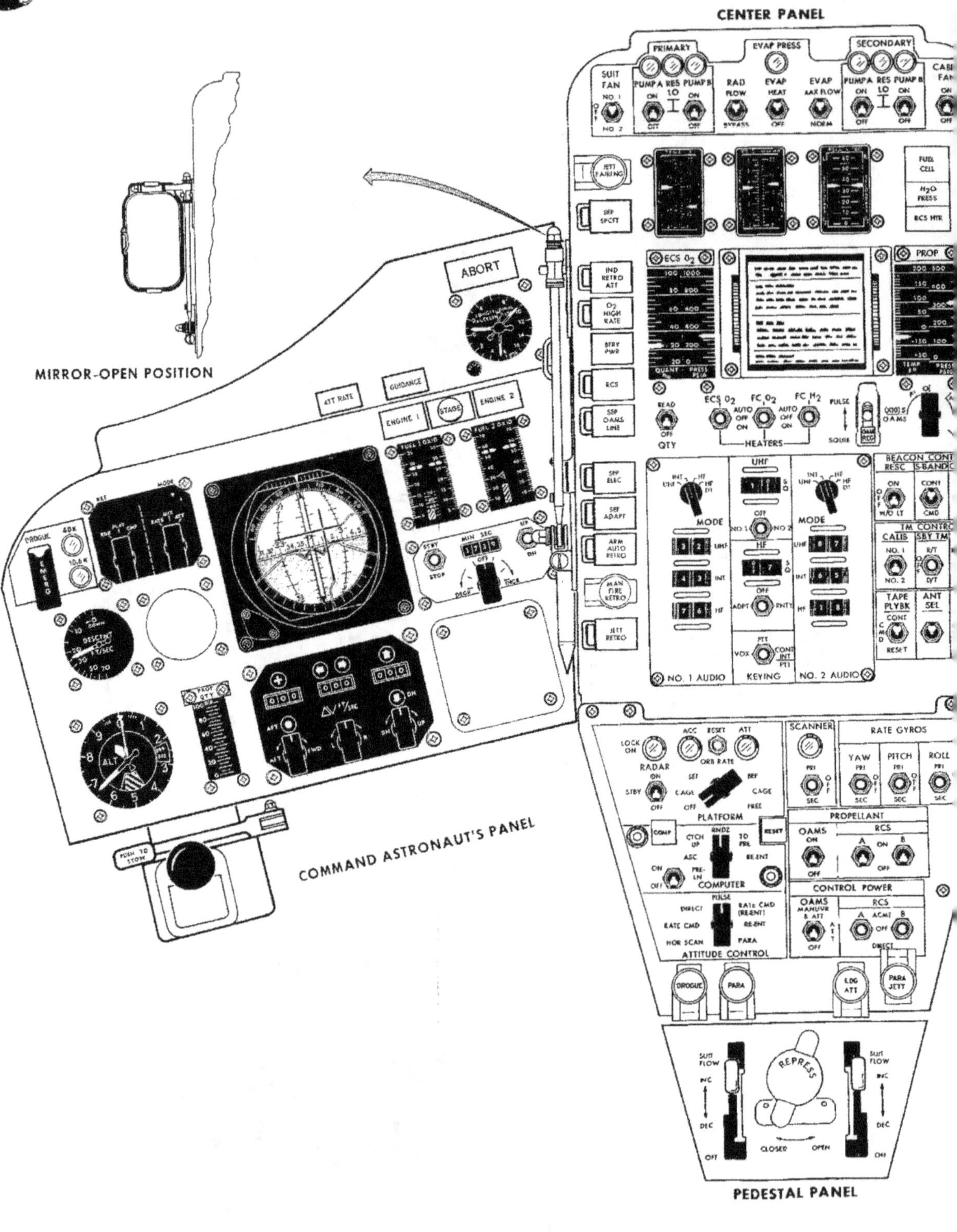

SEDR 300

 **PROJECT GEMINI**

Figure 3-5 Instrument Panels and Controls (Sheet 3 of 5) (S/C 4)

GEMINI

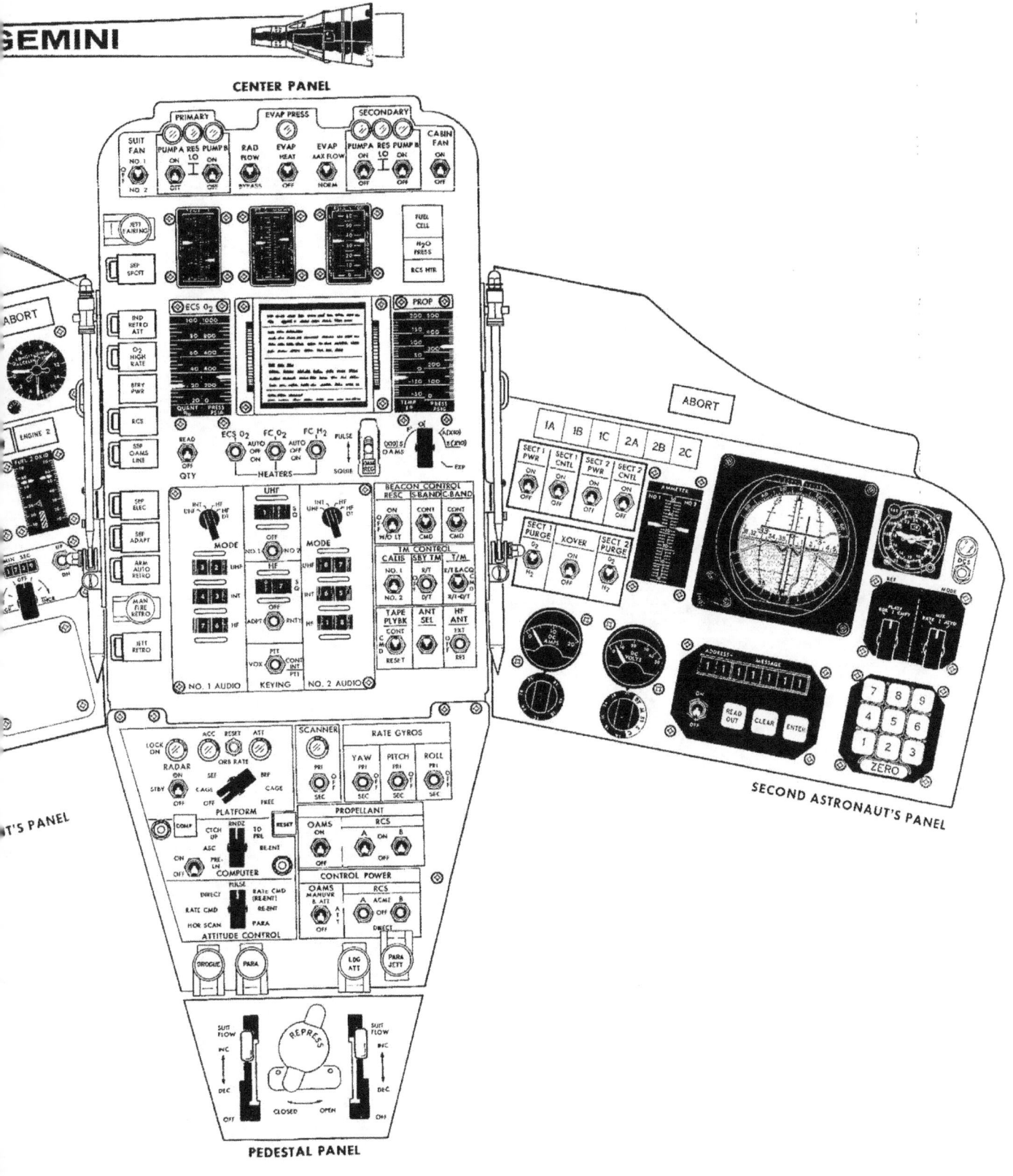

Panels and Controls (Sheet 3 of 5) (S/C 4)

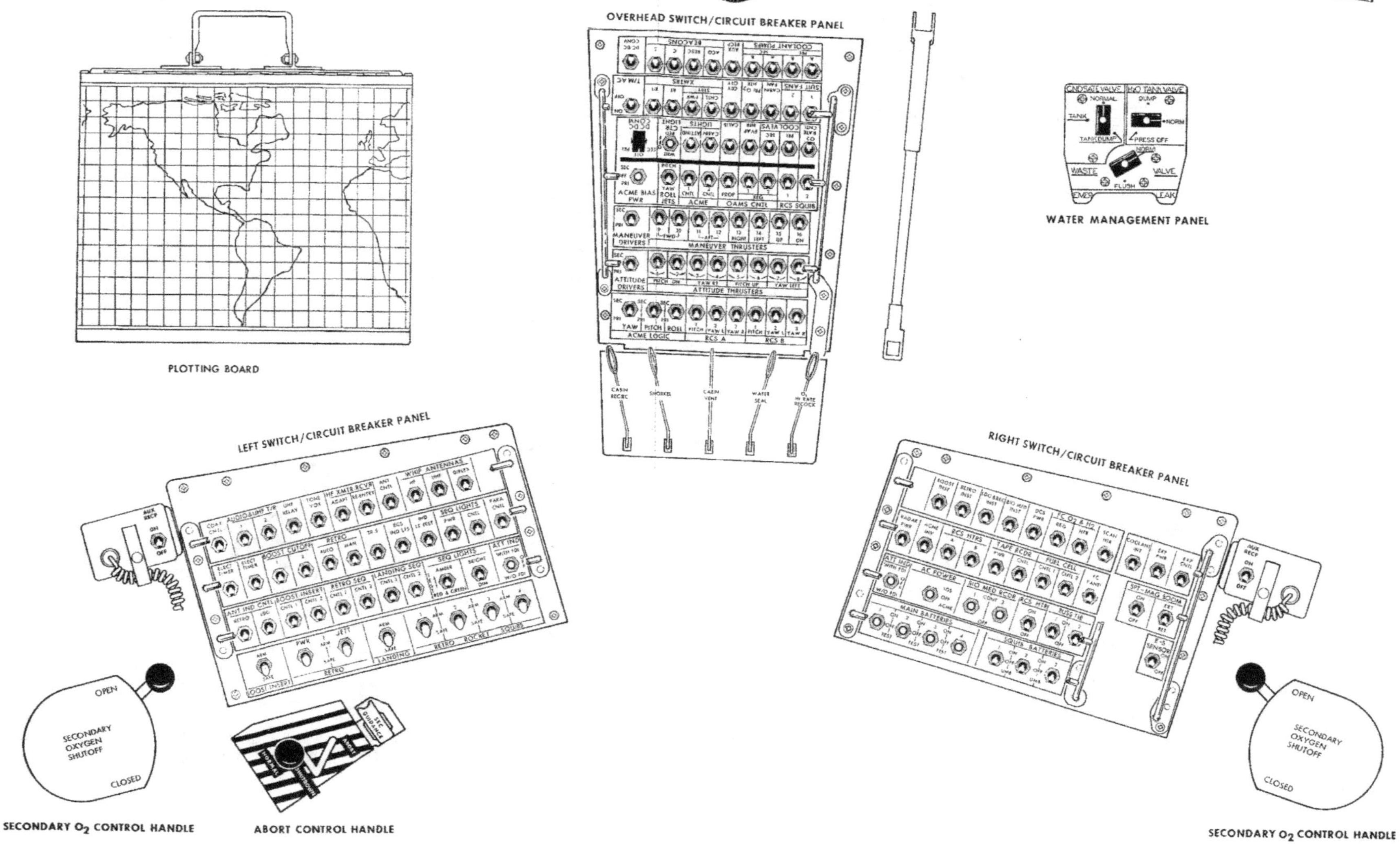

Figure 3-5 Instrument Panels and Controls (Sheet 4 of 5) (S/C 4)

OVERHEAD SWITCH/CIRCUIT BREAKER PA

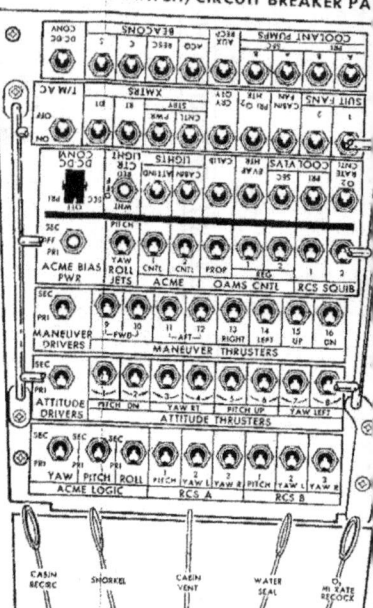

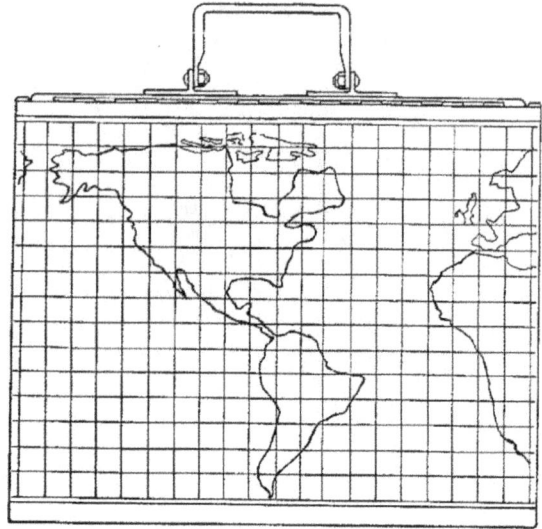

PLOTTING BOARD

LEFT SWITCH/CIRCUIT BREAKER PANEL

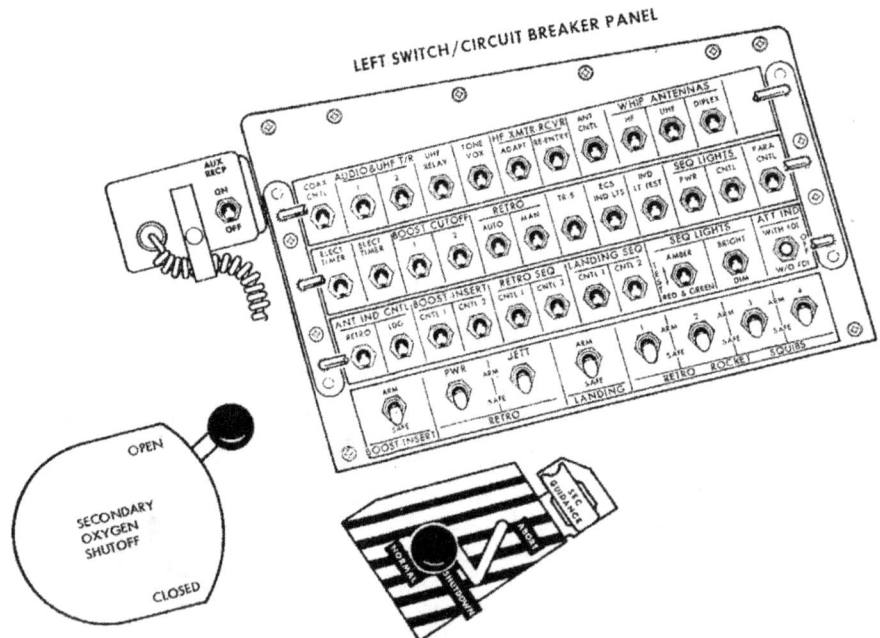

SECONDARY O₂ CONTROL HANDLE   ABORT CONTROL HANDLE

Figur

SEDR 300
PROJECT GEMINI

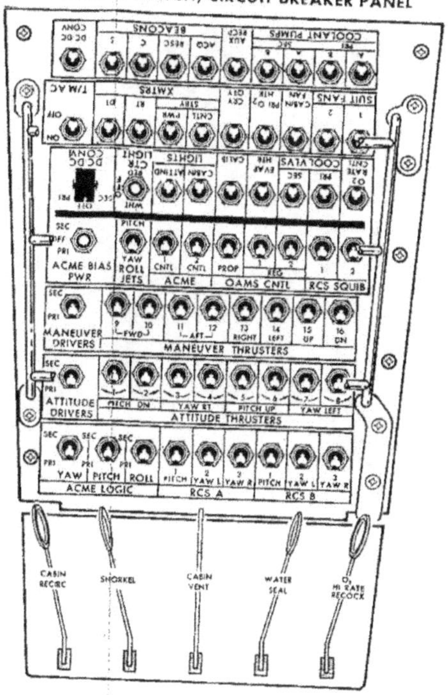

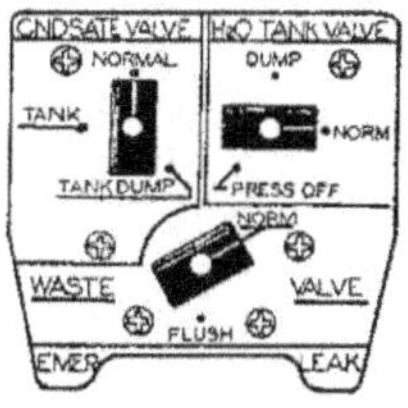

Figure 3-5 Instrument Panels and Controls (Sheet 4 of 5) (S/C 4)

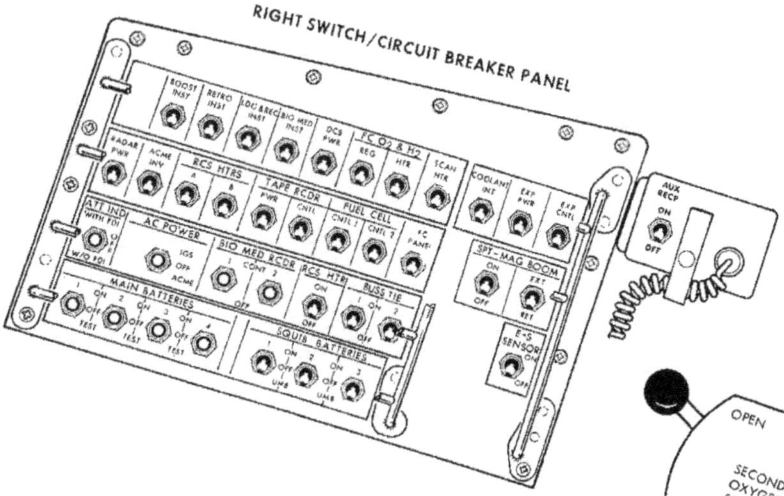

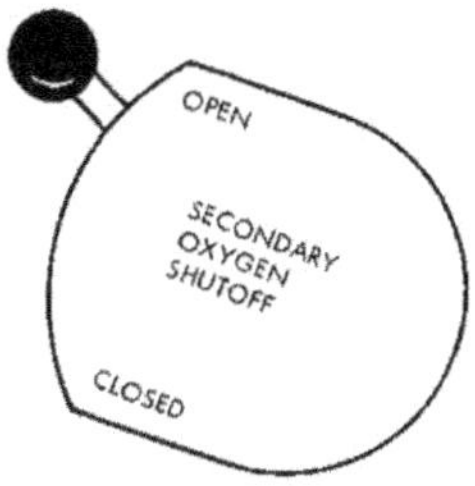

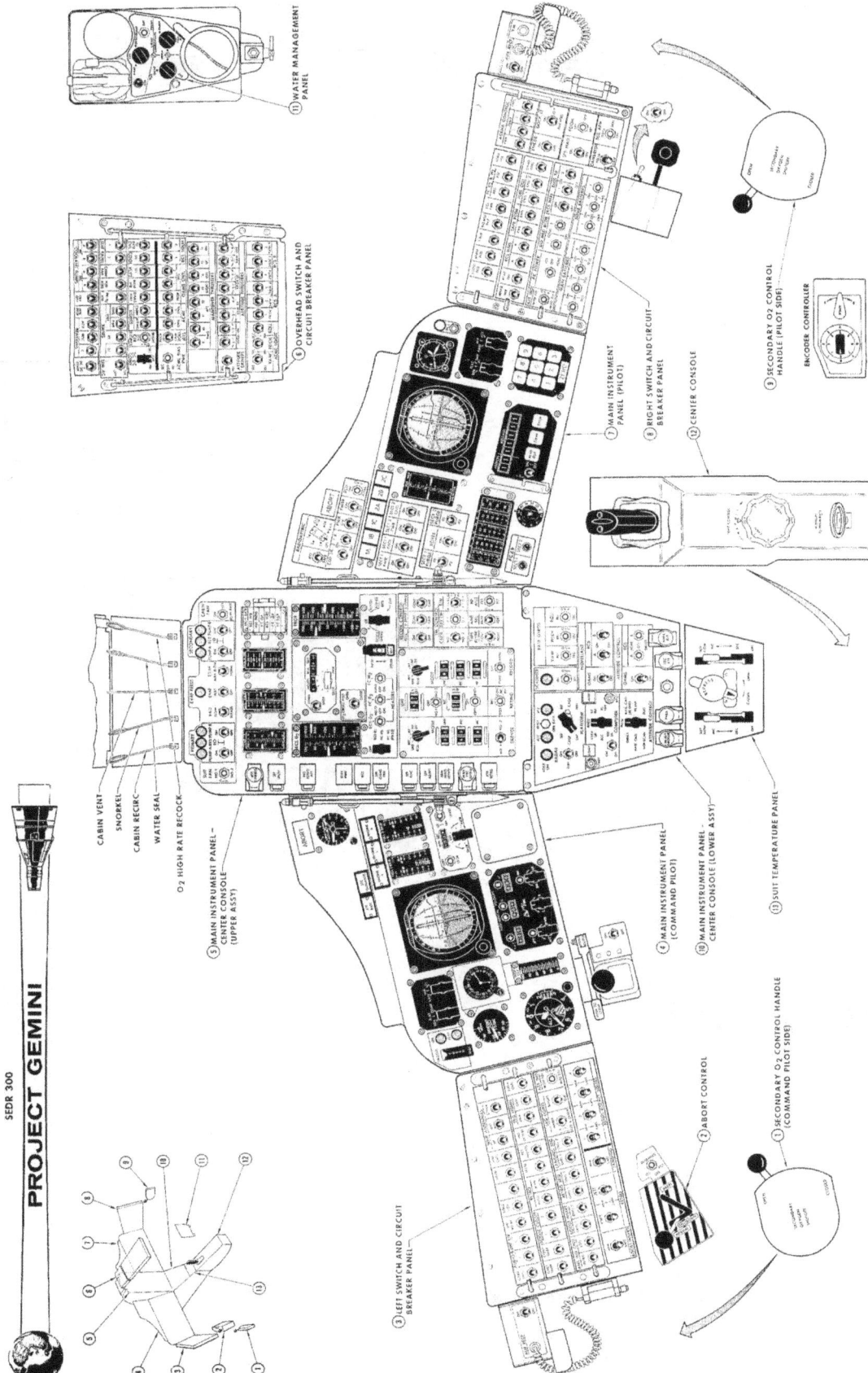

Figure 3-5 Instrument Panels and Displays- S/C 7 (Sheet 5 of 5)

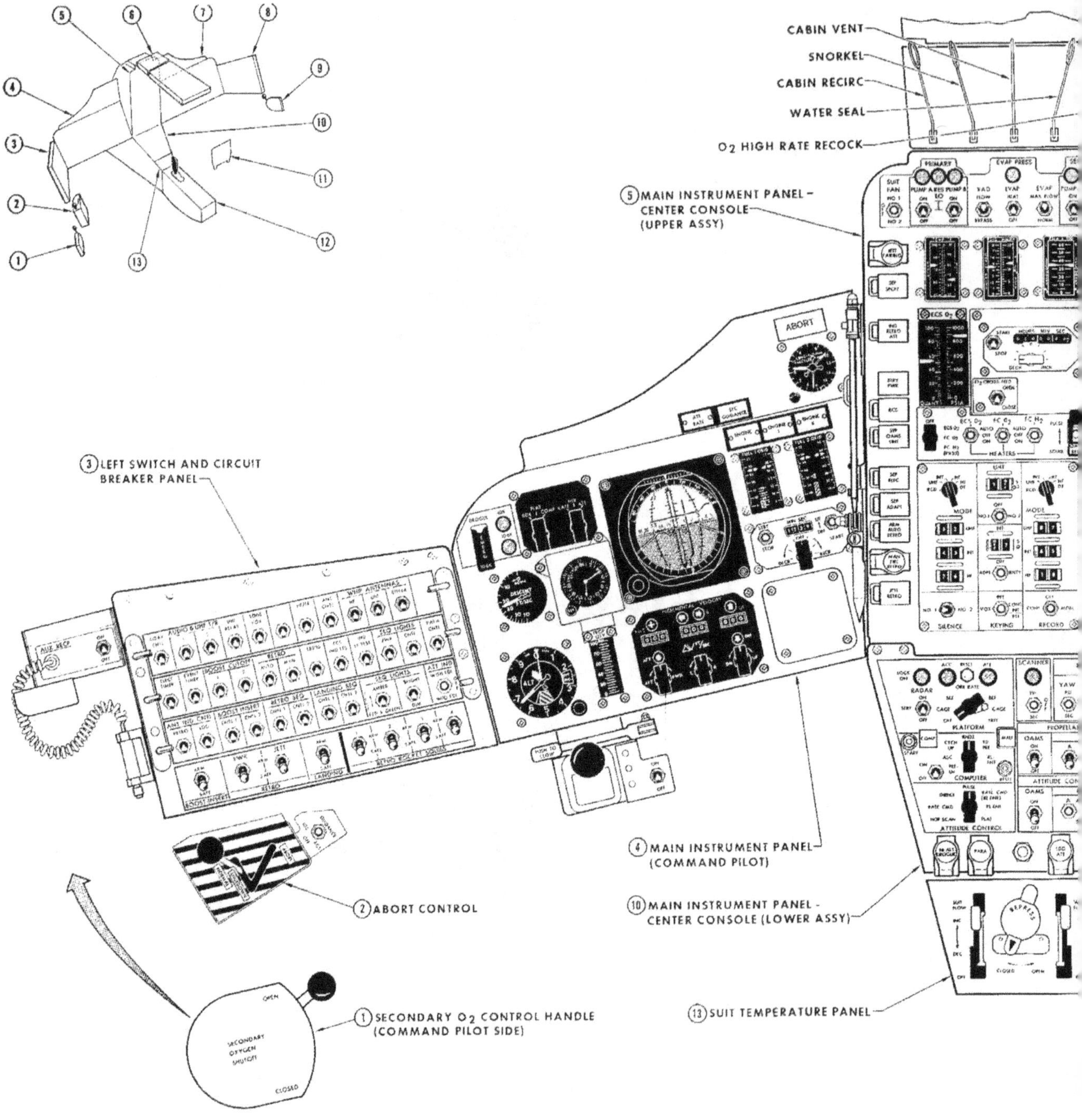

Figure 3-5 Instrument Panels and Displays- S/C 7 (Sheet 5 of 5)

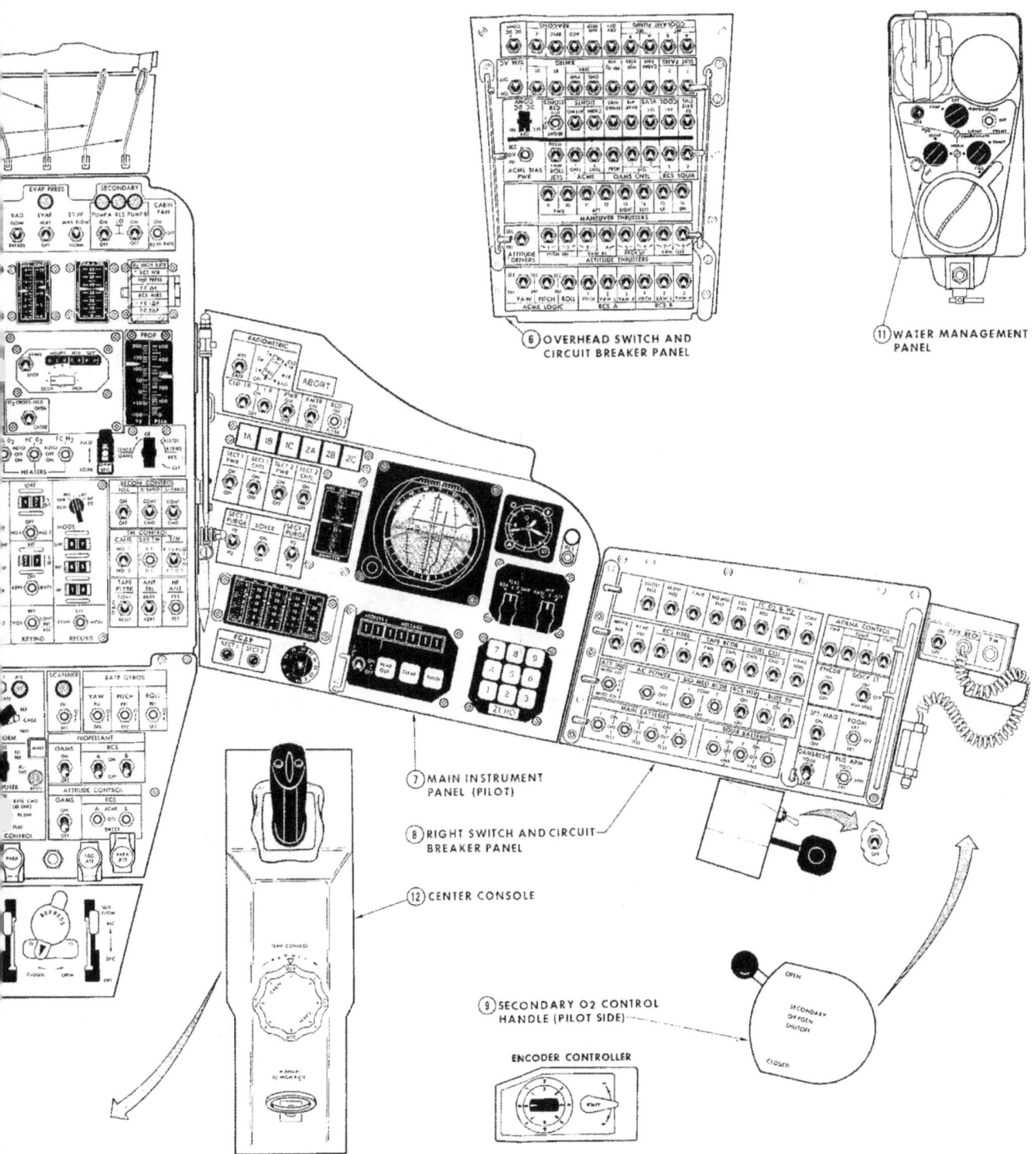

3-25

light leakage. Fingertip lights are provided on the gloves of the NASA furnished pressure suits. Mechanically dimmed white utility lights are stowed in quick release mounts on each cabin wall immediately aft of the switch/circuit breaker panels. These lights are powered from the spacecraft utility electrical outlets through a spiral retracting cord which is stowed when not in use.

ELECTRICAL OUTLETS

The two receptacles, powered by the spacecraft electrical system, are installed on brackets immediately aft of the left and right switch/circuit breaker panels. These receptacles are controlled by adjacent on-off switches and are used for powering the utility light or other electrical equipment.

STATIC SYSTEM

The static pressure system is employed to operate the rate of descent indicator, altimeter, and to supply pressure to the static pressure transducer for instrumentation. The static system is also utilized to provide a differential pressure for the cabin pressure transducer. The static ports (Figure 3-6), used for atmospheric pressure pick-up, are located in the small end of the spacecraft conical section. The static port (Figure 3-6), used for differential pressure pick-up, is located on the forward surface of the small pressure bulkhead.

FOOD, WATER AND EQUIPMENT STOWAGE

Containers to left, right and aft of pilots (Figure 3-7) are provided for equipment and food storage. Although minor changes in storage containers are dictated by mission requirements, the main containers are as follows: Center line stowage box, used for delicate instruments; right aft stowage box, used to stow easily

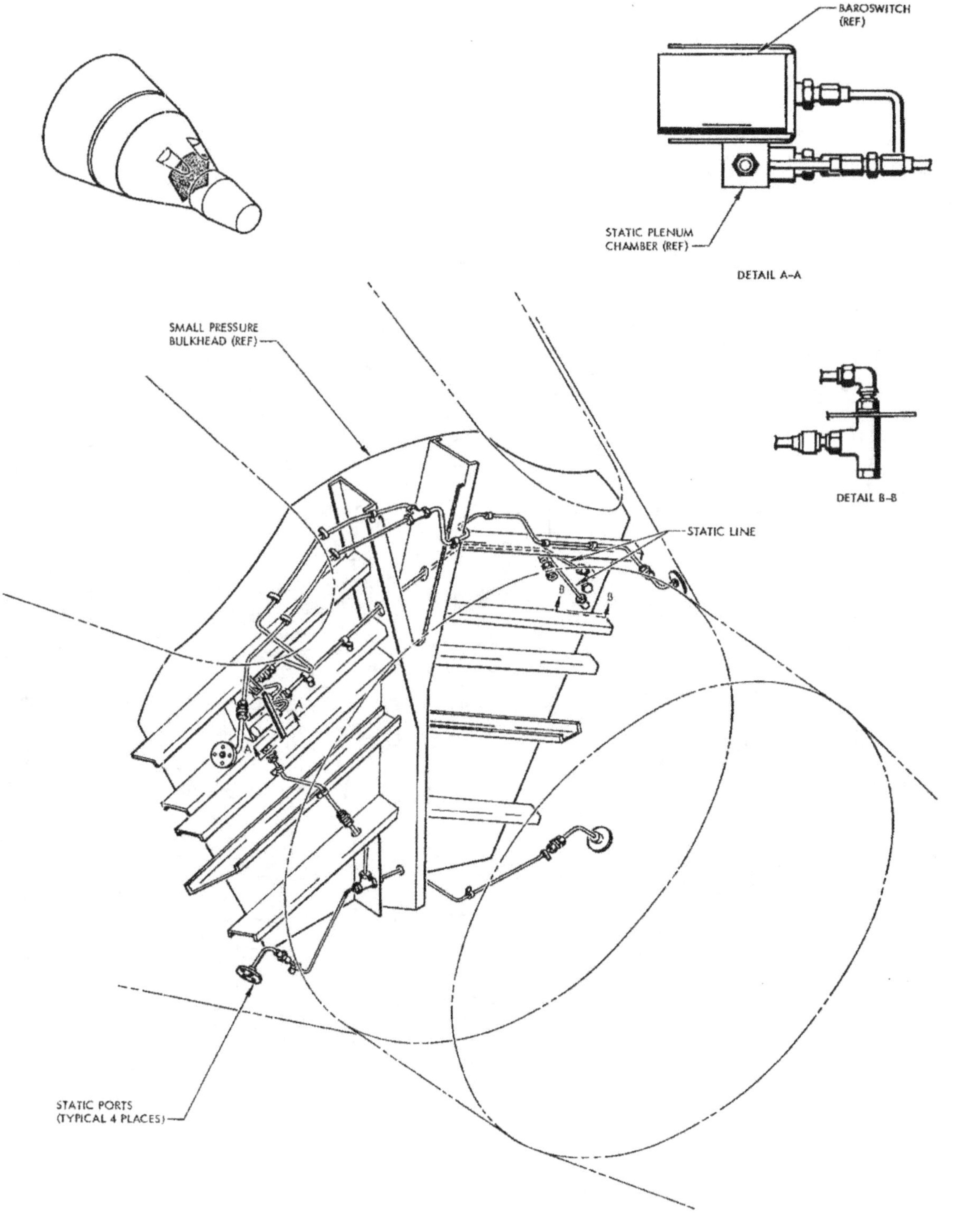

SEDR 300
PROJECT GEMINI

Figure 3-6 Static System

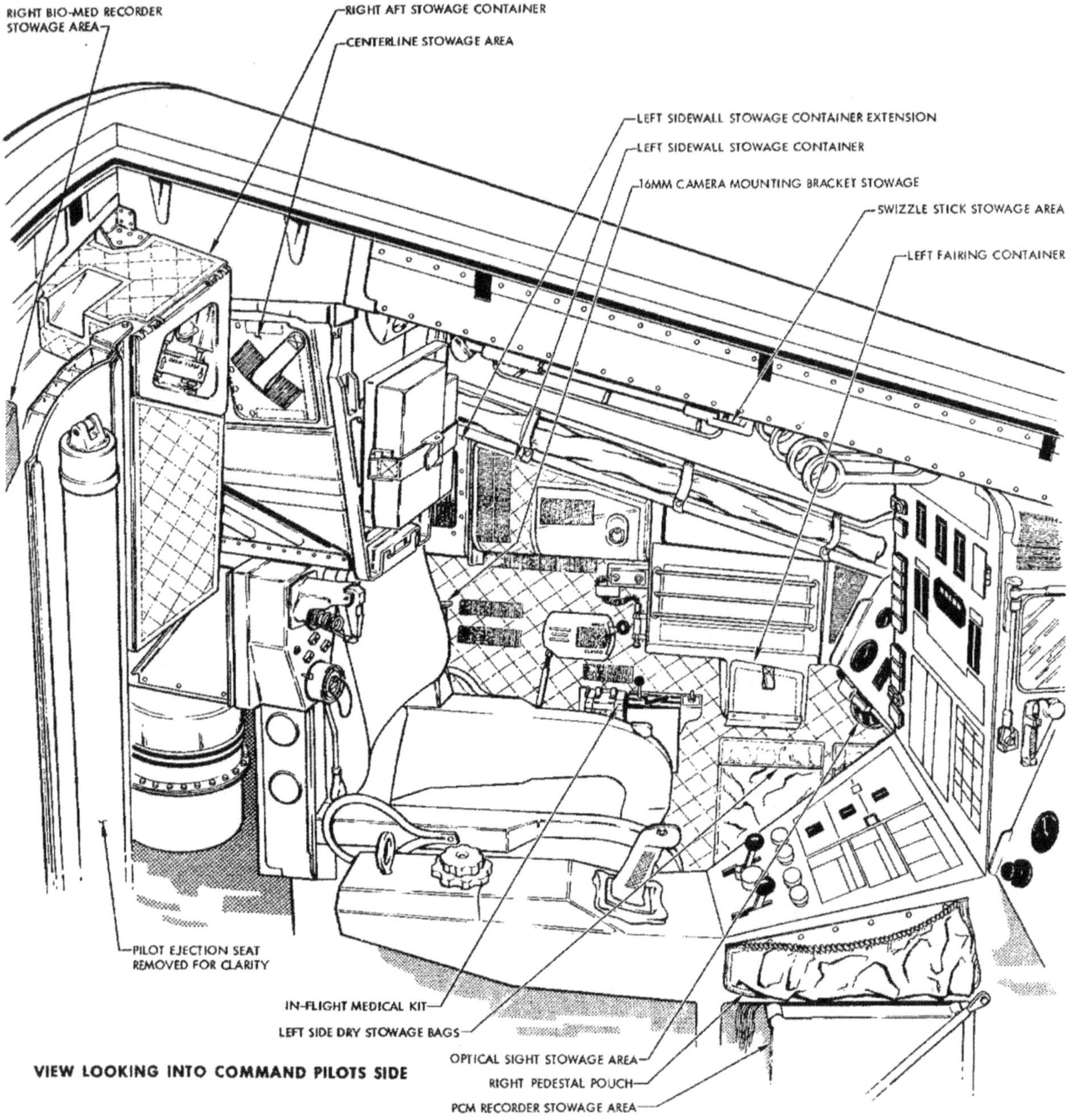

SEDR 300

PROJECT GEMINI

Figure 3-7 Spacecraft Interior Stowage Areas-S/C 7 (Sheet 1 of 2)

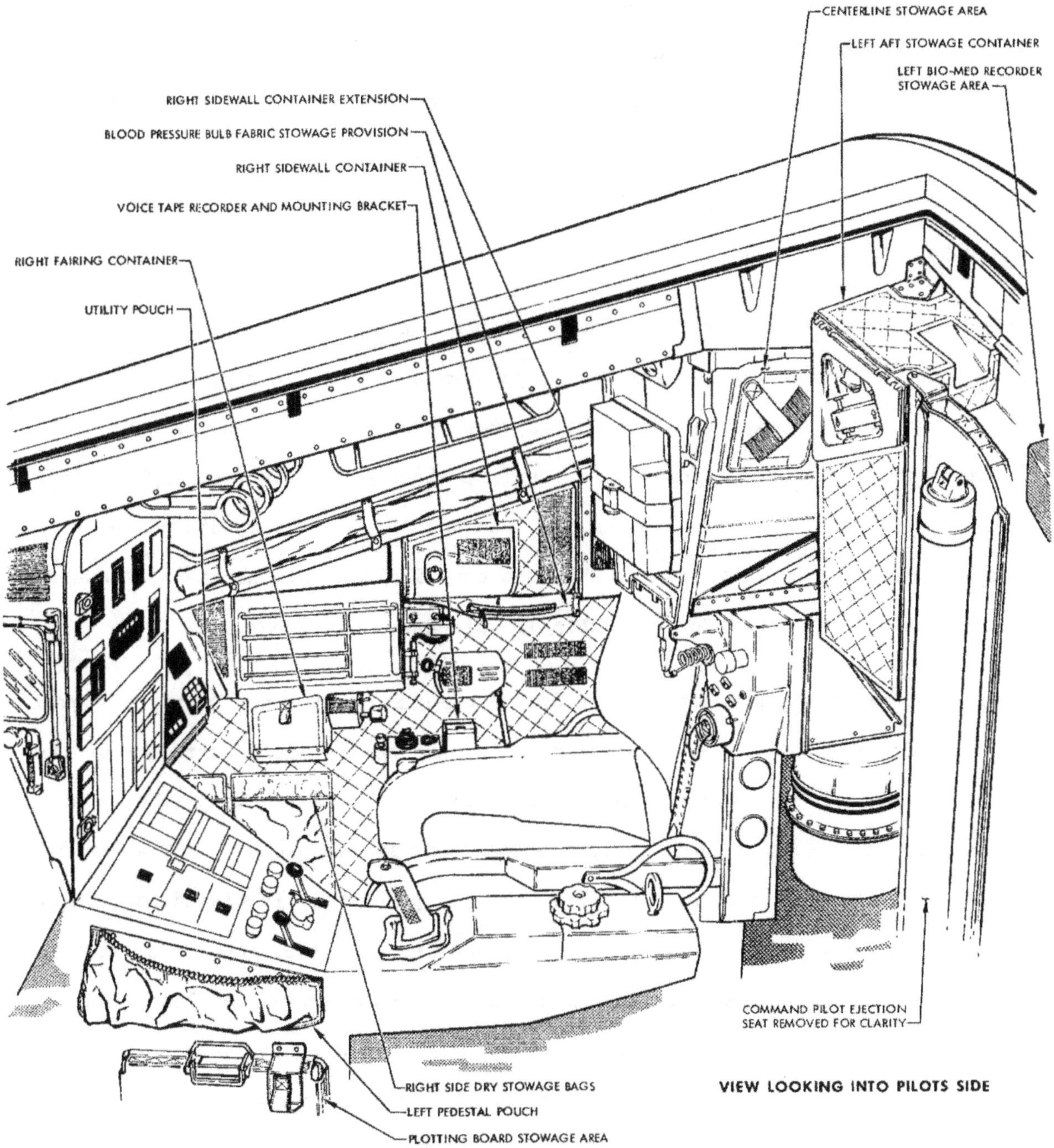

SEDR 300
PROJECT GEMINI

Figure 3-7 Spacecraft Interior Stowage Areas - S/C 7 (Sheet 2 of 2)

packaged equipment; left aft stowage box, used to stow food packages; right and left sidewall stowage boxes, used to stow small pieces of equipment; left and right fabric covered sidewall stowage boxes, used to stow lightweight head sets and sidewall stowage box extensions. The sidewall stowage box extensions are not required for mission equipment at this time. Equipment stowed in the above boxes may change with each mission.

Larger pieces of equipment, emergency equipment or equipment used on every flight have special stowage brackets or fabric pouches positioned throughout the interior of the spacecraft. Examples of specific stowage brackets are as follows: adapter mirror assembly, stowed on centerline stowage box door; inflight medical kit, stowed aft of abort control handle; and the optical sight, stowed under command pilot's instrument panel. Without counting the food packages, stowage facilities are furnished for more than 125 pieces of equipment.

During flight, the right aft stowage box is cleared and the equipment is stowed with velcro tape on the spacecraft sidewalls, and on the inside surfaces of the egress hatch. As debris accumulates during flight, it is placed in the right aft stowage box. Prior to descent, the equipment is re-stowed. Only a general rule can be applied to stowage at this time. Exposed film is placed in insulated containers, previously occupied by cameras and lens, in the center line stowage box. The left aft stowage box is filled and the remainder of the loose equipment is divided among the sidewall stowage boxes on a planned basis. The right aft stowage box, with the flight debris, is seldom used to stow equipment for re-entry.

 **SEDR 300**

PROJECT GEMINI

A water storage container, with a 16-pound capacity, is located forward of the aft pressure bulkhead, between the seats. As the water is used from the main storage container, it is replenished by the water stowed in the adapter section. Drinking is accomplished by means of a tube and manual valve system. Control of the manually operated water system is by means of a 3-knob panel located between the pilots at approximately shoulder level. Food and water will be sufficient for the mission and a postlanding period of 48 hours.

WASTE DISPOSAL

Feces will be collected in a glove-like plastic bag. Provisions are incorporated in the food storage containers for the placement of used and unused feces collector bags. Urine will be collected by means of a horn-shaped receptacle with a self-adjusting opening and directed into an intermediate bellows type container. Urine is disposed of by directing the liquid overboard.

STOWAGE PROVISIONS

Personal stowage facilities are provided for retaining removed portions of the pressure suit and other equipment. These provisions consist of Velcro striped areas on the floor, on the side wall and on the inside surface of the hatch. Items to be stowed in these areas incorporate mating Velcro tape patches.

In addition, several fabric pouches are positioned throughout the cabin interior for stowage of items without Velcro tape patches. These pouches must be kept closed to prevent objects from floating into the cabin. Cabin pouches are kept closed with flaps or elastic bands.

SEQUENCE SYSTEM

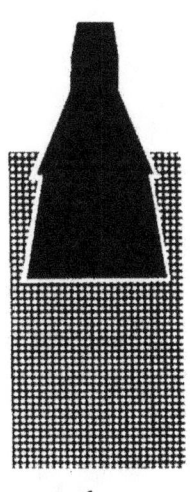

Section IV

TABLE OF CONTENTS

TITLE	PAGE
SYSTEM DESCRIPTION	4-3
SYSTEM OPERATION	4-4
SYSTEM UNITS	4-41

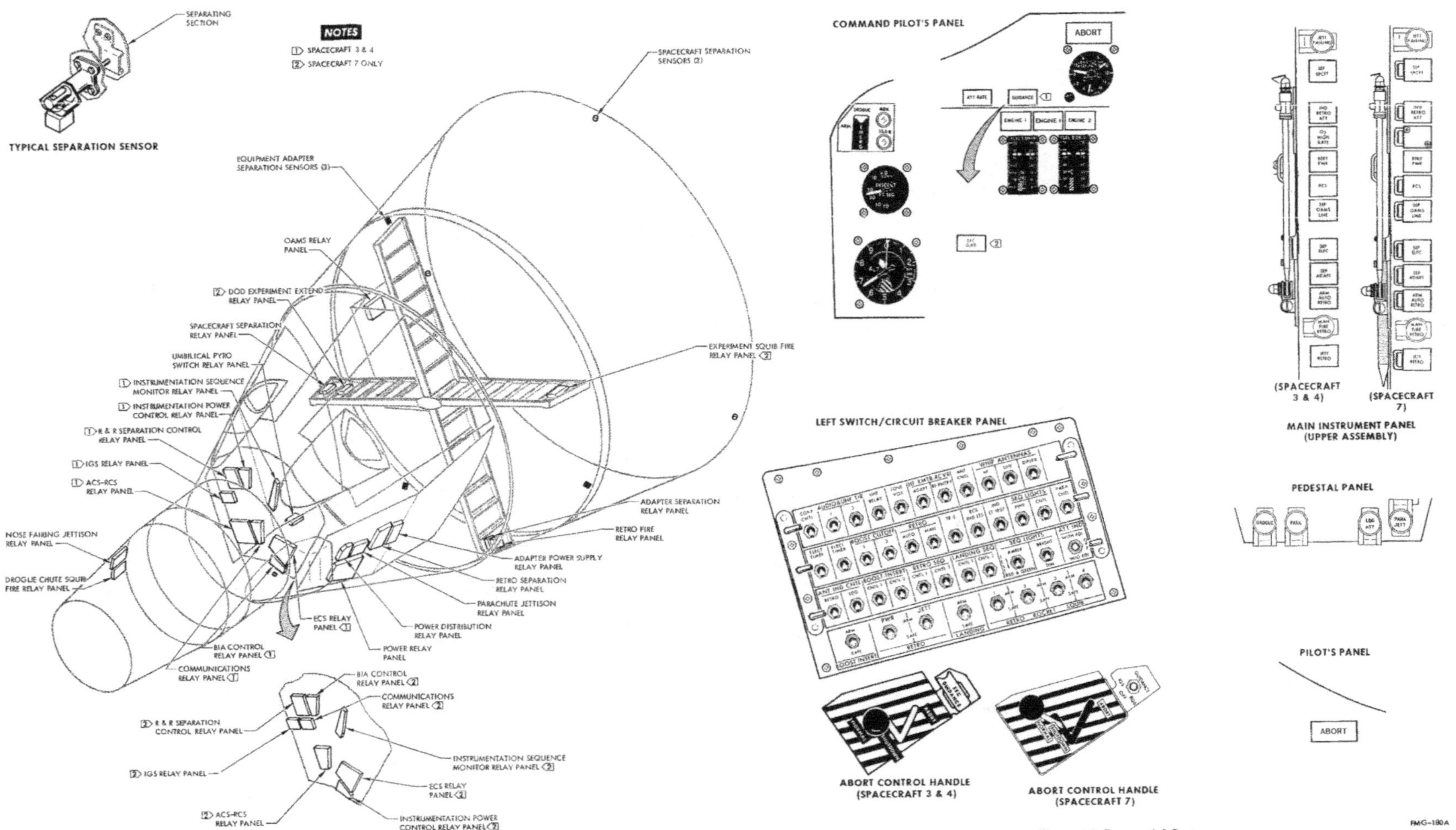

Figure 4-1 Sequential System

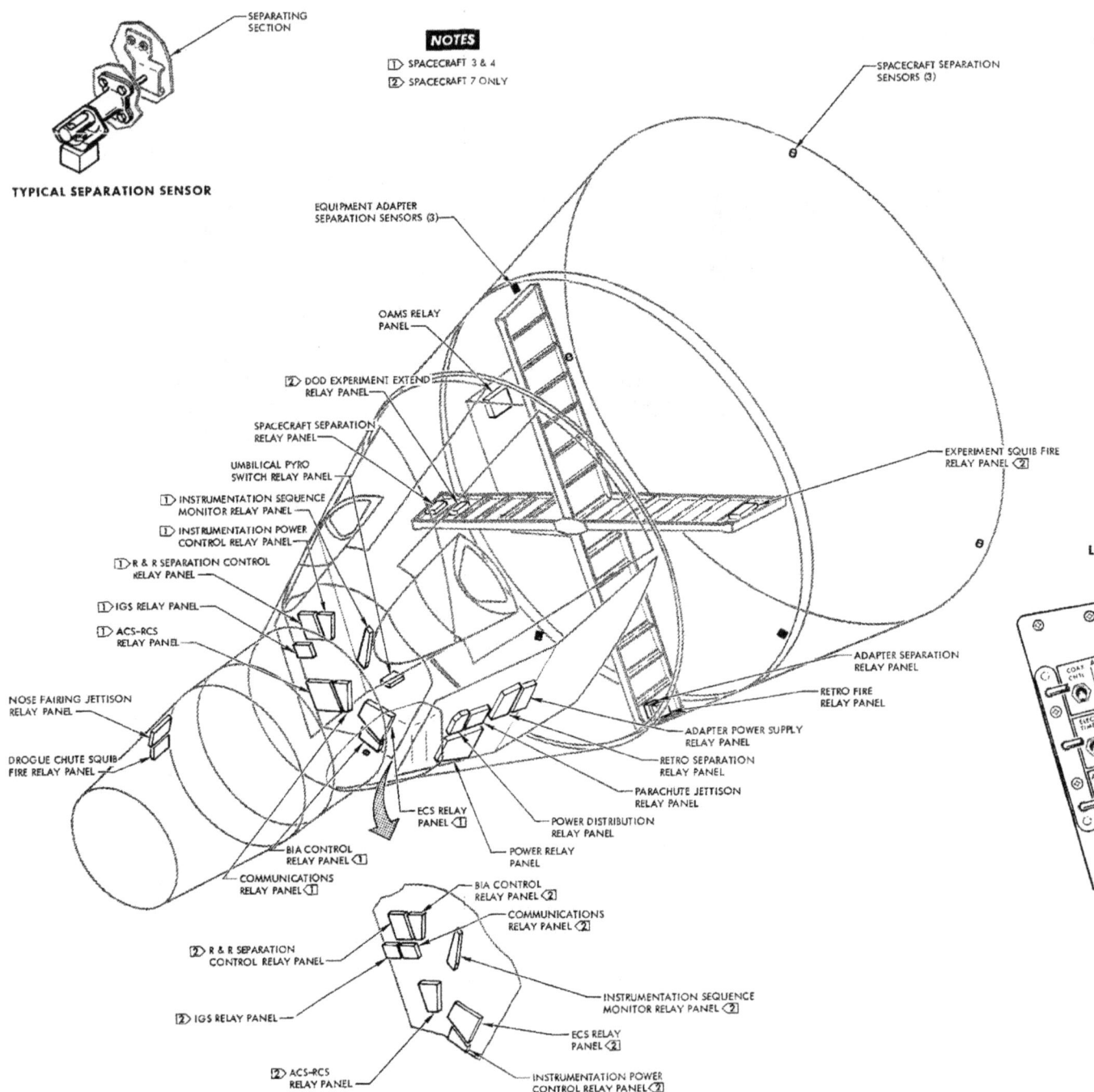

SEDR 300
PROJECT GEMINI

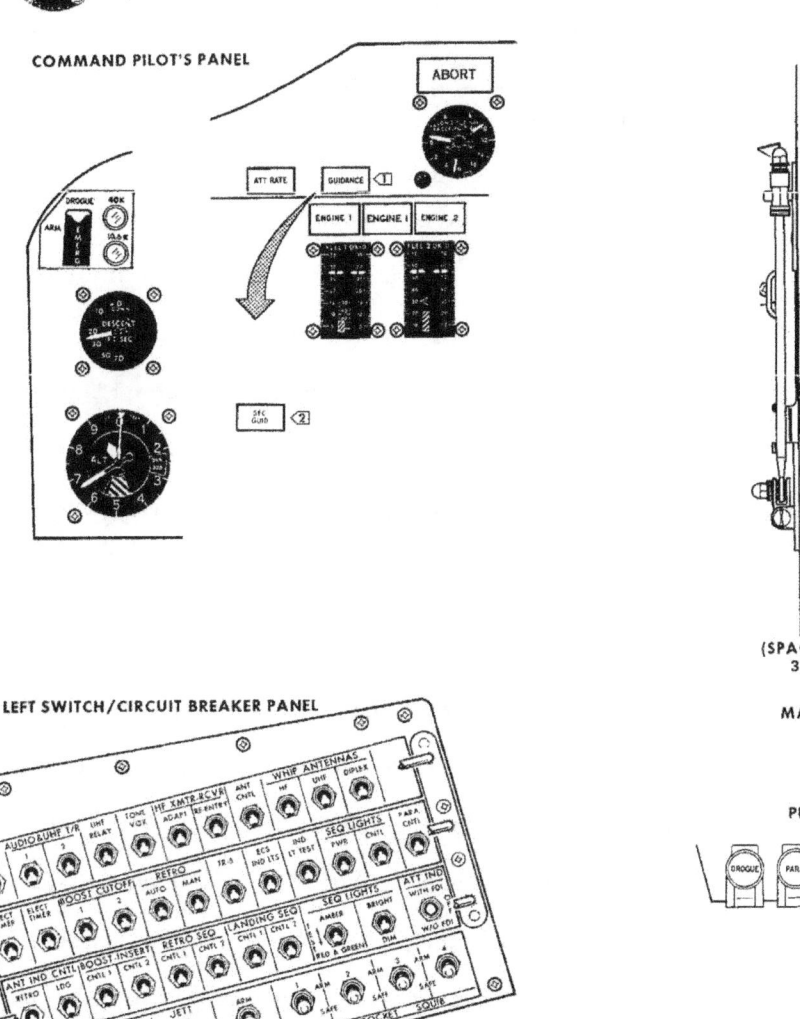

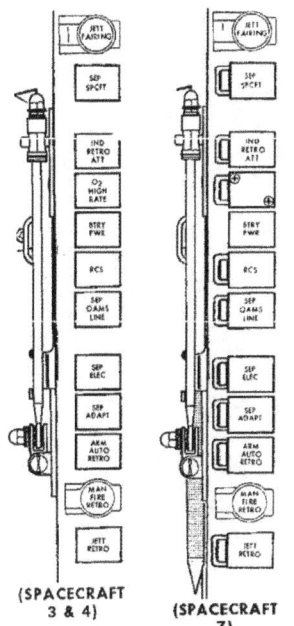

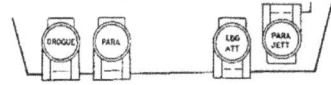

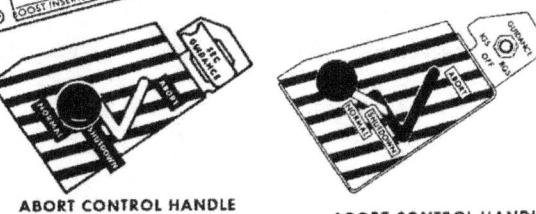

Figure 4-1 Sequential System

SEDR 300

PROJECT GEMINI

SECTION IV SEQUENTIAL SYSTEM

SYSTEM DESCRIPTION

The sequence (or sequential) system of Gemini Spacecrafts 3, 4, and 7 comprises those controls, indicators, relays, sensors and timing devices which provide semiautomatic control of the spacecraft and/or launch vehicle during the critical control times, but which are not parts of other systems. (See Figure 4-1.) The critical control times are: the time from booster engine ignition thru insertion into orbit; the time to prepare to go to retrograde thru post-landing; and the time to abort.

The Gemini crew do not fly the spacecraft during boost and insertion. The spacecraft is controlled by radio guidance or by the inertial guidance system, on-board computer and digital command system. The crew however do monitor certain cabin indicators to keep informed of the operation of the launch vehicle, to anticipate a crisis if one should develop, and to know if and when mission abort is mandatory. The crew assume more control of the spacecraft after second stage engine cutoff (SSECO). Spacecraft separation is accomplished and the final thrust required for the desired orbit is applied by the Command Pilot.

During orbit, the sequence system is in standby. The electronic timer, however, which is part of the time reference system, is counting down the time to go to retrograde.

At 5 minutes before retrograde time on S/C 3 and 4, or 256 seconds before retrograde on S/C 7, a sequence system relay is actuated, and several sequential system indicators illuminate amber. These indicators provide the crew with cues for necessary actions before going to retrograde. Again at 30 seconds before retrofire time, the crew is cued to separate the adapter section. The sequential system, if properly armed, will initiate retrofire automatically, and the crew redundantly initiates it manually as a safety precaution. During descent, altitude indicators remind the crew to deploy their chutes and extend antennas. After splash down, the sequence system is shut down.

The abort system is part of the sequential system. The abort system comprises the abort indicators, controls, relays, and pyrotechnics. The abort modes are seat ejection, modified retro abort, retro abort, and normal re-entry immediately after insertion. The mode of aborting selected is related to the spacecraft altitude when the command to abort is given.

SYSTEM OPERATION

The Sequence System is divided into twelve parts in order to simplify the explanation. These parts are prelaunch, lift-off, boost and staging, telemetry and guidance, abort, separation and insertion, orbit, prepare to go to retrograde, retrograde, re-entry, landing, and post-landing. Figure 4-2 shows these sequential parts, and the detailed functions which each part includes. This simplified block diagram is explained in the following paragraphs.

Prelaunch, lift-off, boost and staging, and separation and insertion are explained first. Orbit, telemetry, and guidance are discussed elsewhere in this manual, and therefore they are not explained in this Section. Prepare to go to retrograde, retrograde, re-entry, landing and post-landing are discussed next.

PROJECT GEMINI

SEDR 300

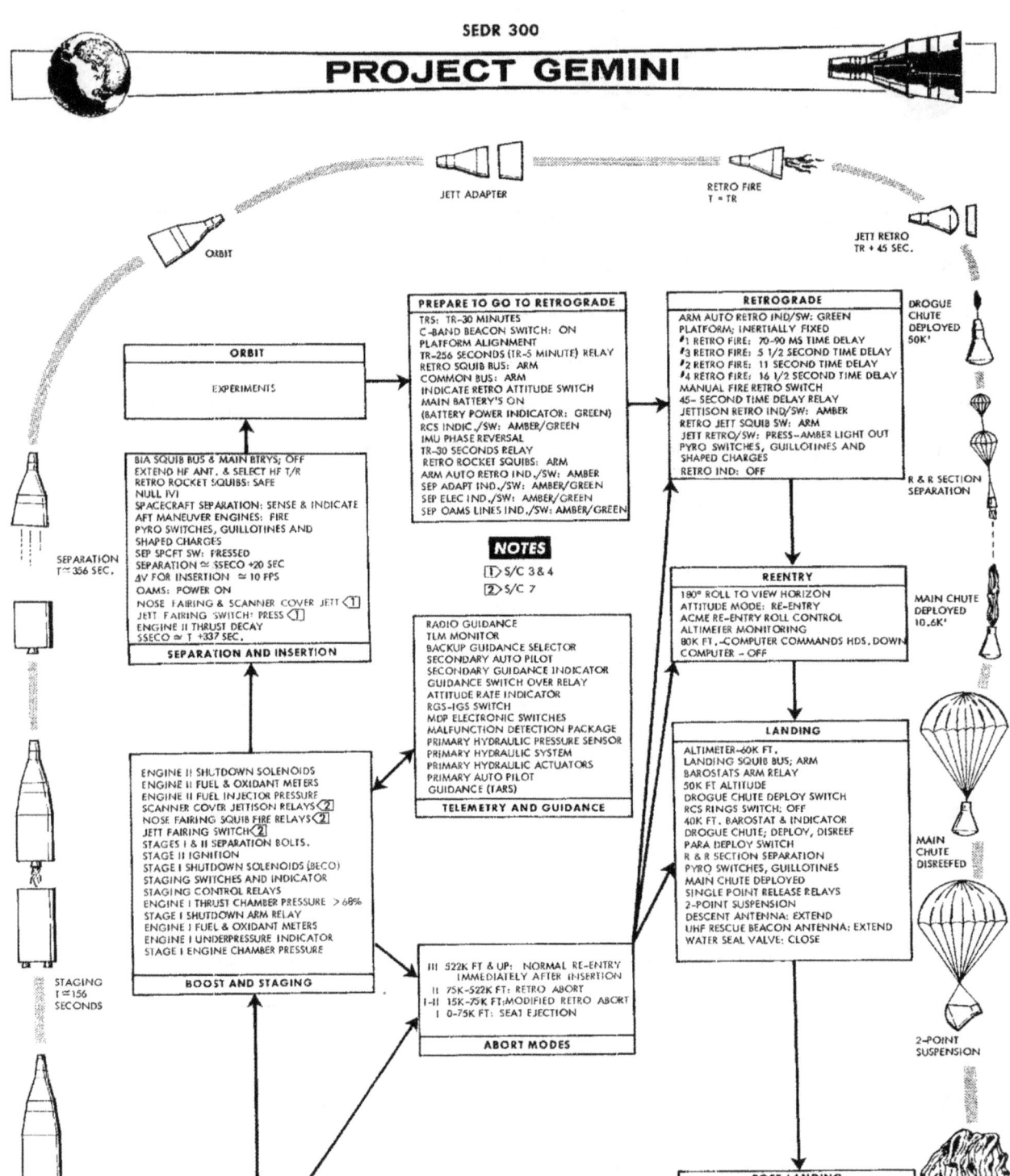

Figure 4-2 Sequential System Simplified Block Diagram

SEDR 300

PROJECT GEMINI

Abort is discussed last. Many of the sequential operations of abort mode II are normal parts of the retrograde sequence. Abort mode III is a normal re-entry maneuver.

PRELAUNCH

The Command Pilot and the Pilot enter the Gemini cabin and take their assigned crew stations. The hatches are closed and locked. The crew checks that both "D" rings are unstowed. The Command Pilot makes sure that the ABORT CONTROL handle is in the NORMAL position; the MANEUVER CONTROLLER is stowed; the altimeter is set; and the Incremental Velocity Indicator is zeroed. He verifies that the Sequence Panel telelights, the two ABORT lights, and the ATT RATE light, the GUIDANCE light, both ENGINE I lights, and the ENGINE II light are extinguished. He places the top three rows of circuit breakers on the Left Switch/Circuit Breaker Panel to the closed (up) position. He places the BOOST-INSERT switch and the RETRO ROCKET SQUIB switches in the bottom row to ARM, and the RETRO and LANDING switches to SAFE. He selects switches for gyro run-up and platform alignment, and performs computer checkout.

The Sequence Panel telelights are tested with the SEQ. LIGHTS TEST switch. The Pilot places the MAIN BATTERIES switches and the SQUIB BATTERIES switches to ON. Both pilots select and check their intercom and UHF communications. The remaining controls and indicators are also monitored or positioned as required. The Command Pilot reports all systems ready for launch.

LIFT OFF

When the prelaunch countdown reaches lift off time (T) minus 4 seconds, the first stage engine ignition signal is given from the blockhouse. Both first stage engines (SA1 and SA2) begin thrust chamber pressure buildup. Both ENGINE I indicators illuminate red but extinguish in about one second. When the thrust chamber pressure (TCP) of these two engines exceeds 77 percent of rated pressure, a two-second time delay is initiated in the blockhouse. If all systems remain "go" during this delay, the holddown-bolt fire command is given and the launch vehicle is committed to flight. First motion sensors detect launch vehicle ascent one and one-half inches off the pad, and energize time-zero relays in the blockhouse and the spacecraft. A 145-second shutdown arm time delay is initiated to prevent accidental booster engine shutdown prior to the scheduled staging time. The umbilical release command is given, disconnecting the environmental control system (ECS), equipment adapter, and re-entry umbilicals. The spacecraft computer is switched from the guidance inhibit mode to the guidance initiate mode and enabled to accept acceleration data. The lift off signal is also applied to the electronic timer and event timer. The electronic timer begins to count down the time to retrograde. The event timer begins to count up the time from lift off.

BOOST AND STAGING

As the missile continues to climb, the crew monitor the boost sequence indicators (Figure 4-3). The two ENGINE I underpressure indicators, the ATTITUDE overrate indicator and both ABORT indicators must remain extingusihed. The ENGINE II indicator illuminates amber. The STAGE 1 FUEL and OXID needles must indicate pressures within the required limits, and the ACCELEROMETER must indicate an increasing acceleration within prescribed limits for the flight time indicated by the event timer. The S/C crew monitor their indicators and report via UHF

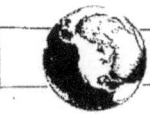

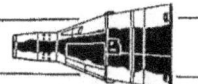

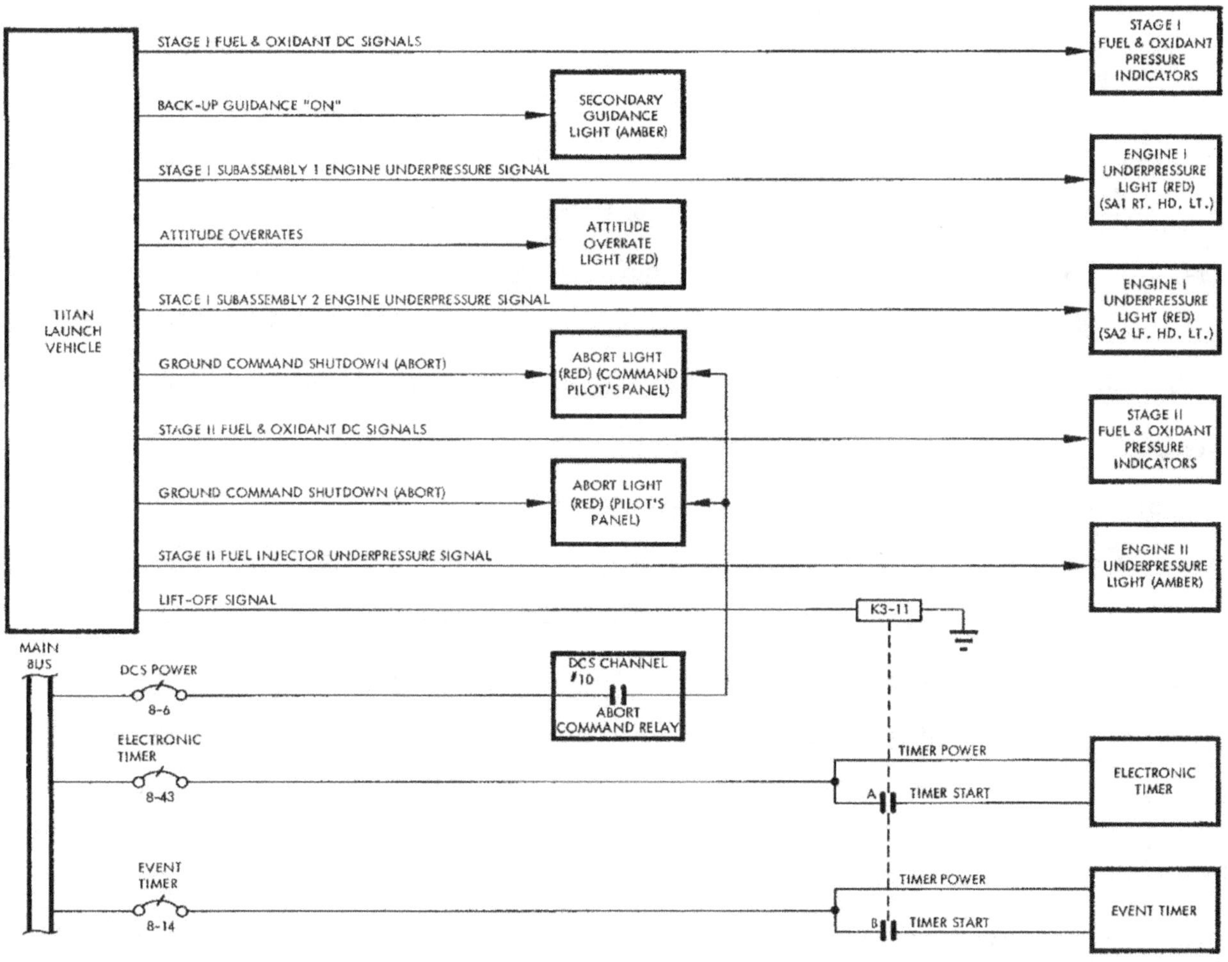

TABLE OF RELAYS			
RELAY	REDUNDANT RELAY	RELAY NOMENCLATURE	RELAY PANEL
K3-11		LIFT OFF	COMMUNICATIONS

Figure 4-3 Boost and Staging Sequence

SEDR 300

PROJECT GEMINI

link to the ground. Abort mode I prevails from lift off to T+50 seconds when abort mode I-II becomes applicable. This mode is replaced by abort mode II at T+100 seconds. Ground stations notify the S/C crew of each change; both pilots acknowledge each change.

At 145 seconds after lift off, when the acceleration has climbed to nearly 6 G, the first stage engine shutdown arm relays are energized. At approximately T+153 seconds, the thrust chamber pressure drops to less than 67 percent. The two ENGINE I indicators illuminate red, and the staging control relays are energized. The staging switches are closed. The stage 1 shutdown solenoids energize and booster engine cut off (BECO) occurs. Acceleration drops sharply to approximately 1.5G. The booster sequential system immediately ignites the second stage engine. The explosive bolts which unite stage 1 and stage 2 are detonated, and the stages separate. Both ENGINE I indicators are extinguished. Fuel injector pressure of the second stage engine rapidly increases above 55 percent, extinguishing the ENGINE II underpressure indicator. The accelerometer begins to climb slowly. The crew reports the results of the staging sequence to the ground station.

The ENGINE II underpressure indicator, the Attitude Overrate (ATT RATE) indicator, and the two ABORT indicators must remain extinguished. The STAGE 2 FUEL and OXID needles must indicate the required pressures, and the ACCELEROMETER must show the required increase.

At approximately T + 310 seconds, the spacecraft has climbed above 522,000 feet and its velocity exceeds 80 percent of orbital velocity. The ground station notifies the crew that abort mode III now replaces abort mode II. Both pilots acknowledge the change of abort modes.

PROJECT GEMINI

SEDR 300

SEPARATION AND INSERTION

At T + 330 seconds, the acceleration has climbed to almost 7G, and the spacecraft has nearly reached orbital velocity and altitude. Approximately 337 seconds after lift off, the blockhouse computer transmits the second engine shutdown command tones to the launch vehicle. The second engine shutdown solenoids energize, second stage engine cutoff (SSECO) occurs, thrust decays, and acceleration falls rapidly. The on-board computer begins to compute the delta-V required for insertion. The Command Pilot places the OAMS PROPELLANT switch to ON, and waits 20 seconds for the launch vehicle thrust decay. Then he depresses and releases the SEP SPCFT telelight/switch on the Sequence Panel shown on Figure 4-1. When the SEP SPCFT switch closes, BIA squib bus #1 power is applied thru the closed BOOST-INSERT CONT 1 circuit breaker to relays K3-22, K3-24, and K3-42. (Refer to Figure 4-4.) K3-22 is the spacecraft shaped charge ignition relay. K3-24 is the launch vehicle/spacecraft wire guillotine relay. K3-42 is the UHF whip antenna extend relay. Redundant contacts of the SEP SPCFT switch energize redundant relays with power from a redundant squib bus. For simplicity's sake, redundant elements are not shown.

Time delays in the relays and pyrotechnics cause the separation events to occur in the following sequence. K3-24, contacts C energize the launch vehicle/spacecraft pyro switch relay K3-26. K3-26, contacts C immediately fire the pyro switch, open-circuiting the wires on the battery side of the guillotine. Next the wire guillotines are fired, severing the launch vehicle spacecraft wires at the interface. Finally the spacecraft shaped charges are ignited, breaking the structural bond between the launch vehicle and the spacecraft. The operation

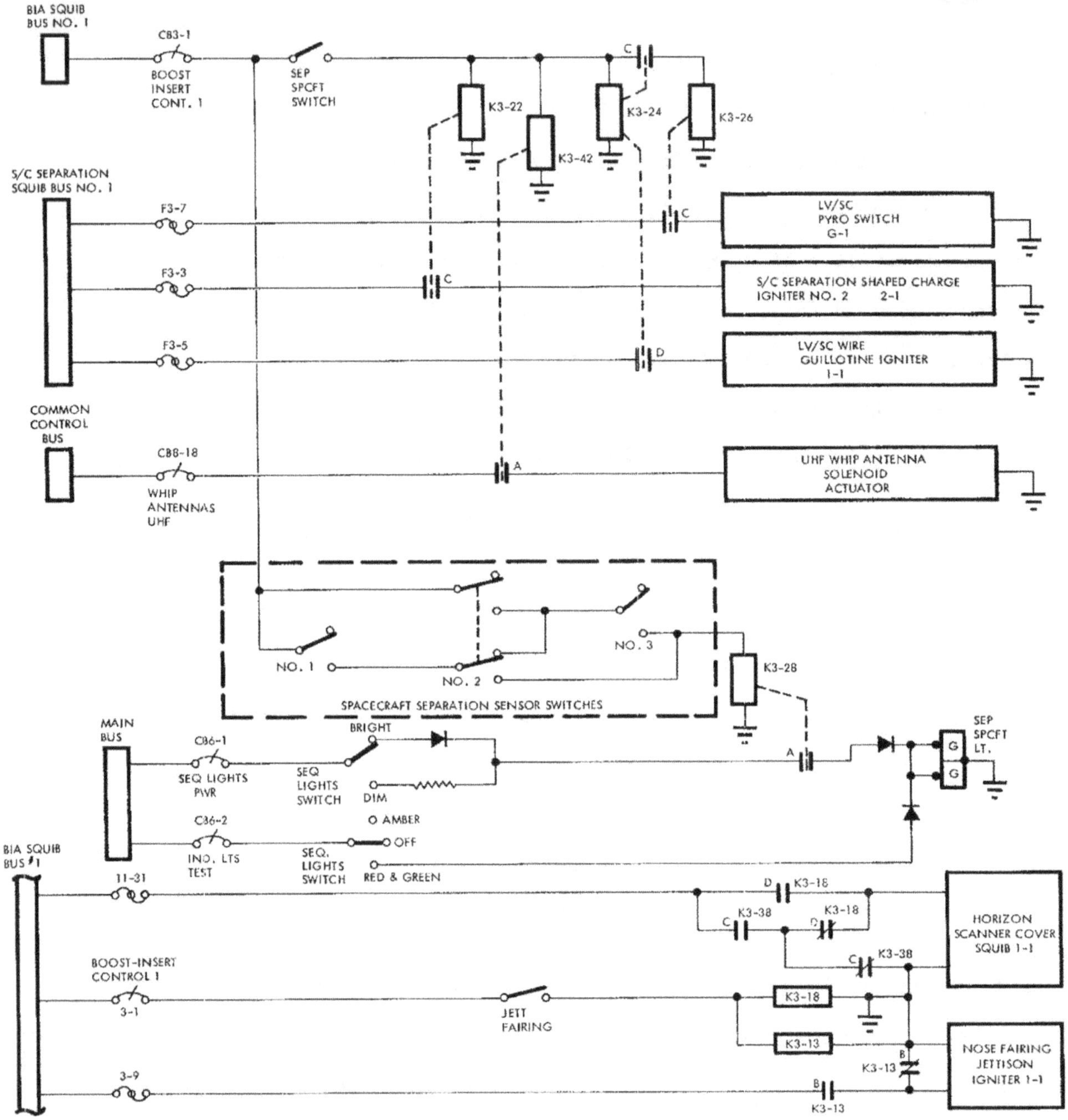

SEDR 300

PROJECT GEMINI

RELAY	REDUNDANT RELAY	NOMENCLATURE	RELAY PANEL
K3-22	K3-23	SPACECRAFT SHAPED CHARGE IGNITION	BIA S/C SEPARATION CONTROL
K3-24	K3-25	LAUNCH VEHICLE/SPACECRAFT GUILLOTINE	BIA S/C SEPARATION CONTROL
K3-26	K3-27	LAUNCH VEHICLE/SPACECRAFT PYRO SWITCH	BIA CONTROL
K3-28	K3-29	SPACECRAFT SEPARATION SENSOR	BIA CONTROL
K3-42	K3-43	UHF WHIP ANTENNA ACTUATOR	COMMUNICATIONS
K4-30	K4-48	LV/SC PYRO SWITCH ABORT	BIA CONTROL
K3-13	K3-17	NOSE FAIRING JETTISON	BIA NOSE FAIRING JETTISON
K3-18	K3-19	SCANNER COVER JETTISON	ACS-RCS
K3-38	K3-39	SQUIB BUS ABORT	POWER DISTRIBUTION

Figure 4-4 Spacecraft Separation Sequence

4-11

 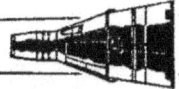

SEDR 300

PROJECT GEMINI

of these and all other pyrotechnics mentioned in this section is explained in Section XI.

The launch vehicle may now separate from the spacecraft, or OAMS thrust may be required to effect separation. Two inches of separation at the interface cause the spacecraft separation sensors to operate. The spacecraft separation sensor relay K3-28 is energized when two of the sensor switches close. Contacts A of K3-28 apply main bus power thru the closed SEQ. LIGHTS PWR circuit breaker and the SEQ. LIGHTS BRIGHT-DIM switch to the telelights. The SEP SPCFT telelight illuminates green.

The Command Pilot fires the aft thrusters of the spacecraft until the IVI is nulled. The spacecraft is in the required orbit. The following switches are placed to these positions: RETRO ROCKET SQUIB to SAFE; BOOST-INSERT SQUIB to SAFE; and MAIN BATTERIES 1, 2, 3, and 4 to OFF. For the communications switches positioned at this time, refer to Section IX.

The nose fairing and horizon scanner cover are jettisoned at this time. To do this, the Command Pilot depresses and releases the JETTISON FAIRING switch on the Sequence Panel. This switch energizes nose fairing jettison relays K3-13 and K3-17 on the boost insert abort (BIA) nose fairing jettison relay panel. The switch also energizes scanner cover jettison relays K3-18 and K3-19 on the attitude control system (ACS) scanner and re-entry system (RCS) squib fire relay panels. These jettison relays ignite the nose fairing squibs and scanner cover squibs, which eject the fairing covers.

SEDR 300

PROJECT GEMINI

ORBIT

During orbit, the crew perform the medical, technical, engineering, scientific, and Department of Defense (DOD) experiments scheduled for the mission.

PREPARE TO GO TO RETROGRADE

Approximately 30 minutes before Retrofire Time (T_R), the crew places the C-Band Beacon switch to CONT and performs the platform alignment procedures.

On S/C 3 and 4 at 5 minutes before retrofire time (T_R-5 minutes), the electronic timer energizes the T_R-5 minute relay K8-16. On S/C 7 at 256 seconds before retrofire (T_R-256 seconds), the electronic timer energizes the T_R-256 second relay K8-16. (See Figure 4-5.) K8-16 energizes K8-17, K8-19, and K8-29. On S/C 3 and 4, K8-16 also energizes K8-18. On S/C 3 and 4, K8-17 is the power system T_R-5 minute relay; on S/C 7, K8-17 is the power system T_R-256 second relay. The A contacts of K8-17 close to illuminate the BTRY PWR sequence telelight amber. K8-19 is the re-entry control system (RCS) amber light relay, and K8-19 illuminates the RCS telelight amber. K8-29 is the indicate retro attitude relay; K8-29 illuminates the IND RETRO ATT telelight amber. On S/C 3 and 4, K8-18 is the environmental control system (ECS) T_R-5 minute relay. K8-18 illuminates the O_2 HI RATE telelight amber.

The amber IND RETRO ATT indicator cues the Command Pilot to place the spacecraft in a retro attitude, and to apply retro bias so that the flight director needles can be used to orient the spacecraft in this mode. Pressing the indicator extinguishes the amber and illuminates the green light of the IND RETRO ATT telelight/switch by actuation of the retro bias relay K12-5. Initiation of

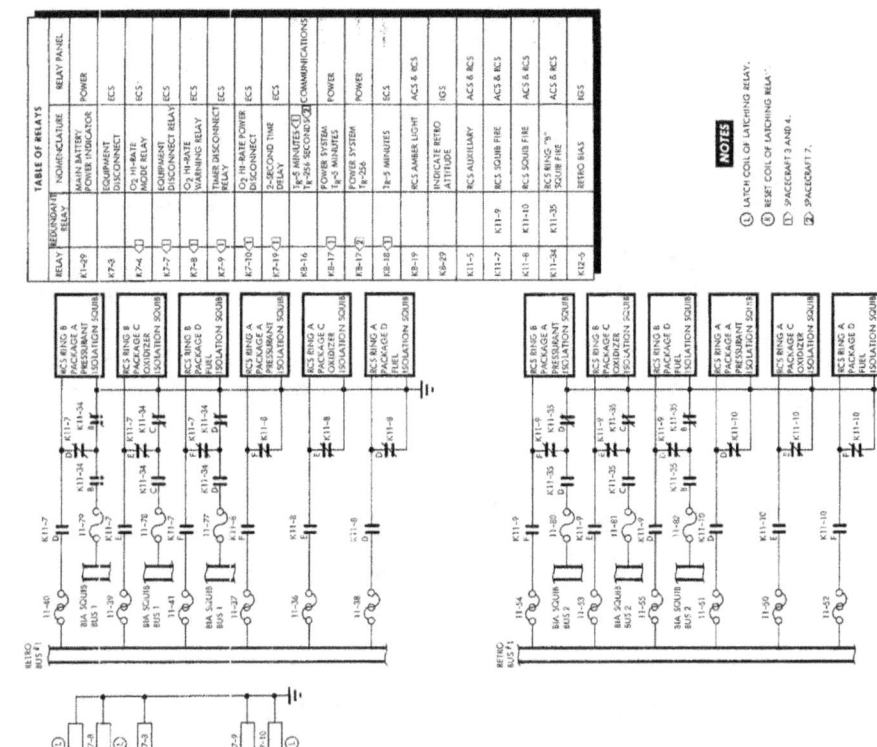

Figure 4-5 Time to Retrograde Minus 5 Minutes (256 Seconds) Sequence

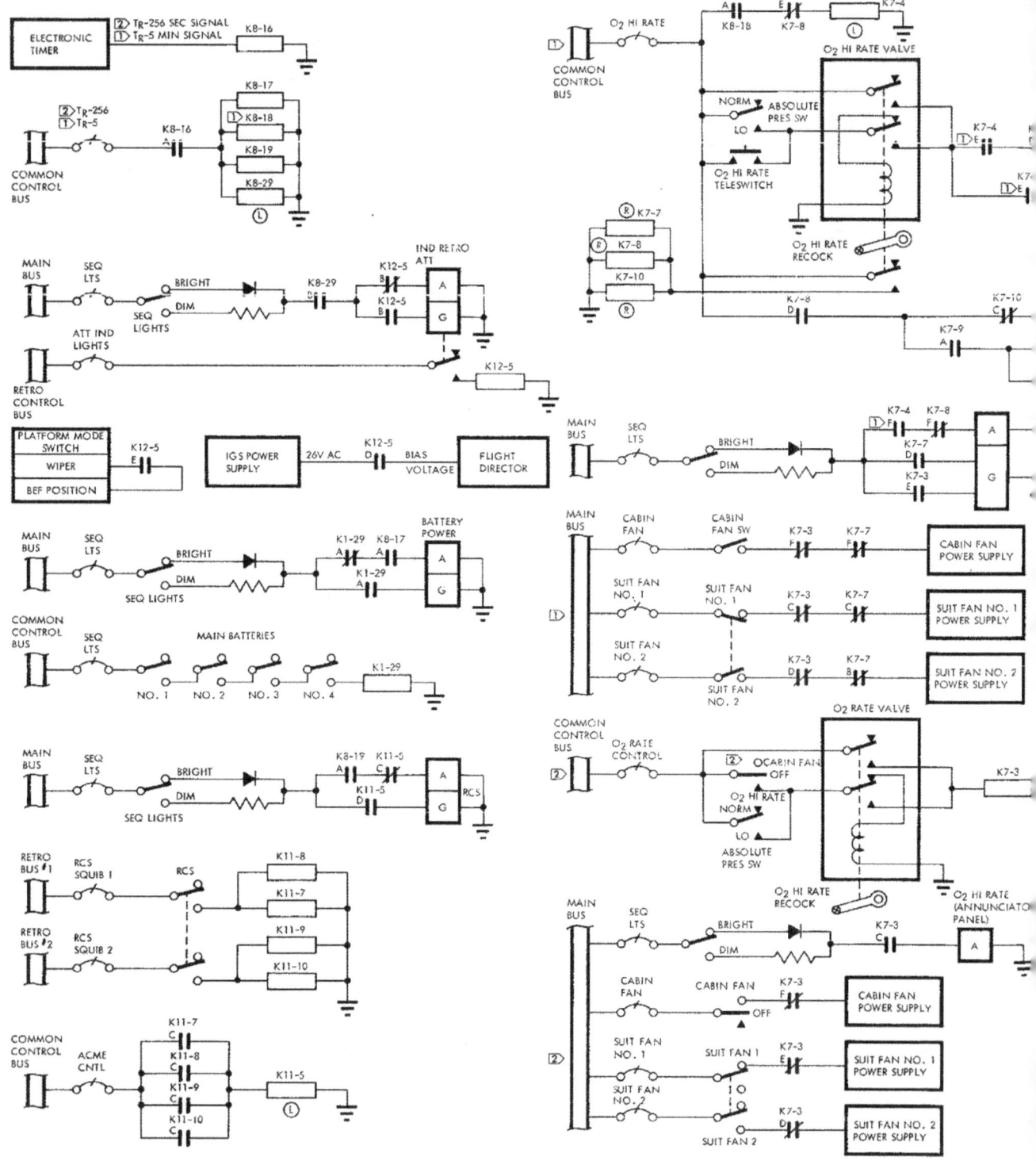

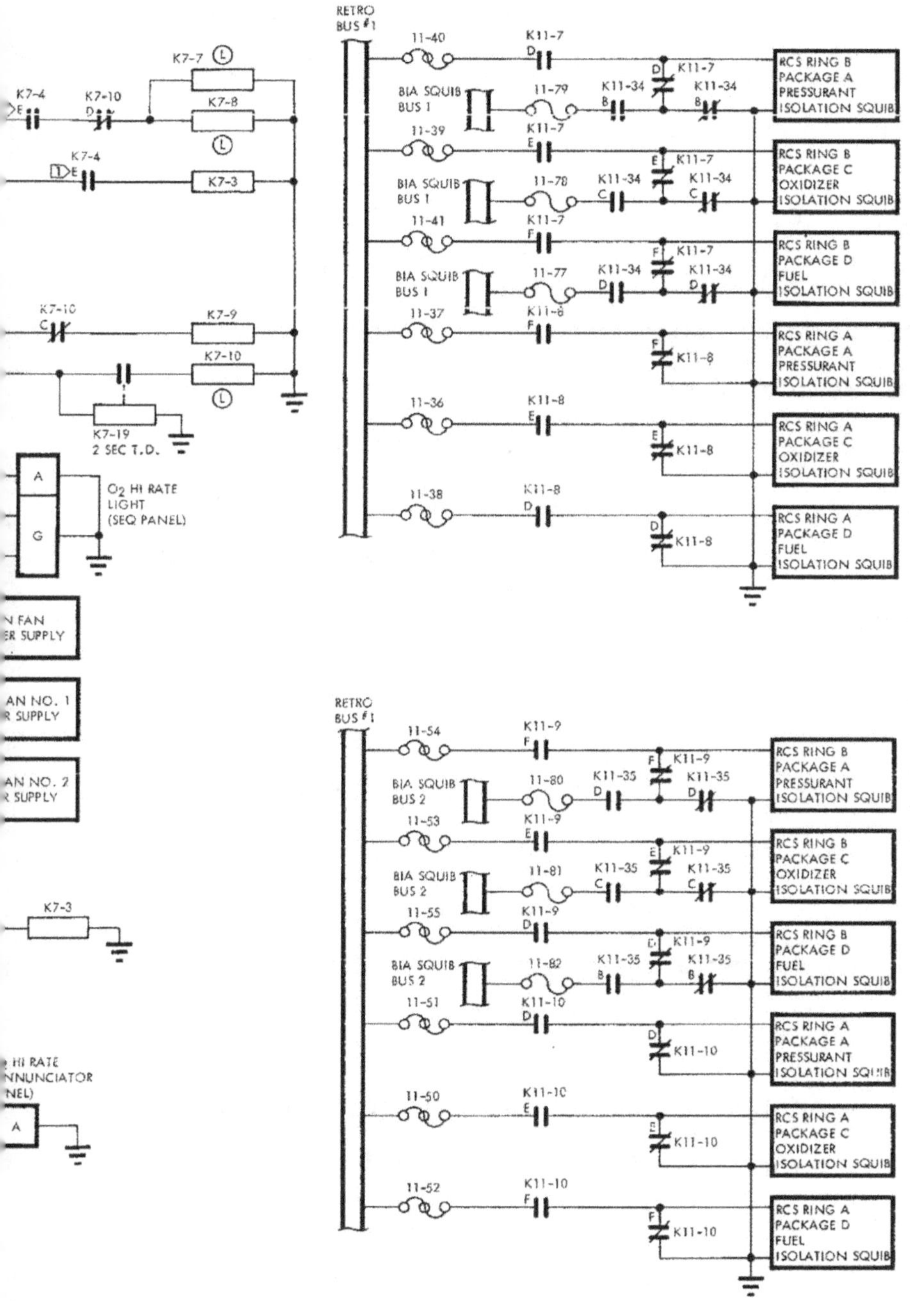

SEDR 300

PROJECT GEMINI

TABLE OF RELAYS			
RELAY	REDUNDANT RELAY	NOMENCLATURE	RELAY PANEL
K1-29		MAIN BATTERY POWER INDICATOR	POWER
K7-3		EQUIPMENT DISCONNECT	ECS
K7-4		O₂ HI-RATE MODE RELAY	ECS
K7-7		EQUIPMENT DISCONNECT RELAY	ECS
K7-8		O₂ HI-RATE WARNING RELAY	ECS
K7-9		TIMER DISCONNECT RELAY	ECS
K7-10		O₂ HI-RATE POWER DISCONNECT	ECS
K7-19		2-SECOND TIME DELAY	ECS
K8-16		T_R-5 MINUTES [1] T_R-256 SECONDS [2]	COMMUNICATIONS
K8-17 [1]		POWER SYSTEM T_R-5 MINUTES	POWER
K8-17 [2]		POWER SYSTEM T_R-256	POWER
K8-18 [1]		T_R-5 MINUTES	ECS
K8-19		RCS AMBER LIGHT	ACS & RCS
K8-29		INDICATE RETRO ATTITUDE	IGS
K11-5		RCS AUXILIARY	ACS & RCS
K11-7	K11-9	RCS SQUIB FIRE	ACS & RCS
K11-8	K11-10	RCS SQUIB FIRE	ACS & RCS
K11-34	K11-35	RCS RING "B" SQUIB FIRE	ACS & RCS
K12-5		RETRO BIAS	IGS

NOTES

(L) LATCH COIL OF LATCHING RELAY.
(R) RESET COIL OF LATCHING RELAY.
[1] SPACECRAFT 3 AND 4.
[2] SPACECRAFT 7.

Figure 4-5 Time to Retrograde Minus 5 Minutes (256 Seconds) Sequence

K12-5 applies a bias voltage to the flight director needles so that they are nulled at the retro attitude and the inertial platform is electrically placed into the blunt end forward (BEF) or retro position.

On S/C 3 and 4, the O_2 high-rate mode relay K7-4 is latched at T_R-5 MIN by the T_R-5 MIN relay K8-18 on the ECS relay panel. The O_2 HI RATE indicator is illuminated amber through the normally closed contacts of K7-8. The Command Pilot then depresses and releases the O_2 HI RATE switch which will latch the O_2 high-rate warning relay K7-8 and the equipment disconnect relay K7-7. Latching relays K7-7 and K7-8 are energized only after the O_2 high-rate valve has been opened. Relay K7-7 removes power from the cabin fan power supply and the two suit fan power supplies, extinguishes the amber lamp and illuminates the green lamp of the O_2 HI RATE indicator. The latching of K7-8 will initiate operation of the time disconnect relay K7-9. K7-9 in turn energizes the two-second time delay relay K7-19. In two seconds K7-19 energizes the O_2 high-rate disconnect relay K7-10. K7-10 removes latching power from K7-7 and K7-8, and deenergizes K7-9. K7-9 deenergizes and removes power from K7-19 and K7-10. K7-7, K7-8, and K7-10 remain latched until reset by the O_2 HI RATE RECOCK lever.

On S/C 7, O_2 high rate flow is initiated after the T_R-256 second sequence at the option of the crew. When the CABIN FAN switch is placed to the O_2 HI RATE position, the O_2 high rate valve is opened, and equipment disconnect relay K7-3 is energized. K7-3 removes power from the cabin fan power supply and both suit fan power supplies, and illuminates the green O_2 HI RATE indicator on the Annunciator Panel.

SEDR 300
PROJECT GEMINI

The amber BTRY PWR indicator cues the Pilot to turn on the main batteries by placing the four MAIN BATTERIES switches to the ON position. Relay K1-29 is energized through the on position of the four battery switches. The BTRY PWR indicator illuminates green.

Depressing the RCS telelight/switch energizes the four RCS squib fire relays K11-7, K11-8, K11-9 and K11-10. Relays K11-7 and K11-8 are energized from retro bus #1 while K11-9 and K11-10 are energized from retro bus #2. When any of the four RCS squib fire relays energize, the RCS auxiliary relay K11-5 is latched, changing the RCS indicator from an amber to a green indication. Relays K11-7 and K11-9 both fire the package A, C and D pressure isolation, oxidizer isolation, and fuel isolation squibs of ring B. Relays K11-8 and K11-10 fire the package A, C and D pressure isolation, oxidizer isolation, and fuel isolation squibs of ring A. The RCS RING A and RING B switches are now placed to ACME, and the Attitude Controller is operated to fire and check the RCS thrusters.

TIME TO RETROGRADE MINUS 30 SECONDS SEQUENCE
After the T_R-5 sequence (on S/C 3 and 4) or the T_R-256 sequence on S/C 7, communications are also selected, as discussed in Section IX.

At thirty seconds prior to retrofire (T_R-30 sec), the electronic timer initiates a contact closure. This closure energizes the retro T_R-30 second relay K4-46, which illuminates the SEP OAMS LINE, SEP ELEC, SEP ADAPT, and ARM AUTO RETRO indicators amber. Figure 4-6 shows a logic presentation of the T_R-30 second sequence. Some of the sub-sequences shown in Figure 4-6 such as SEP OAMS LINES, SEP ELECT, and SEP ADAPT sub-sequences are performed redundantly. However,

4-16

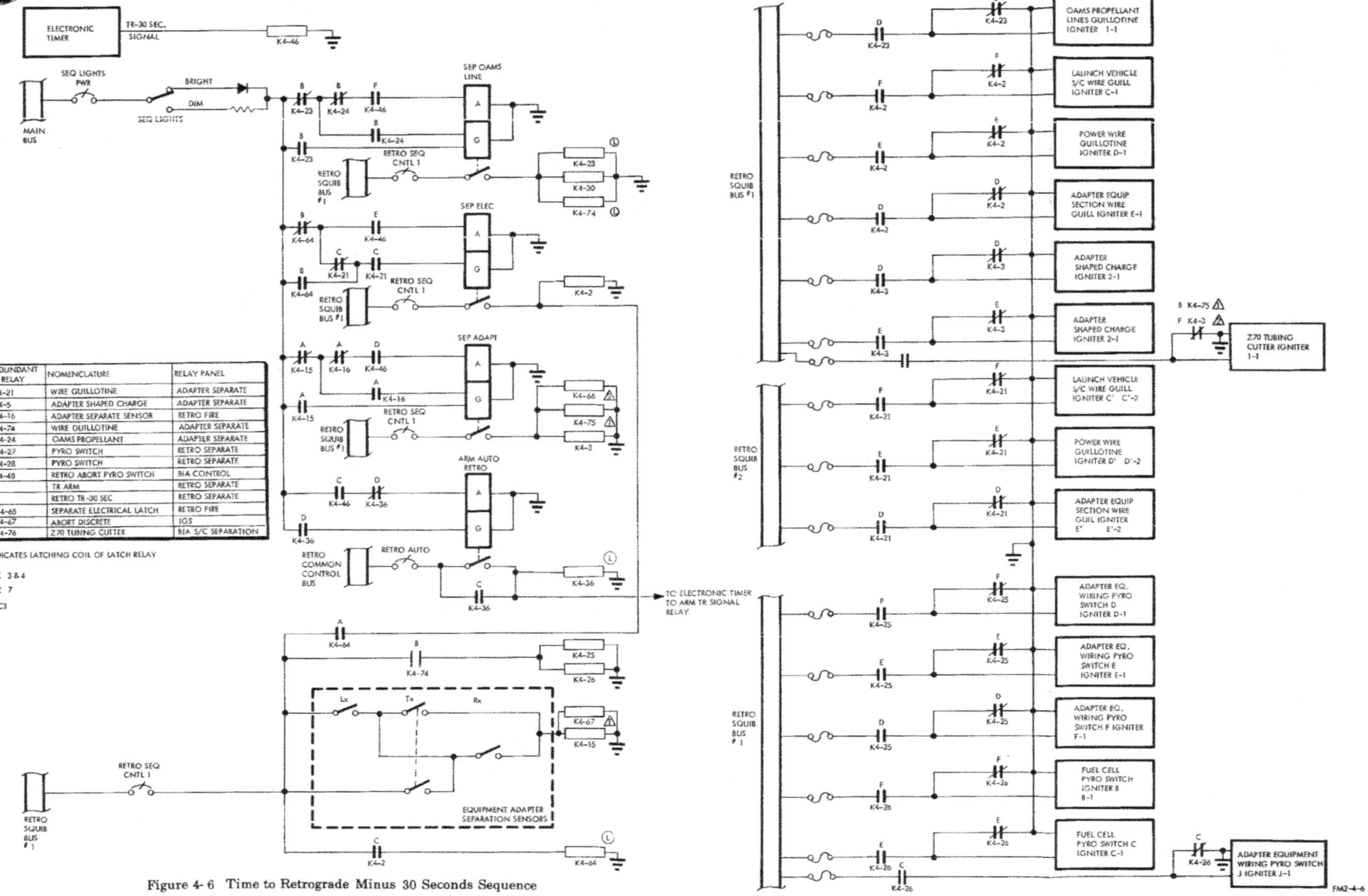

RELAY	REDUNDANT RELAY	NOMENCLATURE	RELAY PANEL
K4-2	K4-21	WIRE GUILLOTINE	ADAPTER SEPARATE
K4-3	K4-5	ADAPTER SHAPED CHARGE	ADAPTER SEPARATE
K4-15	K4-16	ADAPTER SEPARATE SENSOR	RETRO FIRE
K4-22	K4-74	WIRE GUILLOTINE	ADAPTER SEPARATE
K4-23	K4-24	OAMS PROPELLANT	ADAPTER SEPARATE
K4-25	K4-27	PYRO SWITCH	RETRO SEPARATE
K4-26	K4-28	PYRO SWITCH	RETRO SEPARATE
K4-30	K4-48	RETRO ABORT PYRO SWITCH	BIA CONTROL
K4-36		TR ARM	RETRO SEPARATE
K4-46		RETRO TR-30 SEC	RETRO SEPARATE
K4-64	K4-65	SEPARATE ELECTRICAL LATCH	RETRO FIRE
K4-66	K4-67	ABORT DISCRETE	IGS
K4-75 (3)	K4-76	Z70 TUBING CUTTER	BIA S/C SEPARATION

(L) INDICATES LATCHING COIL OF LATCH RELAY

(1) S/C 3 & 4

(2) S/C 7

(3) S/C3

Figure 4-6 Time to Retrograde Minus 30 Seconds Sequence

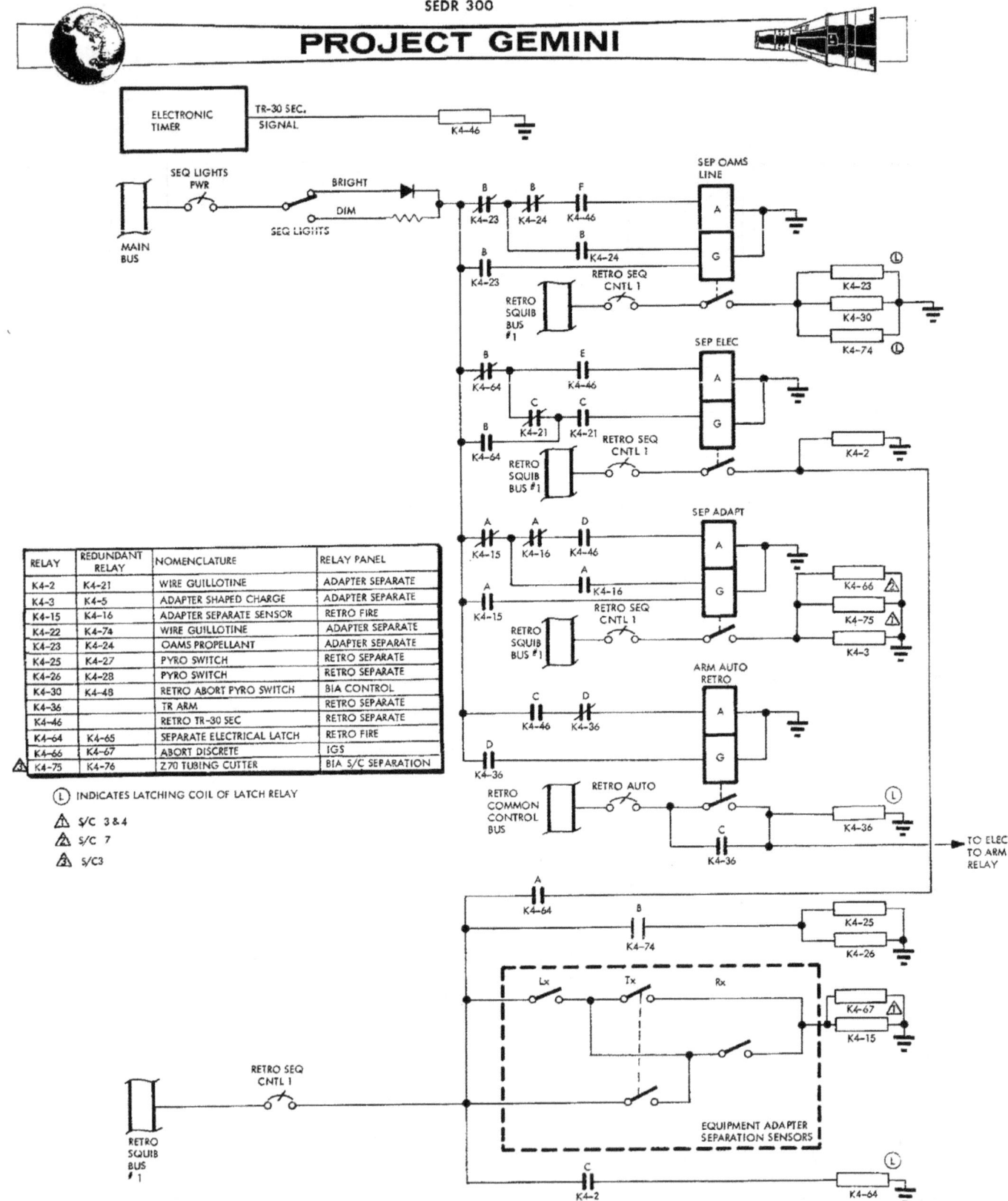

Figure 4-6 Time to Retrograde Minus 30 Seconds Sequence

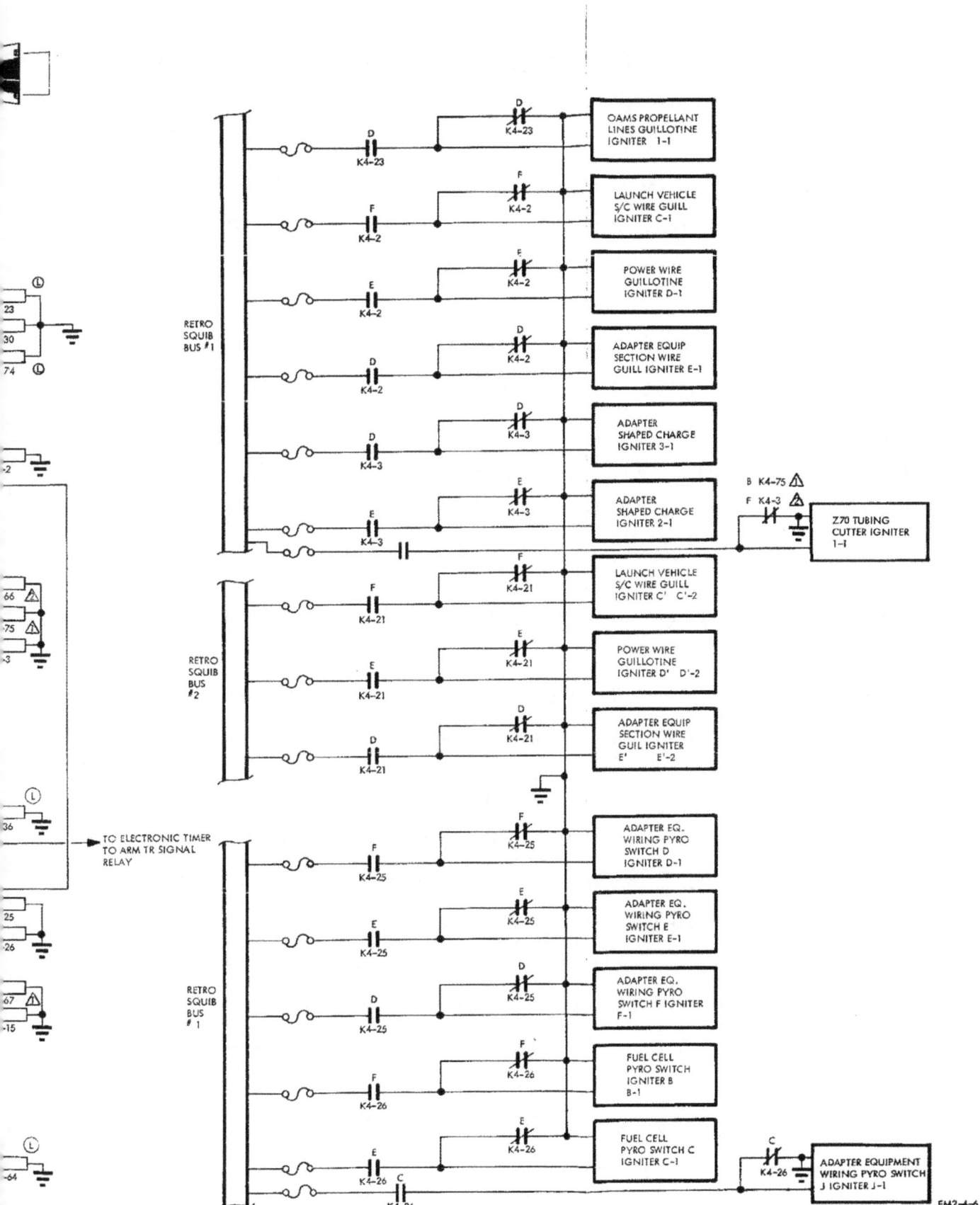

for simplicity only the sequence powered from retro squib bus #1 is shown. The redundant sequence not shown in Figure 4-6 is powered from retro squib bus #2. Since the redundant functions are identical, only the one shown in the referenced figure is described. As soon as the Command Pilot observes that the four indicators have illuminated amber, he depresses and releases the SEP OAMS LINE switch-indicator. This switch closure energizes the OAMS propellant line guillotine relay K4-23 and the wire guillotine relay K4-74. K4-23 changes the SEP OAMS LINE indication from amber to green, and fires the OAMS propellant lines guillotine 1-1. K4-74 energizes pyro switch relays K4-25 and K4-26. Relays K4-25 and K4-26 fire pyro switches B,C,D,E,F, and J.

Next the Command Pilot depresses and releases the SEP ELEC switch-indicator which energizes wire guillotine relay K4-2. K4-2 ignites wire guillotines C,D, and E and energizes the separate electrical latch relay K4-64. When K4-64 energizes, the SEP ELEC indicator changes from amber to green.

Next, the Command Pilot initiates the equipment adapter separation sequence by depressing and releasing the SEP ADAPT switch-indicator. Closure of the SEP ADAPT switch energizes the adapter shaped charge relay K4-3. K4-3 detonates Z70 tubing cutter igniter 1-1 and shaped charge igniters 2-1 and 3-1. The adapter equipment section separates, and separation is sensed by three toggle sensor switches. The switches close when the physical separation is 1-1/2 inches. The closure of any two switches energizes the adapter separate sensor relay K4-15. K4-15 changes the SEP ADAPT indicator from amber to green. The green SEP ADAPT light informs the crew that the adapter equipment section has been jettisoned from the spacecraft.

Lastly the Command Pilot depresses and releases the ARM AUTO RETRO switch-indicator. The ARM AUTO RETRO switch latches the T_R arm relay K4-36. This relay changes the indication from amber to green and arms the electronic timer for the T_R relay contact closure. The four RETRO ROCKET SQUIB switches are now moved to ARM.

RETROGRADE SEQUENCE

A logic diagram of the retrograde sequence is shown in Figure 4-7. As discussed previously, whenever a sequence is initiated from retro squib bus #1, there is an identical redundant sequence initiated from retro squib bus #2. The retrograde sequence is initiated by the T_R signal from the Electronic Timer. The same sequence is initiated redundantly and manually by the Command Pilot.

At time to retrograde T_R, the electronic timer latches the T_R signal relay K4-34. The T_R signal relay in the latched condition energizes the retro rocket auto fire relay K4-7. K4-34 also energizes the 45-second time delay relay K4-4, initiates a 5.5-second, 11.0-second, and a 16.5-second time delay, and deactivates the IGS platform "free" mode. The retro rocket auto fire relay redundantly fires retro rocket #1 from retro squib buses #1 and #2. At the end of the 5.5-second time delay, the retro rocket auto fire relay K4-9 is energized. K4-9 ignites retro rocket #3 from retro squib buses #1 and #2. Retro rocket #2 is redundantly ignited from retro squib buses #1 and #2 when the retro rocket auto fire relay K4-11 energizes at the end of the 11.0 second time delay. Retro rocket auto fire relay K4-13 is energized at the end of the 16.5 second time delay. K4-13 redundantly fires retro rocket #4 from retro squib buses #1 and #2.

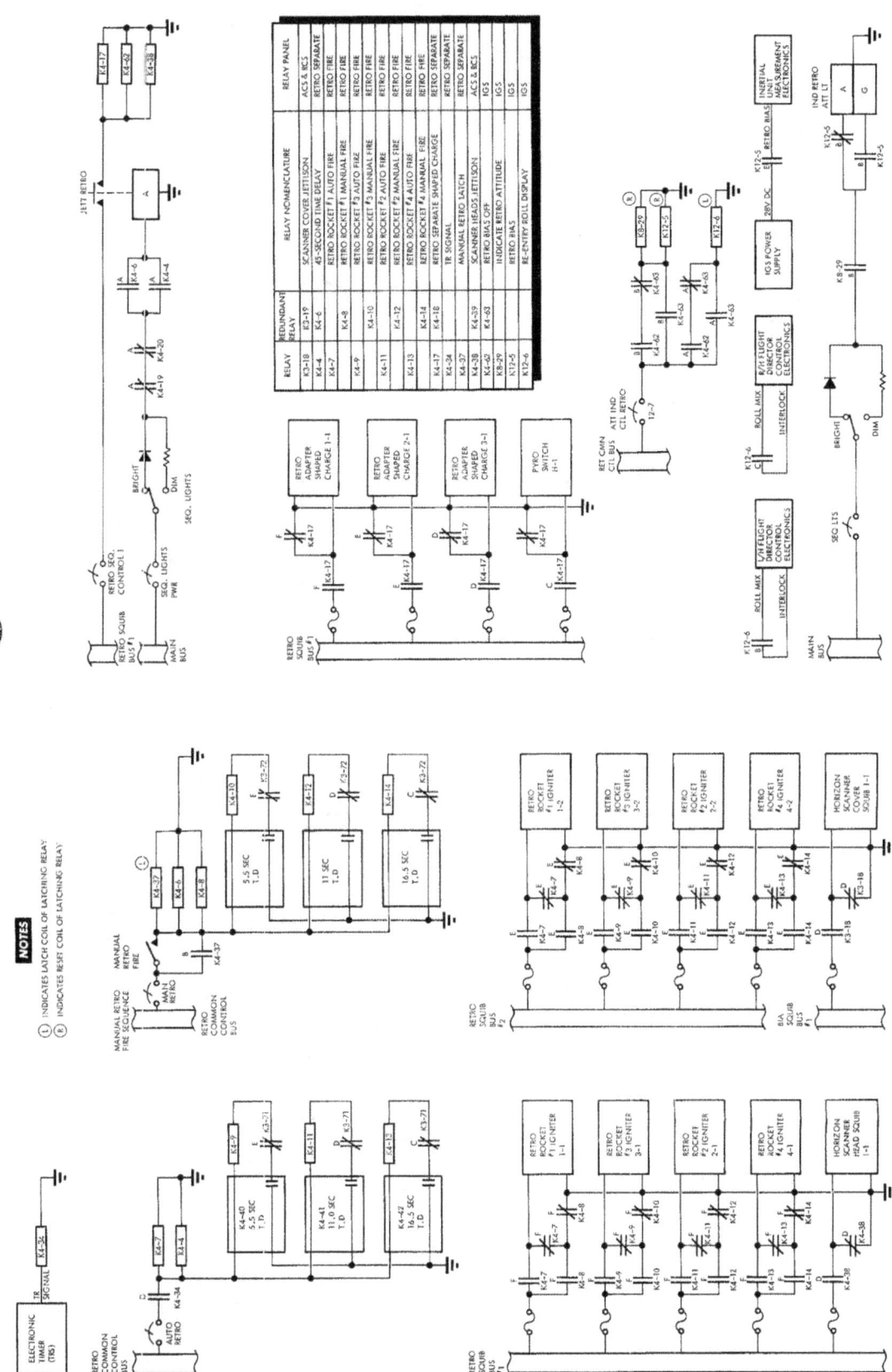

Figure 4-7 Retrograde Sequence

NOTES
(L) INDICATES LATCH COIL OF LATCHING RELAY
(R) INDICATES RESET COIL OF LATCHING RELAY
ELECTRONIC TIMER (TRS)
TR SIGNAL
K4-34
RETRO COMMON CONTROL BUS
AUTO RETRO
K4-7
K4-4
K4-40 5.5 SEC T.D
K4-9
K3-71
K4-41 11.0 SEC T.D
K4-11
K4-42 16.5 SEC T.D
K4-13
MANUAL RETRO FIRE SEQUENCE
MANUAL RETRO FIRE
MAN RETRO
K4-37
K4-6
K4-8
5.5 SEC T.D
K4-10
K3-72
11 SEC T.D
K4-12
16.5 SEC T.D
K4-14
RETRO SQUIB BUS #1
RETRO ROCKET #1 IGNITER 1-1
RETRO ROCKET #3 IGNITER 3-1
RETRO ROCKET #2 IGNITER 2-1
RETRO ROCKET #4 IGNITER 4-1
K4-38
HORIZON SCANNER HEAD SQUIB 1-1
RETRO SQUIB BUS #2
RETRO ROCKET #1 IGNITER 1-2
RETRO ROCKET #3 IGNITER 3-2
RETRO ROCKET #2 IGNITER 2-2
RETRO ROCKET #4 IGNITER 4-2
BIA SQUIB BUS #1
K3-18
HORIZON SCANNER COVER SQUIB 1-1
MAIN BUS
K12-6

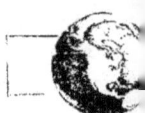

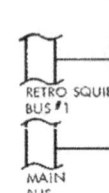

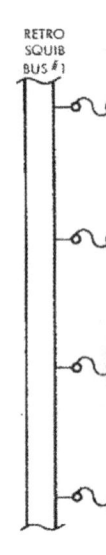

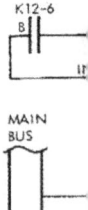

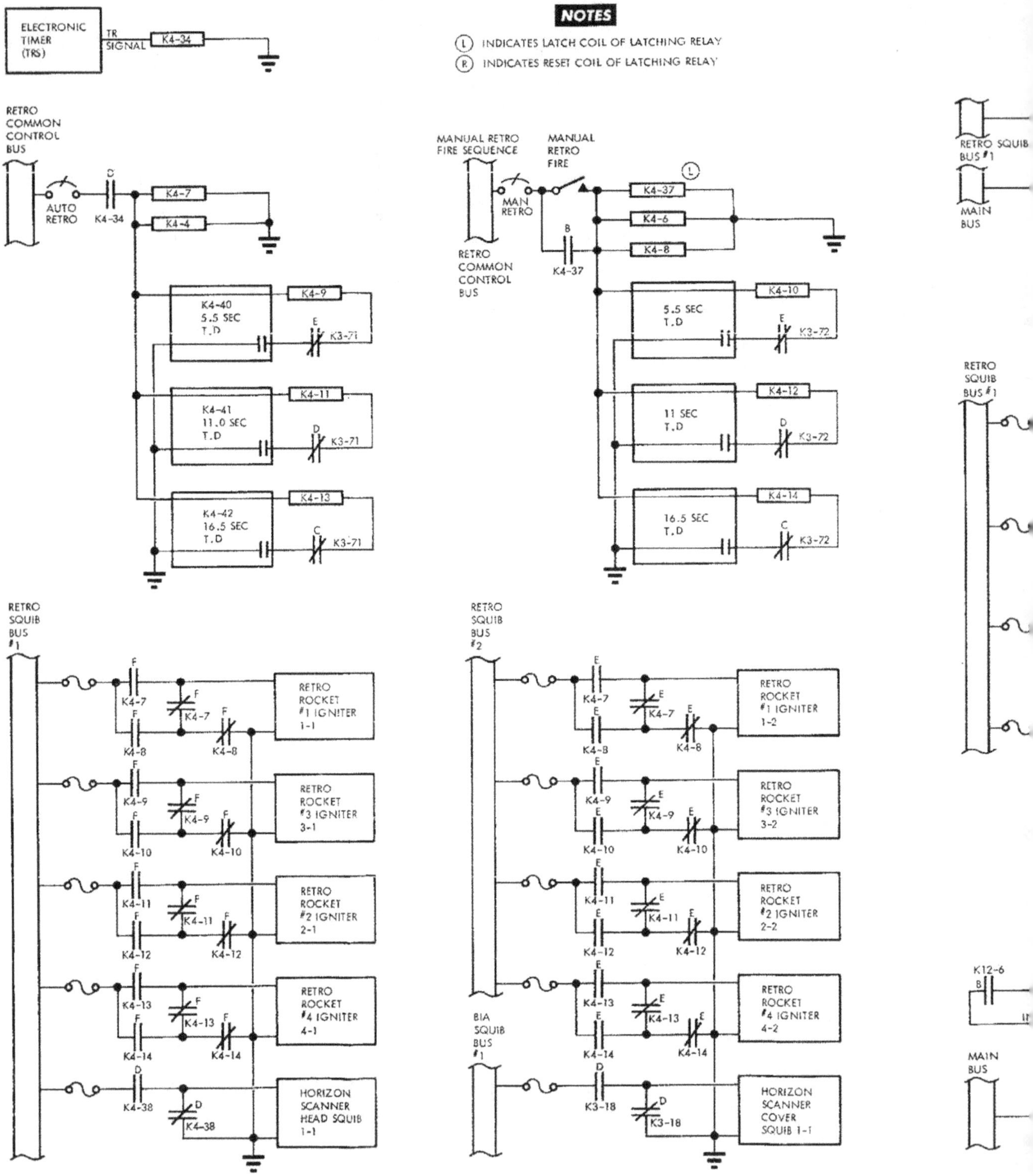

 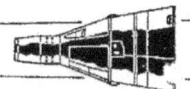

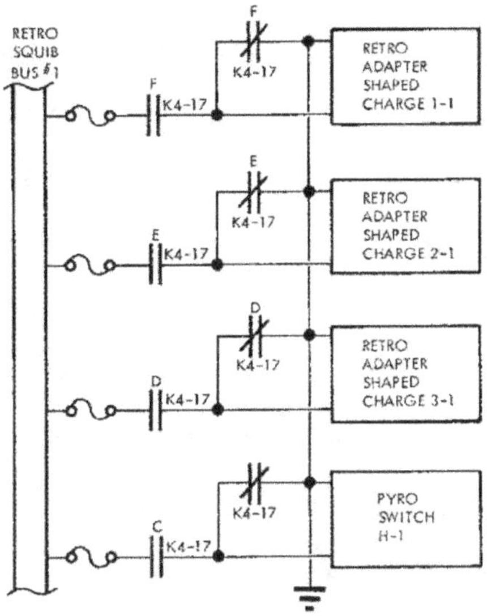

PROJECT GEMINI

SEDR 300

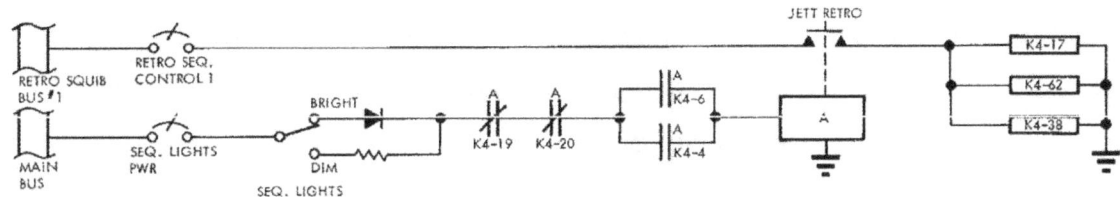

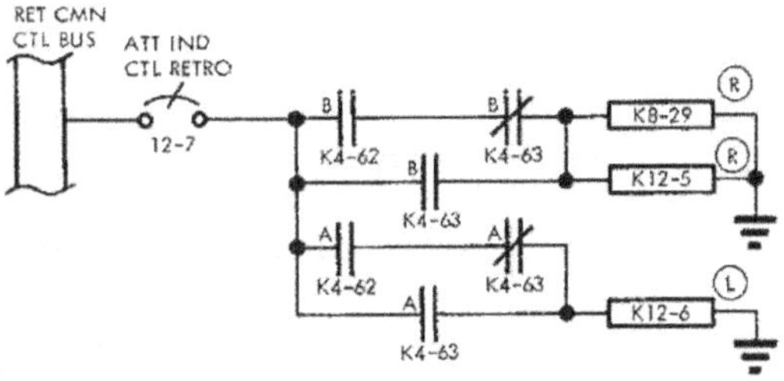

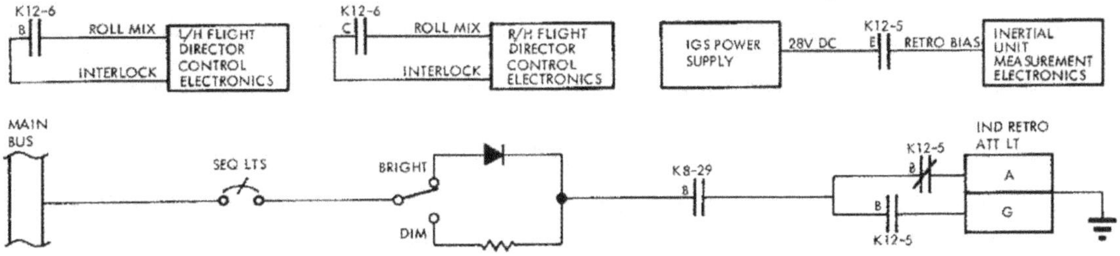

Figure 4-7 Retrograde Sequence

SEDR 300
PROJECT GEMINI

In order to assure retro fire, the Command Pilot depresses and releases the MAN FIRE RETRO switch-indicator approximately one second after automatic retro fire initiation. The MAN FIRE RETRO switch latches the manual retro latch relay K4-37, energizes retro rocket manual fire relay K4-8, and initiates the 45 second time delay relay K4-6. This switch also initiates the 5.5-second, 11-second and 16.5-second time delays. The 5.5, 11 and 16.5-second time delays energize retro rocket manual fire relays K4-10, K4-12 and K4-14 respectively which in turn fire retro rockets #3, #2, and #4 respectively. Retro rocket #1 is fired by K4-8. As in auto retro rocket fire, each retro rocket is fired from retro squib buses #1 and #2. Either one of the two 45-second time delay relays (K4-4 or K4-6) when they time out, illuminate the JETT RETRO indicator of the main instrument panel amber. About 22 seconds after retrofire began, the last retro rocket ceases firing, and the Command Pilot places the JETT RETRO SQUIB ARM switch to ARM. Forty-five seconds after retrofire began, K4-4 or K4-6 energizes and illuminates the JETT RETRO indicator.

As soon as the Command Pilot observes that the JETT RETRO indicator is amber, he depresses and releases the switch-indicator. The switch energizes the retro shaped charge ignition relay K4-17, the retro bias off relay K4-62, and the scanner heads (cover abort) jettison relay K4-38. K4-17 fires the retro adapter shaped charge igniters 1-1, 2-1, and 3-1 and pyro switch H. K4-62 latches the re-entry roll display relay K12-6, removing roll mix interlock from the flight director controller. K4-62 also resets the latched retro bias relay K12-5 and the latched indicator retro attitude relay K8-29. K8-29 when reset extinguishes the IND RETRO ATT indicator (Figure 4-5). K12-5 when reset

4-21

PROJECT GEMINI

removes retro bias voltage from the inertial measurement unit (IMU) electronics. K4-38 ignites the horizon scanner heads squib, jettisoning the heads.

RE-ENTRY

After the retro adapter and scanner heads have been jettisoned, the Command Pilot places the RETRO PWR and RETRO JETT squib switches to SAFE. Using the attitude controller and the flight director indicator needles he rolls the spacecraft 180° so that the horizon is visible in the upper portion of his cabin window. He changes the ATTITUDE CONTROL mode select switch on the Pedestal Panel from PULSE to RATE CMD (RE-ENT). The Command Pilot uses ACME and the Attitude Controller to control the roll attitude during approximately the next 10 minutes in which the altitude diminishes to 400,000 feet. At this altitude the FDI roll needles start to move, the Computer START light illuminates, and the computer begins to calculate the point of impact. The Command Pilot changes the ATTITUDE CONTROL mode select switch from RATE CMD (RE-ENT) to RE-ENT. The computer now computes the roll attitude from optimum re-entry lift and also automatically controls the roll attitude. During approximately the next 10 minutes, the altitude decreases to 100,000 feet. At this altitude, the altimeter indicator begins to come off the peg. At 80,000 feet, the computer commands the spacecraft to assume the best attitude for drogue deployment. Then the Command Pilot places the COMPUTER switch to OFF.

LANDING SEQUENCE

After de-energizing the on-board computer the Command Pilot performs the

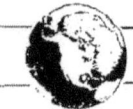 

SEDR 300

PROJECT GEMINI

various landing sequence functions. Figure 4-8 shows a logic diagram of the landing sequence.

At approximately 50,000 feet the Command Pilot deploys the drogue chute by depressing the HI ALT DROGUE switch. Closure of the drogue switch energizes the drogue chute relays K5-83 and K5-84. These relays are energized redundantly from landing squib bus #1 and #2. Actuation of K5-83 and K5-84 ignites the drogue chute mortars from both buses and fires the cabin air inlet door squibs from a separate relay.

In case the Command Pilot did not depress the drogue switch at 50,000 feet, a barostat switch illuminates a 40,000-foot altitude indicator to cue him to deploy the drogue chute.

In the event the drogue parachute did not deploy, or deployed improperly, the Command Pilot depresses and releases the EMERG 10.6K DROGUE switch at an altitude of approximately 11,000 feet. Closure of this switch energizes the emergency chute deploy relays K5-85 and K5-86. This switch also fires the pilot chute apex guillotine and the pilot chute mortar, energizes the drogue chute disconnect relays K5-87 and K5-88, and latches the main chute deploy relays K5-89 and K5-90.

The drogue chute disconnect relays fire the three (L_X, B_Y, T_Y) drogue chute disconnect guillotines while the main chute deploy relays energize the chute deploy 2.5-second time delay relays K5-91 and K5-92.

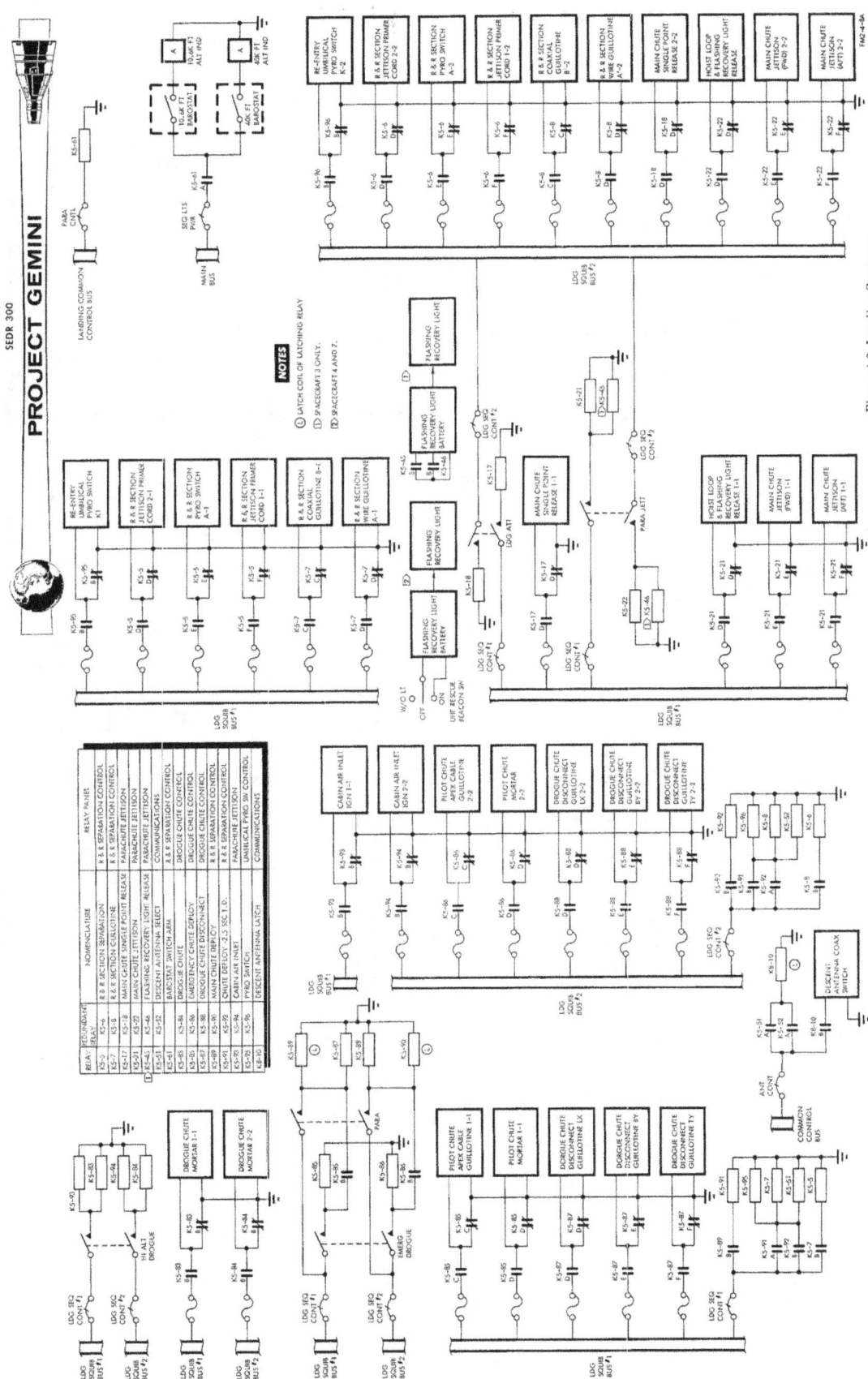

Figure 4-8 Landing Sequence

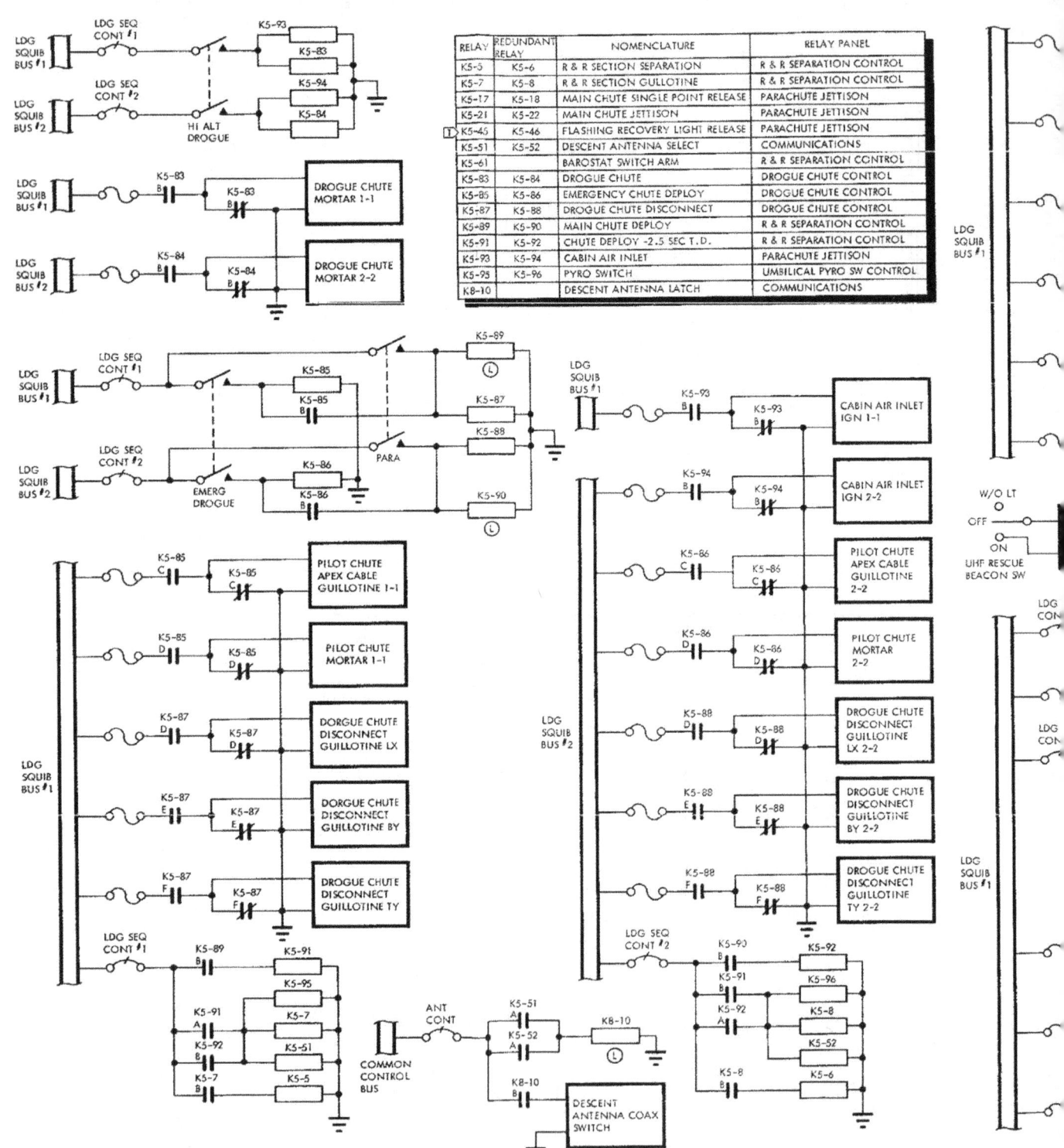

4-24

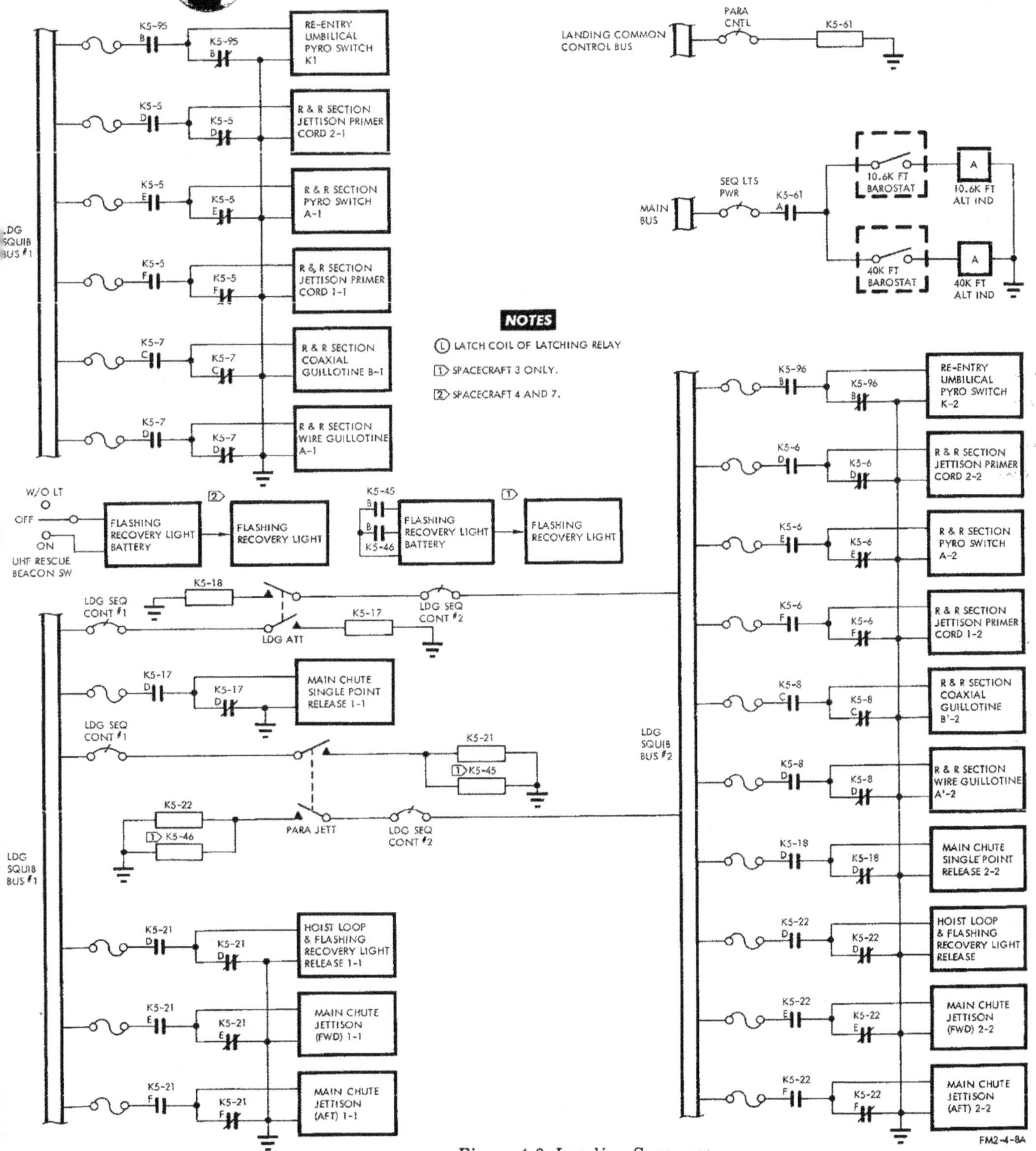

SEDR 300

PROJECT GEMINI

NOTES
- (L) LATCH COIL OF LATCHING RELAY
- [1] SPACECRAFT 3 ONLY.
- [2] SPACECRAFT 4 AND 7.

Figure 4-8 Landing Sequence

SEDR 300

PROJECT GEMINI

When K5-91 and K5-92 energize after 2.5 seconds, they energize the R & R section guillotine relays K5-7 and K5-8, the descent antenna select relays K5-51 and K5-52 and the umbilical pyro switch relays K5-95 and K5-96. The R & R section guillotine relays fire the R & R section coaxial guillotine B, the R & R section wire guillotine A, and energize the R & R section separation relays K5-5 and K5-6. Relays K5-5 and K5-6 redundantly connect landing squib buses #1 and #2 to the two pyro switch igniters and the four jettison primer cord igniters at the R & R section interface. The descent antenna relays latch the descent antenna relay K8-10 which in turn actuates a coaxial switch connecting the output of the quadriplexer to the descent antenna.

In the event that the Command Pilot did not utilize the emergency drogue chute deploy mode, he depresses the PARA switch at 10,600 feet. The 10,600-foot altitude indicator which is illuminated by a 10,600-foot barostat cues the Command Pilot to depress this switch.

The PARA switch energizes the drogue chute disconnect relays K5-87 and K5-88, and latches the main chute deploy relays K5-89 and K5-90. These relays then perform the functions described previously under the emergency mode.

Following main parachute deployment, the Command Pilot selects two-point suspension by depressing the LDG. ATT switch. The LDG. ATT switch energizes the main chute single point release pyrotechnics. At the time the main parachute aft bridle is pulled out of the bridle trough, the UHF recovery and UHF descent antennas are extended. Before landing, AC power is turned off.

SEDR 300
PROJECT GEMINI
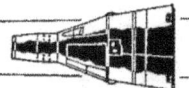

POST LANDING

After splash down, the Command Pilot jettisons the main parachute by depressing the PARA JETT switch. This switch energizes the main chute jettison relays K5-21 and K5-22. The main chute jettison relays fire the hoist loop and flashing recovery light release pyrotechnics and the main chute jettison (forward and aft) pyrotechnics. The UHF recovery beacon is turned on without lights, if rescue is carried out under daylight conditions. On S/C 3, the flashing recovery light relays energize the flashing recovery light.

The Command Pilot extends the HF whip antenna and establishes either HF or UHF voice communications with recovery forces. Spacecraft instrumentation is turned off.

ABORT MODES

An abort is an unscheduled termination of the spacecraft mission. An abort may be initiated at any time during the spacecraft mission. In all cases the actual abort sequence has to be initiated by the crew after an abort command has been received. An abort indication consists of illumination of the ABORT indicators located on either side of the Center Panel. The ABORT indicator may be illuminated by three different methods. During pre-launch prior to umbilical disconnect, the abort indicator may be illuminated from the blockhouse via hardline through the launch vehicle tail plug. After umbilical release, both of the abort indicators may be illuminated by ground command to the spacecraft via some channel of the Digital Command System, or by ground command to the launch vehicle to shutdown the booster.

The abort system is part of the sequential system. The abort system comprises the abort indicators, controls, relays, and pyrotechnics. The part of the abort system which the crew use is determined by the abort mode in effect at the time when the abort command is received or the decision to abort is made. The abort mode to be used at any time during the mission is determined by calculations made on the ground and depends on the altitude and velocity attained by the spacecraft. The critical abort altitudes are 15,000 feet, 75,000 feet, and 522,000 feet. The spacecraft reaches 15,000 feet approximately 50 seconds after lift off, 75,000 feet approximately 100 seconds after lift off, and 522,000 feet approximately 310 seconds after lift off. Below 15,000 feet, seat ejection (Mode I) is used. Between 15,000 and 70,000 feet, seat ejection (Mode I) or modified retro abort (Mode I - II) is used at the option of the Command Pilot. Between 75,000 and 522,000 feet, retro abort (Mode II) is used. Above 522,000 feet, normal re-entry (Mode III) is used, except that the spacecraft electronic timer does not illuminate the sequential lights amber when the time to press them occurs unless the timer is updated by ground command. Figure 4-9 presents a simplified block diagram of the abort sequences in each of the three modes.

Abort Mode I

When an abort becomes necessary during pre-launch, it is accomplished by using abort mode I. The abort command is given from the blockhouse by

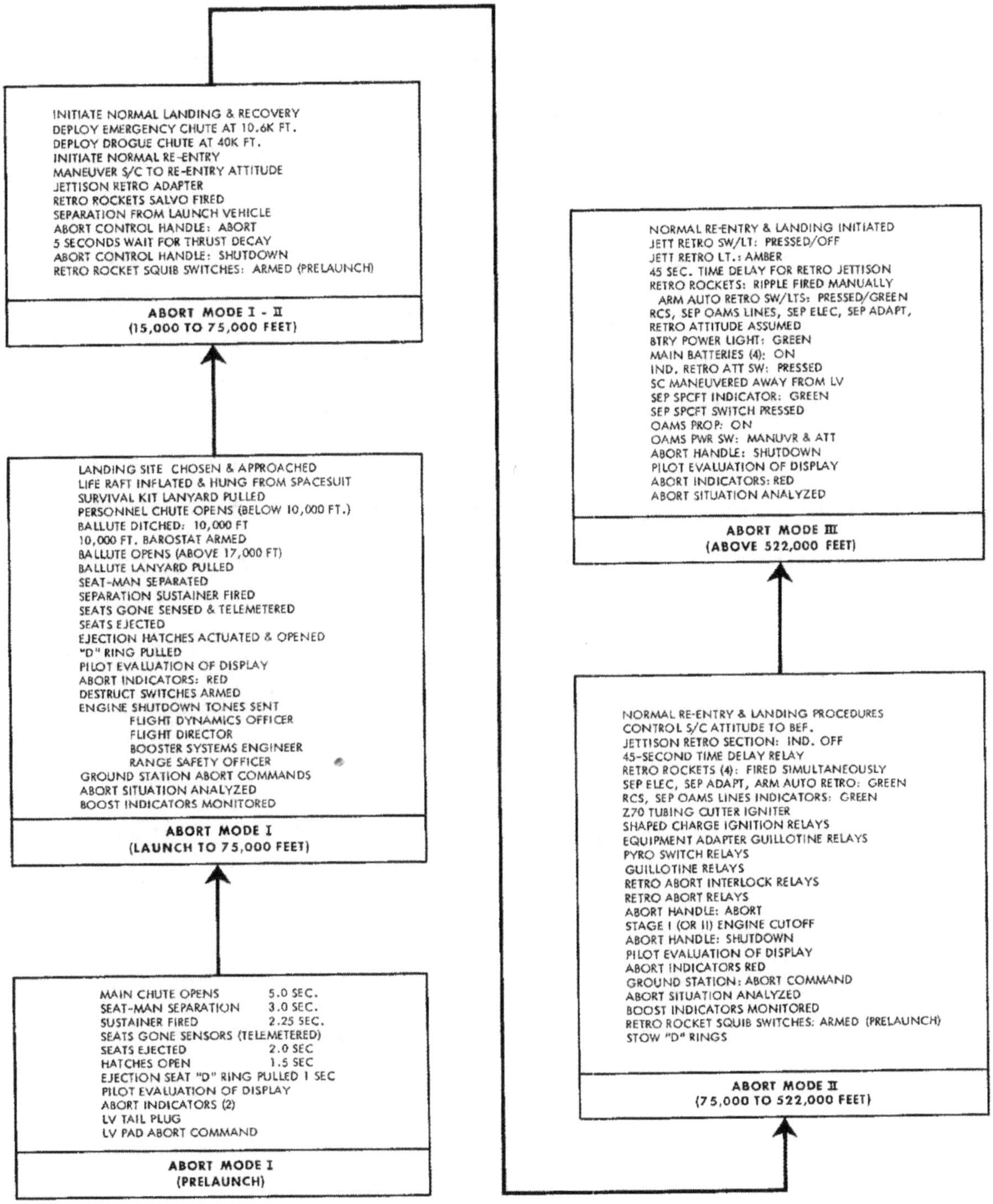

PROJECT GEMINI

SEDR 300

```
                                          ┌──────────────────────────────────┐
                                          │ NORMAL RE-ENTRY & LANDING INITIATED │
                                          │ JETT RETRO SW/LT: PRESSED/OFF    │
┌──────────────────────────────────┐      │ JETT RETRO LT.: AMBER            │
│ INITIATE NORMAL LANDING & RECOVERY│      │ 45 SEC. TIME DELAY FOR RETRO JETTISON│
│ DEPLOY EMERGENCY CHUTE AT 10.6K FT.│     │ RETRO ROCKETS: RIPPLE FIRED MANUALLY│
│ DEPLOY DROGUE CHUTE AT 40K FT.   │      │   ARM AUTO RETRO SW/LTS: PRESSED/GREEN│
│ INITIATE NORMAL RE-ENTRY         │      │ RCS, SEP OAMS LINES, SEP ELEC, SEP ADAPT,│
│ MANEUVER S/C TO RE-ENTRY ATTITUDE│      │ RETRO ATTITUDE ASSUMED           │
│ JETTISON RETRO ADAPTER           │      │ BTRY POWER LIGHT: GREEN          │
│ RETRO ROCKETS SALVO FIRED        │      │ MAIN BATTERIES (4): ON           │
│ SEPARATION FROM LAUNCH VEHICLE   │      │ IND. RETRO ATT SW: PRESSED       │
│ ABORT CONTROL HANDLE: ABORT      │      │ SC MANEUVERED AWAY FROM LV       │
│ 5 SECONDS WAIT FOR THRUST DECAY  │      │ SEP SPCFT INDICATOR: GREEN       │
│ ABORT CONTROL HANDLE: SHUTDOWN   │      │ SEP SPCFT SWITCH PRESSED         │
│ RETRO ROCKET SQUIB SWITCHES: ARMED (PRELAUNCH)│ OAMS PROP: ON              │
└──────────────────────────────────┘      │ OAMS PWR SW: MANUVR & ATT        │
         ABORT MODE I - II                │ ABORT HANDLE: SHUTDOWN           │
         (15,000 TO 75,000 FEET)          │ PILOT EVALUATION OF DISPLAY      │
                                          │ ABORT INDICATORS: RED            │
                                          │ ABORT SITUATION ANALYZED         │
                                          └──────────────────────────────────┘
                                                    ABORT MODE III
                                                  (ABOVE 522,000 FEET)
```

```
┌──────────────────────────────────┐      ┌──────────────────────────────────┐
│ LANDING SITE CHOSEN & APPROACHED │      │ NORMAL RE-ENTRY & LANDING PROCEDURES│
│ LIFE RAFT INFLATED & HUNG FROM SPACESUIT│ CONTROL S/C ATTITUDE TO BEF.    │
│ SURVIVAL KIT LANYARD PULLED      │      │ JETTISON RETRO SECTION: IND. OFF │
│ PERSONNEL CHUTE OPENS (BELOW 10,000 FT.)│ 45-SECOND TIME DELAY RELAY      │
│ BALLUTE DITCHED: 10,000 FT       │      │ RETRO ROCKETS (4): FIRED SIMULTANEOUSLY│
│ 10,000 FT. BAROSTAT ARMED        │      │ SEP ELEC, SEP ADAPT, ARM AUTO RETRO: GREEN│
│ BALLUTE OPENS (ABOVE 17,000 FT)  │      │ RCS, SEP OAMS LINES INDICATORS: GREEN│
│ BALLUTE LANYARD PULLED           │      │ Z70 TUBING CUTTER IGNITER        │
│ SEAT-MAN SEPARATED               │      │ SHAPED CHARGE IGNITION RELAYS    │
│ SEPARATION SUSTAINER FIRED       │      │ EQUIPMENT ADAPTER GUILLOTINE RELAYS│
│ SEATS GONE SENSED & TELEMETERED  │      │ PYRO SWITCH RELAYS               │
│ SEATS EJECTED                    │      │ GUILLOTINE RELAYS                │
│ EJECTION HATCHES ACTUATED & OPENED│     │ RETRO ABORT INTERLOCK RELAYS     │
│ "D" RING PULLED                  │      │ RETRO ABORT RELAYS               │
│ PILOT EVALUATION OF DISPLAY      │      │ ABORT HANDLE: ABORT              │
│ ABORT INDICATORS: RED            │      │ STAGE I (OR II) ENGINE CUTOFF    │
│ DESTRUCT SWITCHES ARMED          │      │ ABORT HANDLE: SHUTDOWN           │
│ ENGINE SHUTDOWN TONES SENT       │      │ PILOT EVALUATION OF DISPLAY      │
│     FLIGHT DYNAMICS OFFICER      │      │ ABORT INDICATORS RED             │
│     FLIGHT DIRECTOR              │      │ GROUND STATION: ABORT COMMAND    │
│     BOOSTER SYSTEMS ENGINEER     │      │ ABORT SITUATION ANALYZED         │
│     RANGE SAFETY OFFICER         │      │ BOOST INDICATORS MONITORED       │
│ GROUND STATION ABORT COMMANDS    │      │ RETRO ROCKET SQUIB SWITCHES: ARMED (PRELAUNCH)│
│ ABORT SITUATION ANALYZED         │      │ STOW "D" RINGS                   │
│ BOOST INDICATORS MONITORED       │      └──────────────────────────────────┘
└──────────────────────────────────┘                ABORT MODE II
         ABORT MODE I                            (75,000 TO 522,000 FEET)
         (LAUNCH TO 75,000 FEET)
```

```
┌──────────────────────────────────┐
│ MAIN CHUTE OPENS        5.0 SEC. │
│ SEAT-MAN SEPARATION     3.0 SEC. │
│ SUSTAINER FIRED         2.25 SEC.│
│ SEATS GONE SENSORS (TELEMETERED) │
│ SEATS EJECTED           2.0 SEC  │
│ HATCHES OPEN            1.5 SEC  │
│ EJECTION SEAT "D" RING PULLED 1 SEC│
│ PILOT EVALUATION OF DISPLAY      │
│ ABORT INDICATORS (2)             │
│ LV TAIL PLUG                     │
│ LV PAD ABORT COMMAND             │
└──────────────────────────────────┘
         ABORT MODE I
         (PRELAUNCH)
```

Figure 4-9 Abort Modes Simplified Block Diagram

FM2-4-9

4-28

 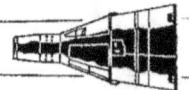

PROJECT GEMINI

hardline thru the launch vehicle tail plug connector. The command lights both ABORT indicators on the Command Pilot and Pilot's Panels. When the pilots see this display, they immediately pull the "D" rings attached to their ejection seats. When one "D" ring is pulled, both ejection systems are energized. One second after the ABORT indicators light, the "D" rings have been pulled. One-half second later, the hatches are open, and one-half second after that the seats have been ejected. Sensors detect the ejection of the seats and notify the blockhouse that the pilots are out of the spacecraft. One-quarter second after the seats are ejected, a sustainer rocket under each seat is fired, which extends the distance between the pilots and the launch vehicle. Then a pyrotechnic ignites and separates the ejection seat from the pilots. Two seconds after sustainer ignition, the main chutes have opened and the pilots are lowered safely to the ground. For illustrations and fuller descriptions of the equipment used for seat ejection abort, refer to Section III of this manual.

After normal lift off, and before the Gemini-Titan reaches an altitude of 15,000 feet, an abort condition could develop. The crew monitor their booster indicators so that they are aware at all times of the manner in which the flight is proceeding. Booster operation is telemetered to the ground for analysis and interpretation. The range safety officer, the booster systems engineer, the flight director, or the flight dynamics officer, who are on the ground, may decide that danger is imminent and an abort mandatory.

A channel of the Digital Command System is used to send the abort command to the spacecraft and ground commands are sent to the launch vehicle to shutdown the booster engines. When the engine shutdown tones are received, the destruct switches of the launch vehicle are armed. The two ENGINE I indicators and both ABORT indicators illuminate red. The Command Pilot and Pilot evaluate these displays and pull the "D" rings. The hatches open and the pilots in their seats are ejected. Refer to Section III for a description of the remainder of this sequence.

Abort Mode I - II

Abort mode I - II is the modified retro abort mode. It is effective at altitudes between 15,000 and 75,000 feet approximately 50 seconds to 100 seconds after lift-off. Abort mode I - II is used when a mode I abort is inadvisable and when a delay to permit entry into the mode II conditions is impractical. The crew however has the option to eject or to ride-it-out depending upon their assessment of the abort conditions. Therefore the D-rings are not stowed during the I - II mode.

Abort mode I - II begins during stage I boost approximately 50 seconds after lift-off. If an abort condition develops, and the crew elect to ride-it-out, the Command Pilot moves the abort control handle from NORMAL to SHUTDOWN. He waits approximately 5 seconds for booster thrust to decay, then moves the handle from SHUTDOWN to ABORT.

The squib bus relays are energized. These relays arm the buses needed for

SEDR 300
PROJECT GEMINI

abort action. The retrograde common control bus is armed from the common control bus. Retro squib buses #1 and #2 are armed from OAMS squib buses #1 and #2. On S/C 3 and 4, S/C separation squib buses #1 and #2 are armed from BIA squib buses #1 and #2. Two parallel circuits are used for redundancy. This arming of buses by means of relays eliminates the motion of throwing the switch ordinarily required to arm the buses. Then, in rapid succession, RCS activate relays, wire guillotine relays, pyro switch relays, and shaped charge igniter relays are energized. The relays ignite the pyrotechnics at the equipment adapter/retrograde adapter mating line, and the vehicles separate. Simultaneously, the four retro rockets are salvo fired and the spacecraft thrusts away from the launch vehicle.

If the abort altitude is between 15,000 and 25,000 feet, the retro adapter is jettisoned 7 seconds after retro rocket salvo fire is initiated. If the abort altitude is between 25,000 and 75,000 feet, the retro adapter is jettisoned 45 seconds after salvo fire.

After retro adapter jettison, the spacecraft is maneuvered to the re-entry attitude. If the abort altitude is above 40,000 feet, the drogue chute is deployed at 40,000 feet, and the main chute at 10,600 feet. If the drogue chute fails or has not been deployed before the spacecraft descends to 10,600 feet, the emergency sequence is used to deploy the main chute.

If one of the two first stage engines should fail and the launch vehicle is above 40,000 feet, the pilots may elect to remain with the spacecraft until the operating engine has boosted them to 75,000 feet. At this altitude, abort mode II would become effective.

Abort Mode II

Abort mode II becomes effective above 75,000 feet. At approximately 100 seconds after lift off on a normal mission, the launch vehicle has boosted the spacecraft to an altitude of 75,000 feet. Ground station computers calculate the time for changeover from abort mode I - II to abort mode II. The ground station notifies the crew via the UHF Communications link of the change to abort mode II. Both the Command Pilot and Pilot acknowledge the change via the same link, and stow the ejection seat handles (D-rings). Initiation of abort mode I above 75,000 feet could be disastrous.

Abort mode II begins during stage 1 boost before BECO and ends during stage 2 boost before SSECO. The crew continue to monitor the booster indicators. If they should notice an abort situation developing, they analyze it. The decision to abort may be theirs or it may come from the ground. If a ground station sends the command to abort, both ABORT indicators illuminate red. In abort mode II, the Command Pilot must act. He moves the ABORT handle to the SHUTDOWN position. The operating engine is cut off. Since launch vehicle destruct is imminent and escape from the fireball is urgent, he moves the ABORT handle to abort. The spacecraft is separated from the launch vehicle at the equipment adapter-retrograde adapter mating line. The retro rockets, armed by four

PROJECT GEMINI

SEDR 300

RETRO ROCKET SQUIB switches during prelaunch checkoff, are salvo fired, propelling the spacecraft away from the launch vehicle.

Since orbital velocity could not have been reached below 522,000 feet, the spacecraft immediately begins a re-entry trajectory. The spacecraft is maneuvered to the retro (BEF) attitude, the retrograde section is jettisoned, and normal landing procedures are initiated.

Abort Mode III

At approximately 310 seconds after lift off, the launch vehicle reaches the altitude of 522,000 feet and a velocity of approximately 21,000 feet per second. The ground station commands a change from abort mode II to abort mode III via the UHF radio.

If an abort after this time should become necessary, the ABORT indicators would be illuminated red. The Command Pilot responds and moves the ABORT handle to the SHUTDOWN position. The shutdown command is thus given to the second stage engine. The ABORT HANDLE remains in the SHUTDOWN position. The Command Pilot then presses the SEP SPCFT telelight/switch on the Sequence Panel. This switch fires the shaped charges and severs the wiring at the launch vehicle/spacecraft mating line as described earlier. OAMS thrust is applied to put distance between the second stage and the spacecraft. The crew perform the T_R-256 seconds and the T_R-30 seconds procedures, using the sequence panel telelight/switches. After retrofire has been initiated manually, normal re-entry, landing, and post-landing procedures are followed.

SEDR 300
PROJECT GEMINI

ABORT SEQUENCE

The abort sequence to be described occurs in abort modes II and I - II. The description covers the series of events which the Abort Control Handle initiates. Figure 4-1 shows the configurations of the Abort Control Handle and Figure 4-10 shows the electrical circuits which cause the abort sequence to occur. Figure 4-10 includes the switches, circuit breakers, buses, relays, and pyrotechnic igniters. A table on Figure 4-10 gives the names, reference designations and relay panel locations of the relays and redundant relays of the abort sequence. The redundant relays, their buses, fuses, and squibs (with a few exceptions) are not shown, since the circuitry and end results are identical with those shown. The omission is made to maintain clarity and simplicity.

Abort mode I, the seat ejection mode, is not covered here. The events of this mode are discussed in another Section of this Manual.

Abort mode III is executed by performing a L/V engine shutdown, a S/C separation sequence and a retrograde sequence. Separation and retrograde in abort mode III differs from normal separation and retrograde in that the abort sequence is performed without cues from the telelights on the Main Instrument Panel. The electrical circuits however are identical with those shown in the shutdown sequence (Figure 4-10), the S/C separation (Figure 4-4), the T_R-5 minutes (or T_R-256 seconds) sequence (Figure 4-5), the T_R-30 seconds sequence (Figure 4-6), and the retrograde sequence (Figure 4-7).

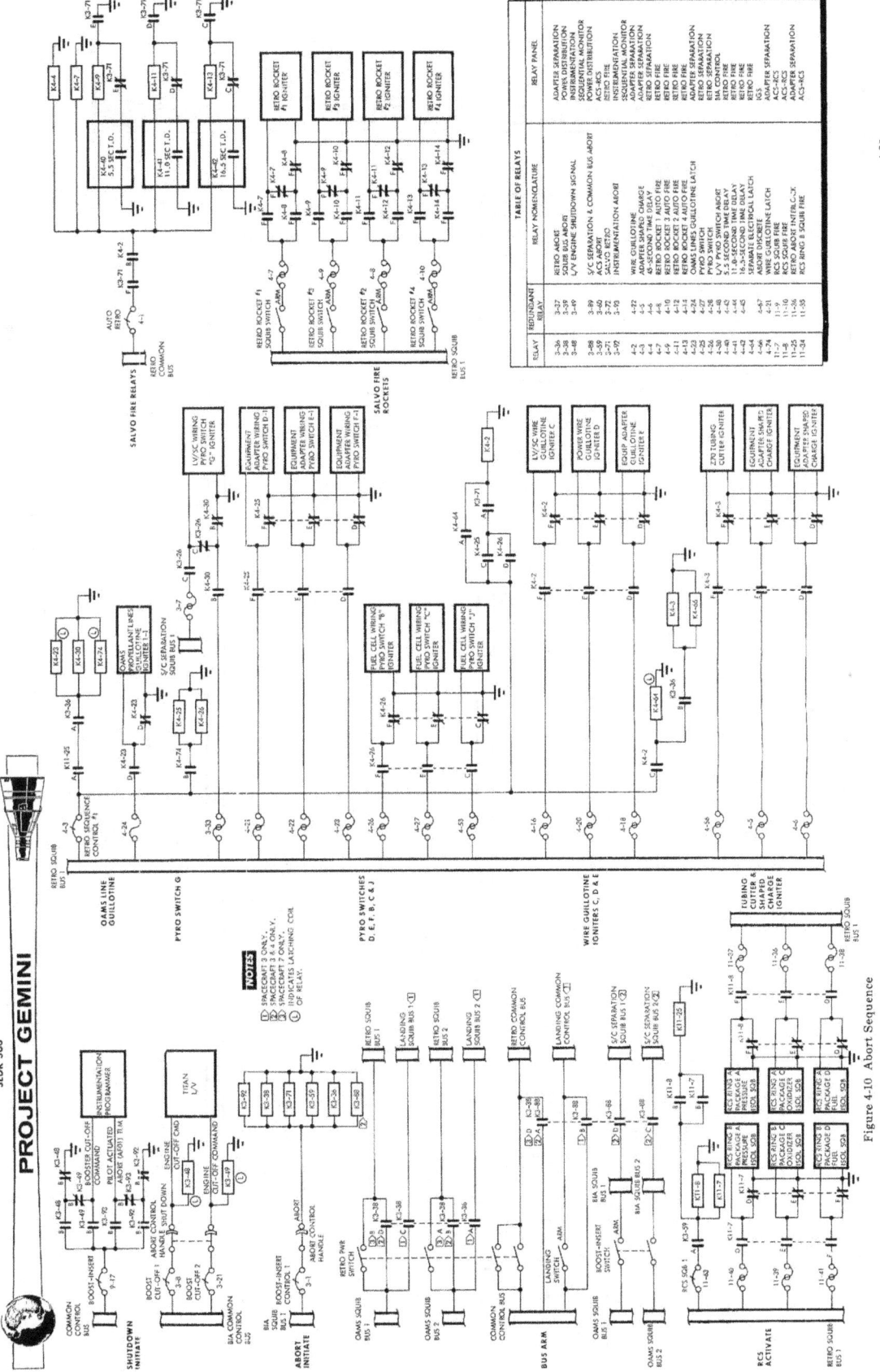

Figure 4-10 Abort Sequence

Figure 4-10 Abort Sequence

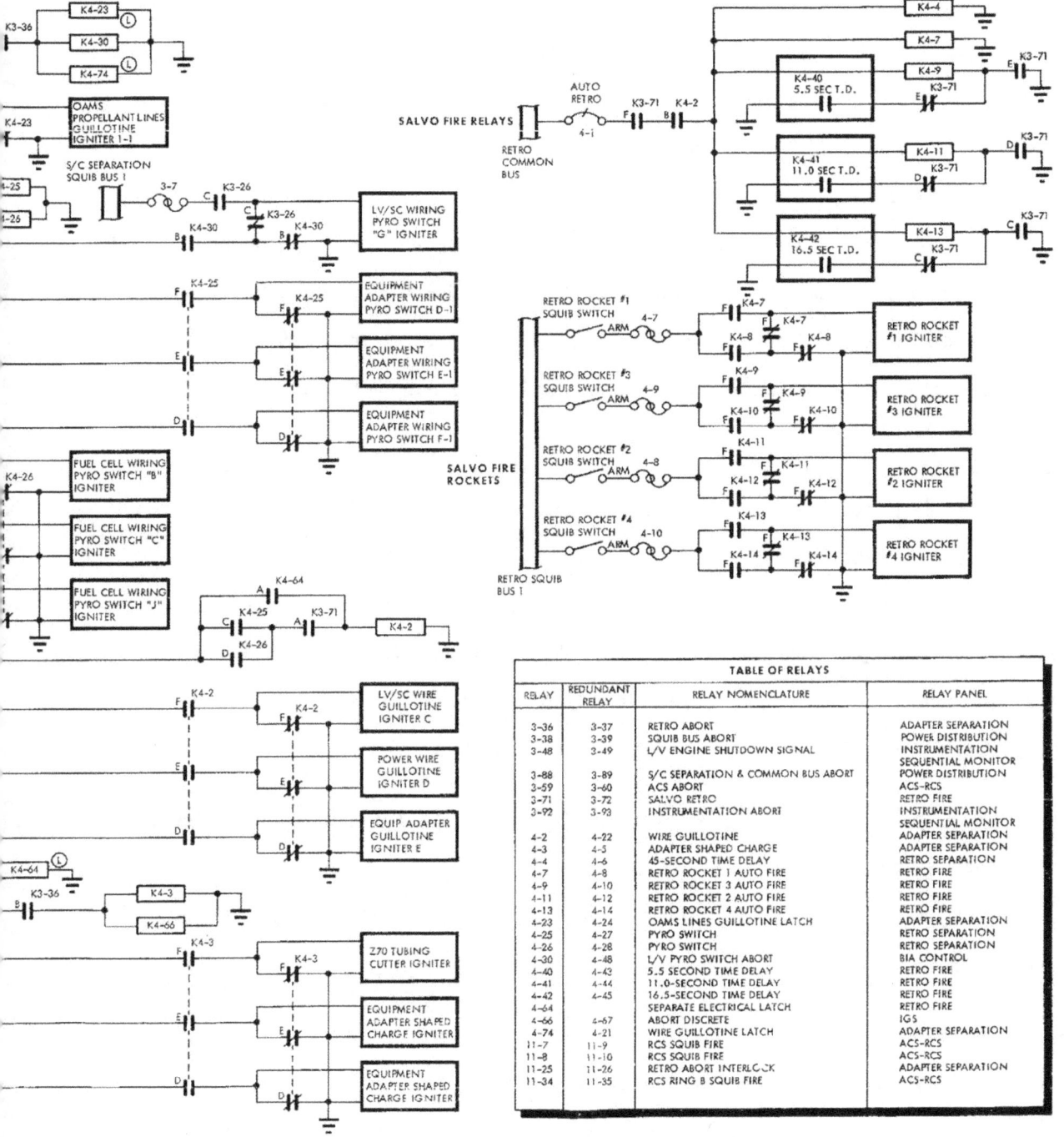

4-35

Shutdown

When the Command Pilot moves the Abort Control Handle to SHUTDOWN, the SHUTDOWN switch is closed. Boost-Insert-Abort (BIA) common control bus power is applied to the launch vehicle (L/V) engine shutdown signal relays K3-48 and K3-49. This power is also applied to the engine shutdown relays in the Titan L/V. The operating engine(s) are cut off. As K3-48 and K3-49 energize, common control bus power is applied thru their B contacts to the S/C instrumentation programmer. The programmer encodes the voltage from this bus as the booster cut-off command signal for telemetry transmission to the ground tracking station.

Abort Initiate

When the Command Pilot moves the Abort Control Handle to ABORT, numerous relays are energized, as shown on Figure 4-10. However five of these relays are key relays in that they control the principal abort operations. These operations are: (1) telemetry of the abort action to the ground; (2) arming of the retro buses; (3) activation of the re-entry control system (RCS); (4) separation of the S/C from the L/V; and (5) salvo firing of the retro rockets.

The relays which control these operations are: (1) the instrumentation abort relay, K3-92; (2) the squib bus abort relay K3-38; (3) the attitude control system (ACS) abort relay K3-59; (4) the retro abort relay K3-36; and (5) the salvo retro relay K3-71.

Abort Telemetry

When the instrumentation abort relay K3-92 is energized by the abort switch, its B contacts connect common control bus power to the S/C instrumentation

SEDR 300
PROJECT GEMINI

programmer. The programmer encodes this signal as the pilot actuated abort signal for telemetry transmission to the ground.

Abort Squib Bus Arming

Abort, if it occurs, requires that power for the circuit used in the retrograde phase of the mission become immediately available. On S/C 3, 4, and 7, the abort switch arms the retro squib buses 1 and 2 and the retro common control bus. On S/C 4, power for firing pyro switch G comes from the S/C separation buses; on S/C 7, from the retro buses. On the S/C 3 mission, power for the landing phase and the S/C separation phase as well as the retrograde phase are all made available. S/C 7 uses only one bus arm relay K3-38; S/C 3 and 4 use two: K3-38 and K3-88.

When the abort switch is closed, BIA squib bus power is applied to K3-38. K3-38 arms the retro squib buses 1 and 2 on S/C 3, 4, and 7. On S/C 3, K3-38 also arms the landing squib buses 1 and 2. On S/C 7, K3-38 also arms the retro common control bus.

On S/C 3 and 4 the abort switch applies BIA squib bus power to K3-88. K3-88 arms the retro common control bus and the S/C separation squib buses 1 and 2. On S/C 3, K3-88 also arms the landing common control bus.

Re-entry Control System (RCS) Activation

Re-entry immediately and automatically follows an abort. Re-entry requires the use of the RCS for control of the S/C during this phase. Hence the RCS is activated. Activation involves opening and pressurizing the RCS fuel and oxidant lines. This is done by firing the squibs of the fuel, oxidant, and

SEDR 300
PROJECT GEMINI

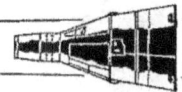

pressurant packages.

In operation, the abort switch applies BIA squib bus power to the Attitude Control System (ACS) abort relay K3-59. K3-59 applies retro squib bus power to RCS (ring A) squib fire relay K11-8 and to RCS (ring B) squib fire relay K11-7. K11-8 applies retro squib bus power to package A, C, and D igniters of RCS ring A. The squibs thus fired open the ring A fuel and oxidant lines and pressurize them. K11-7 applies retro squib bus power to similar igniters of RCS ring B with similar results.

The B contacts of K11-7 and K11-8 energize the retro abort interlock relay K11-25. K11-25, contact A initiates the station Z70 separation sequence.

OAMS Lines and Lower Wires Guillotine

Since the retro rockets are to be fired in the abort modes controlled by the abort switch, the S/C must separate from the L/V at station Z70. Z70 is on the mating line between the S/C retro section and the equipment adapter section. To make separation complete, the Orbit Attitude Maneuver System (OAMS) propellant lines which cross this station must be sealed and guillotined.

The abort switch energizes the retro abort relay K3-36 which arms K4-23, the OAMS lines guillotine latch relay; K4-30, the retro abort pyro switch relay; and K4-74, the wire guillotine relay. When K11-25 is energized, it energizes K4-23, K4-30, and K4-74. The D contacts of K4-23 apply power to the OAMS propellant lines guillotine igniter. The guillotine now seals and cuts the lines. Pyro switch G fires, opening the LV/SC interface circuits. The lower wire bundles are guillotined. The first step toward launch vehicle-spacecraft (LV/SC)

separation has been taken.

Pyro Switch Ignition

The second step in LV/SC separation is the removal of power from the hot wires crossing station Z70. These wires like the propellant lines, must also be guillotined, and the guillotine blade could cause a short circuit of the S/C power. Pyro switches B, C, D, E, F, G and J must be operated to remove power from the wires to be guillotined.

K3-36 and K11-25 apply power to L/V pyro switch abort relay K4-30 and to wire guillotine latch relay K4-74, initiating pyro switch ignition. K4-30 applies power to LV/SC wiring pyro switch G igniter, opening pyro switch G. K4-74 energizes pyro switch relays K4-25 and K4-26. K4-25 ignites equipment adapter pyro switches D, E and F. K4-26 ignites fuel cell wiring pyro switches B, C and J. With the operation of the pyro switches, the second step in LV/SC separation has been taken.

Upper Wire Guillotine Ignition

The third step in LV/SC separation is the cutting the upper wires that cross station Z70. This is accomplished by actuating the wire guillotines. Three wire guillotines igniters must be fired: the LV/SC wire guillotine igniter C, the power wire guillotine igniter D, and equipment adapter wire guillotine igniter E.

When K4-25 and K4-26 energize, they apply power thru the A contacts of K3-71 to wire guillotine relay K4-2. K4-2 fires the wire guillotine igniters C, D and E, cutting the station Z70 wires. On S/C 7, K4-2, contact C energizes the separate electrical latch relay K4-64 and the adapter shaped charge relay K4-3. On S/C 3 and 4, the abort discrete relay K4-66 is energized by the equipment adapter separation sensors; on S/C 7, by K4-2. K4-64, contact A latches K4-2 in the energized position. K4-66 signals the computer to accept re-entry data if it is in the ascent mode. K4-3 prepares the way for the fourth step of LV/SC separation.

Tubing and Structural Bond Cutting

The fourth and final step is to sever the adapter skin at station Z70 and break the LV/SC structural bond.

When K4-2 causes K4-3, the adapter shaped charge relay, to energize, K4-3 fires the Z70 tubing cutter igniter and the equipment adapter shaped charge igniters. These pyrotechnics complete the task of LV/SC separation.

Retro Rocket Salvo Fire

The retro rockets are salvo fired at the same time that the tubing and structural bond is cut. To salvo fire the retro rockets, power must be applied simultaneously to the retro rocket auto fire relays and thus to the retro rockets. Therefore the 5.5, 11.0, and 16.5-second time delay relays must be bypassed. Contacts C, D and E of K3-71 bypass the time delay relays. When K4-2 energizes, retro common bus power simultaneously energizes the retro rocket auto fire relays K4-7, K4-9, K4-11 and K4-13. As these relays energize, retro squib bus power is applied to the igniters of retro rockets 1, 3, 2 and 4. Salvo burn lasts approximately 5.5 seconds.

SEDR 300

 **PROJECT GEMINI**

Retro Section Jettison

When the retro rocket auto fire relays are energized by K4-2, the 45-second time delay relay K4-4 is also energized. When K4-4 energizes after 45 seconds, it illuminates the JETT RETRO telelight as shown on Figure 4-7. The JETT RETRO telelight-switch is then pressed, and the retro section is jettisoned in a mode II abort. However, in a mode I - II abort when the altitude is between 15,000 and 25,000 feet, the telelight switch is pressed seven seconds after the retro rockets begin firing. After the retro section has been jettisoned, normal re-entry and landing procedures are initiated.

SYSTEM UNITS

The Sequence System as shown in Figure 4-1 comprises the following units:

Left switch/circuit breaker panel, consisting of three rows of circuit breakers and one row of switches.

Boost and staging indicators, consisting of seven lights and three meters on the top of the Command Pilot and Pilot's Panels.

Sequence panel, consisting of two pushbutton switches, eight telelight/switches, and one telelight located on the left side of the Center Panel.

Re-entry and landing switches and indicators, consisting of four switches on the Pedestal Panel and one switch, two lights, and two meters on the Command Pilot's Panel.

4-41

Abort controls, consisting of two "D" rings on the ejection seats and one abort control handle on the left hand side of the cabin.

Relay panels, comprising thirteen to fourteen in the re-entry module and six to eight in the adapter and retrograde sections.

Separation sensing devices, consisting of three each in the equipment adapter section and the retrograde section.

The components of the sequence system are described below.

LEFT SWITCH/CIRCUIT BREAKER PANEL

The switches and circuit breakers on the Left Switch and Circuit Breaker Panel perform important functions in the operation of the sequential system. The top row of circuit breakers however pertain largely to communications. The second row of circuit breakers perform functions related to the operation of the sequential system. Their functions are as follows:

ELECTRONIC TIMER Circuit Breaker

The electronic timer circuit breaker CB8-43 applies main bus power to the electronic timer to energize it and to contact A of lift off relay K3-11 which is associated with the timer. When the lift off signal energizes K3-11, closed A contacts start the electronic timer. The timer begins counting the time to go to retrograde.

SEDR 300
PROJECT GEMINI

EVENT TIMER Circuit Breaker

The event timer circuit breaker CB8-14 applies main bus power to the event timer and to contact B of lift off relay K3-11 which is also associated with the event timer. When the lift off signal energizes K3-11, the B contacts start the event counter counting the time since lift off began.

BOOST CUTOFF 1 Circuit Breaker

The boost cutoff 1 circuit breaker CB3-8 applies BIA common control bus #1 power to the booster shutdown switch on the Abort Control Handle and to the Secondary Guidance (RGS-IGS) switch. This circuit breaker arms the booster shutdown circuit.

BOOST CUTOFF 2 Circuit Breaker

The boost cutoff 2 circuit breaker CB3-21 applies BIA control bus #2 power redundantly to the booster shutdown and secondary guidance switches.

RETRO AUTO Circuit Breaker

The retrofire automatic circuit breaker CB4-1 applies retro common control bus power to the ARM AUTO RETRO switch, and to contacts on T_R arm relay K4-36 and T_R signal relay K4-34. If CB4-1 were not closed, the electronic timer T_R contact closure could not automatically fire the retro rockets.

RETRO MAN Circuit Breaker

The retro manual circuit breaker CB4-2 applies retro common control bus power to contacts on the T_R-30 second relay in the electronic timer, to the MAN FIRE RETRO switch, and to contacts on manual retro latch switch relay K4-37. CB4-2

4-43

SEDR 300
PROJECT GEMINI

must be closed before the MAN FIRE RETRO sequential switch can manually fire the retro rockets.

T_R-5 Circuit Breaker

On S/C 3 and 4, the time to retrograde minus five minutes (T_R-5) circuit breaker CB8-16 applies common control bus power to relay contacts in the electronic timer and contacts of the T_R-5 minute relay. CB8-16 enables the T_R-5 minute signal to illuminate amber the IND RETRO ATT, O_2 HI RATE, BTRY PWR, and RCS telelights on the Sequence Panel.

T_R-256 Circuit Breaker

On S/C 7, the time to retrograde minus 256 (T_R-256) seconds circuit breaker CB8-16 applies common control bus power to the normally open A contacts of the T_R-256 second relay in the Electronic Timer. CB8-16 enables the T_R-256 second signal to illuminate amber the IND RETRO ATT, BTRY PWR, and RCS telelights on the Sequence Panel.

SEQ. LIGHTS POWER Circuit Breaker

The sequence lights power circuit breaker CB6-1 applies main bus power to the Sequence Lights (BRIGHT-DIM) switch and to open contacts on barostat switch arm relay K5-61. (See Figure 4-8.)

SEQ. LIGHTS CONTROL Circuit Breaker

The sequence lights control circuit breaker CB1-13 applies common control bus power through the four MAIN BATTERIES switches to relay K1-29. When the main battery power indicator relay K1-29 is energized, the BTRY PWR indicator on the Sequential Panel is illuminated green.

SEDR 300

PROJECT GEMINI

PARA CNTL Circuit Breaker

The parachute control circuit breaker CB5-80 applies the landing common control bus power to the barostat switch arm relay K5-61. K5-61 when energized applies main bus power from CB6-1 (SEQ. LTS. POWER) to the 40K and 10.6K feet indicators. CB5-80 must be closed, or the barostat switches cannot illuminate the indicators.

The third row of circuit breakers on the Left Switch/Circuit Breaker Panel perform functions related to the sequential system. These functions are the following.

ATT IND CNTL RETRO Circuit Breaker

The attitude indicate control retro circuit breaker CB12-7 applies retro control bus power to the IND RETRO ATT switch, and to contacts of retro bias off relays K4-62 and K4-63. Power from CB12-7 energizes retro bias relay K12-5 when the retro jettison switch is pressed.

ATT IND CNTL LDG Circuit Breaker

The attitude indicate control landing circuit breaker CB12-8 applies common control bus power to the attitude control mode switch. In the PARA position of this mode switch, the bus power is connected to the right and left hand attitude display balls as pitch bias.

BOOST-INSERT CONTROL 1 Circuit Breaker

The boost-insert control 1 circuit breaker CB3-1 connects power to the circuits needed in the boost-insert phase. CB3-1 connects BIA squib bus #1 power to the Abort switch, the jettison fairing switch, the separate spacecraft switch, and the spacecraft separation sensors.

SEDR 300

PROJECT GEMINI

BOOST-INSERT CONTROL 2 Circuit Breaker

The boost-insert control 2 circuit breaker CB3-11 connects BIA squib bus #2 power redundantly to the same switches to which CB3-1 connects power.

RETRO SEQ. CNTL 1 Circuit Breaker

The retro sequence control 1 circuit breaker CB4-3 connects the retro squib bus #1 to the separate OAMS lines switch, the separate adapter switch, the separate electrical switch, and the jettison retro switch.

RETRO SEQ. CNTL 2 Circuit Breaker

The retro sequence control 2 circuit breaker CB4-28 connects the retro squib bus #2 redundantly to the same components to which the retro sequence control 1 circuit breaker connects power.

LANDING SEQ. CNTL 1 Circuit Breaker

The landing sequence control 1 circuit breaker CB5-2 applies landing squib bus #1 power to the EMERG DROGUE switch, the HI ALT DROGUE switch, the PARA switch, the PARA JETT switch, and the LDG. ATT. switch.

LANDING SEQ. CNTL 2 Circuit Breaker

The landing sequence control 2 circuit breaker CB5-33 applies landing squib bus #2 power redundantly to the same switches to which CB5-2 applies power.

SEQ. LIGHTS TEST (AMBER-OFF-RED & GREEN) Switch

The sequence lights test switch connects main bus power to all amber-colored sequence lights in the AMBER position, and to all red or green sequence lights in the RED & GREEN position. The test switch does not apply power to any sequence lights in the center (OFF) position.

SEQ. LIGHTS (BRIGHT-DIM) Switch

The sequence light bright-dim switch is a single-pole, double-throw toggle switch. It connects the main bus thru a diode to all sequence light circuits in the BRIGHT position. It connects the bus thru a resistor to the same circuits in the DIM position.

The fourth row on the Left Switch/Circuit Breaker Panel contains eight switches. These switches arm or safety the various squib buses used by the sequential system. Their functions are as follows.

BOOST-INSERT (ARM-SAFE) Switch

The boost-insert squib bus arm-safe switch is a four-pole, double-throw toggle switch. In the ARM position, one pole of this switch connects the OAMS squib bus #1 to Boost-Insert-Abort (BIA) squib bus #1 and the SPCFT SEP switch. Another pole of this switch connects OAMS squib bus #2 to BIA squib bus #2 and the SPCFT SEP switch. The third and fourth poles are connected together; they connect the common control bus to the BIA common control bus.

RETRO PWR (ARM-SAFE) Switch

The retro power squib bus arm-safe switch is a four-pole, double-throw switch. In the safe position it removes power from the retro jettison switch, the retro common control bus, the retro squib buses, and the retro rocket squib arm switches. In the arm position, this switch connects OAMS squib buses #1 and #2 to the RETRO JETT ARM-SAFE switch and to retro squib buses #1 and #2. The retro power switch connects common control bus power to the retro common control bus.

RETRO JETT (ARM-SAFE) Switch

The retro jettison squib bus arm-safe switch is a two-pole double-throw toggle switch. In the safe position, it removes power from retro jettison squib buses #1 and #2. In the arm position it applies power to these squib buses. When the retro jettison relays are energized, these squib buses detonate the retro section shaped charges and the wiring pyro switch.

LANDING (ARM-SAFE) Switch

The landing squib bus arm-safe switch is a four-pole double-throw switch. In the safe position, it prevents deploying the drogue, pilot, and main chutes, and jettisoning the chutes and the rendezvous and recovery section. In the arm position, this squib bus switch connects common control bus power to the landing common control bus from which the barostat switches are armed. The squib bus switch also connects the OAMS squib buses #1 and #2 to landing squib buses #1 and #2, respectively. These squib buses supply power to the HI ALT DROGUE, EMERG. DROGUE, PARA, and PARA JETT switches, to the relays they control, and to their associated pyrotechnics.

RETRO ROCKET SQUIB 1, 2, 3, 4 (ARM-SAFE) Switches

The four retro rocket squib arm switches apply the voltages which ignite the four retrofire rockets to open contacts of the retro rocket automatic and manual fire relays. In the safe positions of these four switches, the ignition voltage is removed from the relays. When the four RETRO ROCKET SQUIB arm switches are placed to the ARM position, the RETRO squib buses #1 and #2 are connected redundantly to the retro rocket fire relays.

BOOST-INSERT-ABORT CONTROLS AND INDICATORS

Seven telelight/switches, three meters and four controls are provided for the boost-insert-abort phase of S/C 3 and 4; six telelight/switches and the same number of meters and controls are used for S/C 7.

ENGINE I Indicators

The two ENGINE I indicators are provided on the Command Pilot's Panel to indicate thrust chamber underpressure of the first stage booster engines. Each indicator illuminates red when the thrust chamber pressure of the engine is 68 percent of rated pressure or less. Both indicators illuminate red at stage 1 ignition but extinguish 0.91 to 1.25 seconds later as the pressure increases above 68 percent. Both indicators illuminate red at booster engine cutoff and extinguish quickly at staging.

ENGINE II Indicator

The ENGINE II indicator on the Command Pilot's Panel illuminates amber to indicate the fuel injector underpressure (or off) condition of the second stage engine. The critical pressure for engine II is 55 percent of rated value. The indicator illuminates when the first stage engine is ignited and stays amber through first stage boost. Approximately one second after both ENGINE I indicators extinguish, the ENGINE II indicator also extinguishes, indicating normal staging and engine II fuel injector pressure build up.

ATT RATE Indicator

The attitude rate indicator on the Command Pilot's Panel indicates an evaluation of the launch vehicle attitude rates during the boost phase. The indicator is extinguished if the attitude rates remain within acceptable limits, but illuminates red if the rates exceed these limits. Acceptable limits for stage 1 flight differ somewhat from those for stage 2 flight. For stage 1, limits in pitch are + 2.5°/second nose up and - 3.0°/second nose down; in yaw, + 2.5°/second right and -2.5°/second left; and in roll, + 20°/second clockwise and -20°/second counterclockwise. For stage 2, pitch limits are + 10°/second and - 10°/second and yaw limits are + 10°/second and - 10°/second. Roll limits for stages 1 and 2 are the same.

GUIDANCE (SEC GUID) Indicator

The GUIDANCE indicator on the Command Pilot's Panel on S/C 3 and 4 and the SEC GUID indicator on S/C 7 indicates the guidance system that is in operation. The indicator is extinguished to indicate that primary guidance is being used. The indicator illuminates amber to indicate that secondary guidance has been selected.

ABORT Indicators

Two ABORT indicators are provided, one for each pilot. Both indicators illuminate red when the abort command is transmitted. When the ABORT indicator is illuminated immediate and appropriate action is imperative. The indicator signals the crew to initiate immediately the abort mode appropriate for the altitude and velocity of the spacecraft. These modes are described under Sequential System Operation. During boost phase, the crew has been reminded via the UHF communications link of the abort mode in effect.

SEDR 300
PROJECT GEMINI

STAGE 1 FUEL/OXID Meters

The stage 1 fuel and oxidizer meters on the Command Pilot's Panel are provided to the crew to monitor the instantaneous conditions and progress of the boost phase, and to permit them to anticipate an abort condition if one should develop. These meters indicate the gas pressures in PSIA of the stage 1 fuel and oxidizer tanks. Dual indicating needles are provided for redundancy. The range of the stage 1 indicators is 35 to 5 PSIA. A time-versus-pressure scale near the bottom of the meter shows the minimum required pressure at 20, 40 and 60 seconds after lift off. Critical fuel tank pressure is indicated by a shaded column at the low end of the scale. After staging when no signals are applied to the meters, the needles reset at the maximum PSIA position.

STAGE 2 FUEL/OXID Meters

The stage 2 fuel and oxidizer meters on the Command Pilot's Panel indicate stage 2 fuel and oxidizer tank pressures over a 70 to 10 PSIA range. Redundant pointers are used. Critical fuel tank pressures are indexed by a shaded column at the low end of the scale. The S-flag at the 30-PSIA mark indicates the minimum acceptable stored pressure in the tank before pressurization. After S/C separation the meters indicate maximum PSIA.

LONG ACCEL Meter

The longitudinal accelerometer on the Command Pilot's Panel indicates the rate in G's at which the launch vehicle engines are changing the velocity of the spacecraft. The range of the accelerometer is -6G to 16G. The meter has positive and

4-51

SEDR 300
PROJECT GEMINI

negative memory pointers. The accelerometer enables the crew to monitor the effectiveness of the engines. It is a secondary indicator of staging.

RGS-IGS Guidance Switch

The guidance switch above the abort control handle permits the Command Pilot to manually change from primary guidance (RGS) to secondary backup guidance (IGS). When backup guidance has been selected either manually or automatically during stage 1 boost, and the ground station determines that primary guidance is feasible during stage 2 boost, primary guidance can be selected again by momentarily placing the guidance switch to the RGS position.

"D" Rings

A "D" ring is provided on the ejection seat of each astronaut. These rings are pulled to initiate Mode 1 Abort at altitudes below 75,000 feet. Refer to Section III of this volume for the location and operation of these devices.

ABORT CONTROL Handle

The Abort Control Handle is located on the Command Pilot's side of the cabin. It is used for spacecraft re-entry in Abort Modes I-II, II and III. These modes are effective above 25,000 feet. The three positions of this handle are NORMAL, SHUTDOWN and ABORT. In NORMAL, the handle is inoperative. When the handle is moved to SHUTDOWN, the engine cutoff command is sent to the operating launch vehicle engine. When the ABORT handle is moved to ABORT, the spacecraft is separated from the launch vehicle at the adapter-retrograde interface, and the retrofire rockets are simultaneously fired.

SEDR 300

PROJECT GEMINI

SEQUENTIAL PANEL CONTROLS AND INDICATORS

The switches, telelights, and telelight/switches on the Main Instrument Panel have the following nomenclature, place in the mission sequence, and functions.

JETT FAIRING Pushbutton Switch

The jettison fairing switch is used in the launch phase of the mission sequence. The Command Pilot presses the switch after SSECO on S/C 3 and 4, and after staging on S/C 7, ejecting the nose fairing and the scanner head cover.

SEP SPCFT Telelight/Switch

The separate spacecraft telelight/switch is used in the separation-insertion phase of the sequence. The Command Pilot presses the switch approximately 20 seconds after second stage engine cutoff when the Incremental Velocity Indicator displays the delta-V required for insertion. Pressing the switch causes several things to happen. Primarily it detonates pyrotechnic devices which separate the spacecraft from the launch vehicle. Secondarily, it extends the UHF and diplexer antennas, and readies the acquisition aid beacon for use. As the spacecraft moves away from the launch vehicle, two of three separation sensors close and energize the spacecraft separation relays. The relays illuminate the telelight green.

IND RETRO ATT Telelight/Switch

The indicate retro attitude telelight is illuminated amber when the electronic timer energizes the T_R-5 minute (or T_R-256 second) relay in the prepare-to-go-to-retrograde phase. The amber light reminds the crew to press the telelight/switch at this time. When pressed, the switch changes the FDI electronics to permit orienting the spacecraft in the retro (BEF) attitude

4-53

to the FDI needles in the same manner as in normal (SEF) flight. The telelight also changes from amber to green, indicating that the FDI can be used in the retro attitude.

O_2 HI RATE Telelight/Switch

The oxygen high rate telelight on S/C 3 and 4 is illuminated amber by the T_R-5 minute relay in the prepare-to-go-to-retrograde sequence. The amber light reminds the Command Pilot to start the oxygen high flow rate for the re-entry phase. When pressed, the telelight/switch opens the secondary oxygen high rate valve, and the telelight changes from amber to green. A somewhat different arrangement for S/C 7 has been made, as explained below.

BTRY PWR Telelight

The battery power telelight is illuminated amber by the T_R-5 minute relay on S/C 3 and 4 or the T_R-256 second relay on S/C 7. The amber light reminds the Command Pilot to turn off the adapter power supply and place the MAIN BATTERIES switches to ON. This change must be made because the adapter section will be jettisoned at retrograde (T_R). When all of the main battery switches are on, the telelight changes from amber to green.

RCS Telelight/Switch

The re-entry control system telelight is illuminated amber by the T_R-5 minute (or T_R-256 second) relay. The amber light cues the Command Pilot to activate the re-entry control system by firing the fuel, oxidant, and pressurant isolation squibs. Pressing the telelight/switch energizes relays which fire the squibs. The telelight changes from amber to green, indicating that the RCS has been activated.

SEP OAMS LINES Telelight/Switch

The separate OAMS lines telelight is illuminated amber by the T_R-30 second relay in the prepare-to-go-to-retrograde phase. The amber light cues the Command Pilot to seal and sever the OAMS lines before jettisoning the adapter. Pressing the telelight/switch energizes relays which ignite the pyrotechnics used to seal and sever the propellant lines and to guillotine the lower wires at Z70. The relays also fire pyro switches which open-circuit the "hot" wires crossing the adapter-retro mating line. The telelight changes from amber to green.

SEP ELEC Telelight Switch

The separate electrical telelight is also illuminated amber by the T_R-30 second relay. The amber light cues the Command Pilot to sever all the wiring at the adapter/retro mating line. Pressing the telelight/switch energizes the upper wire guillotine relays. The pyrotechnics detonate and the wiring is cut. The telelight changes from amber to green to indicate that electrical separation has been accomplished.

SEP ADAPT Telelight/Switch

The separate adapter telelight is illuminated amber by the T_R-30 second relay. The amber light cues the Command Pilot to jettison the adapter section. Pressing the telelight/switch causes the adapter shaped charge and the Z70 tubing cutter pyro to be detonated, and the adapter severed. Separation of the adapter is sensed by two of three toggle switch sensors. Two closed sensors energize the sensor relay and change the telelight from amber to green.

ARM AUTO RETRO Telelight/Switch

The arm automatic retrofire telelight is illuminated amber by the T_R-30 second relay. The amber light cues the Command Pilot to arm the automatic retrofire circuits so that when the electronic timer closes the T_R contacts at T_R time, the retro rockets will fire automatically. Pressing the telelight/switch completes the path from the retro common control bus to the timer T_R contact, and also energizes the T_R arm relay. The relay changes the light from amber to green. Contact closure at T_R time energizes the T_R signal relay. The signal relay energizes the 45-second time delay relay, fires the retro rockets at 5.5-second intervals, and puts the platform in the free mode.

MAN FIRE RETRO Pushbutton Switch

The manual fire retro rockets switch connects the retro common control bus to the manual retro latch relay. Contacts of this relay do several things. They energize the 45-second time delay, fire the retro rockets at 5.5 second intervals, and place the platform in the free mode of operation.

JETT RETRO Telelight/Switch

The jettison retro section telelight is illuminated amber by the 45-second time delay relay 45 seconds after retrofire begins. The amber light cues the Command Pilot to jettison the retro section. Pressing the telelight does this and several other things besides. It fires the pyrotechnic devices which disconnect and guillotine the wires at the retro/re-entry module mating line. It fires the shaped charges which sever the retro section from the re-entry module. It removes the retro attitude signals applied to the flight director needles at T_R-5 minutes (or T_R-256 seconds). It switches the FDI roll channel to the mix mode for re-entry. Finally it extinguishes the IND RETRO ATT, SEP OAMS LINES, SEP ELECT, SEP ADAPT, and ARM AUTO RETRO green telelights. It jettisons the scanner heads.

The telelight is extinguished 80 milliseconds after actuation of the switch.

O_2 HI RATE Indicator

On S/C 7, the O_2 high rate indicator is on the Annunciator Panel, which is located in the upper right corner of the Center Console of the Main Instrument Panel. The indicator is normally off when the cabin fan or the suit compressors are in use. The indicator lights with a green color after the CABIN FAN-O_2 HI RATE switch has been placed momentarily to the O_2 HI RATE position. No automatic lighting of this indicator has been provided on S/C 7. The indicator reminds the pilots that their secondary O_2 supply is now in use, and that the suit compressors and cabin fan cannot be turned on. The indicator remains on until the O_2 high rate valve has been recocked and closed with the O_2 HI RATE RECOCK lever.

CABIN FAN-O_2 HI RATE Switch

On S/C 7, the cabin fan-O_2 high rate switch combines two functions which were separated on S/C 3 and 4. In the CABIN FAN position, the switch turns on the cabin fan. When placed momentarily to the O_2 HI RATE position, the switch causes oxygen from the secondary O_2 supply to flow into the space suits of both pilots. It also causes the suit compressors and cabin fan to be disconnected. In the OFF position, the switch can turn off the cabin fan if it was on, but it cannot stop the flow of secondary O_2. O_2 high rate flow is terminated, and the suit compressors and cabin fan circuits are reconnected, when the O_2 high rate recock lever is operated to close the O_2 high rate valve.

 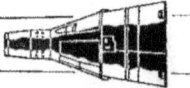

SEDR 300

PROJECT GEMINI

LANDING SEQUENCE SWITCHES AND INDICATORS

Four switches on the Pedestal Panel and one switch on the Command Pilot's Panel control the landing sequence events. Two indicators and two meters on the Command Pilot's Panel provide the crew with important descent data. These components and their functions are as follows.

HI ALT DROGUE Switch

The high altitude drogue chute deploy switch applies landing squib bus #1 and #2 power to drogue chute relays K5-83 and K5-84, and to cabin air inlet relays K5-93 and K5-94. The drogue chute relays apply landing squib bus #1 and #2 power to the drogue chute mortar igniters which upon ignition deploy the drogue chute.

PARA Switch

The parachute deploy switch applies landing squib bus #1 and #2 power to main chute deploy relays K5-89 and K5-90, to drogue chute disconnect relays K5-87 and K5-88. The main chute deploy relays start the 2.5-second time delay chute deploy relays K5-91 and K5-92. The drogue chute disconnect relays fire three drogue chute disconnect guillotines. The 2.5-second time delay relays ignite the pyro switches, the R & R section wire guillotine pyrotechnics, and the R & R section jettison primer cord igniters.

LDG ATT Switch

The landing attitude switch applies the landing squib bus #1 and #2 power to the main chute single-point release relays K5-17 and K5-18. These relays fire the single-point release igniters. The single-point release pyrotechnics initiate the two-point suspension sequence, and also extend the UHF descent antenna and the UHF rescue beacon antenna.

SEDR 300

PROJECT GEMINI

PARA JETT Switch

The parachute jettison switch connects the landing squib buses #1 and #2 to the main chute jettison relays K5-21 and K5-22 and, on S/C 3, to the flashing recovery light relays K5-45 and K5-46. K5-45 and K5-46 when energized close the circuit from the recovery light battery to the recovery light. K5-21 and K5-22 connect squib bus power to the main chute jettison pyrotechnics and to the hoist loop and flashing recovery light release squibs.

EMERG 10.6K DROGUE Switch

The emergency 10.6K-foot drogue chute switch is pressed in case the drogue chute fails. This switch connects landing squib buses #1 and #2 to the emergency chute relays K5-85 and K5-86. These relays fire the pilot chute mortar and apex cable guillotine pyrotechnics, and deploy the pilot chute.

40K FT Indicator

The 40,000-foot indicator illuminates amber when the re-entering spacecraft has descended to an altitude of 40,000 feet. The indicator is illuminated by the 40K-foot barostat switch, when the SEQ. LIGHTS PWR and PARA CNTL circuit breakers have been previously closed.

10.6K FT Indicator

The 10,600-foot indicator illuminates amber when the SEQ. LIGHTS PWR and PARA CNTL circuit breakers have been previously closed and the spacecraft has descended to this altitude.

DESCENT RATE Meter

The descent rate meter is a conventional pneumatic unit driven from a static pressure source. It visually indicates the vertical velocity of the re-entry module during the landing phase.

Altimeter

The altimeter located on the left side of the Command Pilot's Panel is a standard aircraft altimeter. It is used to monitor the altitude of the re-entry module during the landing phase.

RE-ENTRY VEHICLE RELAY PANELS

Twenty relay panels are installed and used on S/C 3 and 4, and twenty-two on S/C 7. See Figure 4-1. Each relay panel contains from 2 to 20 relays, which are used for switching, sequencing, memory, or time delay. The re-entry vehicle contains 14 relay panels. These panels are described below.

Power Relay Panel

On S/C 3, the power relay panel contains twenty relays sixteen of which pertain to fuel cell operations. These sixteen relays are inoperative because the fuel cells are not installed in this spacecraft. The four remaining relays are used: to arm OAMS squib bus #1 (K1-2); to arm OAMS squib bus #2 (K1-90); to indicate that the four main batteries should be selected at 5 minutes before retrofire (K8-17); and to indicate that the four main batteries have been selected (K1-29).

On S/C 4, there are fifteen relays on the power relay panel, eleven of which pertain to the operation of fuel cells. The fuel cells are not installed in Gemini S/C 4 either, and these relays are inoperative. The four remaining relays are the same relays used in S/C 3, and perform the same functions.

SEDR 300

PROJECT GEMINI

On S/C 7, there are fourteen relays on this panel. All of the relays are operative. Ten pertain to fuel cell operation, two to OAMS squib bus arming, and two to the use of main battery power.

Power Distribution Relay Panel

In the event of an abort, the spacecraft separation, common, and squib buses are armed by means of the four relays of the power distribution relay panel.

IGS Relay Panel

The inertial guidance system (IGS) relay panel contains nine relays to perform the following IGS functions; abort command transfer, re-entry roll display, retro bias application, retro attitude indication, guidance switch over fade in, and flight director indicator (FDI) ascent scale factoring.

BIA Control Relay Panel

The boost-insert-abort (BIA) control relay panel contains six relays to perform spacecraft separation indicator control and launch vehicle/spacecraft pyro switch firing.

Retro Separation Relay Panel

The necessary functions required for adapter retro section separation are performed by the twelve relays of the retro separation relay panel. The relays perform such functions as; pyro switch and shaped charge ignition, T_R-30 second indication, automatic IGS "free" mode selection, and arming of the T_R contacts of the TRS.

Parachute Jettison Relay Panel

The parachute jettison relay panel contains two relays to perform each of the following functions: main chute single point release, main chute jettison, flashing recovery light actuation and cabin air inlet door guillotine ignition.

ACS Scanner and RCS Squib Fire Relay Panel

Re-entry control system (RCS) squib firing, scanner cover and scanner heads jettison, abort interlock, RCS amber light actuation, and RCS ring B squib firing test prior to launch are provided by the eighteen relays of the attitude control system (ACS) scanner and RCS squib fire relay panel.

Communication Relay Panel

The communication relay panel consists of nine relays to perform the following functions: lift-off sensing, T_R-5 minute indication on S/C 3 and 4, T_R-256 second indication on S/C 7, descent antenna select, acquisition aid disable, and UHF whip antenna actuate.

ECS Relay Panel

The various sequentially controlled functions of the environmental control system (ECS) are performed by the seventeen relays of the ECS relay panel. The relay panel performs such functions as O_2 high rate indicator power control, suit and cabin fan power disconnect, and O_2 high rate selection.

R & R Section Separation Control Relay Panel

The required functions for the rendezvous and recovery (R & R) section separation are performed by the R & R section separation control relay panel. The nine relays of the R & R section separation control relay panel perform such functions as guillotine ignition, barostat arming, and R & R section shaped charge firing.

Adapter Power Supply Relay Panel

The adapter power supply contains twelve relays which control the transfer of electrical power from the S/C 3 and 4 adapter batteries or the S/C 7 fuel cell stacks in the equipment adapter.

Instrumentation Sequence Monitor Relay Panel

The instrumentation sequence monitor relay panel contains eleven relays in S/C 3 and 4, and nine in S/C 7 which switch signals representing significant sequence operations into the telemetry transmitter for transmission to tracking stations.

Umbilical Pyro Switch Relay Panel

The umbilical pyro switch relay panel, located in the main landing gear well, contains two relays which operate umbilical pyro switch K during the landing phase of the mission.

ADAPTER SECTION RELAY PANELS

The retrograde adapter section contains the spacecraft separation control, retro fire, and adapter separation relay panels in S/C 3, 4 and 7, plus the DOD equipment extend and experiment squib fire relay panels in S/C 7. The orbit attitude maneuver system (OAMS) squib fire relay panel is located in the equipment adapter section of S/C 3, 4 and 7.

Spacecraft Separation Relay Panel

The spacecraft separation control relay panel contains six relays to perform the functions of shaped charge ignition and launch vehicle/spacecraft guillotine firing.

PROJECT GEMINI

Retro Fire Relay Panel

The retro rockets are either manually or automatically, in salvo or in rotation, fired by the relays of the retro fire relay panel.

Adapter Separate Relay Panel

The adapter separate relay panel contains relays for shaped charge ignition, OAMS propellant lines guillotine firing, and electrical wires guillotine firing.

DOD Equipment Extend Relay Panel

The Department of Defense (DOD) equipment extend relay panel contains relays which control the initiation of some DOD experiments on the S/C 7 mission.

Experiment Squib Fire Relay Panel

The experiment squib fire relay panel contains relays used to initiate a number of experiments performed by the crew during the S/C 7 mission.

OAMS Squib Fire Relay Panel

The adapter equipment section contains the OAMS squib fire relay panel used for OAMS squib firing and controlling of regulator valves.

R & R SECTION RELAY PANELS

The R & R section contains the drogue chute control relay panel and the nose fairing jettison relay panel to perform such functions as drogue chute mortar ignition and nose fairing jettison pyrotechnic ignition.

SEPARATION SENSORS

The sequence system contains the following separation sensors as illustrated in Figure 4-1: three launch vehicle/spacecraft separation sensors and three equipment adapter section separation sensors. The separation sensors are toggle switches that are normally open before the spacecraft is launched. The separating section will close the sensors when it is separated from the spacecraft. The closure of any two of three sensors is sufficient to detect separation.

ELECTRICAL POWER SYSTEM

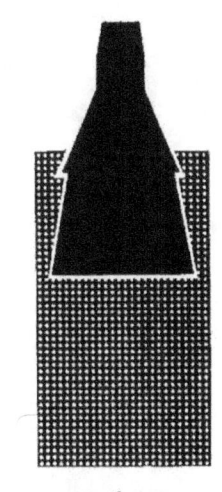

Section V

TABLE OF CONTENTS

TITLE	PAGE
SYSTEM DESCRIPTION	5-3
SYSTEM OPERATION	5-10
PRE-LAUNCH	5-10
ORBIT	5-11
RE-ENTRY	5-12
MONITOR AND DISPLAY	5-14
SYSTEMS UNITS	5-16
SILVER-ZINC BATTERIES	5-16
POWER SYSTEM RELAY PANEL	5-17
ADAPTER POWER SUPPLY RELAY PANEL	5-17
AMMETERS	5-18
VOLTMETER	5-18
POWER SYSTEM MONITOR	5-19
FUEL CELL BATTERIES	5-19
REACTANT SUPPLY SYSTEM	5-27

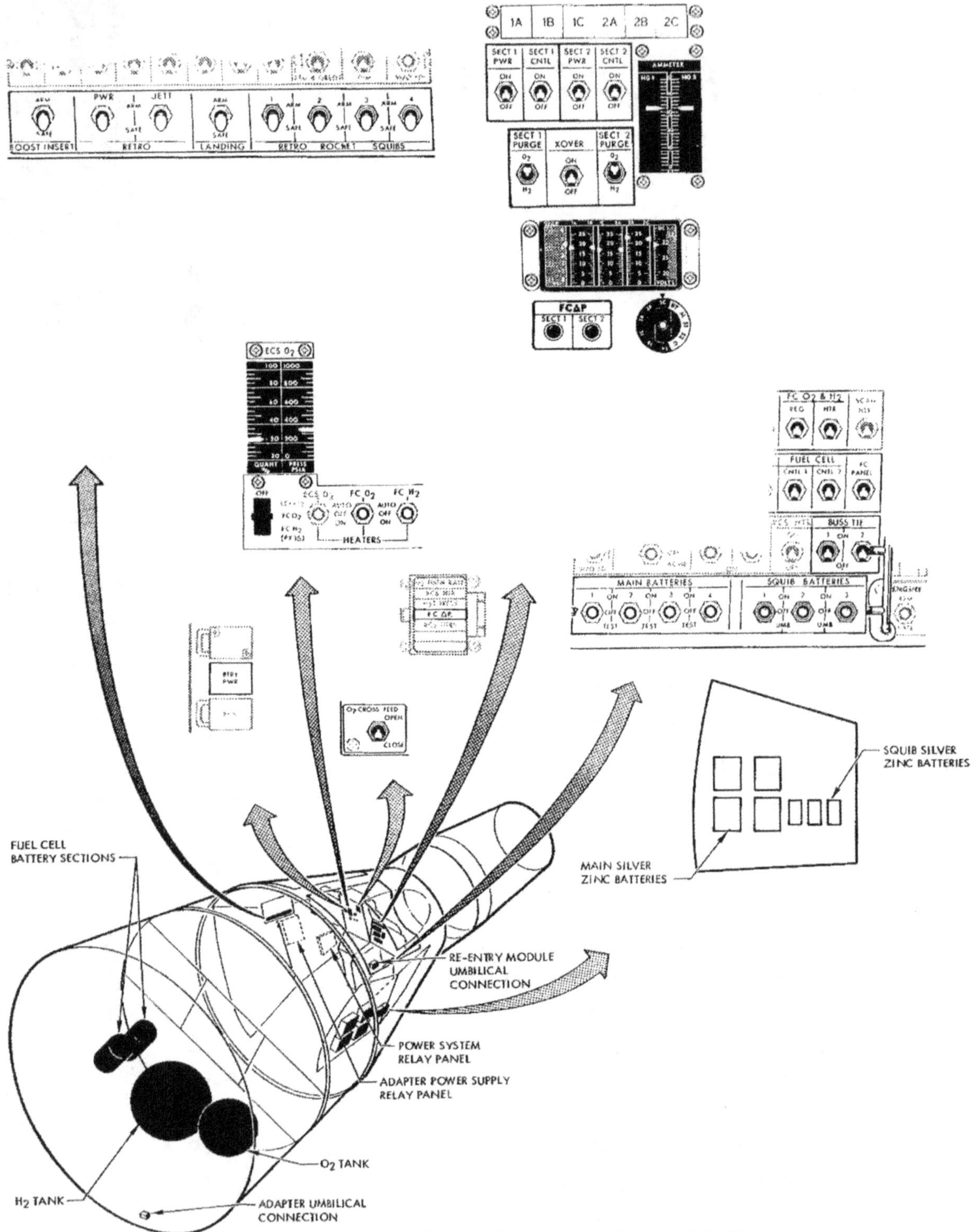

Figure 5-1 Electrical Power System Installation (S/C 7)

SEDR 300
PROJECT GEMINI

SECTION V ELECTRICAL POWER SYSTEM

SYSTEM DESCRIPTION

The electrical power system for the Gemini spacecraft basically consists of two fuel cell battery sections, four silver-zinc main batteries and three silver-zinc squib batteries. Spacecraft 3 (S/C 3) uses three 400 ampere/hour silver-zinc batteries and S/C 4 uses six 400 ampere/hour silver-zinc batteries in lieu of fuel cell batteries. Refer to Figure 5-1 for S/C 7 configuration and Figure 5-2 for S/C 3 and 4 configuration.

The electrical power system includes switches, circuit breakers, relay panels, ammeters, a voltmeter and telelights which provide control, distribution and monitoring for the power system. Also included as a power system sub system, is the reactant supply system (RSS) which provides storage and control of the reactants (H_2 and O_2) used for fuel cell operation (not applicable to S/C 3 and 4). Provisions are made for utilizing external power and remote monitoring of the spacecraft power buses during ground tests and pre-launch operations.

The two fuel cell battery sections and four main batteries provide DC power to the spacecraft main power bus. On S/C 3 and 4, the adapter module batteries and the four main batteries provide DC power to the main bus.

The three squib batteries provide DC power to the common control bus and the two orbital attitude maneuvering system (OAMS) squib buses. The OAMS squib buses in turn distribute DC power to the boost-insert-abort (BIA), retro and landing squib buses via the individual squib bus arming switches. See Figure 5-3 for S/C 7 configuration and Figure 5-4 for S/C 3 and 4 configuration.

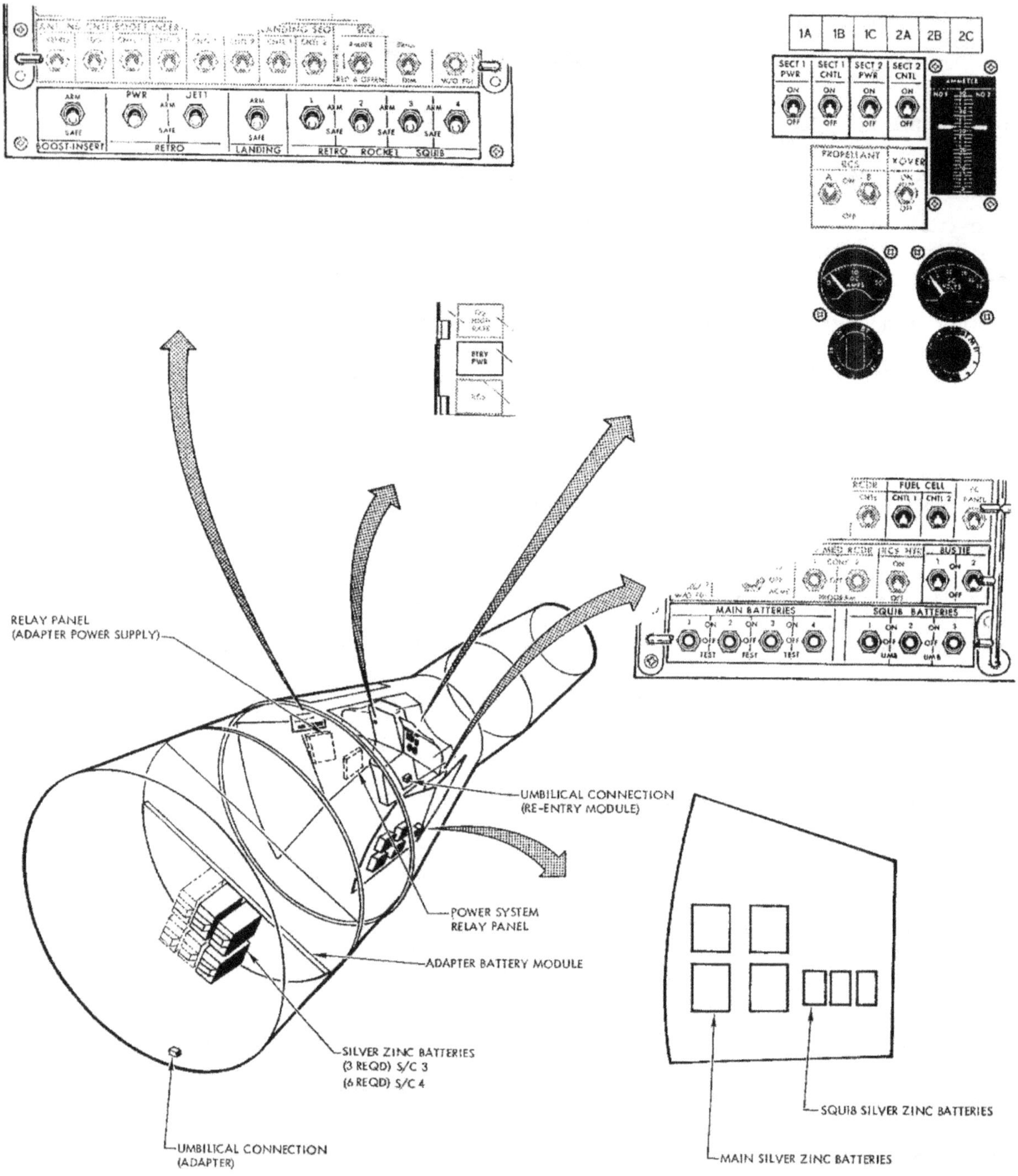

SEDR 300

PROJECT GEMINI

Figure 5-2 Electrical Power System Installation (S/C 3 & 4)

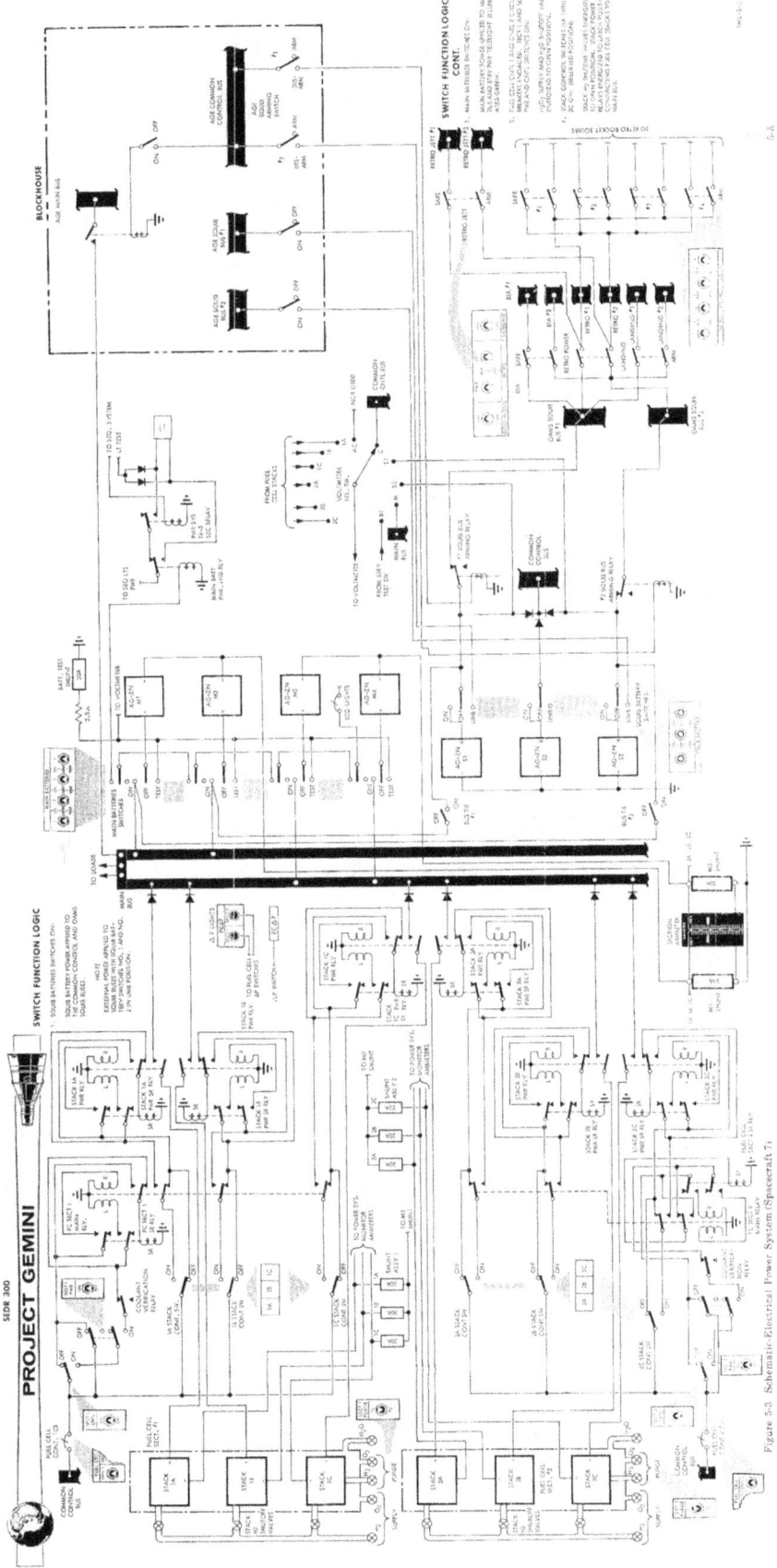

Figure 5-3. Schematic-Electrical Power System (Spacecraft 7)

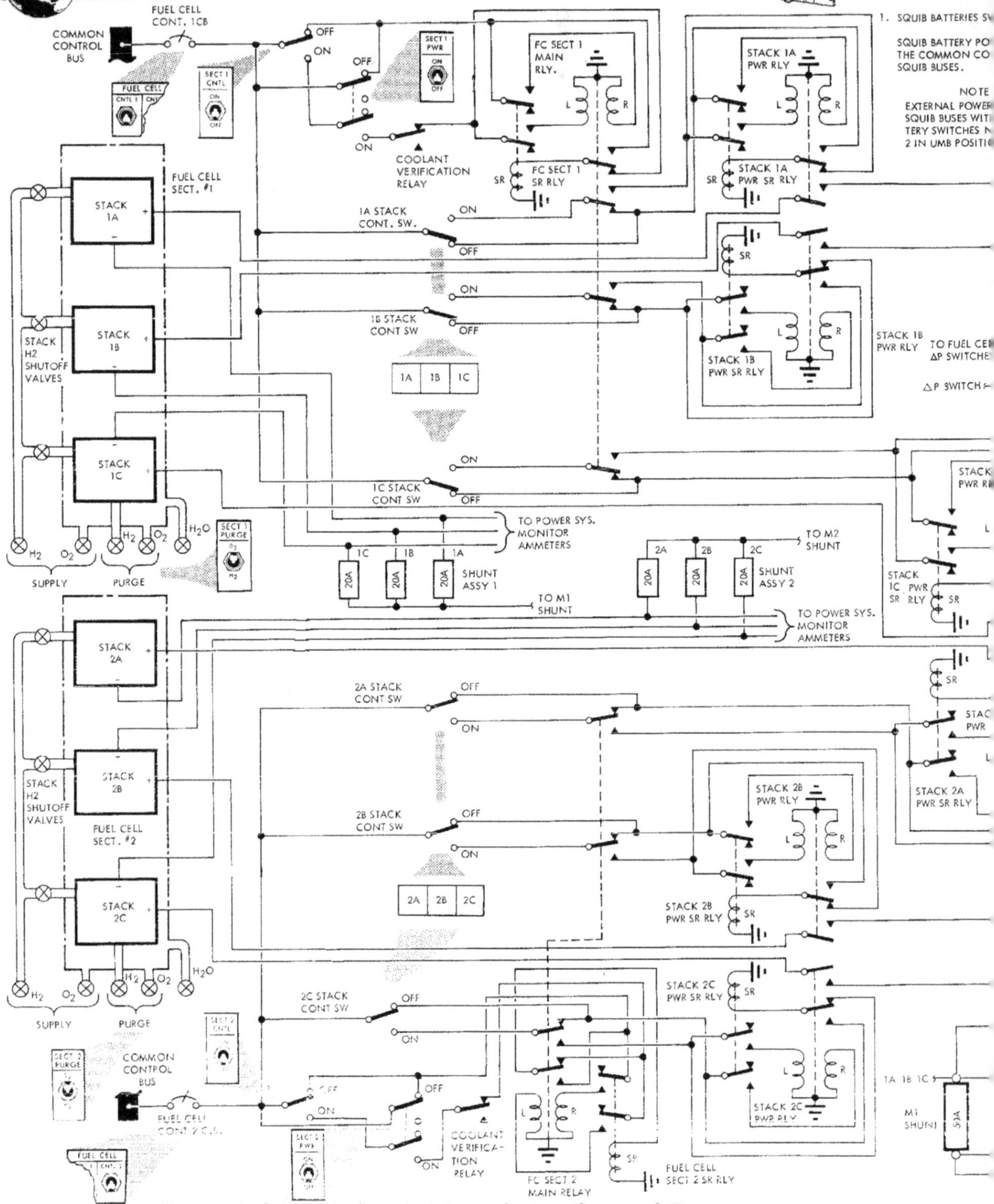

Figure 5-3 Schematic-Electrical Power System (Spacecraft 7)

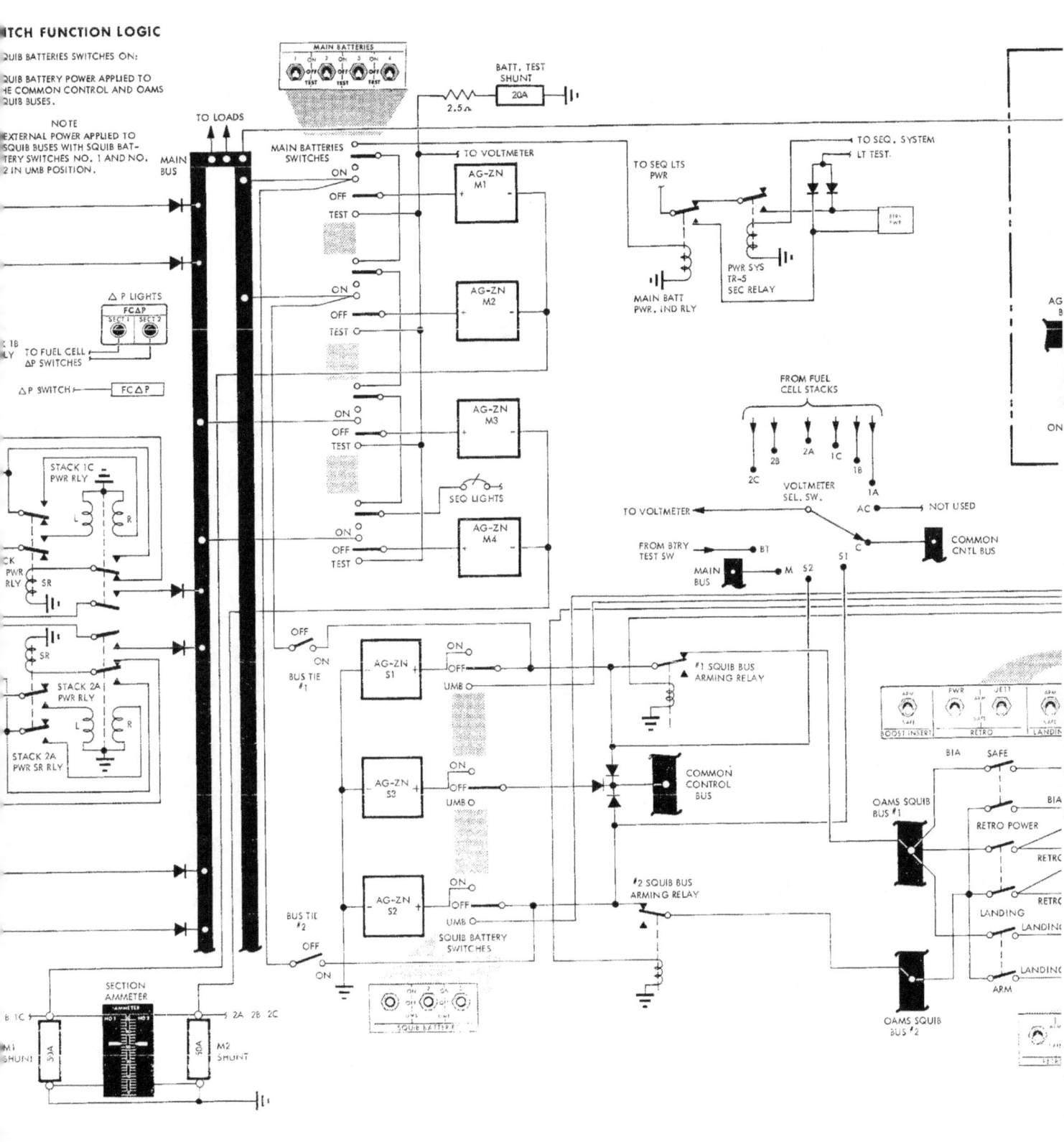

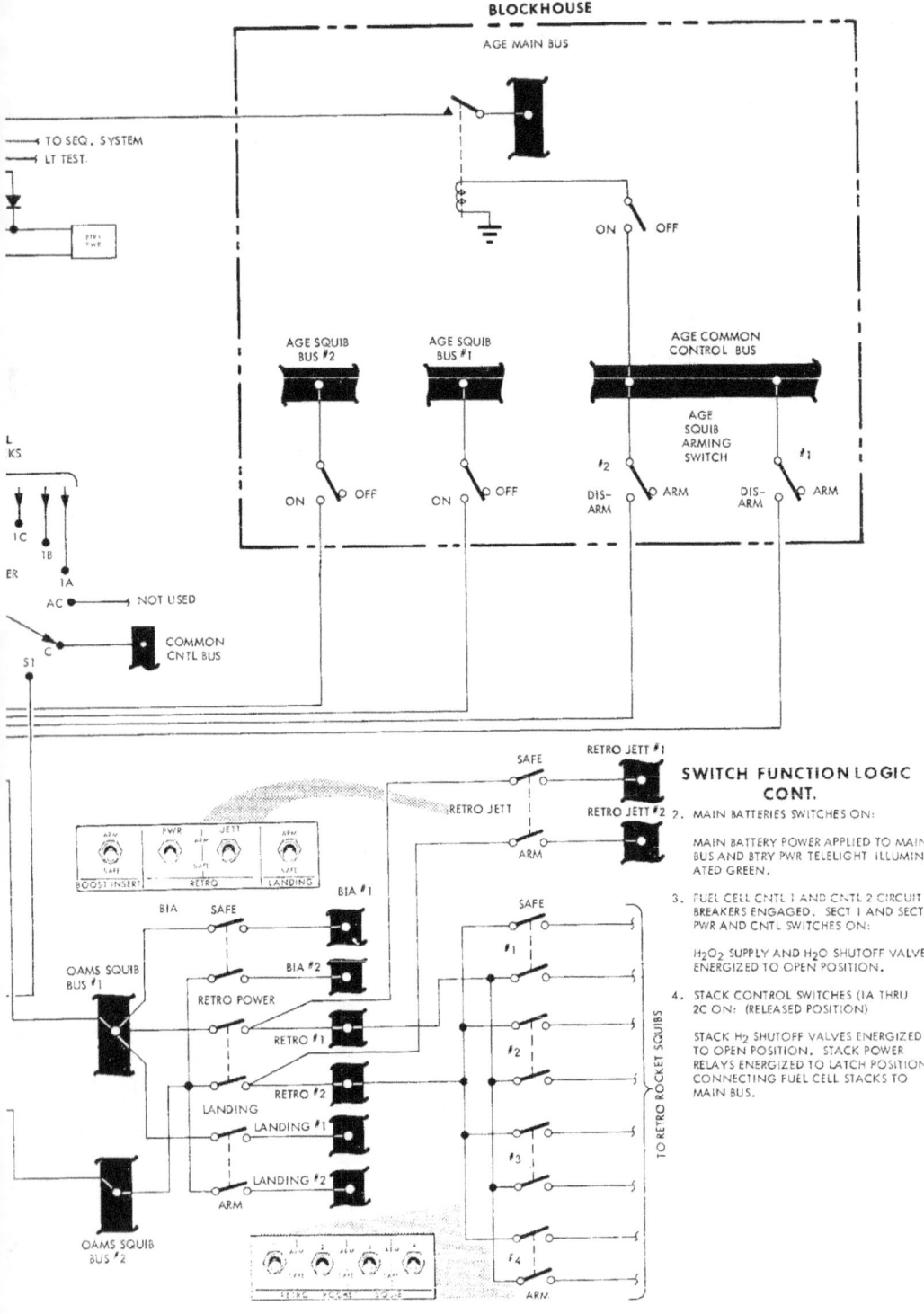

SWITCH FUNCTION LOGIC CONT.

2. MAIN BATTERIES SWITCHES ON:

   MAIN BATTERY POWER APPLIED TO MAIN BUS AND BTRY PWR TELELIGHT ILLUMINATED GREEN.

3. FUEL CELL CNTL 1 AND CNTL 2 CIRCUIT BREAKERS ENGAGED. SECT 1 AND SECT 2 PWR AND CNTL SWITCHES ON:

   H_2O_2 SUPPLY AND H_2O SHUTOFF VALVES ENERGIZED TO OPEN POSITION.

4. STACK CONTROL SWITCHES (1A THRU 2C ON: (RELEASED POSITION)

   STACK H_2 SHUTOFF VALVES ENERGIZED TO OPEN POSITION. STACK POWER RELAYS ENERGIZED TO LATCH POSITION, CONNECTING FUEL CELL STACKS TO MAIN BUS.

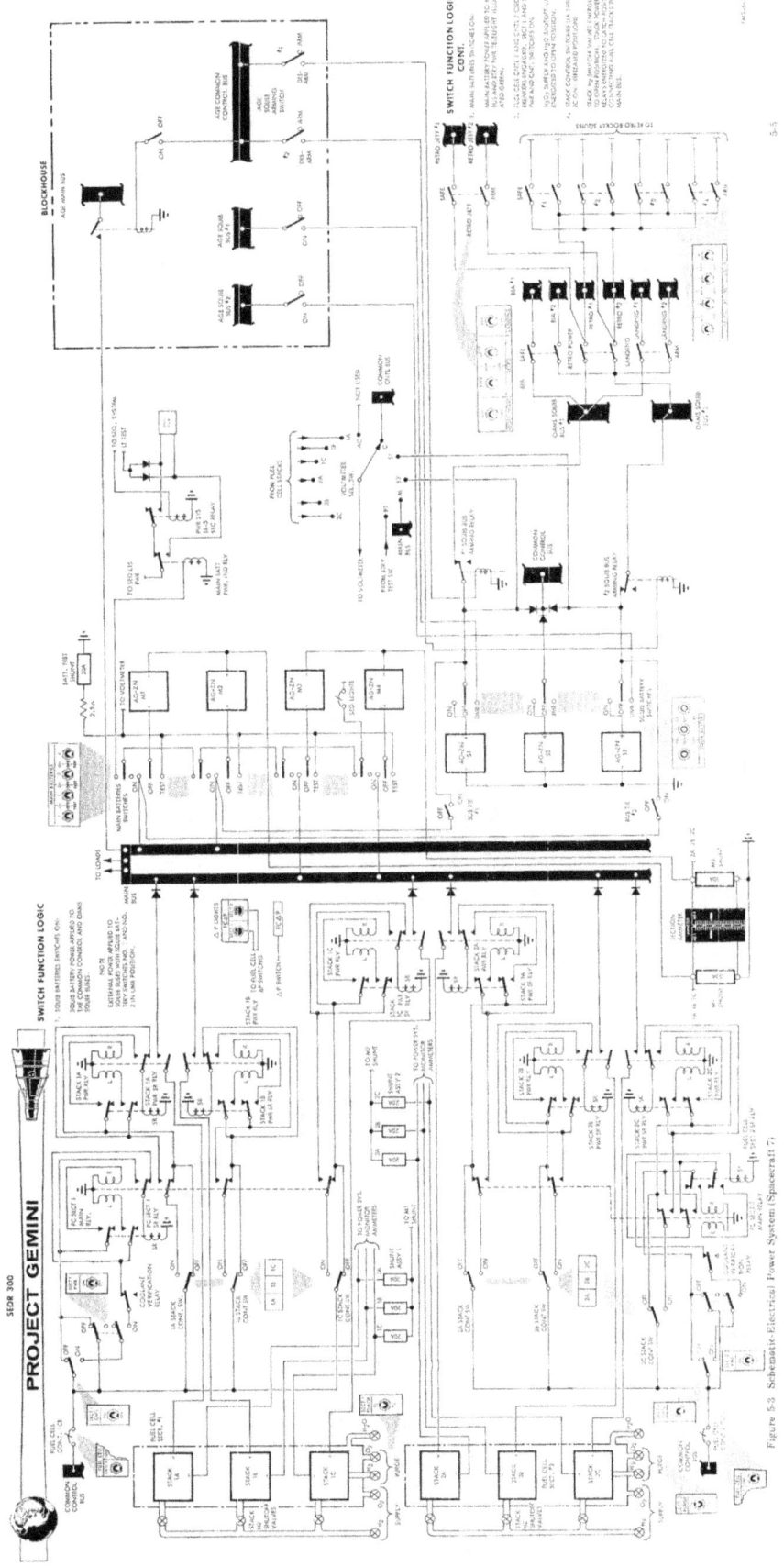

Figure 5-3. Schematic-Electrical Power System (Spacecraft 7)

SWITCH FUNCTION LOGIC

1. SQUIB BATTERIES SWITCHES ON:

   SQUIB BATTERY POWER APPLIED TO THE COMMON CONTROL AND OAMS SQUIB BUSES.

   NOTE

   EXTERNAL POWER APPLIED TO SQUIB BUSES WITH SQUIB BATTERY SWITCHES NO. 1 AND NO. 2 IN UMB POSITION.

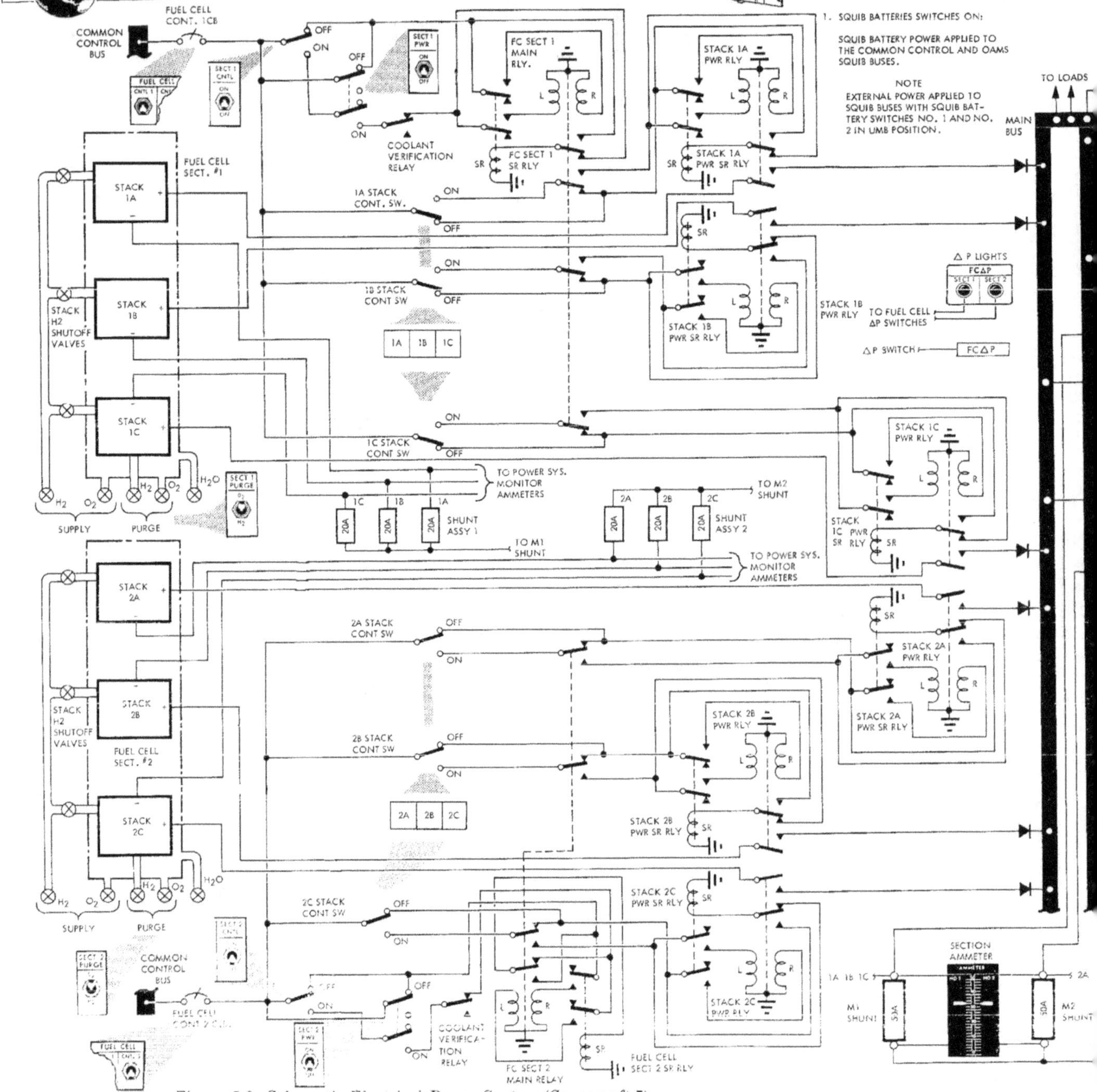

Figure 5-3 Schematic-Electrical Power System (Spacecraft 7)

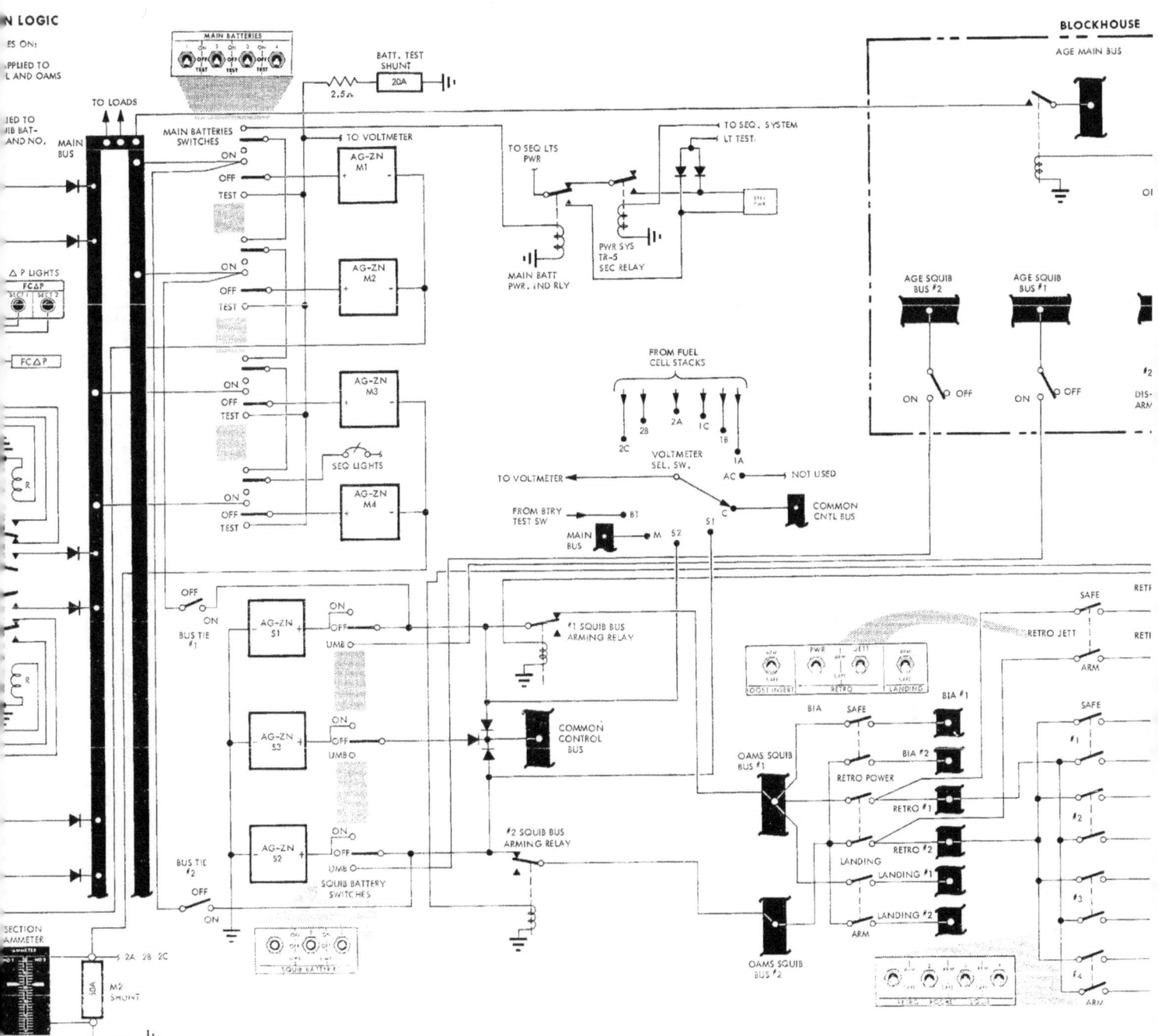

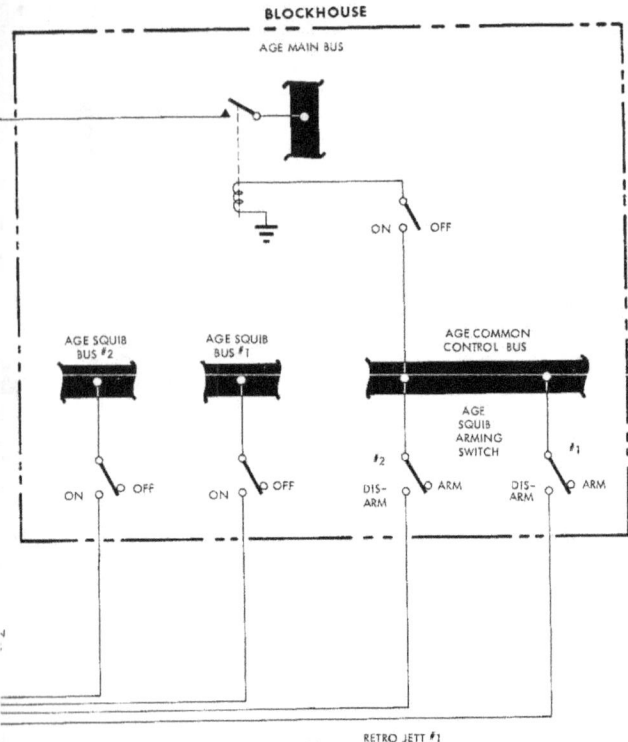
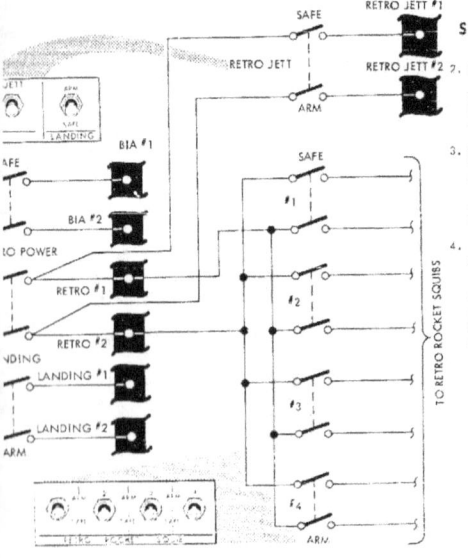

SWITCH FUNCTION LOGIC CONT.

2. MAIN BATTERIES SWITCHES ON:

   MAIN BATTERY POWER APPLIED TO MAIN BUS AND BTRY PWR TELELIGHT ILLUMINATED GREEN.

3. FUEL CELL CNTL 1 AND CNTL 2 CIRCUIT BREAKERS ENGAGED, SECT 1 AND SECT 2 PWR AND CNTL SWITCHES ON:

   H_2O_2 SUPPLY AND H_2O SHUTOFF VALVES ENERGIZED TO OPEN POSITION.

4. STACK CONTROL SWITCHES (1A THRU 2C ON: (RELEASED POSITION)

   STACK H_2 SHUTOFF VALVES ENERGIZED TO OPEN POSITION. STACK POWER RELAYS ENERGIZED TO LATCH POSITION, CONNECTING FUEL CELL STACKS TO MAIN BUS.

SEDR 300

PROJECT GEMINI

On S/C 7, the fuel cell battery sections, along with the required RSS components, are installed in the RSS/Fuel Cell module (Figure 5-5) which is located in the spacecraft adapter section. On S/C 3 and 4, silver - zinc batteries are installed in the adapter battery module (Figure 5-6) which is located in the spacecraft adapter section. The main and squib batteries are installed in the right cabin equipment bay.

The fuel cell SECT 1 and SECT 2 PWR and CNTL switches, stack control switches (1A thru 2C), SECT 1 and SECT 2 PURGE switches and crossover (XOVER) switch are located on the right instrument panel. The fuel cell SECT 1 and SECT 2 power, control and stack control switches are used to control the module battery power on S/C 3 and 4. The PURGE and XOVER switches are inoperative on S/C 3 and 4.

A dual-vertical-readout (main bus section 1 and 2) ammeter is located on the right instrument panel. On S/C 7, a power system monitor, consisting of a delta pressure indicator, three dual-vertical-readout ammeters and an AC/DC voltmeter with associated selector switch, is installed on the right instrument panel. On S/C 3 and 4, a conventional voltmeter and ammeter, with associated selector switches, are located on the right instrument panel.

On S/C 7 two fuel cell delta pressure (FC$\triangle$P) telelights are also located on the right instrument panel.

The MAIN BATTERIES switches, SQUIB BATTERIES switches, BUS - TIE switches, FUEL CELL CNTL 1 and CNTL 2 circuit breakers, FC PANEL circuit breaker and FC O_2 and H_2 REG and HTR (regulator and heater) circuit breakers are located

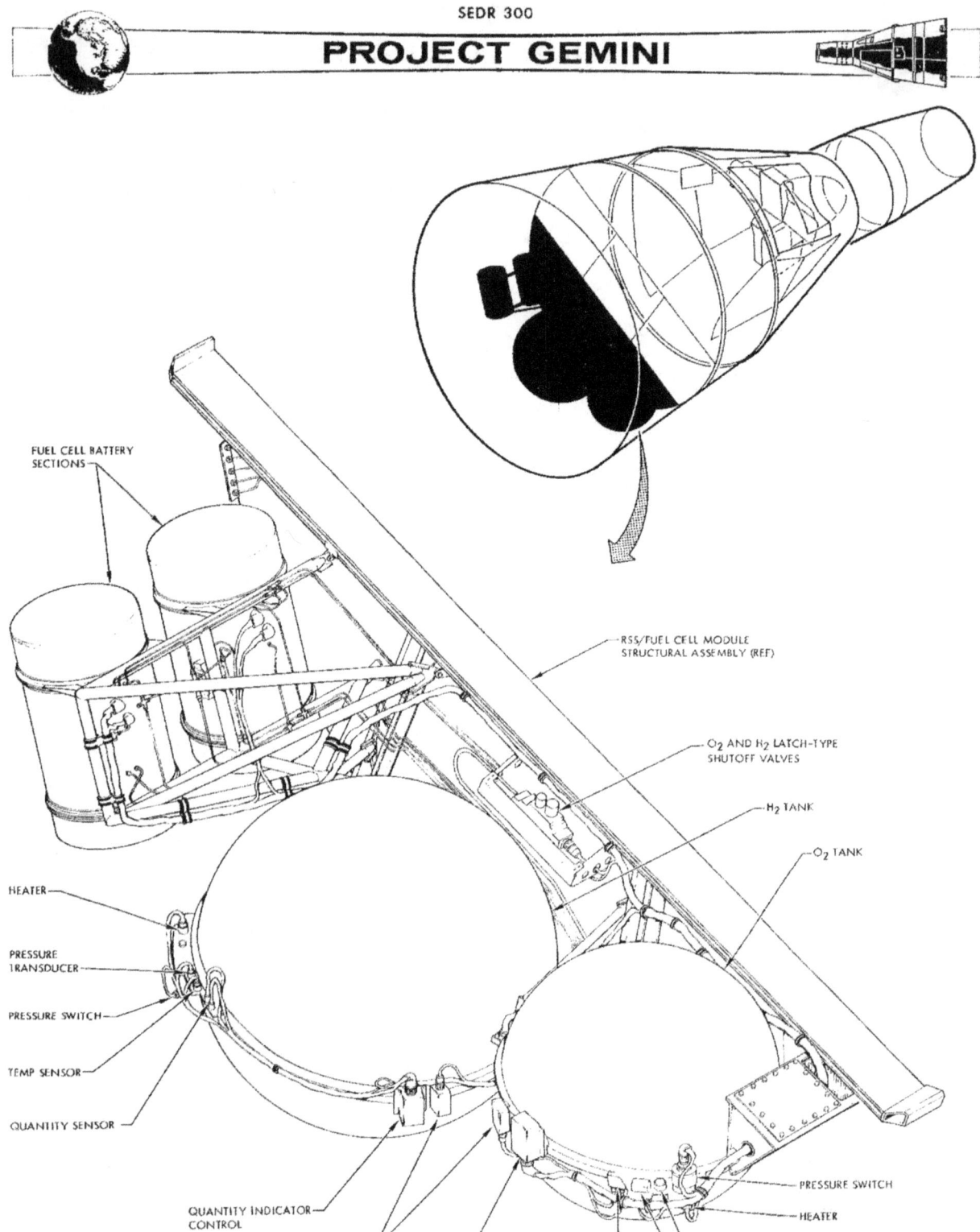

Figure 5-5 RSS/Fuel Cell Module (S/C 7)

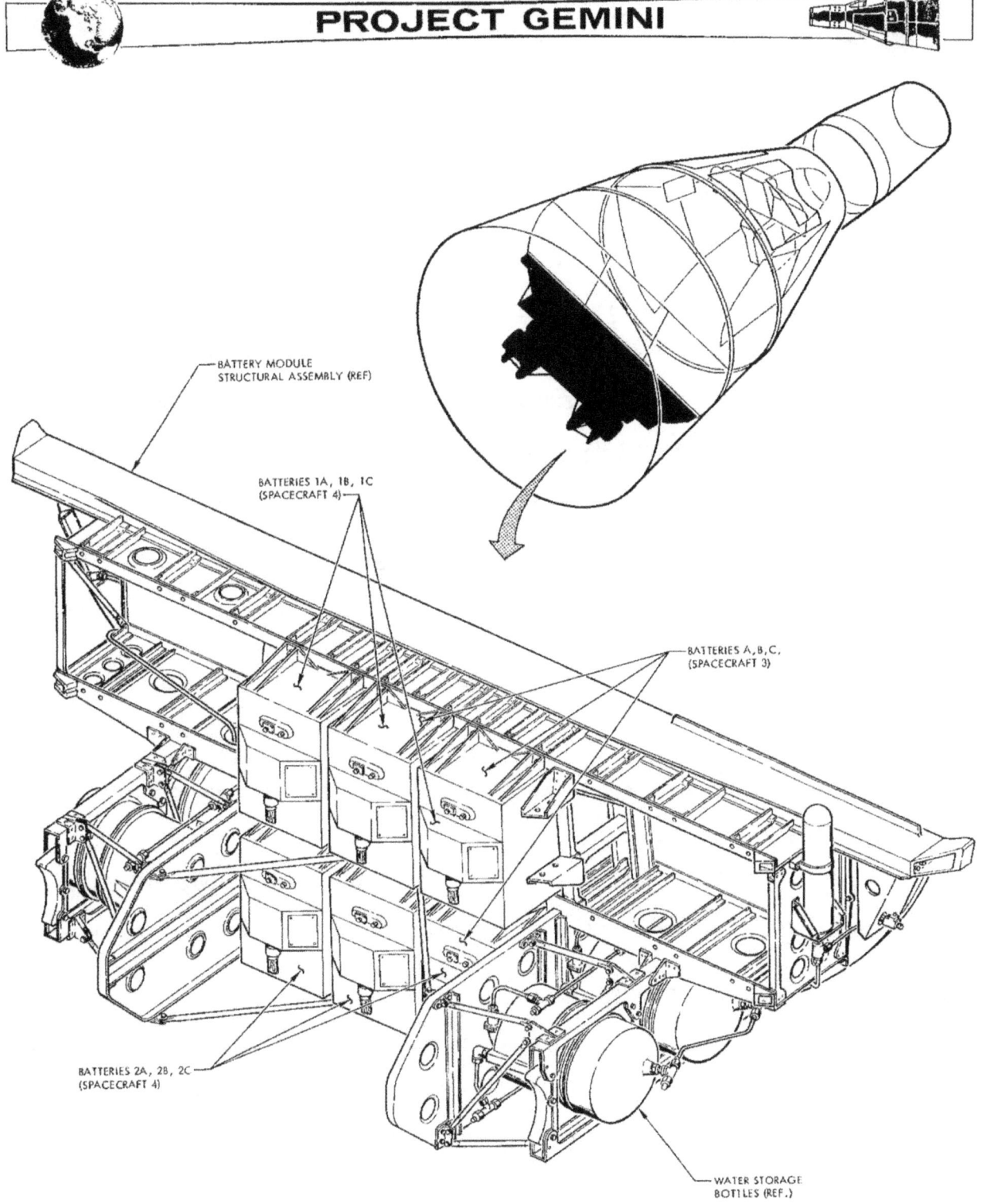

Figure 5-6 Adapter Battery Module (S/C 3 & 4)

PROJECT GEMINI

SEDR 300

on the right switch/circuit breaker panel. The BOOST-INSERT, RETRO, LANDING and RETRO ROCKET squib bus arming switches are located on the left switch/circuit breaker panel.

A BTRY PWR (main batteries) sequence light, FCΔP telelight, O_2 and H_2 heater switches, O_2 and H_2 quantity indicator (integral with ECS O_2 indicator) with selector switch and an O_2 CROSS FEED switch (S/C 7) are located on the center instrument panel. The O_2 and H_2 heater switches, fuel cell O_2 and H_2 quantity indication and fuel cell delta P telelight are inoperative on S/C 3 and 4.

The power system relay panel and adapter power supply relay panel are located in the left equipment area of the cabin.

SYSTEM OPERATION

PRE-LAUNCH

In order to conserve spacecraft battery power, external electrical power is utilized during the pre-launch phase of the mission. External power is supplied to the spacecraft common control, main and squib power buses through umbilical cables connected to the re-entry module and adapter section umbilical receptacles.

No. 1 and No. 2 SQUIB BATTERIES switches must be placed in the UMB (umbilical) position in order to apply external power to the spacecraft squib buses. Remote control of the spacecraft squib bus arming relays and remote monitoring of the spacecraft power buses is also accomplished through the re-entry and adapter umbilical cables.

 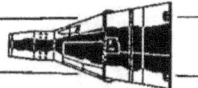

Prior to launch, all MAIN BATTERIES and SQUIB BATTERIES switches, SECT 1 and SECT 2 PWR and CNTL switches and stack control switches (1A thru 2C) are set to the ON position to insure maximum redundancy of the electrical power system during the launch phase of the mission. (Stack control switches are ON in the released position). On S/C 7, the fuel cell batteries are activated in sufficient time, prior to launch, to insure launch readiness of the fuel cell and reactant supply system.

The common control bus and the OAMS squib buses are switched from external power to the squib batteries in sufficient time, prior to launch, to verify the squib battery circuits. The boost-insert-abort (BIA) squib buses are armed prior to launch by setting the BOOST-INSERT ARM/SAFE switch to ARM position.

The re-entry module and adapter section umbilicals are disconnected from the spacecraft at approximately T-2.9 seconds. Normally, umbilical separation is accomplished by an electrical solenoid device. A backup method of separation is also provided by a lanyard initiated mechanism which is actuated by movement of the launch vehicle.

ORBIT

From launch time until booster separation and insertion into orbit both the fuel cell battery sections (module batteries on S/C 3 and 4) and the four main batteries are connected in parallel to the main power bus. After booster separation is accomplished, the MAIN BATTERIES switches are placed in the OFF position, to conserve the main battery power. Also, the pilots will disarm the BIA squib buses by setting the BOOST-INSERT ARM/SAFE switch to SAFE.

The SQUIB BATTERIES switches remain in the ON position throughout the entire mission until landing is accomplished. All three squib batteries are connected to the common control bus through diodes for individual fault protection. Squib batteries No. 1 and No. 2 are connected to the two OAMS squib buses via the de-energized squib bus arming relays.

The BUS-TIE switches remain in the OFF position unless the necessity arises where the pilots must use main bus power to fire the squibs. The BUS-TIE switches provide a method of connecting the main bus to the OAMS squib buses.

On S/C 7, a small percentage of the reactant gases must be purged from the fuel cell system periodically to insure that the impurities contained in the feed gases do not restrict reactant flow to the cells and also to remove any accumulation of product water in the gas lines. This purging function is performed by the pilots manually actuating the O_2 and H_2 PURGE switches.

The O_2 CROSS FEED switch remains in the CLOSED position except in the event of a loss of RSS O_2 tank pressure. The switch controls the O_2 cross feed reactant valve. This valve, when in OPEN position, connects the ECS O_2 supply to the RSS O_2 supply.

RE-ENTRY

At T_R-256 seconds the pilots will arm the retro squib buses by setting the RETRO PWR ARM/SAFE switch and the individual RETRO ROCKET SQUIB No. 1, 2, 3 or 4 ARM/SAFE switches to ARM position. The retro rockets are used according to the mission requirements.

PROJECT GEMINI

SEDR 300

The MAIN BATTERIES switches must be returned to the ON position at T_R-256 seconds to insure continuity of main bus power at the time of separation of the adapter section, containing the RSS/Fuel Cell module (adapter battery module on S/C 3 and 4), from the spacecraft. There is no automatic switching provided for this function.

The stack control switches (1A thru 2C) and the SECT. No. 1 and SECT. No. 2 PWR and CNTL switches are set to OFF position after the main batteries are properly connected to the main bus. (Stack control switches are OFF when in depressed position).

After retro rocket firing has been accomplished, the pilots will set the JETT RETRO ARM/SAFE switch to ARM. This switch provides a method of arming the JETT RETRO (retro section jettison) switch and is effectively an interlock to prevent inadvertant jettisoning of the retro section prior to firing of the retro rockets.

After the adapter and retro sections are separated from the spacecraft, the pilots will disarm the retro squib buses by setting the RETRO PWR ARM/SAFE switch, RETRO JETT ARM/SAFE switch and RETRO ROCKET SQUIBS ARM/SAFE switches to SAFE position. At this time, the landing squib buses are armed setting the LANDING ARM/SAFE switch to ARM position.

In the event of an aborted mission, all the squib buses are armed via the common control bus and squib bus abort relays (sequential system), which effectively bypass the squib bus arming switches.

After landing is accomplished, the pilots will disarm the landing squib buses

 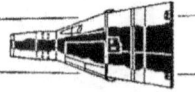

by returning the LANDING ARM/SAFE switch to SAFE. At this time, power will be removed from the common control bus and the OAMS squib buses by setting the SQUIB BATTERIES switches to OFF position. All unnecessary electrical equipment will be deactivated to conserve the spacecraft main batteries for recovery equipment operation. The MAIN BATTERIES switches will remain in the ON position throughout the recovery phase of the mission.

MONITOR AND DISPLAY

Throughout the mission visual displays of bus voltage and current are provided by the system voltmeter and ammeters. On S/C 7, a power system monitor, which consists of one pressure indicator, three dual ammeters and an AC/DC voltmeter, is utilized.

The ammeters monitor individual fuel cell stack current (1A thru 2C). The voltmeter, used in conjunction with a selector switch, displays individual fuel cell stack voltage, common control bus voltage, OAMS squib bus No. 1 and No. 2 voltage, main bus voltage and individual main battery voltage, with the selector switch in battery test (BT) position and a particular MAIN BATTERIES switch in TEST position.

The FC$\triangle$P telelights (center and lower right instrument panel) provide an out of tolerance differential pressure indication (O_2 versus H_2 and O_2 versus H_2O) in the fuel cell battery sections. The FC$\triangle$P telelights are illuminated red when a possible malfunction exists.

In the event the delta P exceeds the prescribed limits, the pilots must evaluate the fuel cell battery performance, and if a malfunction exists, shut down the malfunctioning fuel cell battery section by setting the applicable SECT PWR, CNTL and stack control switches to the OFF position. The delta P telelights

are not operative on S/C 3 and 4.

The reactant (O_2 and H_2) supply quantities are displayed on the ECS O_2 quantity indicator (center instrument panel) when the associated selector switch is set to FC O_2 or FC H_2 positions. (Not applicable to S/C 3 and 4).

The BTRY PWR (main batteries) sequence light, located on the center instrument panel, is illuminated amber at T_R-256 seconds during the mission by action of the T_R-5 relay in the power system relay panel. This informs the pilots that they must return the MAIN BATTERIES switches to the ON position to insure continuity of main bus power due to the impending separation of the spacecraft adapter section containing the adapter power supply (fuel cell battery sections on S/C 7 and module batteries on S/C 3 and 4). With all main batteries properly connected to the main bus, the BTRY PWR sequence light is illuminated green.

The dual-vertical-readout section ammeter provides a display of section No. 1 and No. 2 main bus current. Section No. 1 includes 50 percent of the adapter power supply current plus main batteries No. 1 and No. 2 current. Section No. 2 includes 50 percent of the adapter power supply current plus main batteries No. 3 and No. 4 current.

The stack ammeter (used for battery test ammeter on S/C 3 and 4), with selector switch in 1A, 1B, 1C or 2A, 2B, 2C positions, displays applicable module battery current. (On S/C 3 50 percent battery current and reading must be multiplied by 1.25.) With the selector switch in battery test (BT) position,

SEDR 300
PROJECT GEMINI

the ammeter displays individual main battery test current as the appropriate MAIN BATTERIES switch is set to TEST position.

On S/C 7, the power system monitor ammeters provide a display of individual fuel cell stack (1A thru 2C) current (reading must be multiplied by .8). The power system monitor voltmeter, with selector switch, provides a display of main bus, common control bus and squib bus voltages. The delta pressure indicator and AC portion of the voltmeter are inoperative on S/C 7.

SYSTEMS UNITS

SILVER-ZINC BATTERIES

The four main batteries are 45 ampere/hour, 16 cell, silver-zinc batteries. The three squib batteries are 15 ampere/hour, 16 cell, silver-zinc batteries. The squib batteries are special high-discharge-rate batteries which will maintain a terminal voltage of 18 volts for one second under a 75 ampere load.

On S/C 3, there are three 400 ampere/hour, 16 cell, silver-zinc batteries installed in the adapter battery module. On S/C 4, there are six 400 ampere/hour 16 cell silver-zinc batteries installed in the adapter battery module. These batteries are used in lieu of fuel cell batteries. All of the silver-zinc batteries have an open circuit terminal voltage of 28.8 to 29.9 volts.

The main and squib battery cases are made of titanium. The approximate activated (wet) weight for each squib battery is 8 lbs and each main battery 17 lbs. The adapter module battery cases are constructed of magnesium and the approximate wet weight of each battery is 118 lbs.

The battery electrolyte consists of a 40 percent solution of reagent grade potassium hydroxide and distilled water. The main and squib batteries have a vent valve in each cell designed to prevent electrolyte loss and will vent the cell to atmospheric pressure in the event a pressure in excess of 40 PSIG builds up within the cell.

All of the silver-zinc batteries are equipped with relief valves which maintain a tolerable interior to exterior differential pressure in the battery cases. The batteries are capable of operating in any attitude in a weightless state. Prior to installation into the spacecraft, the batteries are activated and sealed at sea level pressure. All of the batteries are cold plate mounted to control battery temperature.

POWER SYSTEM RELAY PANEL

The power system relay panel contains relays necessary for controlling and sequencing power system functions. The panel contains the control relays for the fuel cell and RSS system, main battery power sequence light relay, T_R-5 relay and the squib bus arming relays.

ADAPTER POWER SUPPLY RELAY PANEL

The adapter power supply relay panel contains relays necessary for controlling adapter module power to the main power bus. The relay panel contains the stack power relays which connect the individual fuel cell stacks to the main bus. On S/C 3 and 4, the stack power relays connect the adapter module batteries to the main bus. The panel also contains diodes used for reverse current protection between the adapter power supply and the spacecraft main power bus.

AMMETERS

The main bus section ammeter is a dual-edge-readout vertical reading meter having a 0-50 ampere range with a total accuracy of two percent. The No. 1 scale displays main batteries No. 1 and No. 2 and 50 percent of the adapter power supply current. The No. 2 scale displays main batteries No. 3 and No. 4 and 50 percent of the adapter power supply current. The ammeter is shunt connected between the main power bus and spacecraft ground.

The fuel cell stack ammeter (used as a battery ammeter on S/C 3 and 4), with associated selector switch, provides a display of individual main battery test current with the selector in battery test (BT) position and a particular MAIN BATTERIES switch in TEST position. With the selector switch in 1A, 1B, 1C, or 2A, 2B, 2C positions, the ammeter displays the applicable adapter module battery current. (50 percent battery current on S/C 3).

The meter has a 0-20 ampere scale. On S/C 3 the meter is connected across a 25 ampere shunt which provides a 0-25 ampere range when the meter reading is multiplied by 1.25. On S/C 4, the meter is connected across a 20 ampere shunt providing a 0-20 ampere range. The meter is read direct on S/C 4.

VOLTMETER

On S/C 3 and 4, the voltmeter, used in conjunction with a selector switch, displays main bus, common control bus and squib bus voltage. Individual main battery voltage may be monitored with the voltmeter selector switch set to battery test (BT) position and a particular MAIN BATTERIES switch set to TEST position. The voltmeter displays applicable adapter module batteries (A, B and C) voltage when the selector switch is set to 1A, 1B, 1C or 2A, 2B, 2C positions. The voltmeter has a 0-50 VDC range.

POWER SYSTEM MONITOR

The power system monitor (not applicable on S/C 3 and 4) consists of five vertical reading indicators; an O_2 delta pressure indicator, three dual-readout ammeters and an AC/DC voltmeter. The delta pressure indicator and the AC portion of the voltmeter are not operative on S/C 7.

The ammeters provide a display of individual fuel cell stack (1A thru 2C) current (reading must be multiplied by .8 on S/C 7). The voltmeter, with selector switch in appropriate position, displays individual fuel cell stack voltage, main bus, squib bus, common control bus voltages and individual main battery voltage (with a particular MAIN BATTERIES switch in TEST position). The voltmeter has an 18-33 volt DC range.

FUEL CELL BATTERIES

Construction

The fuel cell battery, used in the Gemini spacecraft, is of the solid ion-exchange membrane type using hydrogen (H_2) for fuel and oxygen (O_2) for an oxidizer. The fuel cell battery is comprised of two separate sections which are sealed in air tight pressure containers. Each section is made up of three interconnected fuel cell stacks with plumbing for transferring hydrogen, oxygen and product water. (See Figure 5-7).

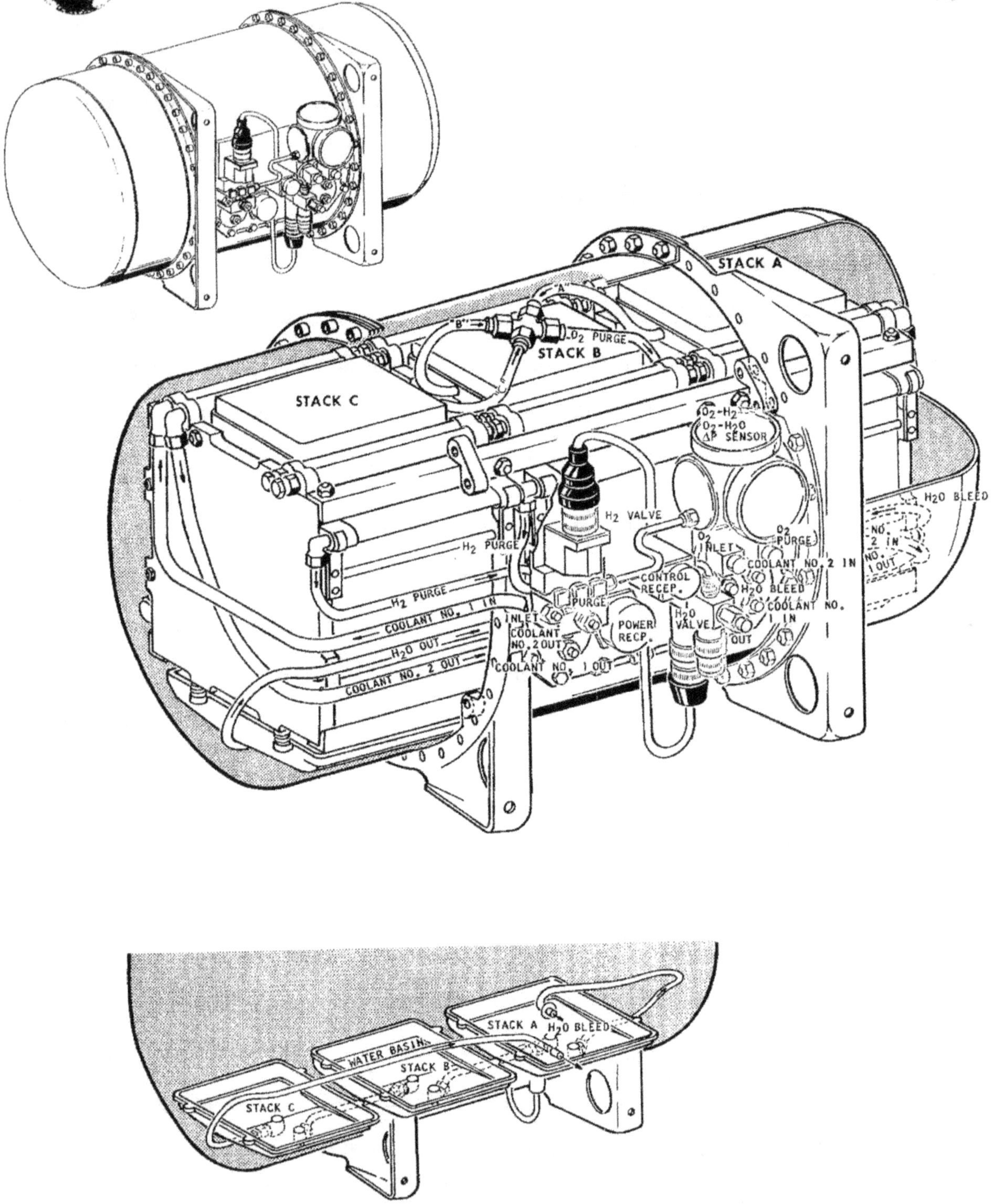

SEDR 300
PROJECT GEMINI

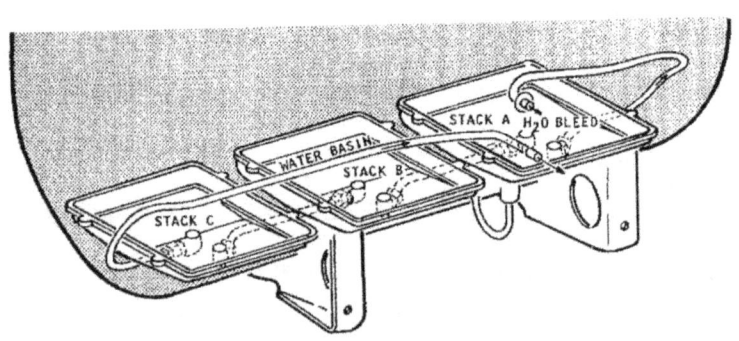

Figure 5-7 Fuel Cell Battery Section

PROJECT GEMINI

Each fuel cell stack consists of 32 individual fuel cells. Each basic fuel cell is made up of two catalytic electrodes separated by a solid type electrolyte in laminated form. (See Figure 5-8 and 5-9).

The electrolyte is composed of a sulfonated styrene polymer (plastic) approximately .010 inches thick. Thin films of platinum catalyst, applied to both sides of the electrolyte, act as electrodes and support ionization of hydrogen on the anode side of the cell and oxidation on the cathode side of the cell.

A thin titanium screen, imbedded into the platinum catalytic electrode, reduces the internal resistance along the current flow path from the electrode to the current collector and adds strength to the solid electrolyte.

On the hydrogen side of the fuel cell, a current collector is attached by means of a glass-cloth-reinforced epoxy frame which assures a tight seal around the edges of the cell, forming a closed chamber. Ribs in the collector are in contact with the catalytic electrode on the fuel cell, providing a path for current flow.

The hydrogen fuel is admitted through an inlet tube in the frame of the current collector and enters each gas channel between the collector ribs by way of a series of slots in the tube. Another tube provides a purge outlet, making it possible to flush accumulated inert gases from the cell. The collector plate is made of approximately .003 inch thick titanium.

On the oxygen side of the cell, a current collector of the same configuration and material as the hydrogen side collector is attached. Its ribs, located at right angles to those of the other collector, provide structural support

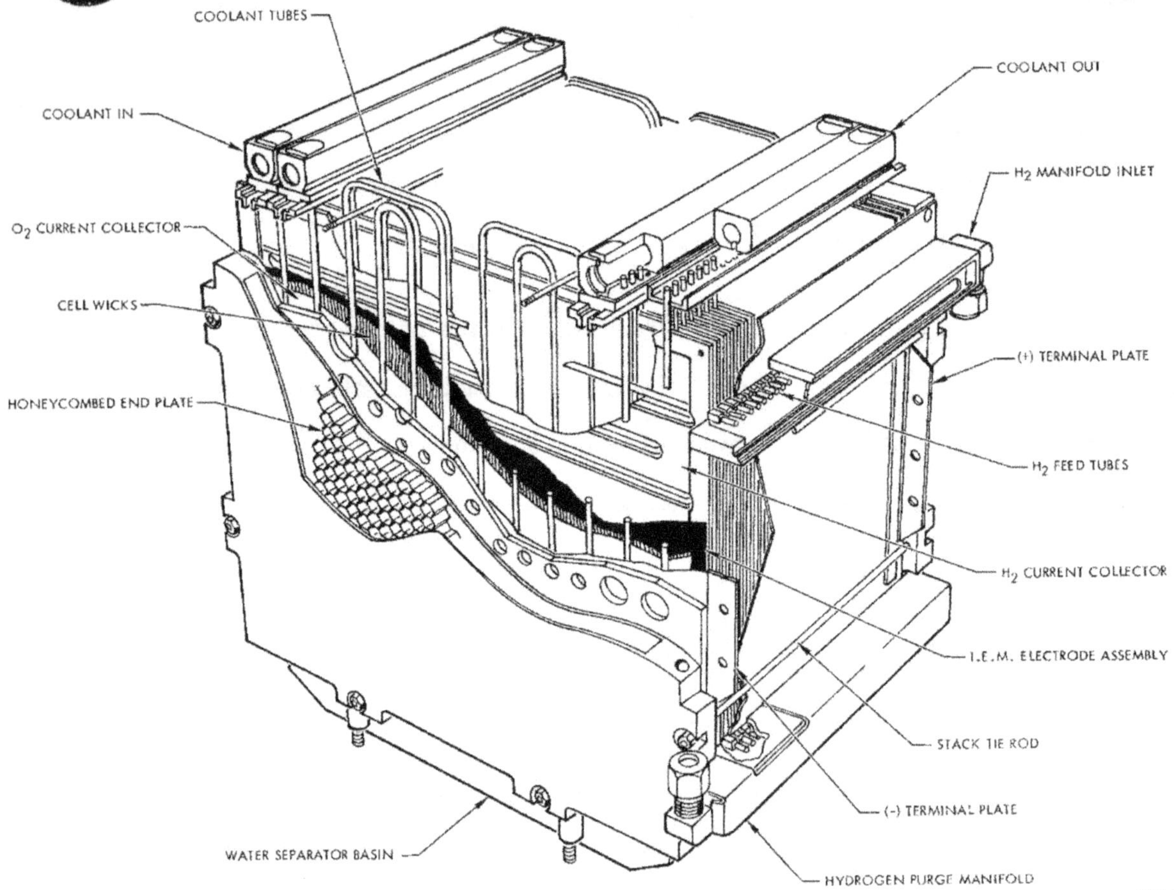

SEDR 300

PROJECT GEMINI

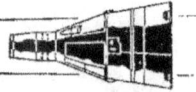

Figure 5-8 Fuel Cell Stack Assembly

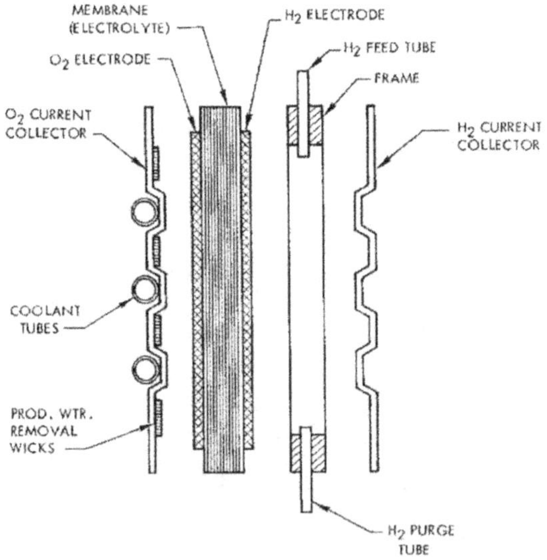

Figure 5-9 Basic Fuel Cell Assembly

to the electrolyte-electrode structure.

A dacron cloth wick, attached between the ribs, carries away the product water through capillary action, by way of a termination bar on one side of the assembly. Oxygen is admitted freely to this side of the fuel cell from the oxygen filled area of the section container.

The cell cooling system consists of two separate tubes bonded in the cavity formed by the construction of the oxygen side current collector and the back side of the hydrogen current collector. Each tube passes through six of the collector ribs and has the cooling capacity to maintain operating temperature. The cooling of the oxygen current collector, which holds the product water transport wicks, provides the cold plate for water condensation from the warmer oxygen electrode.

The individual fuel cell assemblies are arranged in series to form a stack as shown in Figure 5-8. When assembling the cells into a stack, the ribs of the oxygen side current collector contact the solid electrolyte of the fuel cell assembly. Titanium terminal plates are installed on the ends of the two outside cells to which connections are made for the external circuit. End plates, which are honey-comb structures of epoxy-glass laminate 0.5 inch thick, are installed on the outside of the terminal plates.

Stainless steel insulated tie rods hold the stack together and maintain a compression load across the area of each cell assembly. This assures proper contact of the solid electrolyte with the ribs of each current collector. The fuel cell stacks are packaged in a pressure tight container, together with the necessary reactant and coolant ducts and manifolds, water separator for each stack, required electrical power and instrumentation wiring.

The hydrogen inlet line, hydrogen purge line, and the two coolant lines for each cell lead from their respective common manifolds running the length of the stack. The manifolds are made of an insulating plastic material and the individual cell connections are potted in place after assembly to provide a leak-tight seal. The oxygen sides of the cells are open to the oxygen environment surrounding the fuel cell assemblies within the container.

An accessory pad is mounted on the outside of the fuel cell section container. It includes the gas inlet and outlet fittings, purge and shut-off valves, water valve and electrical power and control receptacles. Structurally, the container is a titanium pressure vessel consisting of a central cylinder with two end covers and two mounting brackets. Within the container, the fuel cell stacks are mounted on fiberglass-impregnated epoxy rails by bolts which pass through the stack plates. These rails are in turn bolted to the mounting rings sandwiched between the two flanges on the section container.

The hydrogen manifolds, on each stack within a section, are parallel fed with a hydrogen shut-off valve and check valve in the feed line to each stack. Oxygen is fed into the section container so that the entire free volume of the container contains oxygen at approximately 22.5 PSIA. The coolant reaches the fuel cell battery sections by two separate isolated lines. Any malfunction in the coolant line in one section will not affect the cooling function of the coolant line in the other section.

Each stack in the section has its own water-oxygen separators which are manifolded into a single line coming out of the section container. All hydrogen, oxygen, coolant, electrical and water storage pressure line connections at the section container are fastened to standard bulkhead fittings on the accessory pad.

After the stacks are completely assembled within the container, all void spaces are filled with unicellular foam. The purpose of this foaming is for vibration dampening, accoustical noise deadening and minimizing free gas volume to prevent possible fire propagation. Thin plastic covers are placed over the top and bottom of each stack to manifold oxygen to the stack and to keep the foam material from entering areas around the coolant manifolds and oxygen water separator.

Operation

The basic principle, by which the fuel cell operates to produce electrical energy and water, is the controlled oxidation of hydrogen. This is accomplished through the use of the solid electrolyte ion-exchange membrane. On the hydrogen side of the fuel cell, hydrogen gas disassociates on the catalytic electrode to provide hydrogen ions and electrons. The electrons are provided a conduction path of low resistance by the current collector, either to an external load or to the next series-connected fuel cell.

When a flow of electrons is allowed to do work and move to the oxygen side of the fuel cell, the reaction will proceed. By use and replacement, hydrogen ions flow through the solid electrolyte to the catalytic electrode on the oxygen side of the fuel cell. When electrons are available on this surface, oxygen disassociates and combines with the available hydrogen ions to form water. (See Figure 5-10).

PROJECT GEMINI

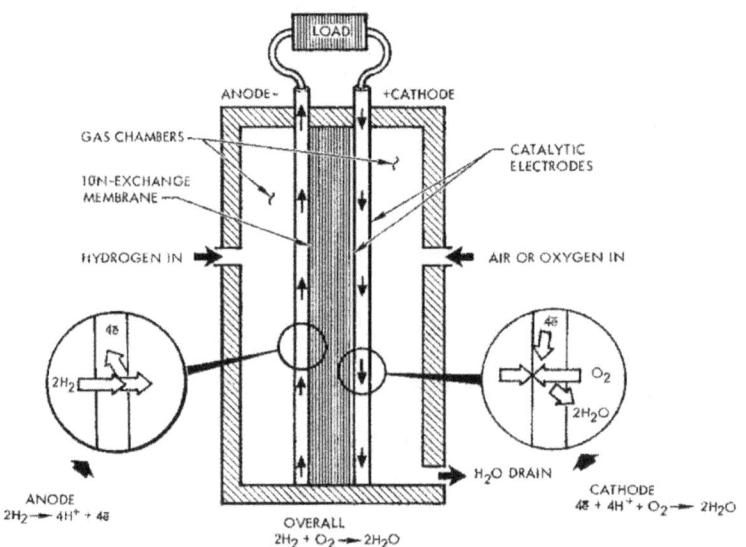

Figure 5-10 Principal of Operation.

The oxygen current collector provides the means of distributing electrons and condensing the product water on a surface to be transported away by the wick system through capillary action. The individual cell wicks are integrated into one large wick which routes the water to an absorbent material that separates the water from the gas.

By using the oxygen outlet pressure as a reference, a small pressure differential is obtained over the length of the water removal system. This pressure is sufficient to push the gas-free water toward the storage reservoir.

Waste heat, generated during the fuel cell battery operation, is dissipated by means of the circulating coolant, provided by the environmental control system (ECS). In addition, the total coolant flow provides the function of pre-heating the incoming reactant gases.

In the spacecraft, the reactant gases (hydrogen and oxygen) are supplied to the fuel cell sections by the reactant supply system (RSS). This system contains the reactant supply tanks, control valves, heat exchangers, temperature sensors and heaters required for management of the fuel cell reactants. See Figure 5-11 for a functional diagram.

REACTANT SUPPLY SYSTEM

The RSS is essentially a sub system for the fuel cell battery sections. The system provides storage for the cryogenic hydrogen and oxygen, converts the reactants to gaseous form and controls the flow of the gases to the fuel cell battery sections. The RSS components are installed in the RSS/Fuel Cell module. See Figure 5-5 for component installation and Figure 5-11 for a functional diagram.

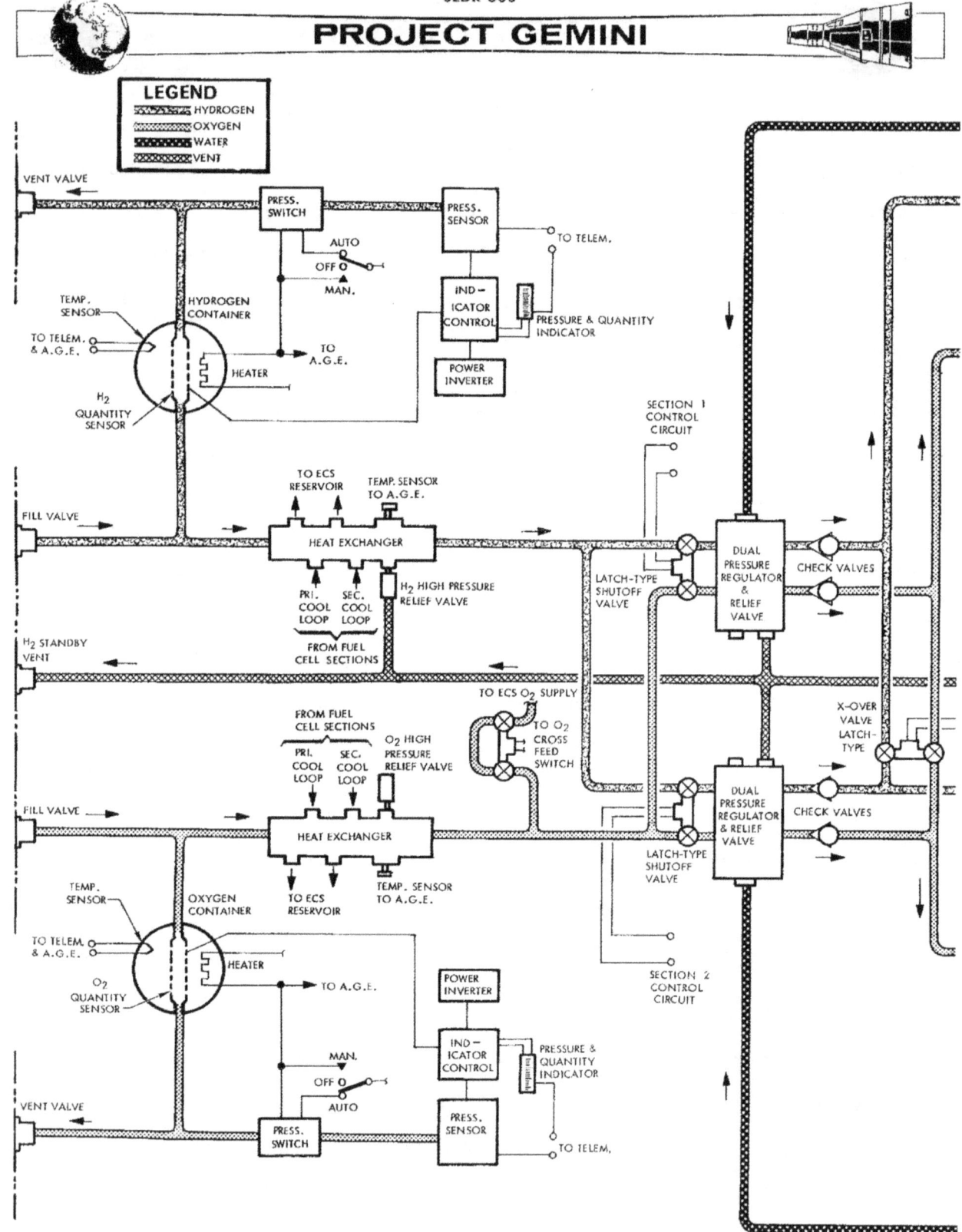

Figure 5-11 RSS/Fuel Cell Cell System Functional Diagram (S/C 7) (Sheet 1 of 2)

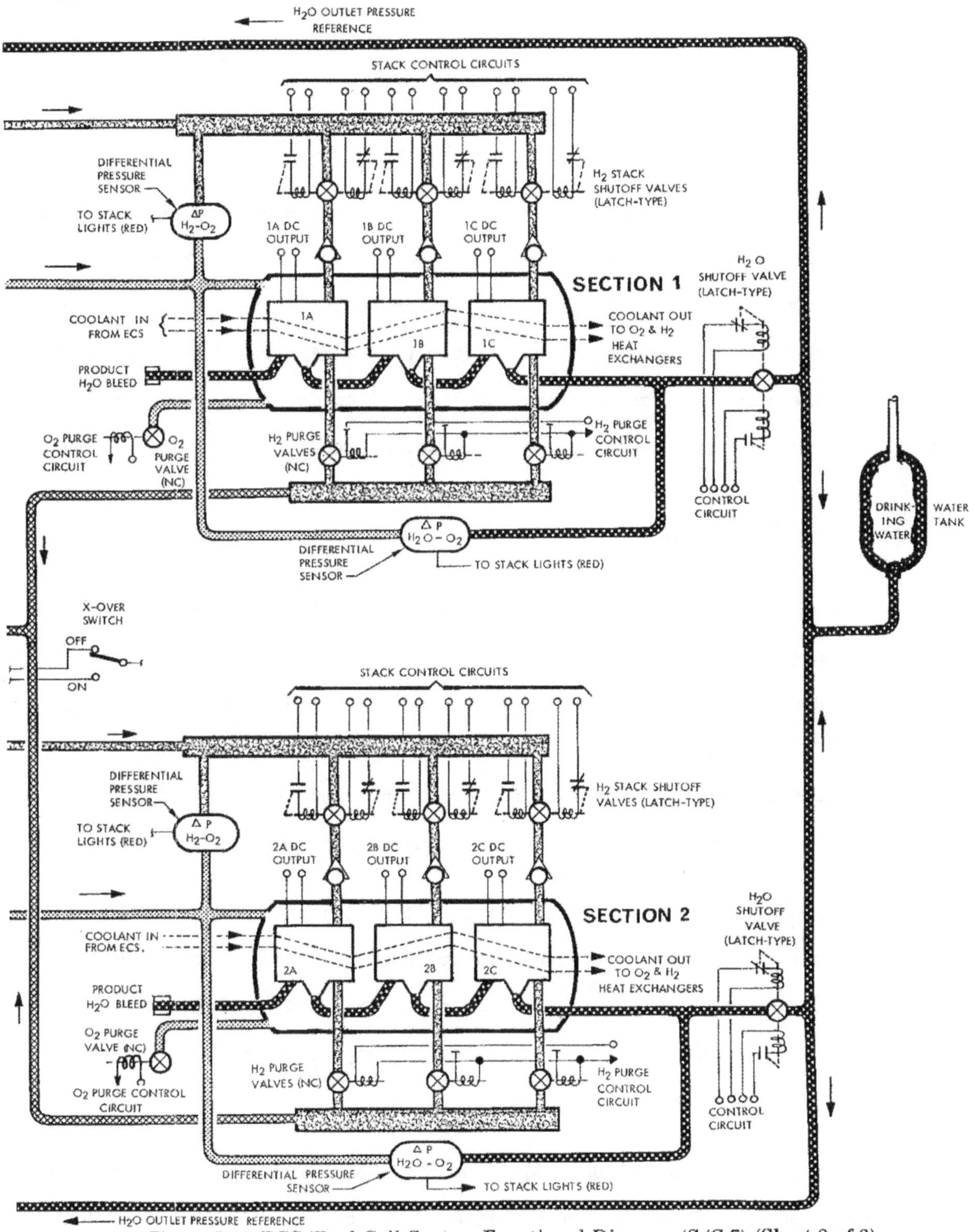

PROJECT GEMINI

Figure 5-11 RSS/Fuel Cell System Functional Diagram (S/C 7) (Sheet 2 of 2)

Components

Reactant Supply Tanks

Two tanks are utilized to separately contain the cryogenic hydrogen and oxygen required for the operation of the fuel cell battery sections. The tanks are thermally insulated to minimize heat conduction to the stored elements which would cause the homogeneous solution to revert to a mixture of gas and liquid. The tanks are capable of maintaining the stored liquids at super-critical pressures and cryogenic temperatures. The total amount of liquid stored in the hydrogen vessel is 22.25 lbs. The total amount of liquid stored in the oxygen vessel is 180 lbs.

The hydrogen vessel is composed of titanium alloy and the oxygen vessel is made of a high strength nickel base alloy. Both vessels are spherical in shape and double walled. A vacuum between the inner and outer vessel (a space of approximately one inch) provides thermal insulation from ambient heat conduction. The inner wall is supported in relation to the outer wall by an insulating material supplemented by compression loading devices.

Each storage tank contains a fluid quantity sensor, a pressure sensor, a temperature sensor and an electrical heater installed in the inner vessel in intimate contact with the stored reactants. The fluid quantity sensor is an integral capacitance unit which operates in conjunction with an indicator control unit containing a null bridge amplifier.

The sensor varies the capacitance (in proportion to fluid level) in a circuit connected to the null bridge amplifier. The amplified signal is then used to drive a servo motor, which in turn operates a visual indicator for quantity indication. Power inverters supply 400 cycle, 26 VAC power to the fluid quantity circuits.

The temperature sensor is a platinum resistance device capable of transmitting a source signal to a balanced bridge circuit. The sensor provides cryogenic fluid temperature monitoring for telemetry and AGE.

The pressure sensor is a dual resistive element, diaphram type transducer. The sensor provides signals for cryogenic fluid pressure monitoring on a spacecraft meter.

The electrical heaters provide a method of accelerating pressure build-up in the reactant supply tanks. The heaters may be operated either in a manual or automatic mode. In the automatic mode, a pressure switch removes power from the heater element when the tank pressure builds up to a nominal 900 PSIG in the oxygen tank and a nominal 250 PSIG in the hydrogen tank. In the manual mode a spacecraft pressure meter indicates proper switch operation.

Fill and Vent Valves

The fill and vent valve provides a dual function in permitting simultaneous fill and vent operations. Quick disconnect fittings are provided for rapid ground service connection to both the storage tank fill check valve and the vent check valve. When fill connections are made, the pressure of the ground service connection against the fill and vent valve poppet shaft simultaneously opens both the fill and vent ports. When ground service equipment is removed, the valve poppet automatically returns to its normally spring loaded-closed position.

PROJECT GEMINI

SEDR 300

The vent check valve is a single-poppet type, spring loaded-closed check valve which opens (when system pressure exceeds 20 PSIG) to relieve through the fill and vent valve vent port.

Heat Exchangers

The supply temperature control heat exchangers are finned heat exchangers in which the supply fluid temperature is automatically controlled by absorbing heat from the recirculating coolant loop fluid of the environmental control system. The special double-pass design precludes freezing of the environmental control system coolant and assures a reactant fluid supply at 50°F minimum and 140°F maximum.

Dual Pressure Regulator and Relief Valves

The dual pressure regulator and relief valve is a normally open poppet-type regulator which controls downstream pressure to the fuel cell sections. The regulator maintains the hydrogen pressure at approximately 21.7 PSIA and the oxygen pressure at approximately 22.2 PSIA. The oxygen side of the regulator is referenced to hydrogen pressure. The hydrogen side of the regulator is referenced to product water pressure.

The relief valve provides overpressurization protection for the regulated pressure to the fuel cell battery sections. This valve is pre-calibrated to operate at a pressure of approximately 10 PSIA above the normal supply level.

High Pressure Relief Valves

The high pressure relief valve is a single poppet-type, spring loaded, normally closed valve which provides system and overpressurization protection. The valve vents system gas to ambient when pressure exceeds the system limits.

Solenoid Shut-off Valves

The solenoid shut-off valves are solenoid operated latching type valves which eliminate fluid loss during the non-operating standby periods. The valves are normally open and are closed only during fill and standby periods by applying power to the solenoids.

Crossover Valve

The crossover valve is a solenoid operated latching type valve which provides the capability of selecting both dual pressure regulators to supply hydrogen and oxygen to the fuel cell battery sections for the purpose of increasing flow rate for more effective purging. The crossover valve is controlled by the XOVER switch on the right instrument panel.

O_2 Cross Feed Reactant Valve

The O_2 cross feed reactant valve is a solenoid operated, latching type valve which provides the capability of pressurizing the RSS oxygen supply with pressure from the ECS oxygen supply. This provides a redundant method of supplying the proper reactant oxygen pressure to the fuel cell sections in the event of a malfunction in the RSS oxygen supply. The cross feed valve is controlled by the O_2 CROSS FEED switch located on the center instrument panel.

Operation

During pre-launch, the two separate reactant supply tanks are serviced (using AGE equipment) with liquid hydrogen and oxygen. After the tanks are filled, in order to accelerate pressure buildup within the tanks, the internal tank heaters are operated, utilizing external electrical power. In approximately one hour, the liquid is converted into a high density, homogeneous fluid at a constant pressure.

During the fill operation, the solenoid shut-off valves, between the storage tanks and the dual pressure regulators, are closed. Once operating pressure is obtained, the solenoid shut-off valves may be opened by applying power to the coil of the valves. The high density, homogeneous fluid will then flow upon demand.

The fluid flows from the supply tanks to the heat exchangers. The fluid temperature, when entering the heat exchangers, is approximately -279°F for the oxygen and approximately -423°F for the hydrogen. The heat exchangers absorb heat from the recirculating coolant fluid of the adapter section cooling system. This heat, applied to the high density fluid, raises the temperature of the reactants to approximately 50°F to 140°F.

The reactants, now in gaseous form, flow through the heat exchangers, past the high pressure relief valves and AGE temperature sensors, to the supply solenoid shut-off valves. During fuel cell battery operation, if the demand on the fluid flow is inadequate to keep tank pressures within limits, the high pressure relief valves will vent, externally, the excess fluid. The AGE temperature sensors on the heat exchangers are used for pre-launch checkout only.

The reactants flow through the supply solenoid shut-off valves to the dual pressure regulator and relief valves. The dual pressure regulators reduce the pressure of the reactants to approximately 21.7 PSIA for the hydrogen and approximately 20.5 PSIA for the oxygen. The gas now flows through the manual shut-off valves and is then directed to the fuel cell battery sections at a flow rate that is determined by both the electrical load applied to the fuel cell battery sections, and the frequency of purging. The flow rate of the gases may be increased for more effective purging by opening the crossover valve.

After launch, the supply tank heaters are operated by spacecraft power. The heaters operate as required to maintain proper system pressures.

ENVIRONMENTAL CONTROL SYSTEM

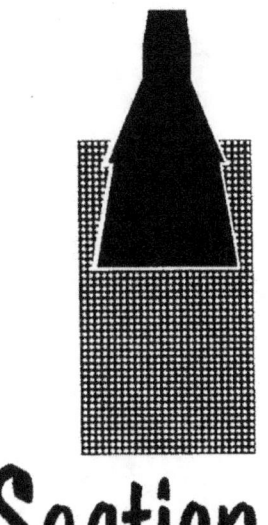

Section VI

TABLE OF CONTENTS

TITLE	PAGE
SYSTEM DESCRIPTION	6-3
SYSTEM DISPLAYS AND CONTROLS	6-27
SYSTEM OPERATION	6-32
SYSTEM UNITS	6-41

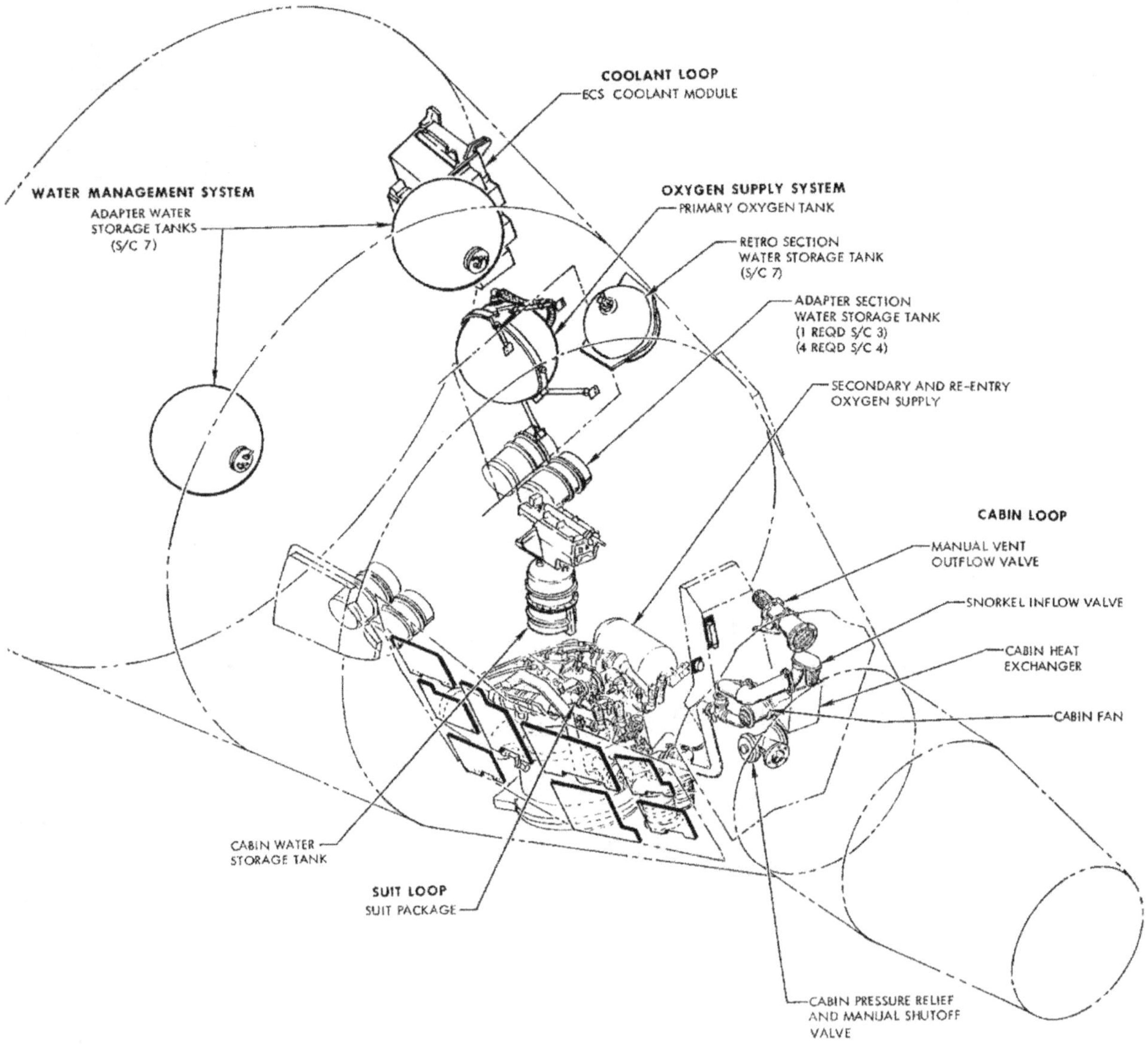

SEDR 300

PROJECT GEMINI

Figure 6-1 Environmental Control System

 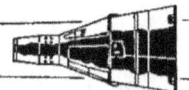

SEDR 300

PROJECT GEMINI

SECTION VI ENVIRONMENTAL CONTROL SYSTEM

SYSTEM DESCRIPTION

The Environmental Control System (E.C.S.) (Figure 6-1, 6-2) may be defined as a system which provides a safe and comfortable gaseous atmosphere for the pilots. The system must perform such tasks as providing fresh oxygen, pressurization, temperature control, water removal and toxic gas removal. In addition to providing atmospheric control for the pilots, the system provides equipment cooling and regulated temperatures for certain pieces of equipment.

For ease of understanding, the Environmental Control System may be separated into five systems or loops which operate somewhat independent of each other. These loops are:

(1) The oxygen supply system.

(2) The cabin loop.

(3) The suit loop.

(4) The water management system.

(5) The coolant system.

OXYGEN SUPPLY SYSTEM

There are three oxygen systems: Primary, Secondary and Egress.

Primary Oxygen (Figure 6-3, 6-4)

This system stores and dispenses oxygen for breathing and for suit and cabin pressurization.

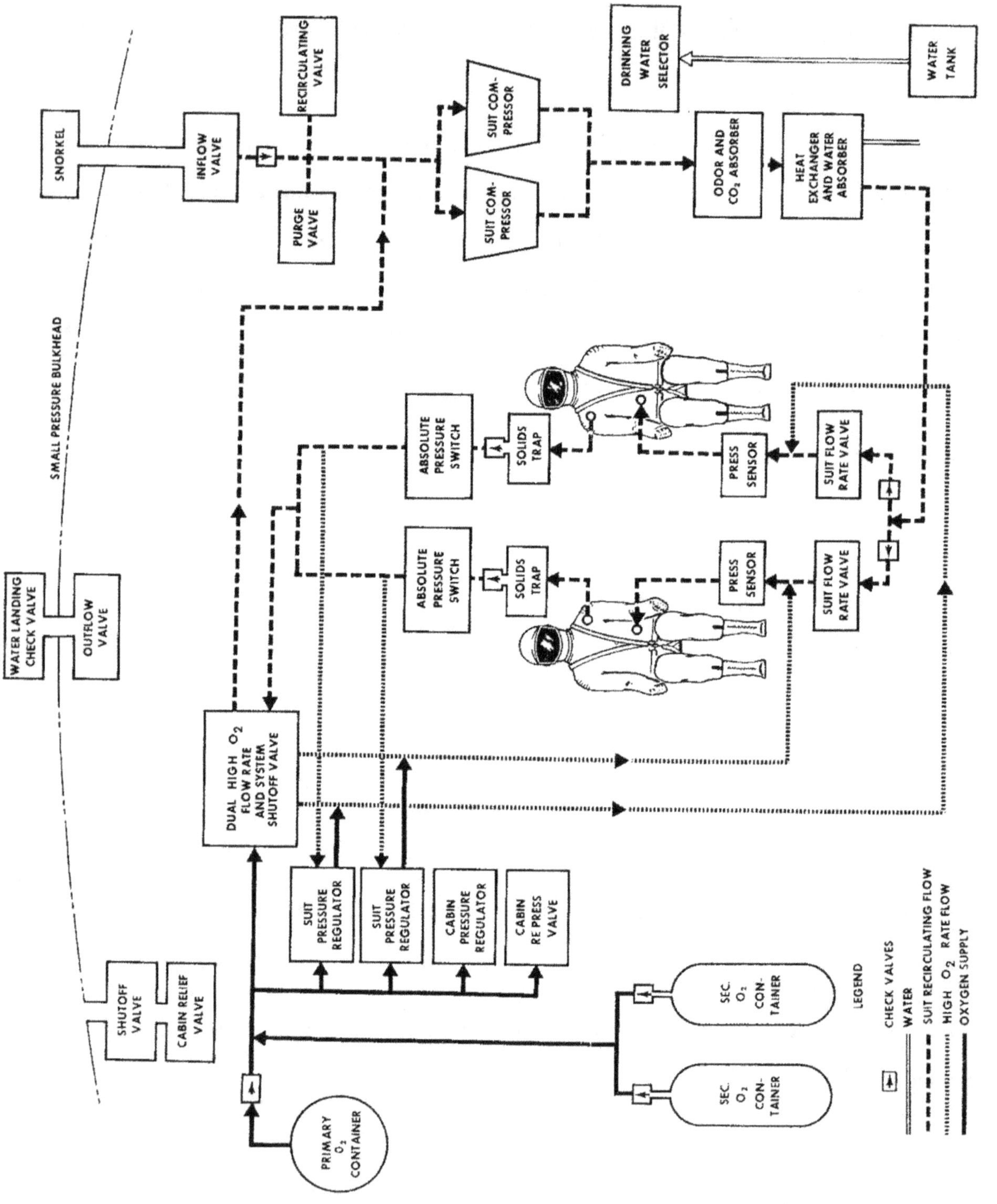

PROJECT GEMINI

SEDR 300

Figure 6-2 ECS Block Diagram

6-4

 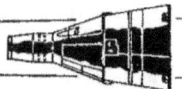

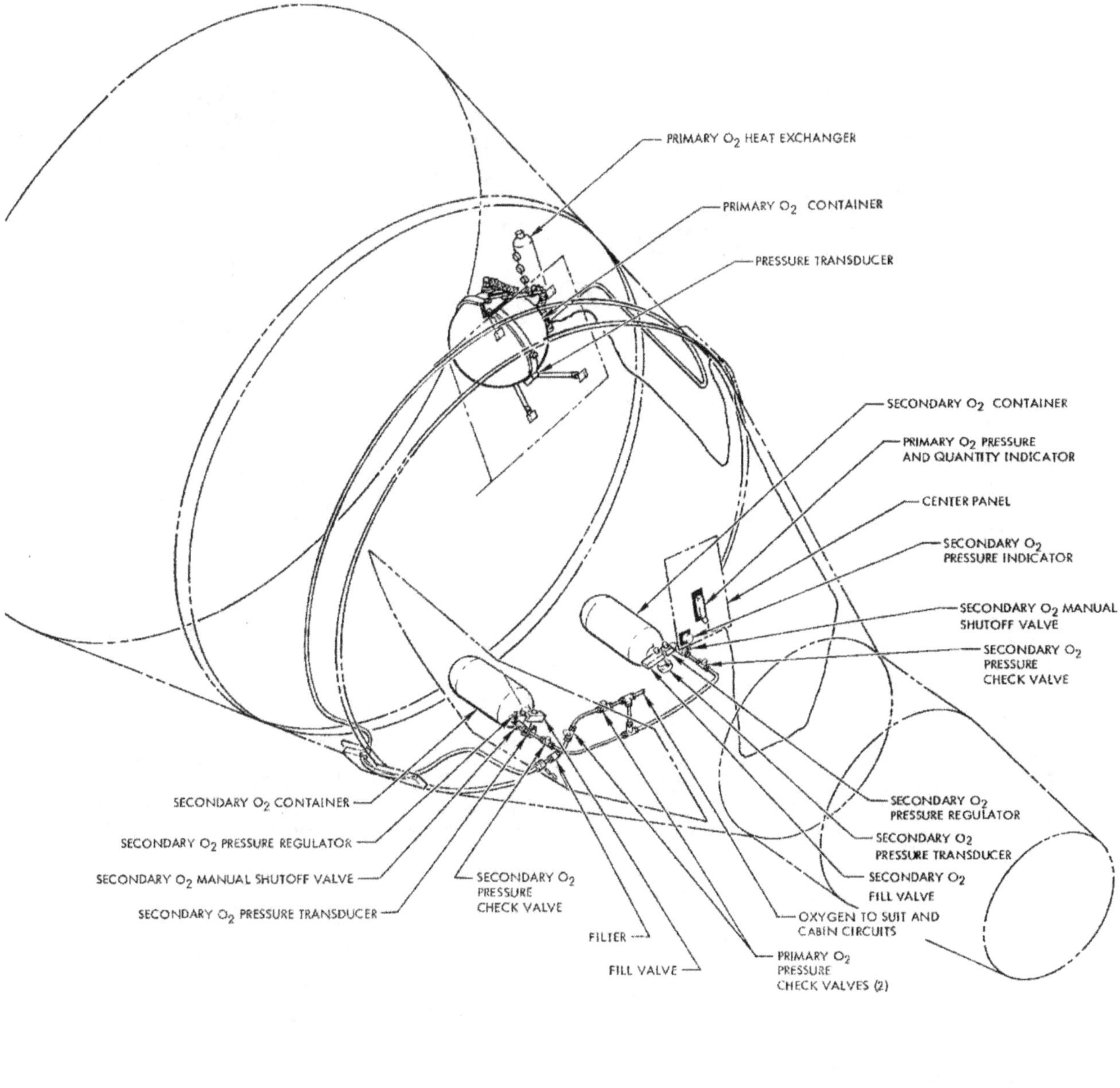

Figure 6-3 Primary and Secondary Oxygen System

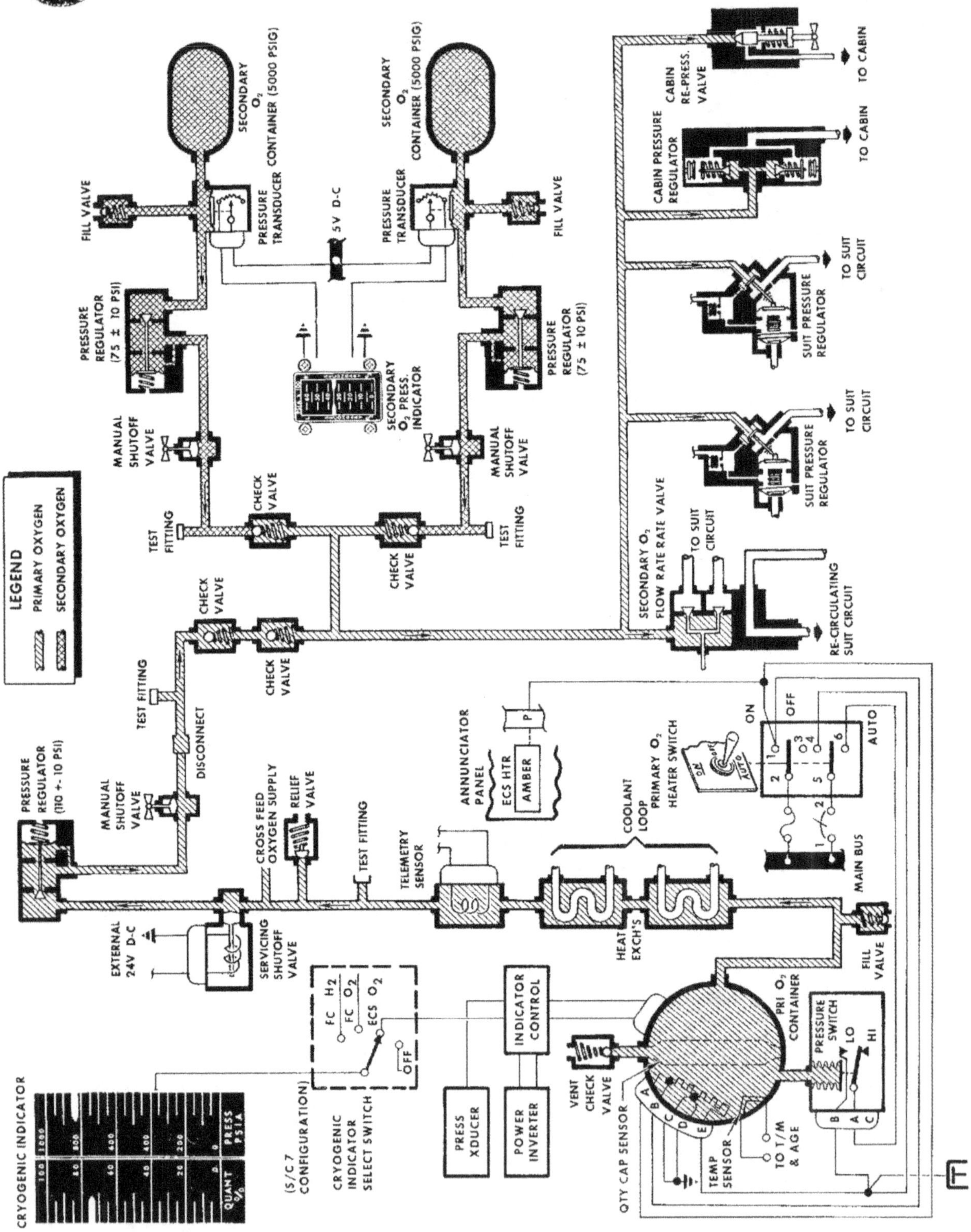

SEDR 300

PROJECT GEMINI

Figure 6-4 Primary and Secondary Oxygen Schematic

6-6

PROJECT GEMINI

SEDR 300

This system provides oxygen during the period commencing two hours prior to launch and terminating with jettison of the adapter section at retrograde.

The primary oxygen supply is stored at supercritical pressure in a cryogenic spherical container in the adapter section of the spacecraft. This container is filled with liquid oxygen at atmospheric pressure. Heat is supplied by thermal leakage through the container insulation and by activation of an electric heater in order to build pressure to the critical point of 736 psia. Above this point liquid oxygen becomes a homogeneous mixture, described for simplicity as a dense supercritical fluid. This fluid is warmed, regulated and filtered before it enters the suit or cabin loop.

The primary loop consists of the following components: primary oxygen container, pressure control switch, pressure transducer, fill and vent valves, temperature discharge sensor, pressure relief valve, pressure regulator, shutoff valve, filter, check valves, and heat exchanger.

Secondary Oxygen (Figure 6-3, 6-4)

The secondary oxygen system is capable of performing the same functions as the primary oxygen system and operates when pressure in the primary system falls below 75 $\pm$10 psi. At retrograde, when the primary oxygen container is jettisoned with the equipment adapter, the secondary oxygen system assumes the duties of the primary oxygen system.

The secondary gaseous oxygen supply is stored in two cylinders located in the re-entry module. Each cylinder contains 6.5 pounds of usable oxygen pressurized to 5000 psig maximum at 70°F. This oxygen supply is then regulated before it enters the suit or cabin loop.

The secondary system consists of two: cylinders, fill valves, transducers, pressure regulators, shutoff valves and check valves.

Egress System (Figure 6-5)

This system provides each pilot with oxygen for breathing and for suit pressurization in the event that they initiate ejection procedures at 45,000 feet or below, during launch or re-entry.

The egress gaseous oxygen supply is stored in a tank located in each seat-mounted egress kit. Each tank contains 0.31 pound of usable pressurized oxygen.

Each egress system consists of a tank, pressure regulator, pressure gage, restrictor, check valve, shutoff valve, and composite disconnects.

CABIN LOOP (Figure 6-6)

Design cabin leakage at ground test conditions is 670 scc/min of nitrogen at 5.0 psig. Makeup oxygen, to maintain cabin pressure at nominal 5.1 psia level, is called for by the cabin pressure regulator. In order to obtain maximum utilization of oxygen, it first passes through the suit loop before it is dumped into the cabin through the suit pressure relief valves.

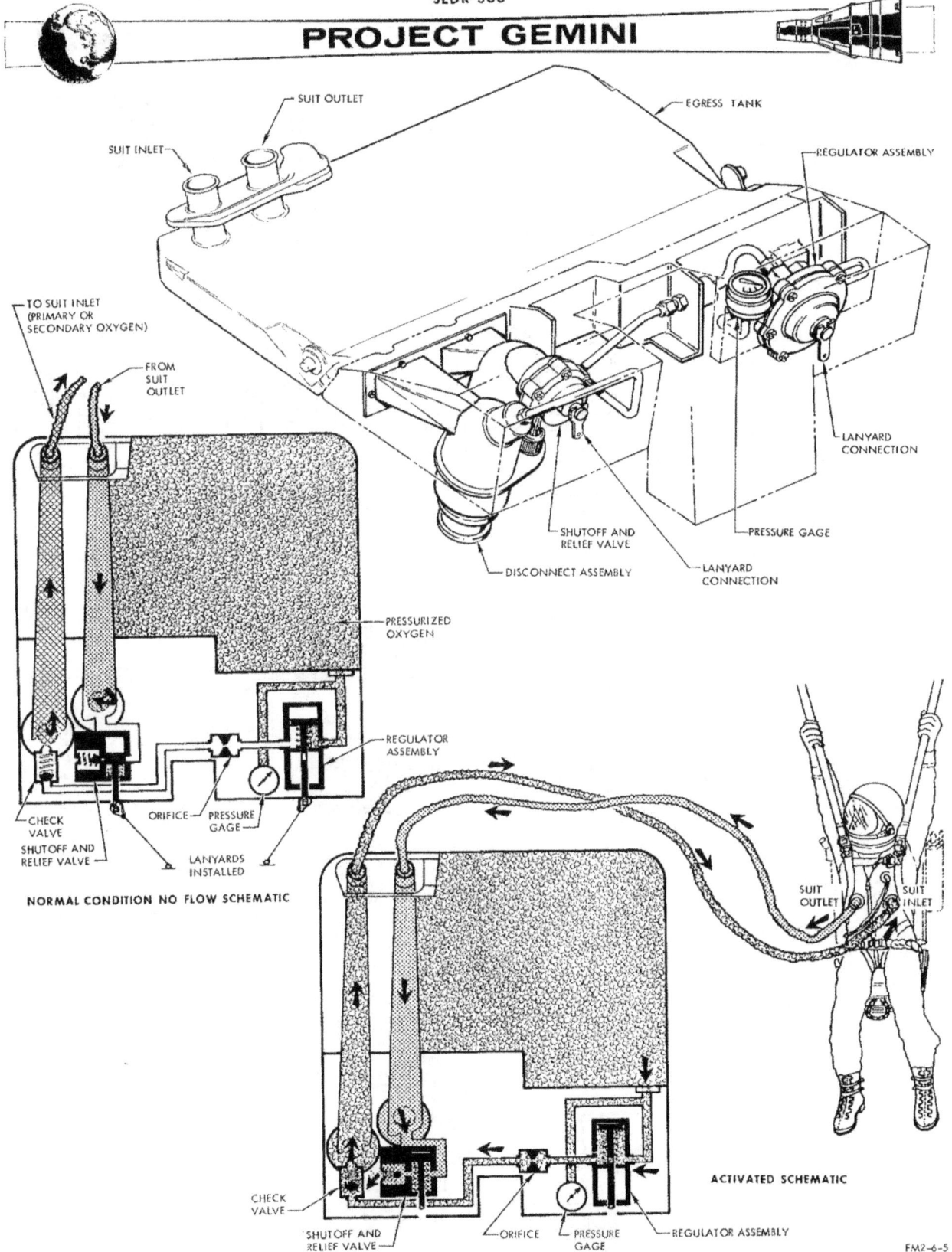

Figure 6-5 Egress Oxygen System

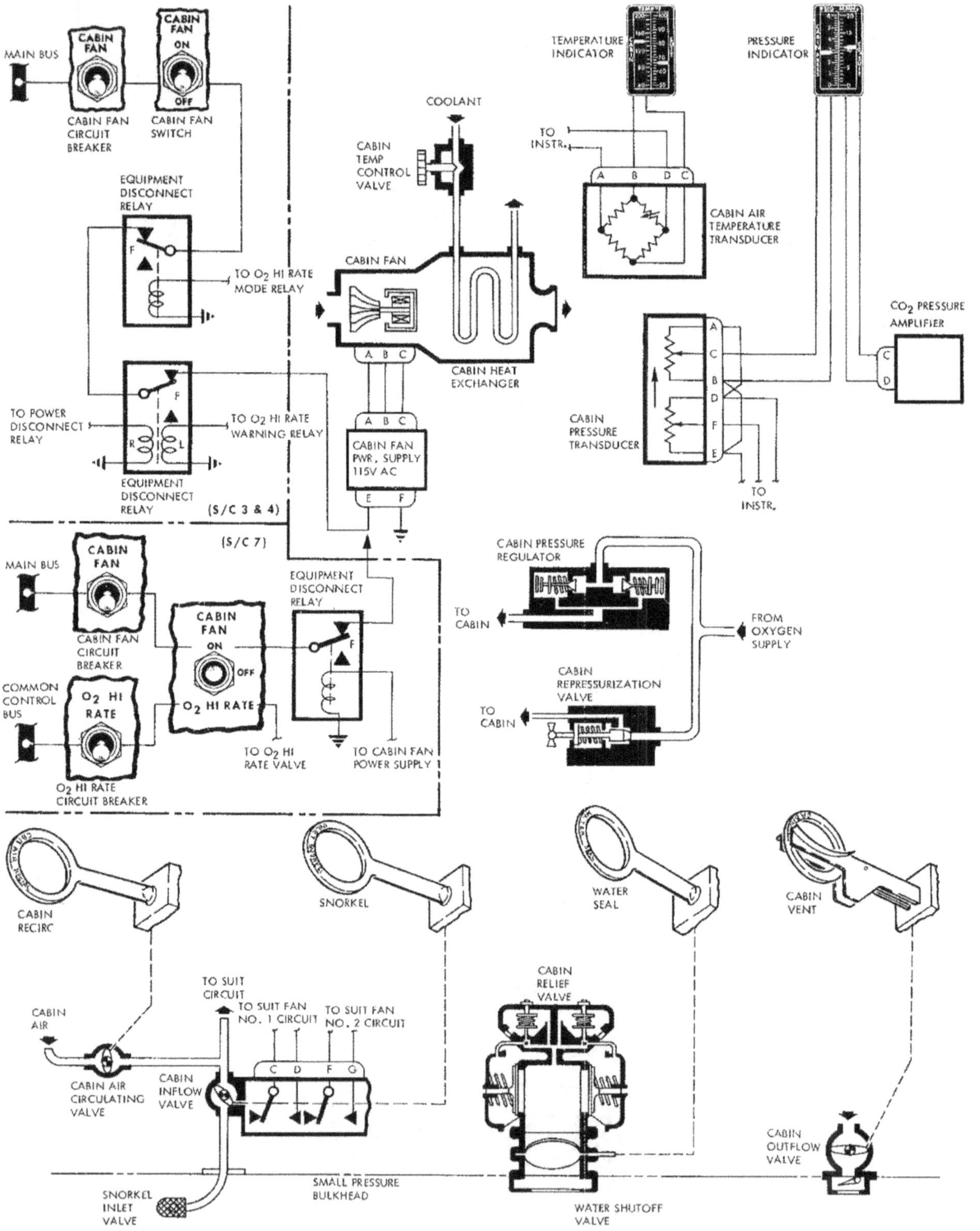

Figure 6-6 Cabin Environmental Control

Primary cabin components are a cabin heat exchanger and a fan. This loop also contains a relief valve for both positive and negative pressure relief, a pressure regulator and manual valves to either dump cabin pressure or repressurize. In the latter operation, oxygen is fed directly into the cabin.

SUIT LOOP (Figure 6-7, 6-8, 6-9)

The pilots are provided with redundant atmospheres by having a closed pressure suit circuit within the pressurized cabin. This suit circuit provides for cooling, pressurization, purification and water removal.

The suit loop is a closed system with two pressure suits operating in parallel. Circulation of oxygen through the suit is provided by a centrifugal compressor. Carbon dioxide and odors are removed by an absorber bed containing lithium hydroxide and activated charcoal. The gases are cooled in a heat exchanger by a liquid coolant, MCS 198, at a point below the dew point temperature. Water condensing within the heat exchanger is dumped overboard or routed to the water evaporator. The reconditioned oxygen is mixed with fresh makeup oxygen.

The suit circuit has two modes of operation, the normal recirculation mode which was discussed in the previous paragraph and the high rate mode which shuts off the recirculation system and dumps oxygen directly into the suit.

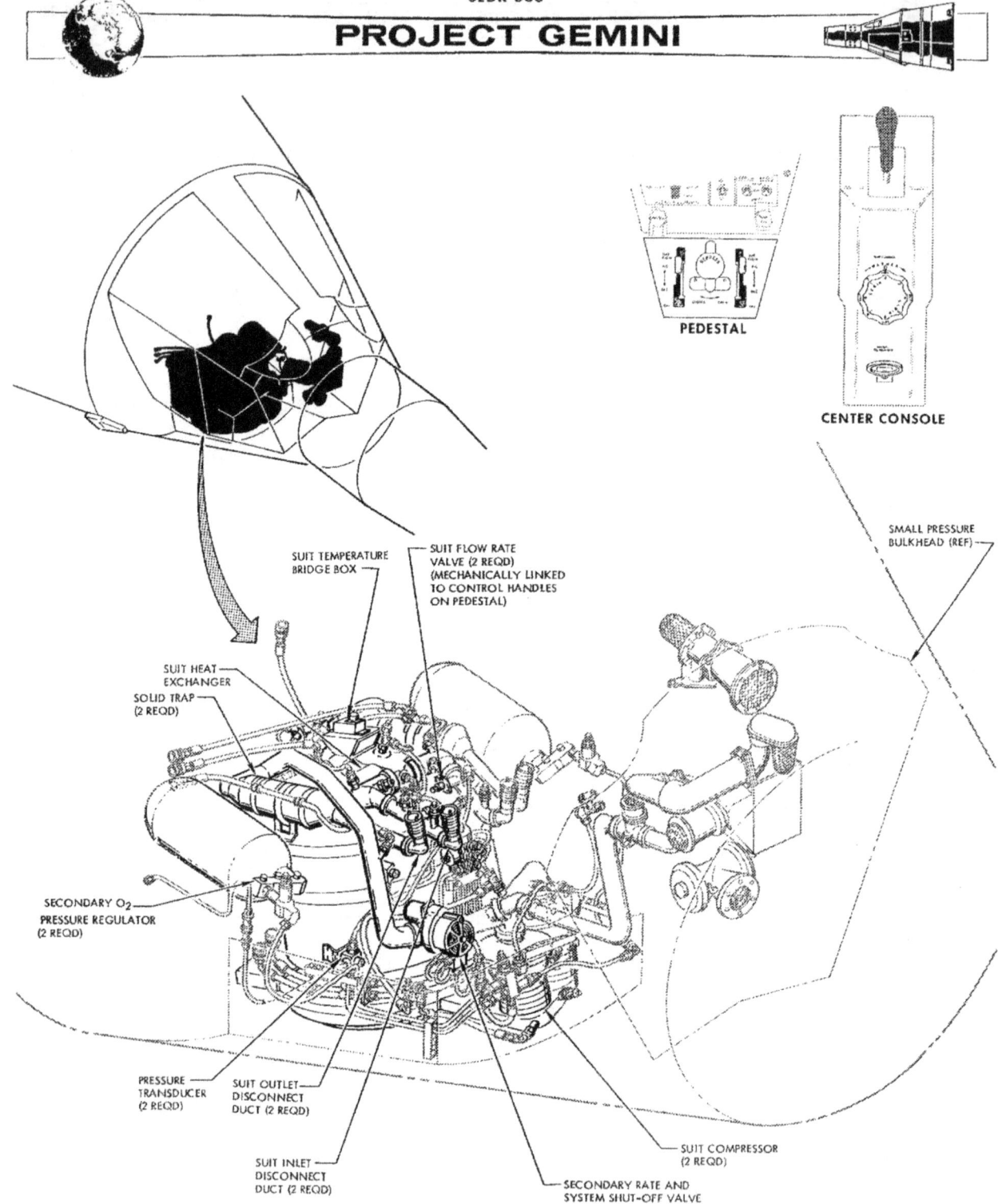

Figure 6-7 Suit Loop (ECS Package) (Sheet 1 of 2)

SEDR 300

PROJECT GEMINI

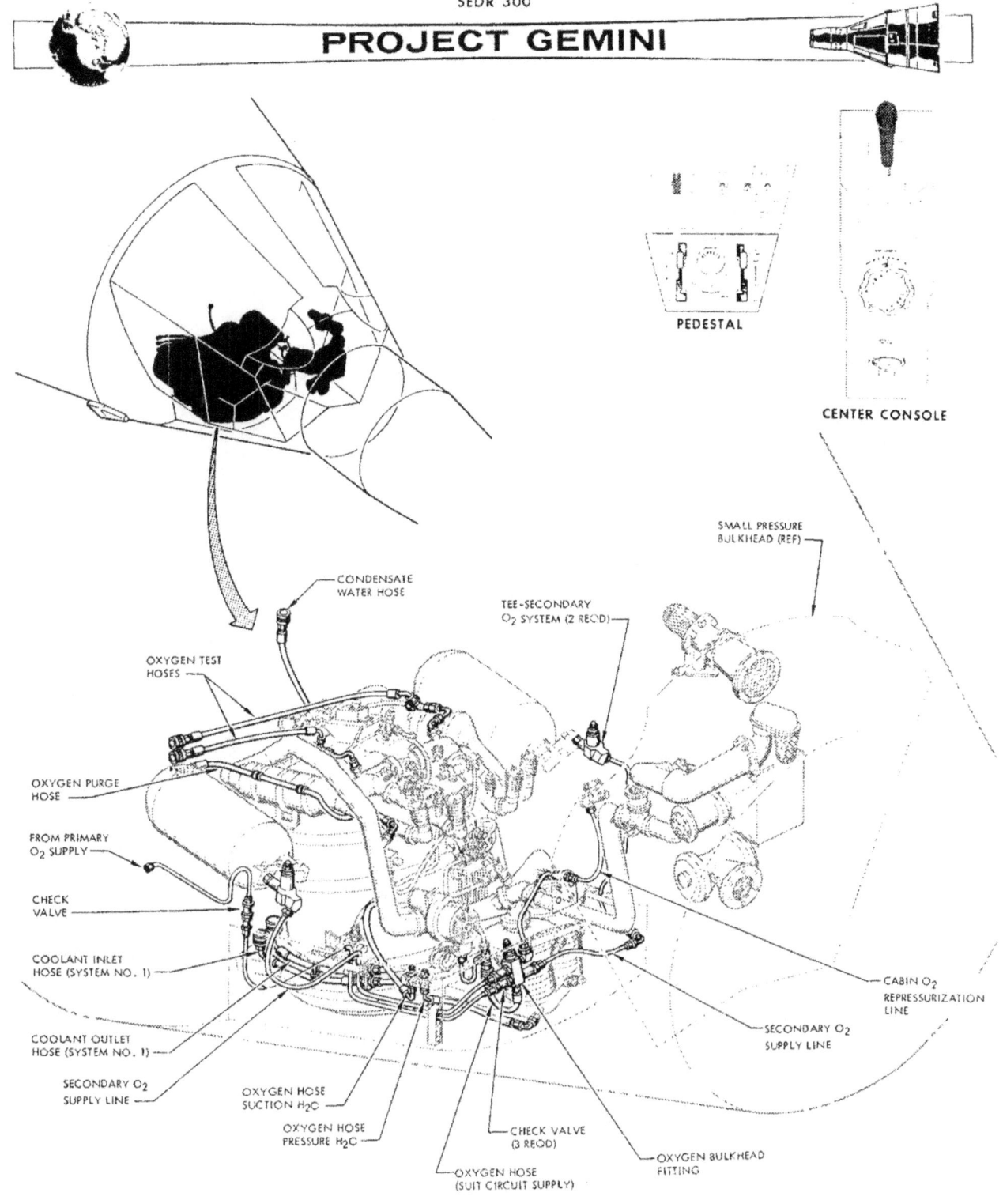

Figure 6-7 Suit Loop (ECS Package) (Sheet 2 of 2)

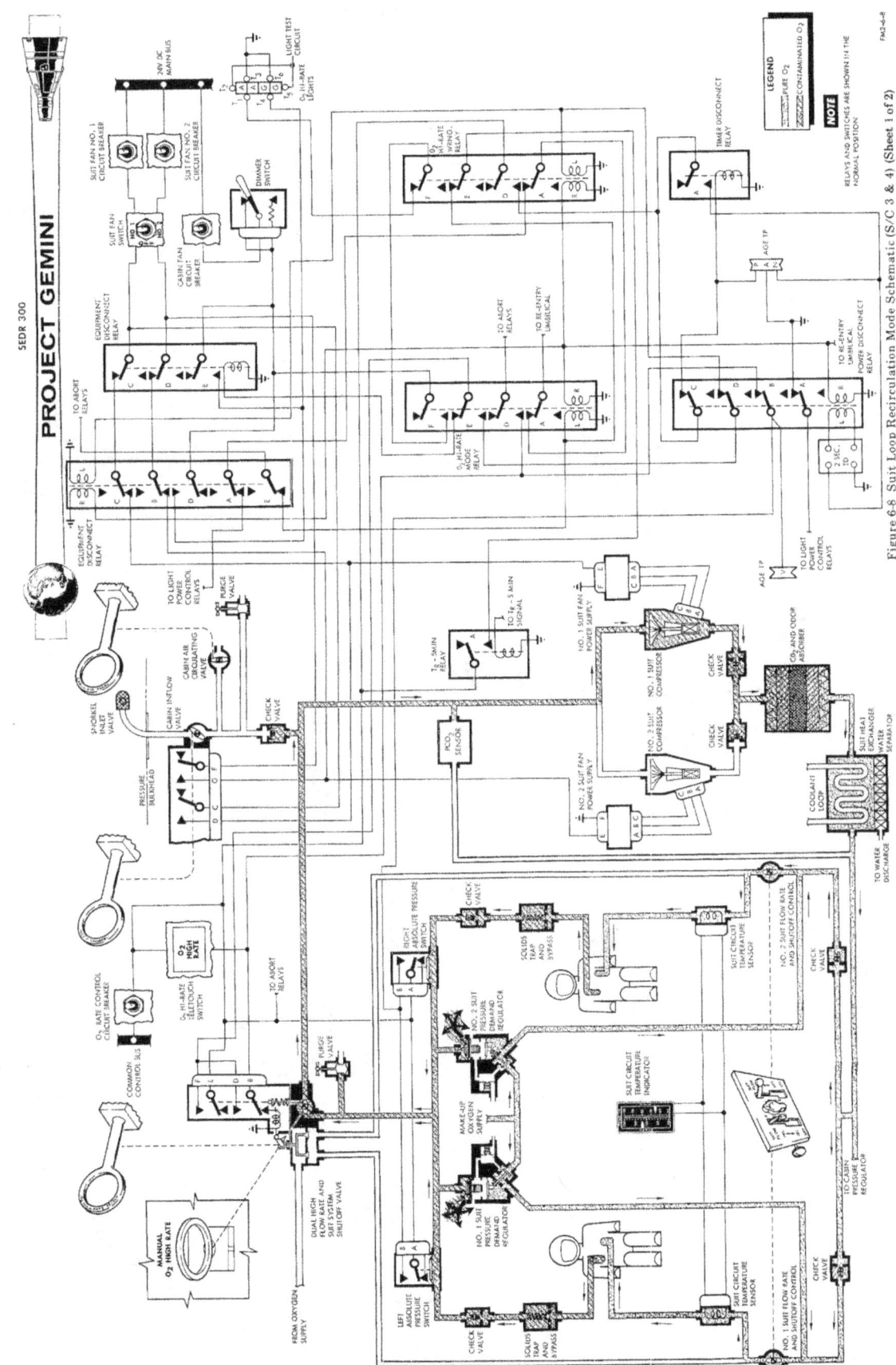

Figure 6-8 Suit Loop Recirculation Mode Schematic (S/C 3 & 4) (Sheet 1 of 2)

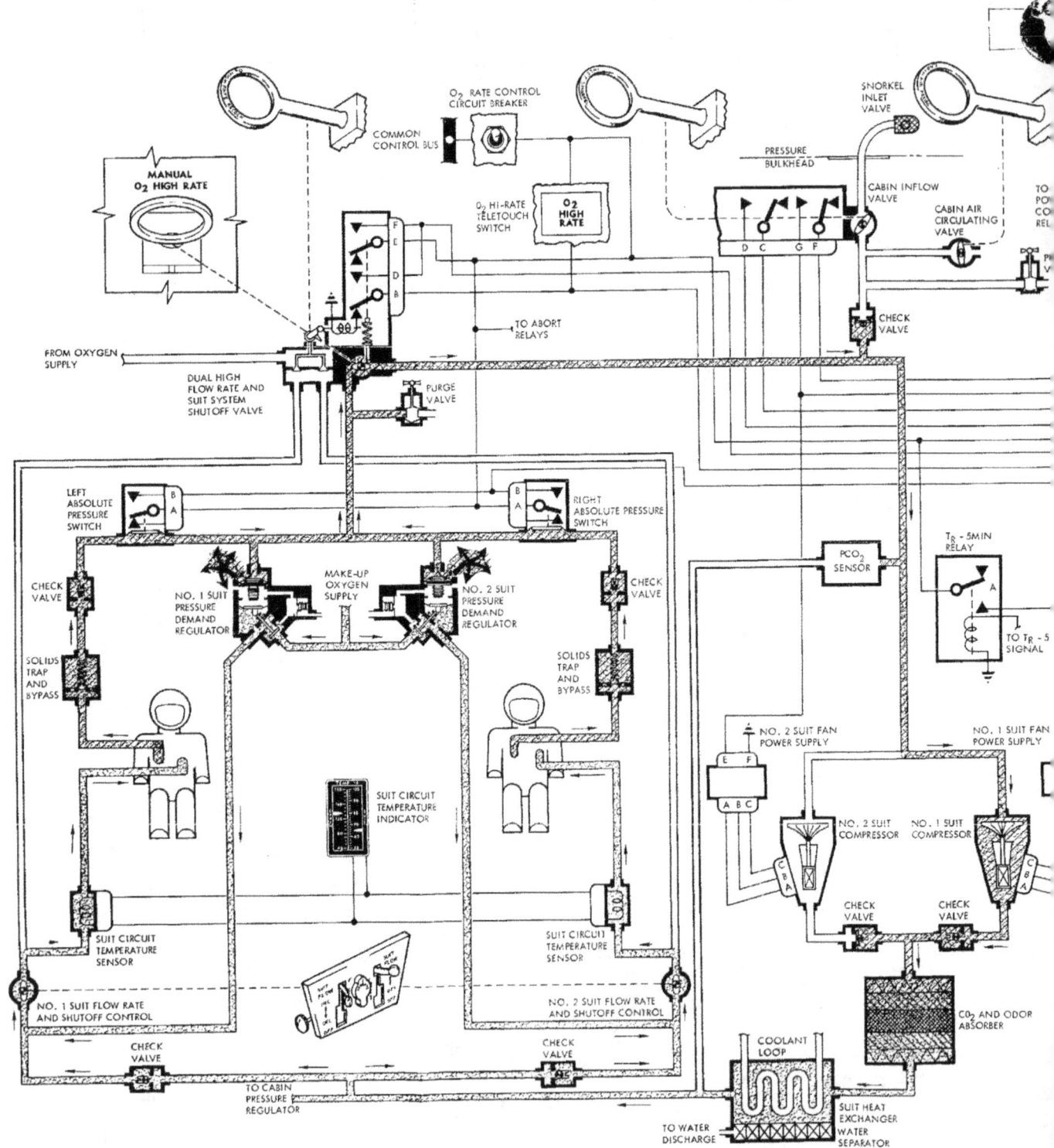

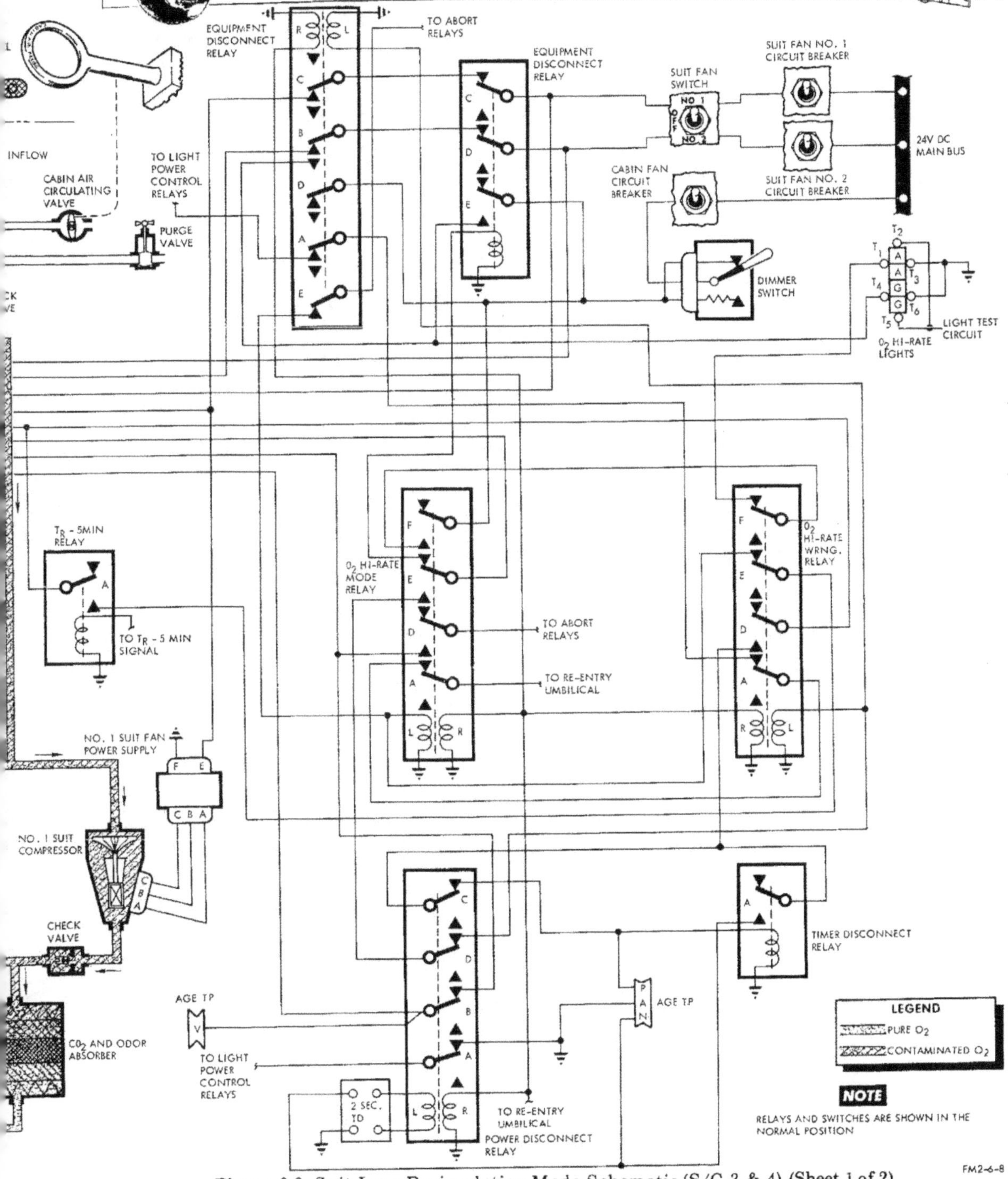

Figure 6-8 Suit Loop Recirculation Mode Schematic (S/C 3 & 4) (Sheet 1 of 2)

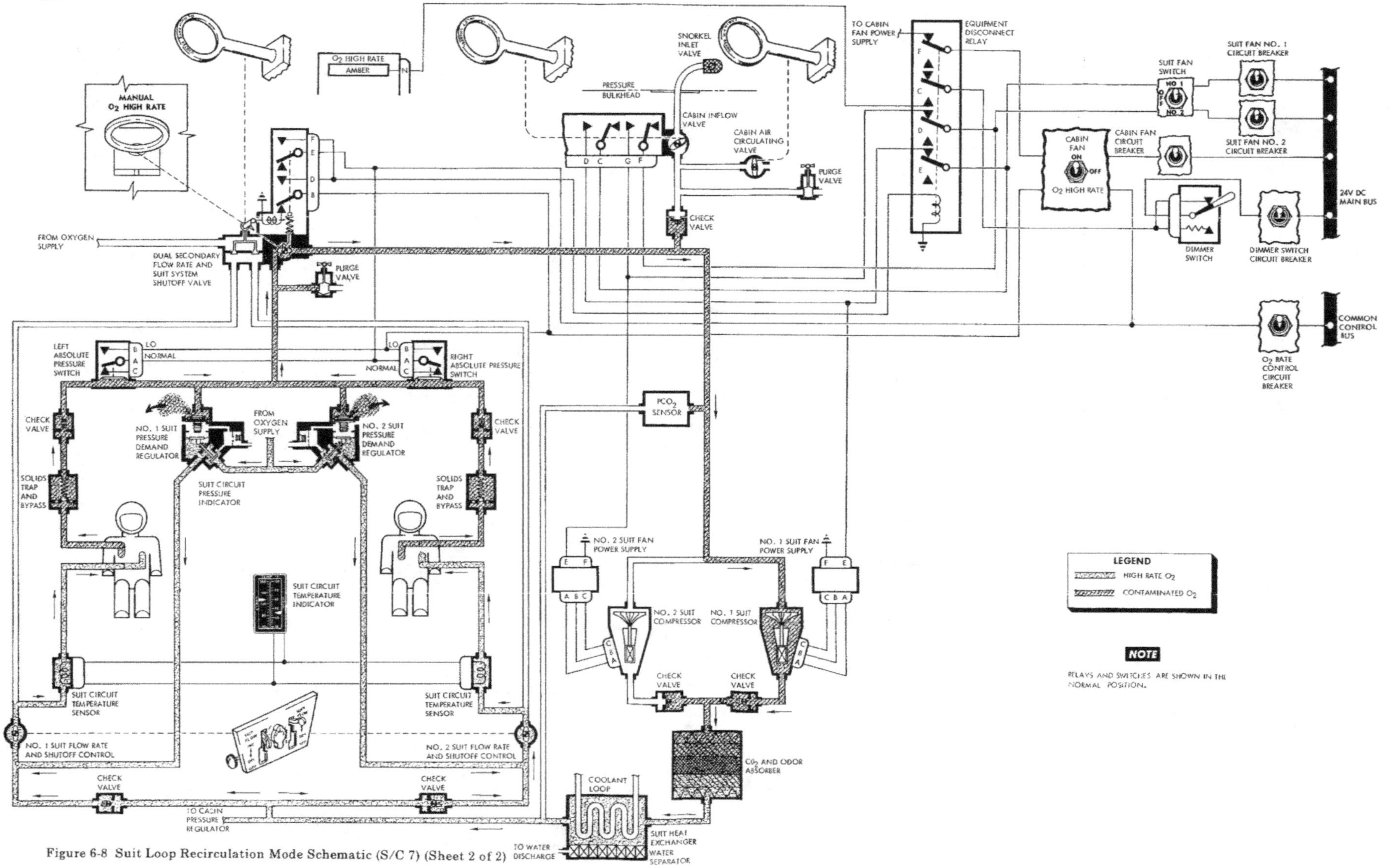

Figure 6-8 Suit Loop Recirculation Mode Schematic (S/C 7) (Sheet 2 of 2)

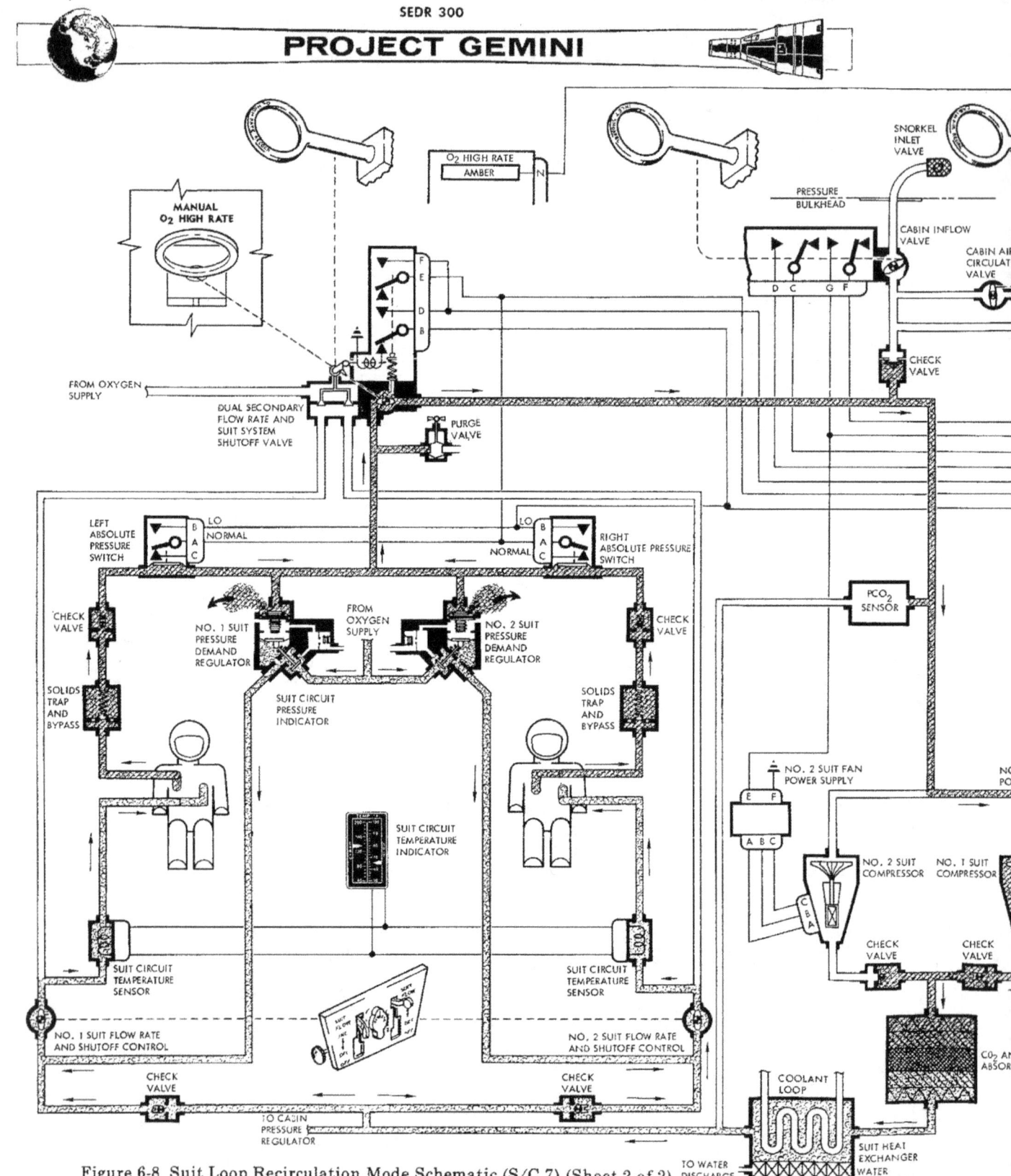

Figure 6-8 Suit Loop Recirculation Mode Schematic (S/C 7) (Sheet 2 of 2)

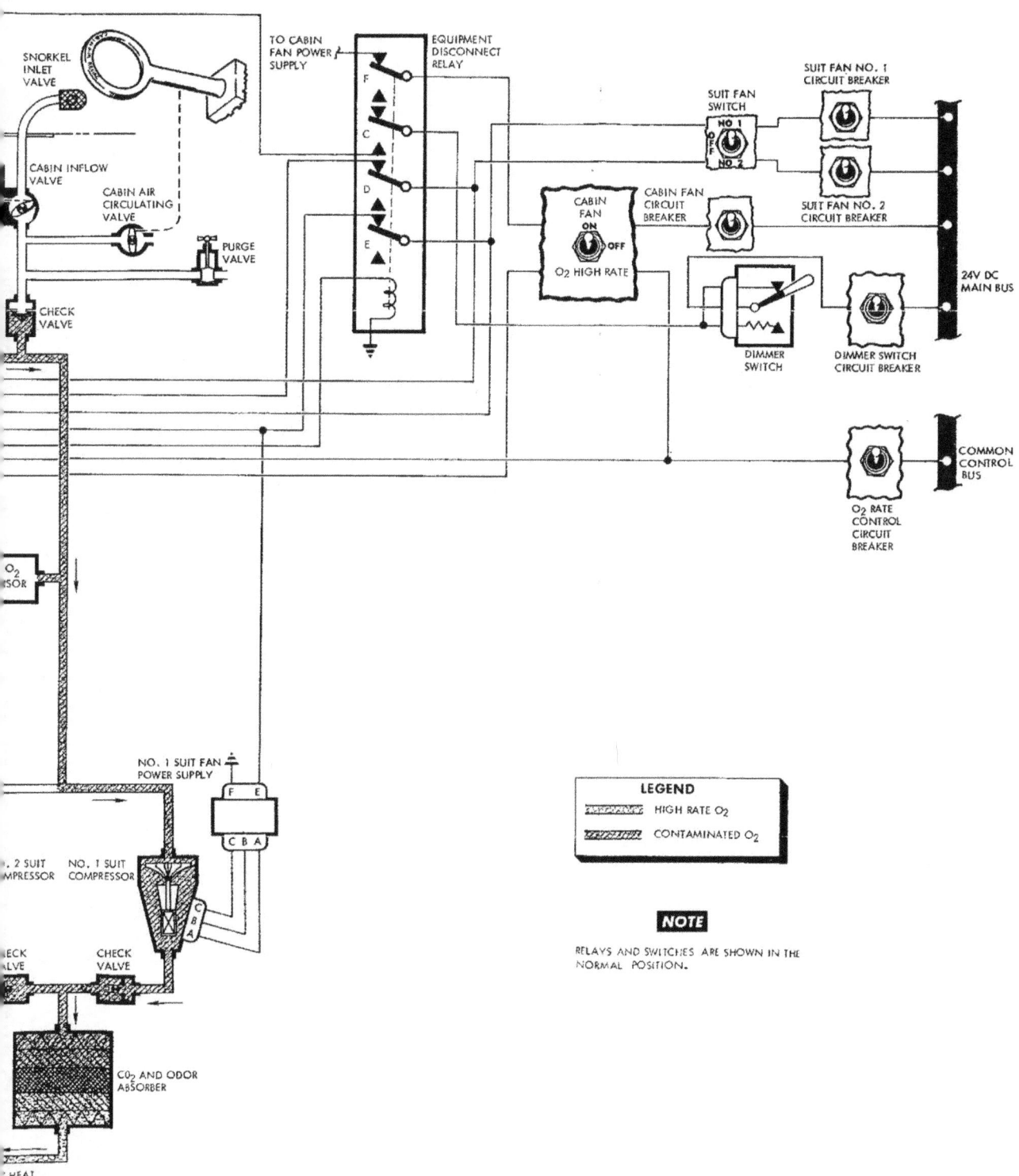

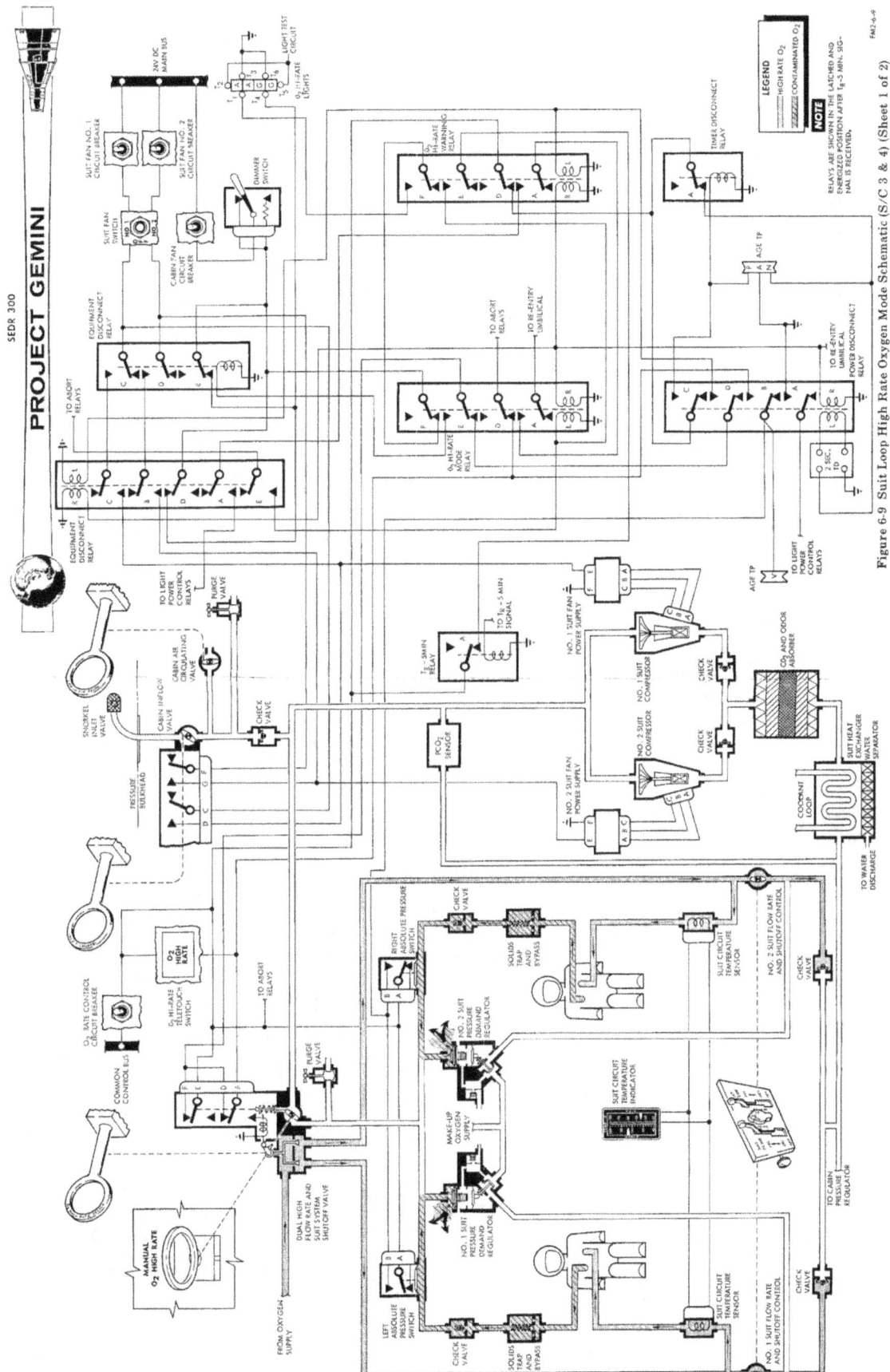

Figure 6-9 Suit Loop High Rate Oxygen Mode Schematic (S/C 3 & 4) (Sheet 1 of 2)

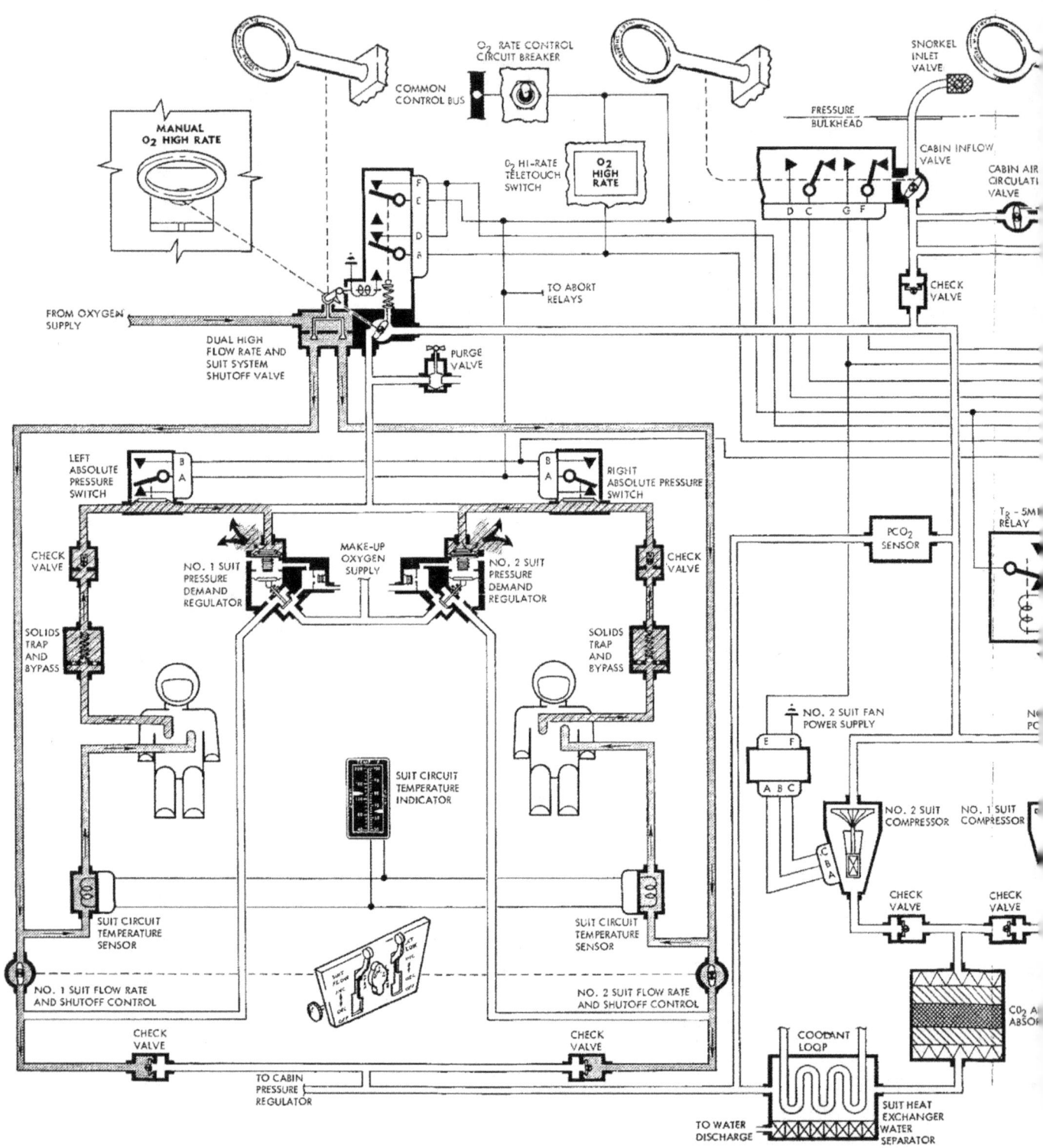

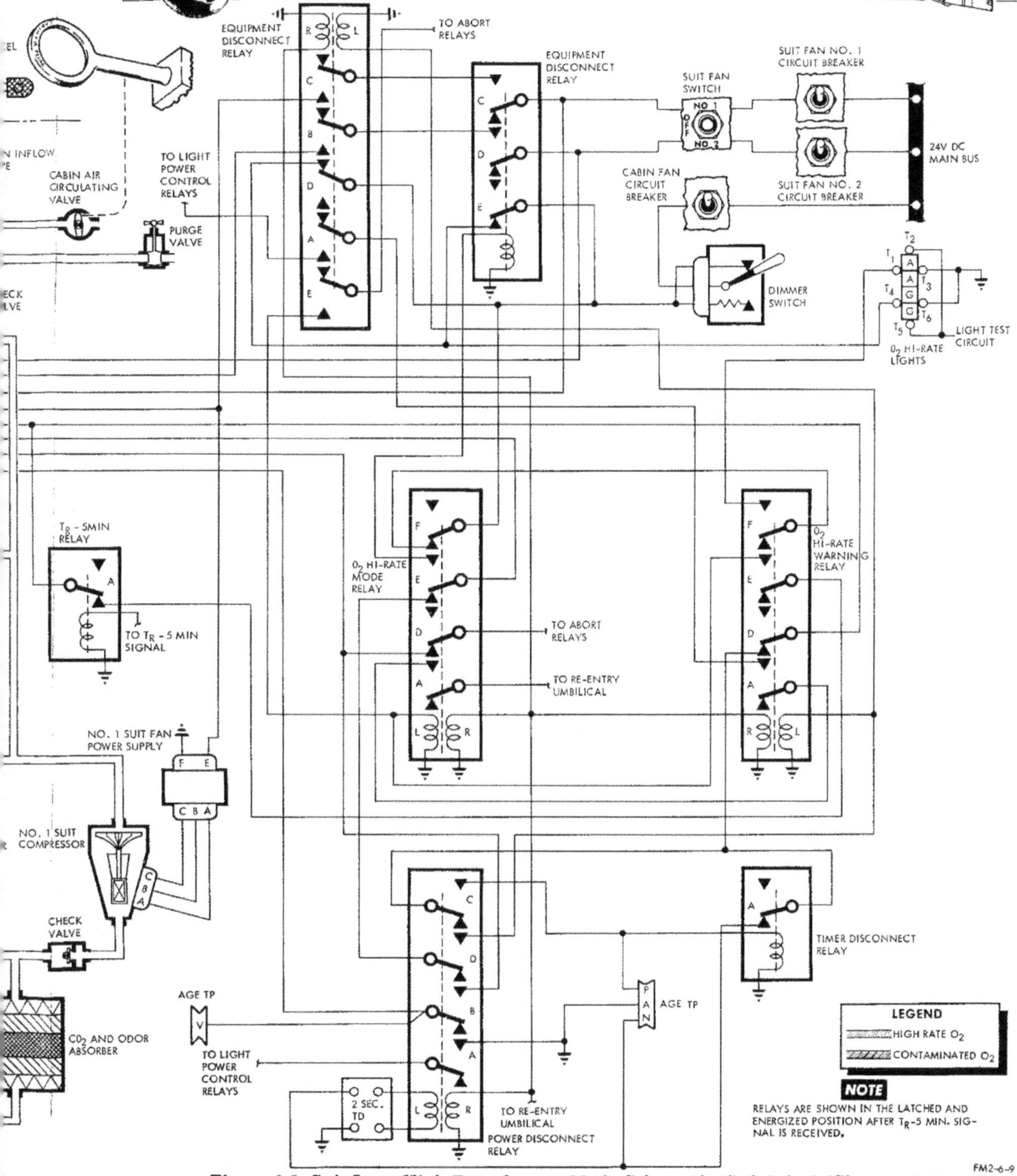

Figure 6-9 Suit Loop High Rate Oxygen Mode Schematic (S/C 3 & 4) (Sheet 1 of 2)

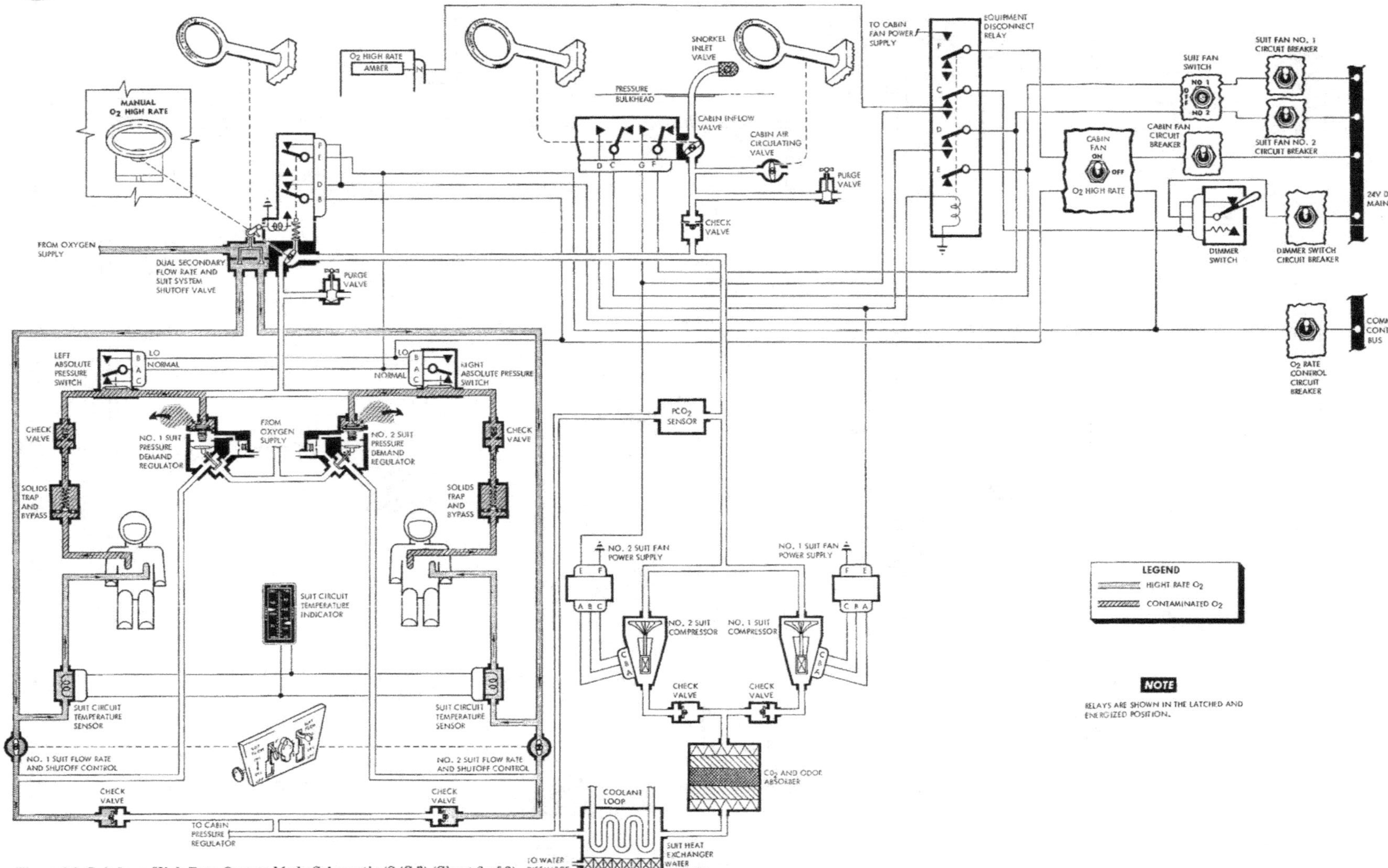

Figure 6-9 Suit Loop High Rate Oxygen Mode Schematic (S/C 7) (Sheet 2 of 2)

6-17

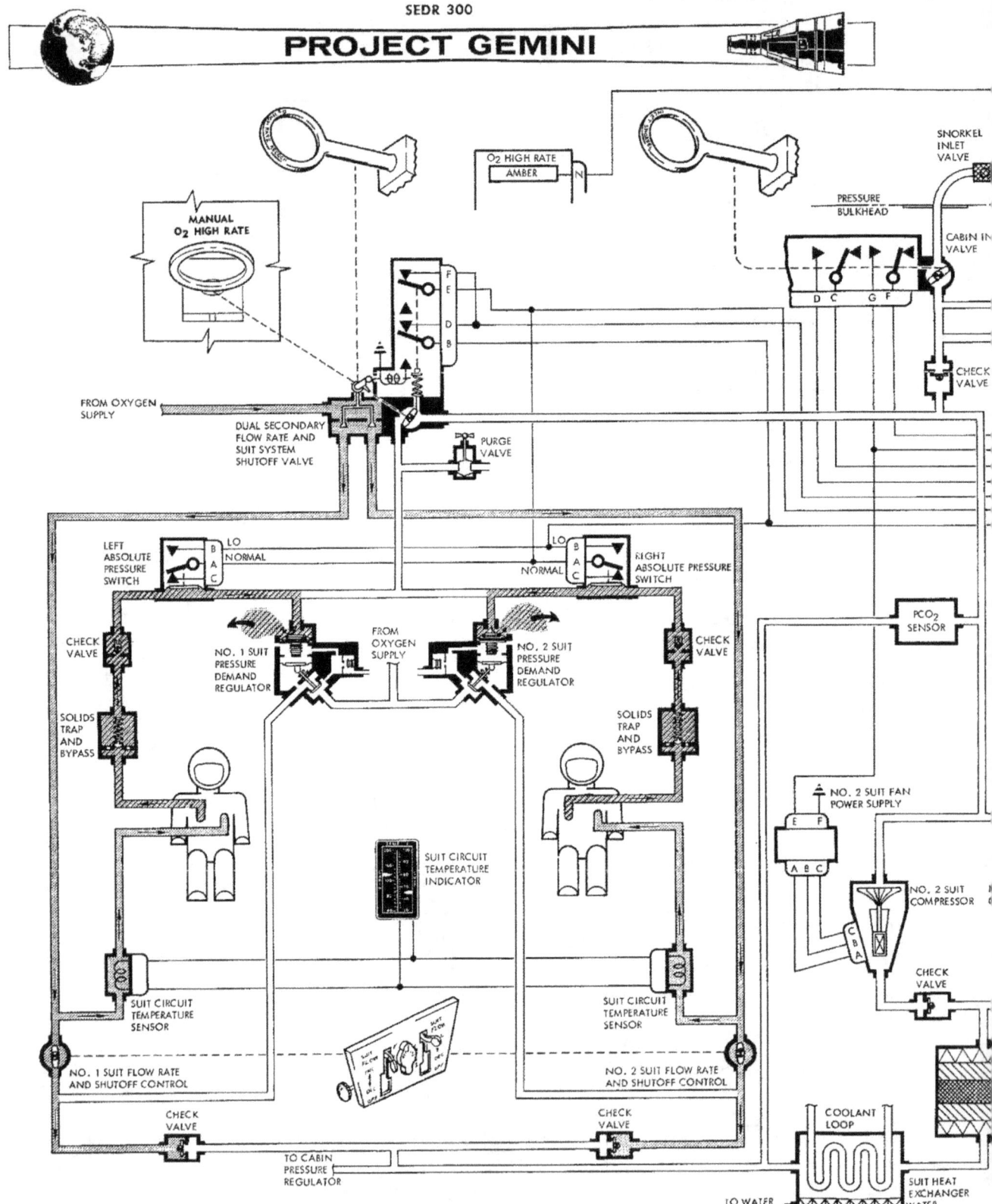

Figure 6-9 Suit Loop High Rate Oxygen Mode Schematic (S/C 7) (Sheet 2 of 2)

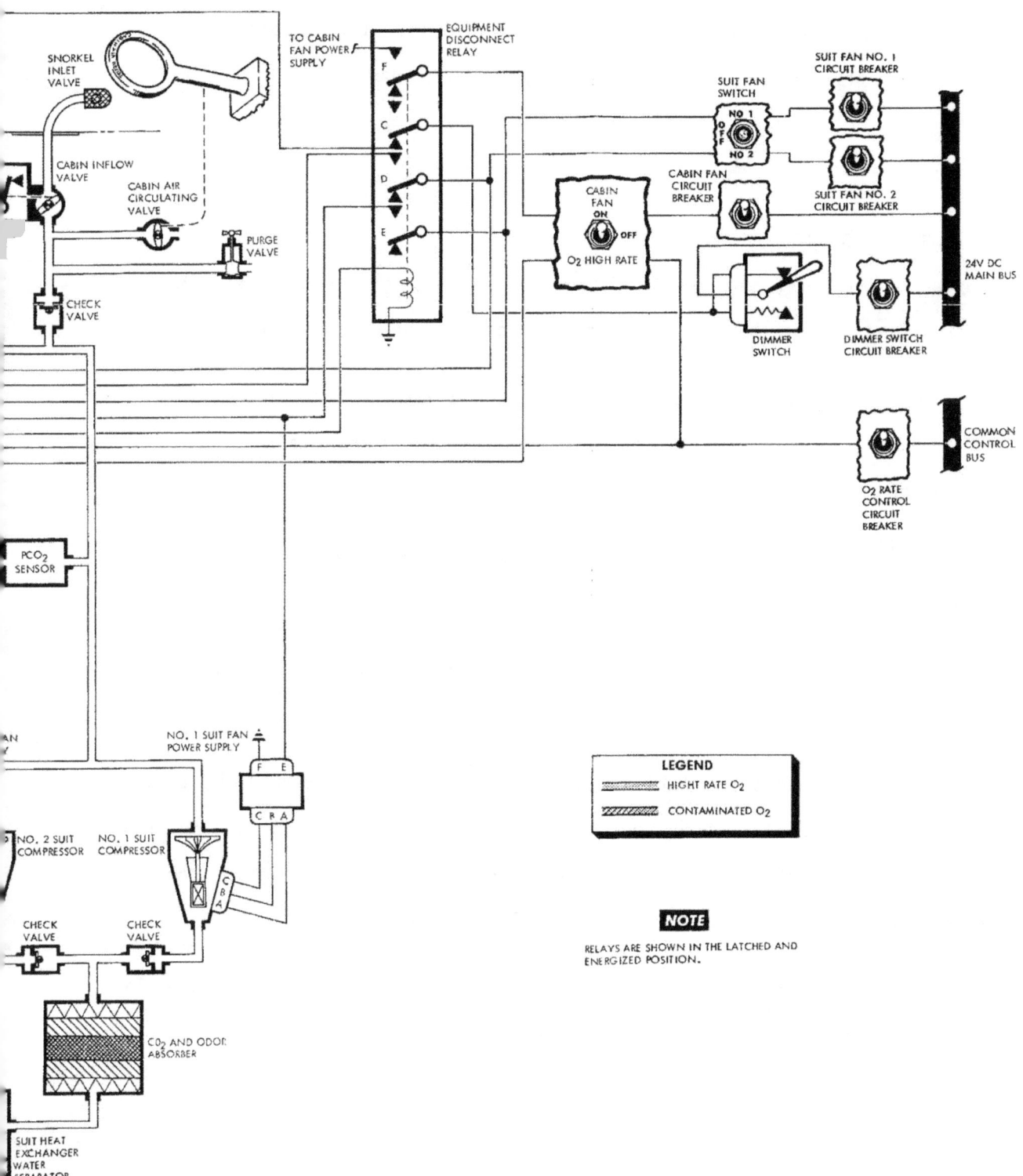

SEDR 300

 **PROJECT GEMINI**

The suit loop consists of two demand regulated and relief valves, four check valves, two throttle valves, two solids traps, a system shutoff and high flow rate valve, two compressors, one carbon dioxide and odor absorber, and a suit heat exchanger.

WATER MANAGEMENT SYSTEM (Figure 6-10, 6-11)
The purpose of the water management system is to store and dispense drinking water, collect and route unwanted water to the evaporator or dump overboard.

Drinking water is stored in a tank or tanks in the adapter. Each tank contains a bladder and is pressurized to supply water to the transparent tank in the re-entry module.

S/C 3 has only one adapter drinking water storage tank and uses oxygen for the pressurant.

S/C 4 utilizes four adapter drinking water storage tanks and uses oxygen for the pressurant.

S/C 7 utilizes three adapter water storage tanks. Tanks A and B hold 149 lbs. of water each. A combination of gas and fuel cell by-product water is used as the pressurant. Tank C holds 25 lbs. of drinking water. Tanks A and C store the drinking water. Tank B stores water from the fuel cell.

Urine and condensated water from the suit circuit heat exchanger are absorbed by the wick in the water boiler or dumped overboard.

6-18

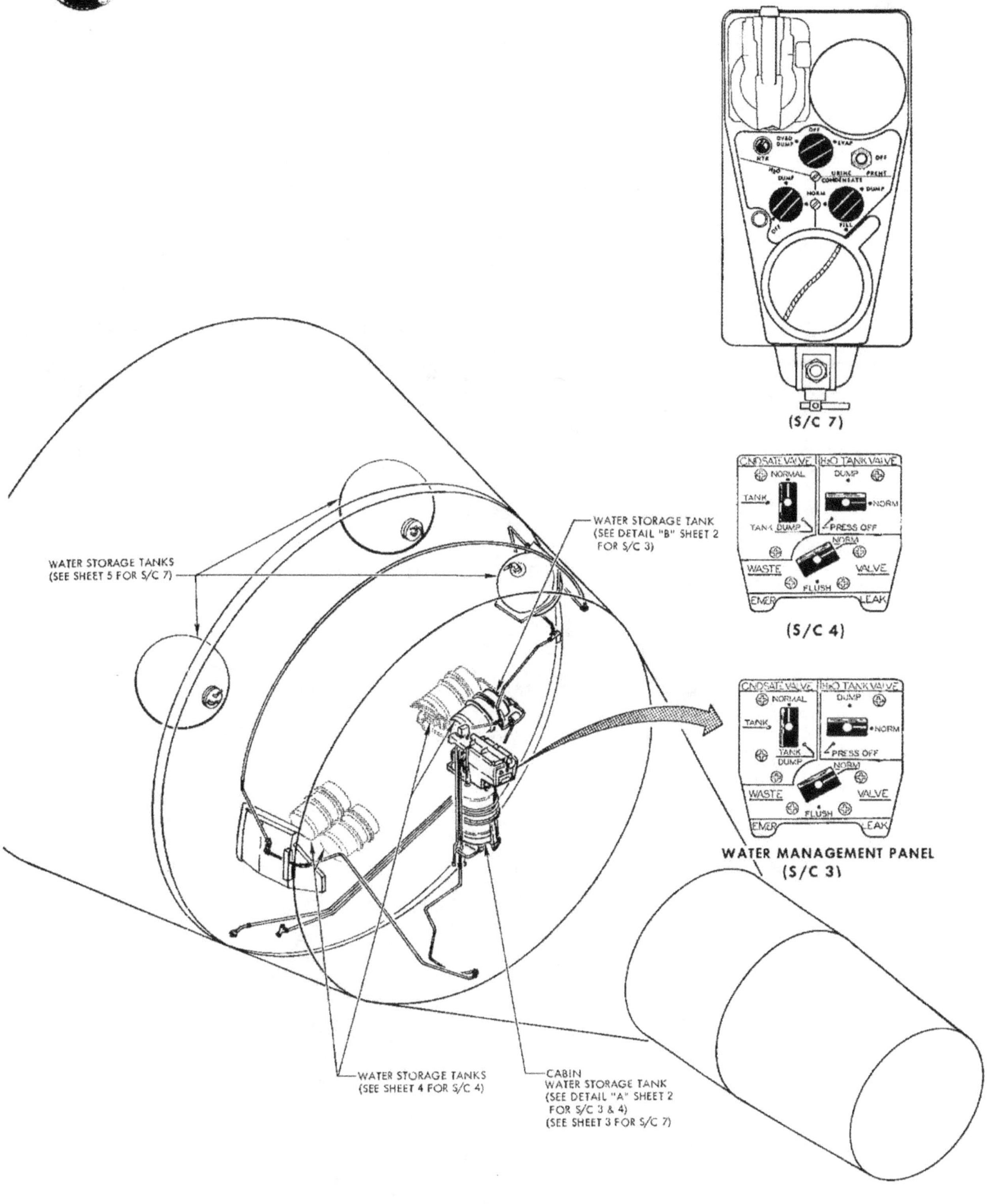

PROJECT GEMINI

SEDR 300

Figure 6-10 Water Management System (Sheet 1 of 5)

DETAIL A
(S/C 3 & 4 CONFIGURATION)

DETAIL B (S/C 3 CONFIGURATION)

Figure 6-10 Water Management System (Sheet 2 of 5)

FMG2-171-2

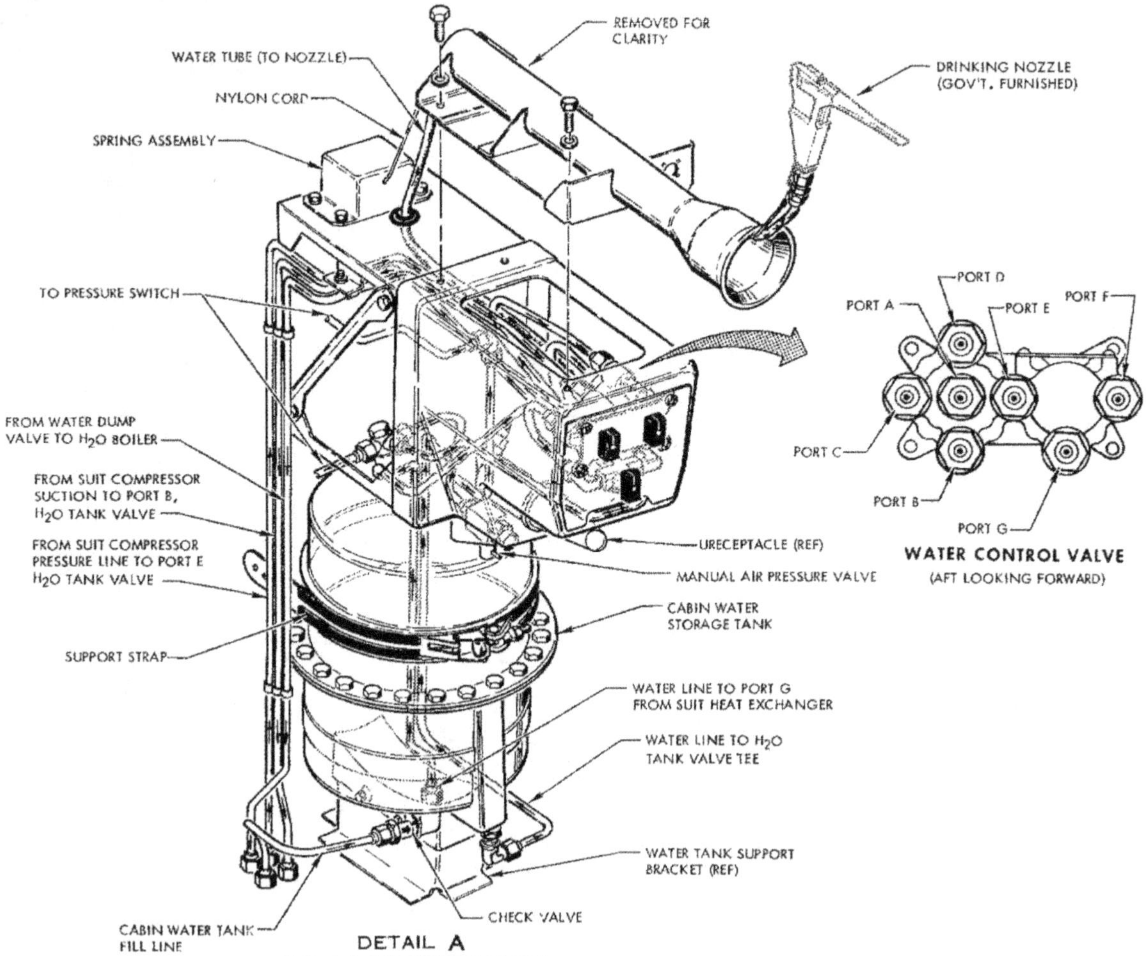

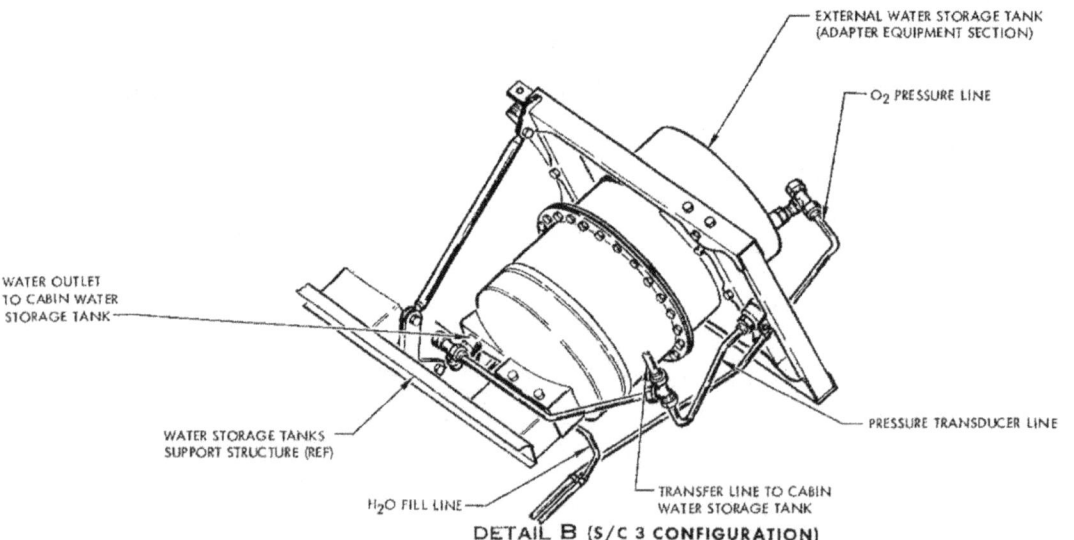

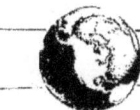

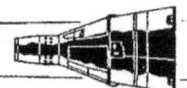

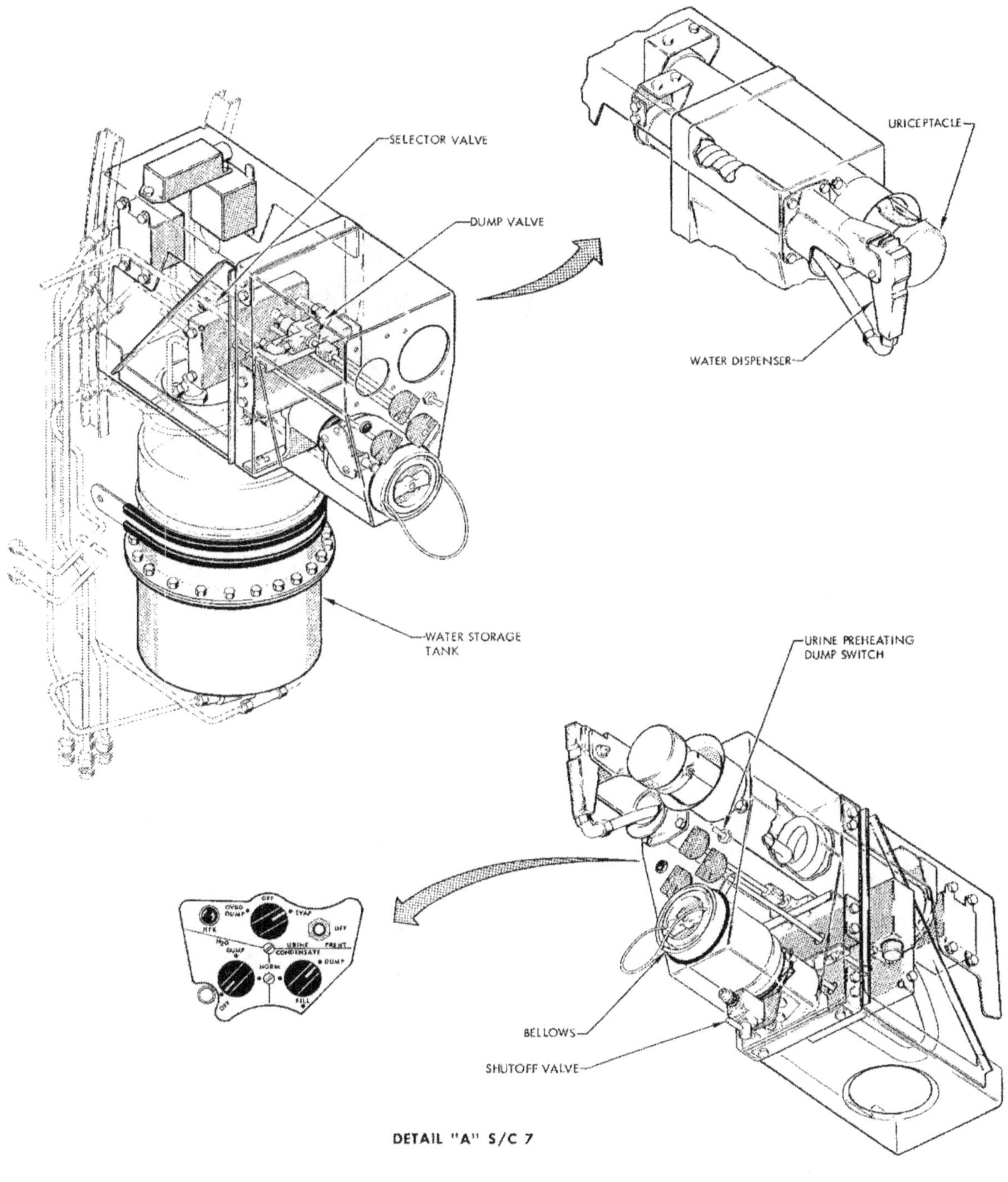

Figure 6-10 Water Management System (Sheet 3 of 5)

FM2-6-10 (4)

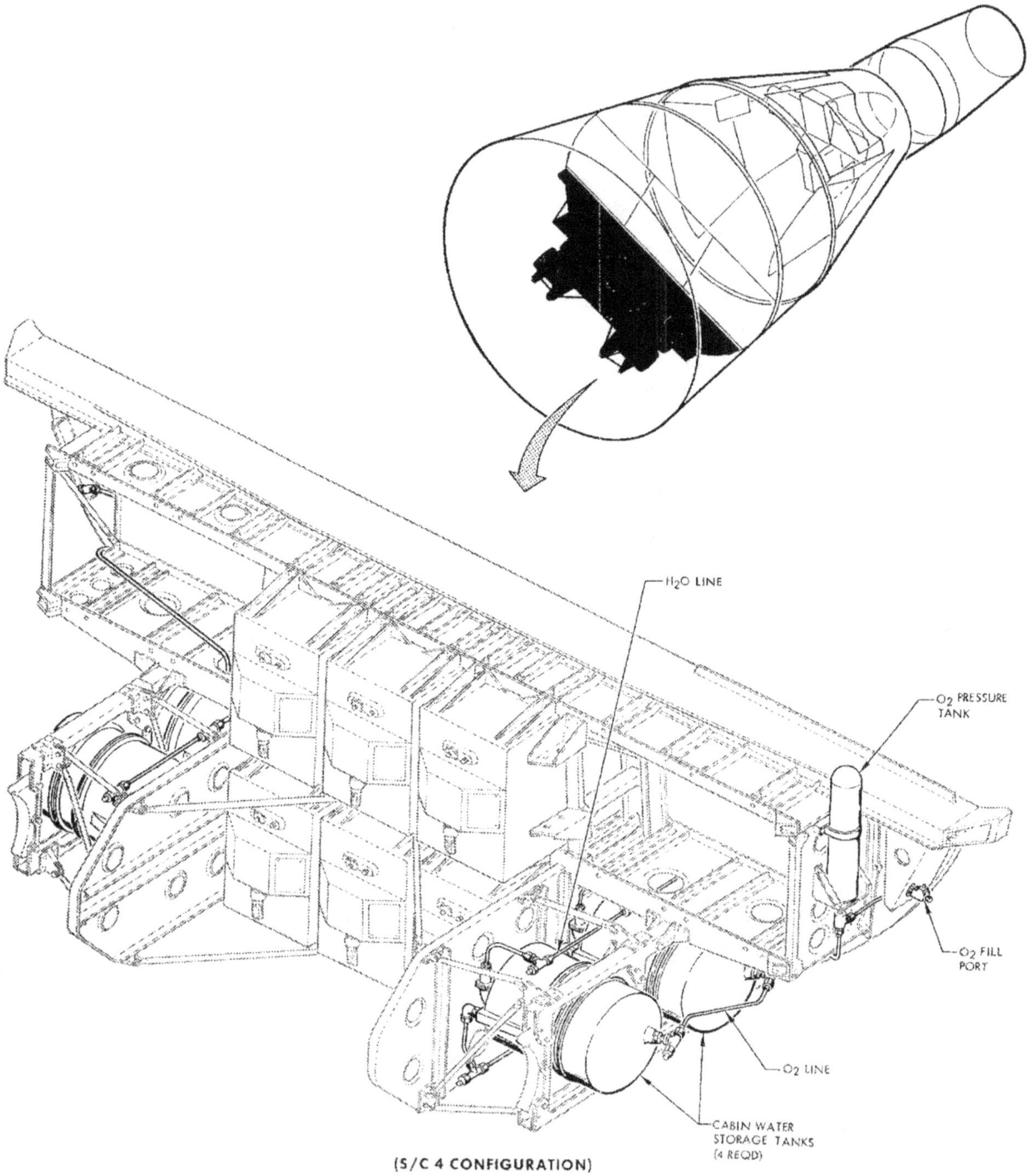

PROJECT GEMINI

SEDR 300

(S/C 4 CONFIGURATION)

Figure 6-10 Water Management System (Sheet 4 of 5)

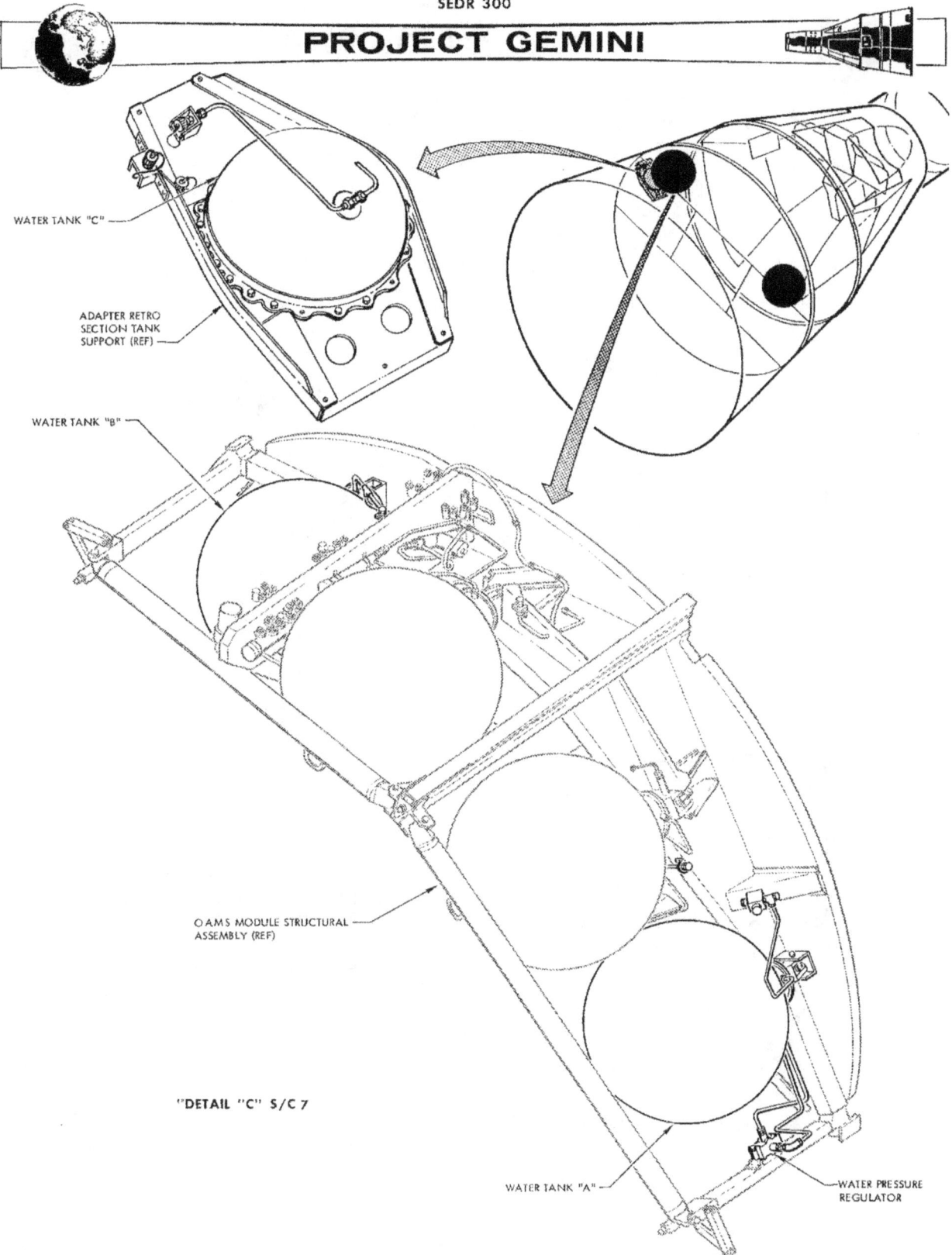

Figure 6-10 Water Management System (Sheet 5 of 5)

 SEDR 300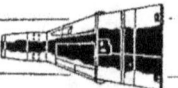

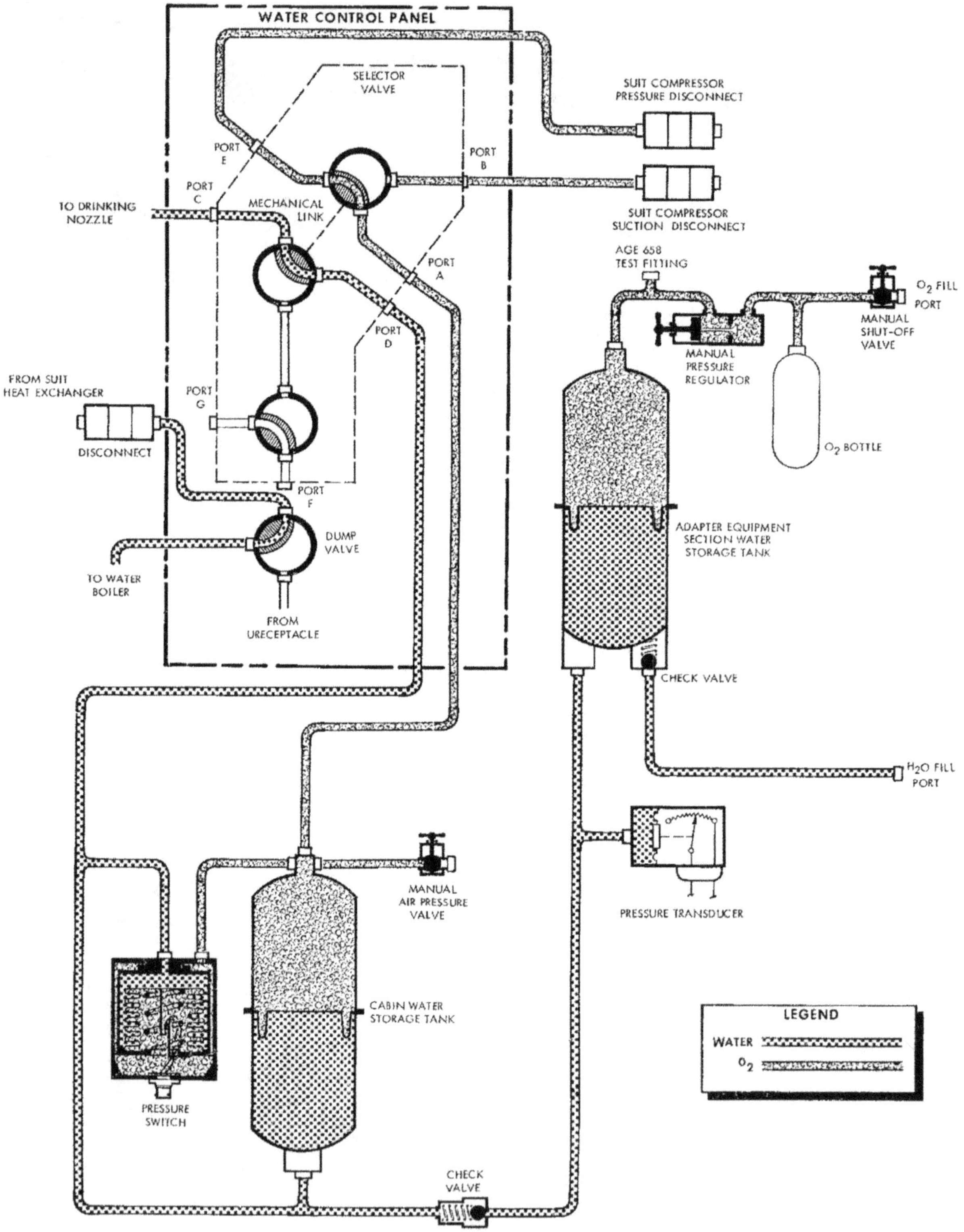

PROJECT GEMINI

Figure 6-11 Water Management Schematic (Sheet 1 of 3)
(S/C 3 ONLY)

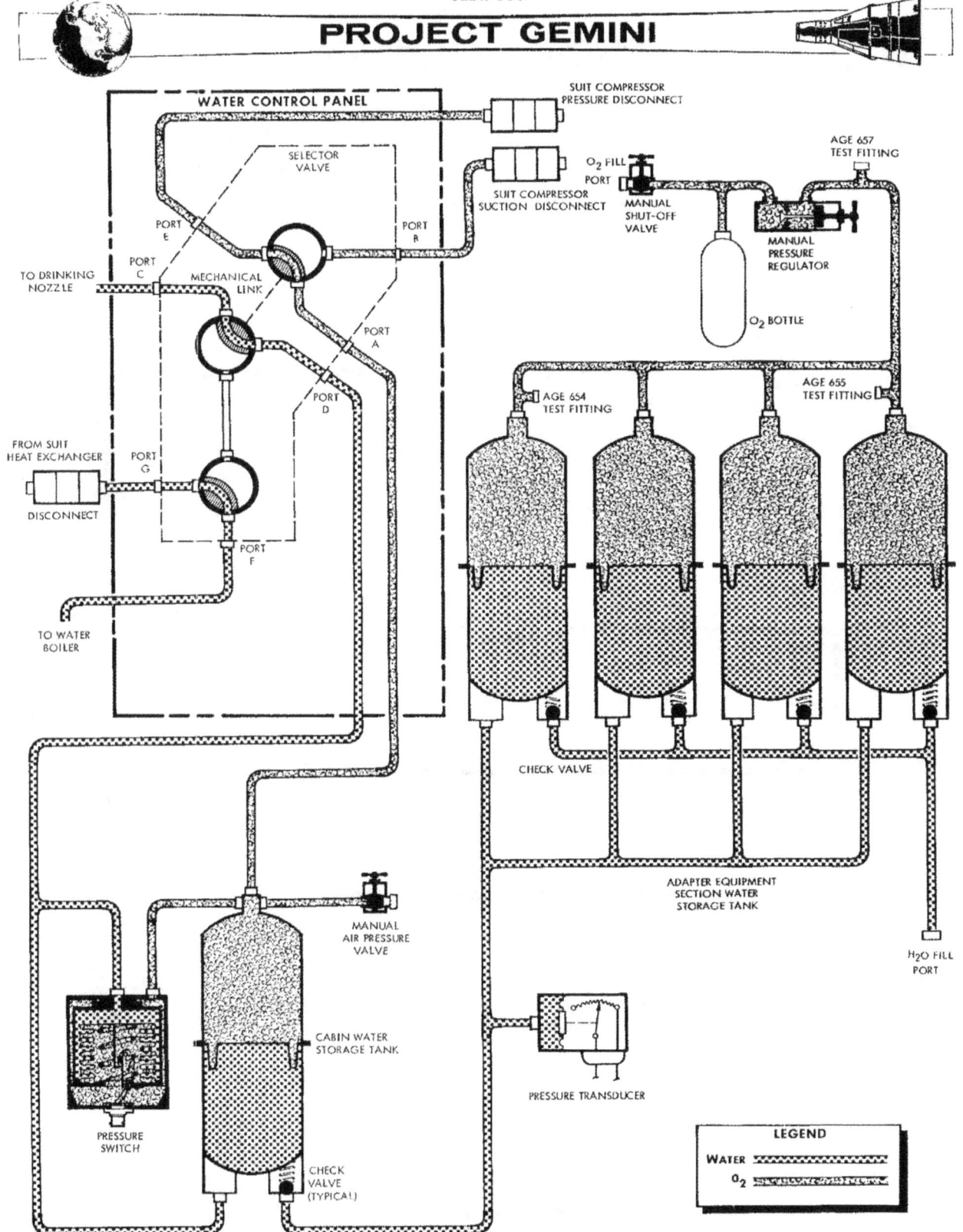

Figure 6-11 Water Management Schematic (Sheet 2 of 3)
(S/C 4 ONLY)

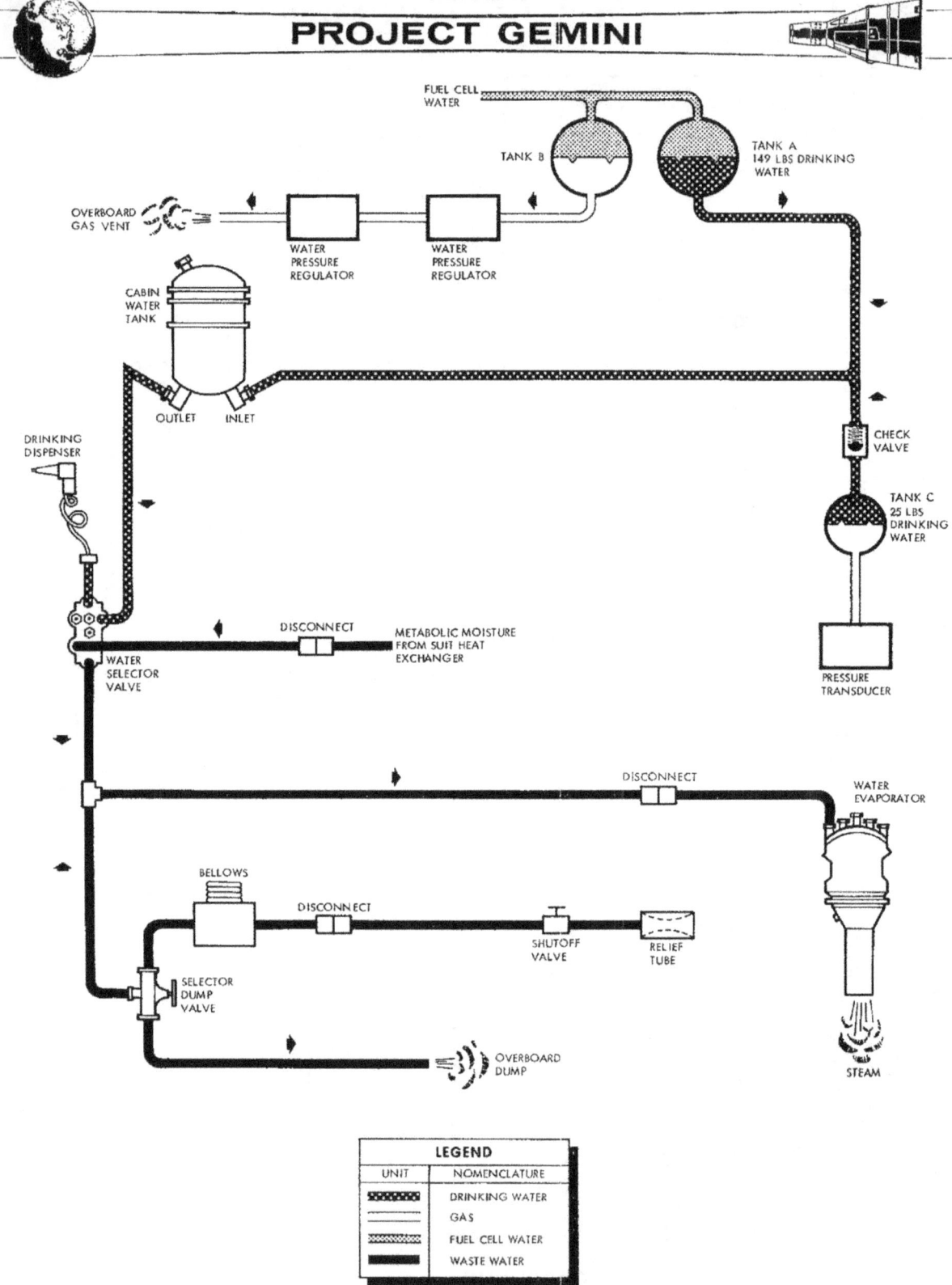

Figure 6-11 Water Management Schematic (Sheet 3 of 3)
(S/C 7 ONLY)

Components of the water management system, in addition to the water tanks, are a water control valve, dump valve, water evaporator, two water pressure regulators and a solenoid valve.

The urine disposal equipment is government furnished equipment (GFE). It includes the urine line, bellows assembly, quick disconnect coupling, and uriceptacle.

COOLANT SYSTEM

See SECTION VII

SYSTEM DISPLAYS AND CONTROLS

The displays and controls for the Environmental Control System are provided in the cabin and function as specified. (Figure 3-6)

SECONDARY OXYGEN SHUTOFF INSTALLATION

A manual secondary oxygen shutoff handle is provided for each member of the flight crew for complete and positive shutoff of each secondary oxygen container. The handles are located aft of the right and left switch/circuit-breaker panels. The position OPEN or CLOSED is noted.

OXYGEN HIGH RATE TELELIGHT/SWITCH

The following paragraph applies only to S/C 3 and 4.

Five minutes prior to retrograde initiation, an amber light in the O_2 HIGH RATE telelight/switch assembly will illuminate as a warning that high flow rate of oxygen should be initiated. After T_R-5, the light will illuminate green when

SEDR 300
PROJECT GEMINI

the high oxygen rate valve is opened either manually or automatically. Oxygen high rate will be available at any time during the mission by depressing the switch. However, the green light will not illuminate prior to $T_{R}-5$.

Spacecraft 7 O_2 HI RATE telelight does not illuminate until the high oxygen rate valve is opened, either manually or automatically; then an amber light illuminates. This telelight is located in the Annunciator Panel.

The O_2 HI RATE switch is connected to the high oxygen rate valve. This is the same switch that activates the cabin fan. It has three positions: CABIN FAN, O_2 HI RATE, and OFF. It is located in the upper right hand corner of the center panel.

CABIN AIR RECIRCULATION HANDLE

This handle controls the recirculation valve which permits entry of cabin air into the suit circuit for removal of odors and carbon dioxide. This procedure will renovate cabin air without cabin decompression and reduces the possibility of carbon dioxide pockets by increasing circulation of the cabin atmosphere.

INLET SNORKEL HANDLE

This handle controls the cabin air inlet valve which provides for ventilation during landing and postlanding phases of the mission.

CABIN VENT HANDLE

This handle controls the operation of the cabin outflow valve to permit emergency decompression in orbit and cabin ventilation during the landing phase.

WATER SEAL HANDLE

This handle provides for watertight closure of the cabin pressure relief valve during a water landing.

OXYGEN HIGH RATE RECOCK HANDLE

This handle provides for the manual return of the oxygen high rate valve to the closed position, thereby restoring normal oxygen flow rate. Actuation of this handle also re-establishes the capability of initiating high rate oxygen flow when necessary.

CABIN PRESSURE AND SUIT CARBON DIOXIDE PARTIAL PRESSURE INDICATOR

A dual indicator provides for monitoring cabin atmospheric pressure and the amount of carbon dioxide at the suit inlet. Cabin atmospheric pressure is calibrated in pounds per square inch. Carbon dioxide partial pressure is calibrated in millimeters of mercury.

SECONDARY OXYGEN PRESSURE INDICATOR

A dual indicator is provided for monitoring pressure in the individual gaseous oxygen containers in the secondary oxygen subsystem. The indicator range is from 0 to 6000 psia, divided into 500-pound increments and numbered at each 1000-pound interval. Readings must be multiplied by 100 to obtain correct values.

CRYOGENIC OXYGEN QUANTITY AND PRESSURE INDICATOR

This indicator provides for monitoring quantity and pressure of cryogenic oxygen in the primary oxygen container. The quantity scale displays from 0 to 100 per cent in 2 per cent increments, numbered at 20 per cent intervals. The

 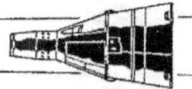

SEDR 300

PROJECT GEMINI

pressure scale ranges from 0 to 1000 psia in 20-pound increments, numbered at 200-pound intervals. Red undermarkings are incorporated on the oxygen meter to indicate the point at which thermal pressurization may be discontinued by de-energizing the heaters.

CRYOGENIC INDICATOR SWITCH

This switch provides for using the same indicator when monitoring the pressure and quantity of cryogen in any of the three cryogenic containers. The three containers are: the ECS primary oxygen supply, the RSS or FC oxygen supply, and the RSS or FC hydrogen supply. It is located below the indicator on the center panel, and only on S/C 7.

ECS O_2 HEATER SWITCH

This switch is connected to the heaters in the ECS primary oxygen container, and has three positions, AUTO, OFF, and ON. It is located below the Flight Plan Display on the center panel.

SUIT FAN SWITCH

This switch is connected to the suit fan power supplies, and has three positions, SUIT FAN NO. 1, OFF, and NO. 1 & 2. It is located in the upper left hand corner of the center panel.

WATER MANAGEMENT PANEL

A three knob panel is provided for managing, replenishing, and dumping waste water overboard.

OXYGEN CROSS-FEED SWITCH

This switch, when in the OPEN position, permits oxygen from the primary oxygen supply module for the ECS to be used in the RSS in the event of a failure in the RSS oxygen module or vice versa.

6-30

PROJECT GEMINI

MANUAL OXYGEN HIGH RATE HANDLE

This handle is located on the console between the members of the flight crew and provides for manual control of the dual high oxygen rate and suit system shutoff valve. Actuation of the handle shall initiate the oxygen high flow rate and de-energize the suit compressor. Resumption of normal system operation shall be effected by actuation of the oxygen high rate recock handle.

SUIT FLOW CONTROL LEVERS

An individual lever is provided for each member of the flight crew for regulation of circulatory oxygen flow through the suit circuits. The levers are located on the lower section of the pedestal and shall provide any selected flow valve setting from fully open to fully closed. A detent provides an intermediate position to prevent inadvertent shutoff of suit flow. This detent may be by-passed by moving the lever outboard.

CABIN REPRESSURIZATION CONTROL

A rotary handle control is provided for cabin repressurization after a decompression has occurred. The control rotates approximately 90° between fully OPEN (repressurize) and fully CLOSED (off) positions. This control is located on the center console between the suit flow control panels.

ECS HEATER TELELIGHT

This telelight, located on the annunciator panel of the center instrument panel, illuminates when the heater in the primary oxygen container has been manually activated.

SEDR 300

PROJECT GEMINI

SYSTEMS OPERATION

The environmental control system (Figure 6-1, 6-2) is semi-automatic in operation and provides positive control in all modes of operation. There are six operational modes:

1. Pre-Launch
2. Launch
3. Orbit
4. Re-Entry
5. Postlanding
6. Emergency

Prior to the pre-launch mode, it is necessary to service and to check the system functionally.

SERVICE AND CHECKOUT

For this operation, it is assumed that the spacecraft has been mated with booster on the launch pad and in the unserviced condition.

1. Fill primary, secondary, and egress oxygen storage tanks.
2. Fill water boiler.
3. Fill drinking water supply tank.
4. Replace cartridge in the suit loop cannister.

PRE-LAUNCH

The pre-launch phase is defined as the period after the servicing has been completed and prior to launch.

SEDR 300

PROJECT GEMINI

Suit Loop

The pilots in their suits, with faceplates open, are connected to the suit circuit. The suit circuit compressor is actuated and the suit temperature control valve is adjusted to satisfy the pilot desiring the cooler temperature. The other pilot becomes comfortable by adjusting his suit flow rate control valve toward the closed position to obtain a warmer setting. A ground supply of pure oxygen is connected to the pressure suit circuit purge fitting. Flow is initiated with the faceplates closed. The suit circuit gas is sampled periodically until an acceptable oxygen content is attained. A suit circuit leakage test is conducted. After satisfactorily completing the suit circuit leakage test, the primary and secondary oxygen manual shutoff valves are opened and the suit circuit purge system is disconnected and removed.

Cabin Loop

The cabin hatches are closed. A ground supply of pure oxygen is connected to the cabin purge fitting, flow is initiated and the cabin is purged. The cabin fan is actuated and the recirculation valve is opened. A cabin leakage test is conducted. After satisfactorily completing the cabin purge and leakage test, the cabin purge system is disconnected and removed and the cabin purge fitting is capped.

Oxygen Loop (Figure 6-3, 6-4)

The primary and secondary oxygen manual shutoff valves are opened.

The liquid oxygen inside the primary supercritical container has been changing from a liquid to a supercritical fluid by thermal leakage and heater activation.

A pressure control switch provides for automatic or manual activation of these heaters. The manual control switch is located on the center control panel.

An indicator also on the center control panel indicates both pressure and quantity from a transducer and control unit that are attached to the container.

The oxygen gas flows from the container and is warmed to approximately 50°F in a heat exchanger. This heat exchanger also contains a relief valve that limits maximum pressure to 1000 psig. This valve opens, permitting full flow and reseats within the range of 945-1000 psig.

A discharge temperature sensor provides an indication, for telemetering only, of the temperature in the primary oxygen line downstream of the heat exchanger.

The oxygen gas is regulated from 1000 psia maximum to 110 ± 10 psig. Flow capacity is 0.35 lb/min with an inlet pressure from 800 to 1000 psia and an inlet temperature of 60°F. This regulator also contains a relief feature that limits downstream pressure to 215 psig in the event of a failed-open condition.

A 10-micron absolute rated filter provides filtration of the primary oxygen supply before it enters the suit or cabin loop.

LAUNCH

Cabin Loop

The cabin pressure relief valve opens to limit the pressure differential between cabin and ambient to $5.5 ^{+.5}_{-.0}$ psi.

SEDR 300

PROJECT GEMINI

<u>Suit Loop</u> (Figure 6-8)

Oxygen is supplied to the suit loop through the suit pressure regulator. The suit pressure is controlled to between 2 and 9 inches of water above cabin pressure by the suit pressure regulator.

Suit circuit oxygen from the suit circuit demand regulator recirculates through the suit compressor, the carbon dioxide and odor absorber, the suit heat exchanger and water separator, the pressure suits, and the suit circuit solids traps. There are two compressors in the circuit. One is an alternate to be used if a compressor failure occurs. The alternate compressor is activated by positioning the SUIT FAN switch on the center panel. The cartridge of lithium hydroxide and activated charcoal remove carbon dioxide and odors of an organic nature that could have any ill effects on the pilots. As suit circuit oxygen flows through the suit heat exchanger, the temperature is controlled as selected by the pilots.

Solids traps, located in the oxygen outlet ducts of both pilots' suits, remove particulate solids, preventing contamination of the suit circuit system. An integral by-pass opens if the traps become choked with collected solids permitting continuous oxygen flow through the suit circuit.

ORBIT

<u>Cabin Loop</u>

Normal cabin leakage allows the cabin pressure to decay to a nominal value of 5.1 psia. The cabin pressure control valve maintains this value automatically.

A dual cabin pressure regulator supplies makeup oxygen through the pilots pressure suits to the cabin on demand, as sensed by two aneroid elements within the regulator. The regulator supplies the makeup oxygen at a controlled pressure between 5.0 to 5.3 psia.

The cabin fan circulates cabin air through the cabin heat exchanger. One or both of the pilots may open their faceplates. The cabin air circulating valve is in the open position to provide for recirculation of the cabin oxygen through the suit circuit.

In the event of spacecraft depressurization, whether intentionally or by spacecraft puncture, the dual cabin pressure regulator closes when cabin pressure decreases to $4.1 ^{+0.2}_{-0.1}$ psia, preventing excessive loss of oxygen.

Suit Loop (Figure 6-8, 6-9)

The suit circuit demand regulator senses cabin pressure and maintains suit circuit pressure at 2.5 to 3.5 inches of water below to 2 to 9 inches of water above cabin pressure. Should cabin pressure decrease below 3.5 psia, the suit circuit demand regulators maintain the suit circuit pressure at $3.5 ^{+.4}_{-.0}$ psia by constant bleed orifices and sensing aneroids within the regulator. When cabin pressure is restored to $5.1 ^{+0.2}_{-0.1}$ psia, the suit circuit demand regulators return to normal operation.

In the event of cabin and suit circuit malfunction, the suit circuit will automatically revert to the high rate of operation when suit circuit pressure decreases below $3.0 ^{+0.1}_{-0.0}$ psia. A suit circuit pressure sensing switch energizes

the solenoid of the dual high flow rate and system shutoff valve. This initiates a high oxygen flow rate of 0.08 ± 0.008 lb/min per man (total flow: 0.16 lb/min). This high flow rate flows directly into the suits by-passing the suit demand regulators. The suit recirculating system is shut off and the suit compressors are de-activated when the solenoid of the dual high flow rate and system shutoff valve has been energized. The O_2 HI RATE light on the center panel illuminates when the suit circuit is on the high flow rate. There is also a manual control for the high flow rate and system shutoff valve located on the center console.

When the suit circuit pressure is restored to a level above $3.0 \, ^{+0.1}_{-0.0}$ psia, the high rate and system shutoff valve is reset manually by using the control marked O_2 HIGH RATE RECOCK located on the center panel. This returns the suit circuit to normal operation by opening the system shutoff valve and closing the high rate valve. The suit compressor is also reactivated.

Water Management System (Figure 6-11)

The drinking water system is pressurized and manually controlled by the pilots.

Water from the adapter supply is used to replenish the cabin tank water supply.

The water tank drink selector valve is set in the NORM position.

The pilots manually operate the drinking dispenser to provide drinking water from the cabin storage tank.

The water separator removes metabolic moisture through a wicking material positioned between the plates of the suit heat exchanger.

 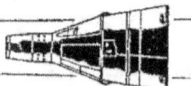

PROJECT GEMINI

The dump selector valve is positioned to route the urine either to the water boiler or dumped overboard. The normal procedure is to dump. Before it is dumped the urine dump system is preheated by positioning its heater switch located on the water management panel. A urine dump heater light is also provided and located on the water management panel. This light illuminates when the heater is activated. The shutoff valve is opened and the bellows operated to remove urine from the system.

RE-ENTRY

Oxygen System

The primary oxygen system is disconnected when the adapter section is separated from the re-entry module. This removes the primary oxygen supply pressure which automatically activates the secondary oxygen supply.

The system shutoff and high rate valve is positioned to the high rate position before the adapter section is jettisoned.

Cabin Loop

The pressure in the suit and cabin remains constant at 5 psia (nominal) until an altitude of approximately 27,000 feet is reached.

As ambient pressure increases during descent, the cabin pressure relief valve admits ambient air into the cabin, preventing high differential pressures. The cabin pressure relief valve begins to open when the ambient pressure is 15.0 inches of water greater than cabin pressure and opens to maximum flow when the pressure differential is 20 inches of water.

SEDR 300
PROJECT GEMINI

At an altitude of 25,600 feet, or below, the pilots manually open the cabin inflow and outflow valves to circulate external air through the cabin and suit circuit.

Maximum negative pressure on the cabin should not exceed 2 psi as controlled by the cabin relief valve.

Suit Loop

Prior to re-entry the faceplates should be closed. The high flow rate of oxygen is flowing directly into the suit circuit.

When the cabin inflow valve is opened it activates the suit compressor and external air is circulated through the suit circuit.

POSTLANDING

Ventilation is provided by the suit compressor as long as electrical power is available (12 hours minimum).

Ambient air is drawn into the vehicle through the snorkel inflow valve, by the suit compressor, circulated through the suit circuit into the cabin, then discharged overboard through the outflow vent valve.

The snorkel inlet valve functions as a water check valve. When the snorkel inlet valve is above water level, the ball check is retained freely in a wire mesh cage, permitting ambient air to enter the suit circuit. Normal oscillations of the spacecraft in the sea may result in the snorkel valve being momentarily submerged. This will cause the ball check to seat and is held there by suction from the suit compressor. Opening the cabin air circulating valve allows the ball to drop from its seat.

To prevent water from entering the cabin through the cabin pressure relief valve, the manual shutoff section of the valve is closed.

EMERGENCY

Cabin Loop

If cabin depressurization becomes necessary due to toxic contaminants or fire, the cabin outflow valve is opened to depressurize the cabin. The cabin regulator will close, stopping the oxygen supply to the cabin, permitting the escape of toxic contaminants and preventing oxygen assistance to combustion in the event of fire. The cabin repressurization valve permits repressurization of the spacecraft cabin.

The control knob for the cabin repressurization valve is located on the lower console and is rotated counterclockwise to open the valve. It is rotated clockwise to close the valve when cabin pressure is between 4.3 and 5.3 psia. Cabin pressure is then automatically controlled at $5.1 ^{+0.2}_{-0.1}$ psia by cabin pressure regulator valve.

Egress Oxygen (Figure 6-5)

Operation of the egress oxygen system is initiated by three of the four lanyards which are pulled when the seat leaves the spacecraft. One lanyard pulls a pin in the composite disconnect allowing it to separate and close the normal suit circuit. Two of the remaining lanyards open the container shutoff valve and circuit relief valve activating the egress oxygen system.

Each of the egress oxygen containers is pressurized to 1800 psig with gaseous oxygen. The oxygen flows from the containers through a pressure regulator, where the pressure is reduced to 40 psia. It then flows through a shutoff valve and a flow restrictor, which allows a flow of .052 to .063 lb/min, then through a check valve to the suit. After leaving the suit, the oxygen flows through the shutoff and relief valve, which dumps the oxygen overboard, as well as controls the suit pressure to $3.5 ^{+.6}_{-0}$ psia if ejection occurs at an altitude above 31,500 feet, and 2 to 8.23 inches of water above ambient at an altitude below 31,500 feet.

SYSTEM UNITS

DUAL HIGH OXYGEN RATE AND SUIT SYSTEM SHUTOFF VALVE (Figure 6-12)

This unit is a combination of a 2-inch-nominal-diameter, spring-loaded-closed, manually opened, spoon-type-butterfly shutoff valve and a spring-loaded-closed poppet valve for high oxygen rate flow. The unit consists basically of: (1) a spoon-type butterfly close element assembled within an aluminum in line housing; (2) a cable-operated valve reset mechanism; (3) a 24-vdc solenoid with a manual override to act as a holding mechanism to retain the butterfly in the open position during normal system operation; and (4) a spring-loaded-closed poppet valve assembly to control high oxygen rate flow.

In normal operation, the shutoff valve is opened by linear motion of the reset cable. This rotates the butterfly arm to an engagement position with the solenoid latch assembly. As long as the solenoid is in the de-energized position, the butterfly remains latched in the open position. When the shutoff valve is in the open position, the high oxygen rate valve is always closed. The main drive, torsion spring, loads the butterfly toward the closed position. Closure of the shutoff valve and initiation of the high oxygen rate flow is accomplished when the

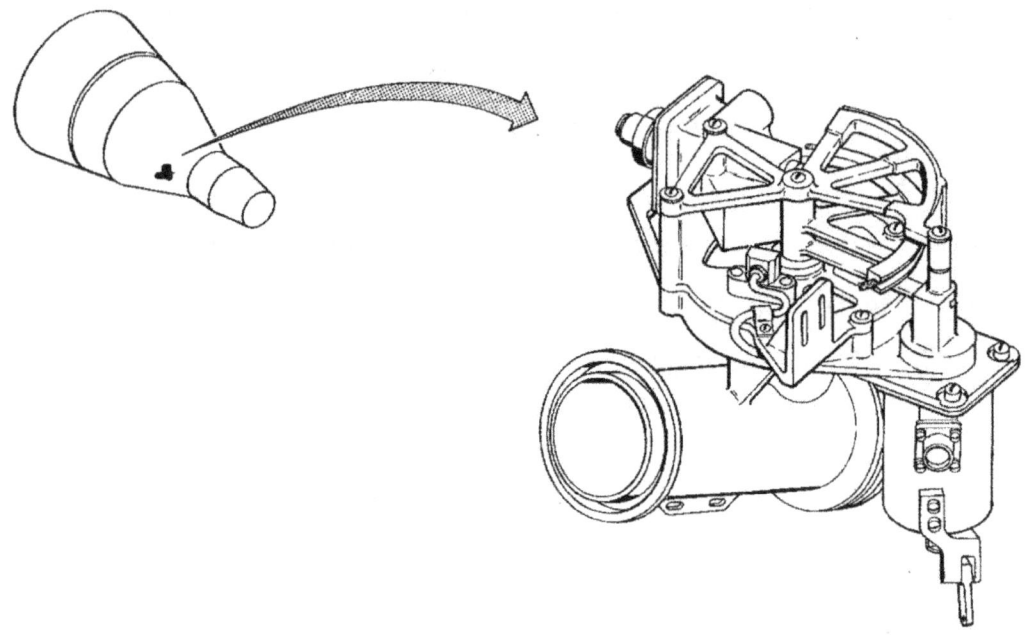
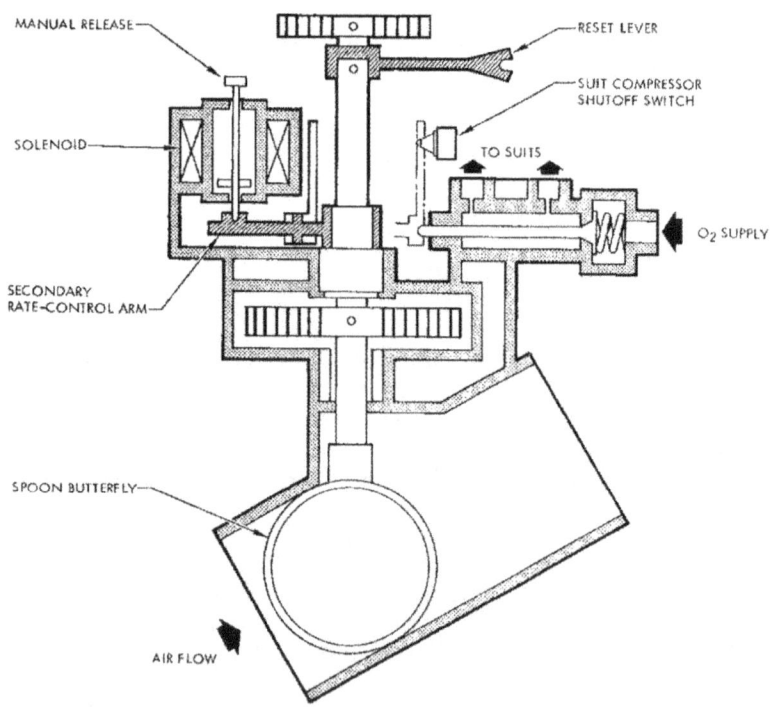

Figure 6-12 Dual High Oxygen Rate and Shutoff Valve

solenoid is disengaged from the butterfly arm. This is accomplished by either of the following actions:

1. Electrical signal from the control panel or the absolute pressure switch that senses suit pressure.

2. Manually disengaging the solenoid by pulling the manual control knob.

As the butterfly closes, the butterfly engages the solenoid cutoff switch, removing power from the solenoid, turn off the suit compressor and cabin fan, and illuminates the O_2 HI RATE lamp. At the same time the butterfly arm engages the high oxygen rate valve poppet, lifting it from its seat against the poppet spring force. Opening the poppet valves allows oxygen to flow to each pilot's suit through fixed orifices at a rate of 0.08 $\pm$0.008 lb/min per man (total flow 0.16 lb/min.)

SUIT OXYGEN DEMAND REGULATOR (Figure 6-13)

The suit oxygen demand regulator controls the oxygen to the suit circuit from the primary or secondary oxygen system and replenishes oxygen used by the pilots or lost by leakage.

Cabin pressure is sensed on one side of the diaphragm and suit pressure is sensed on the opposite side of the diaphragm. The differential pressure across this diaphragm opens or closes a poppet valve admitting or stopping oxygen flow into the suit circuit. With cabin pressure of 5.0 psia, the suit regulator maintains suit pressure at 2.5 to 3.5 inches of water below cabin pressure.

A resilient diaphragm type valve relieves pressure in the suit during ascent and limits excess pressure to between 2.0 and 9.0 inches of water above cabin

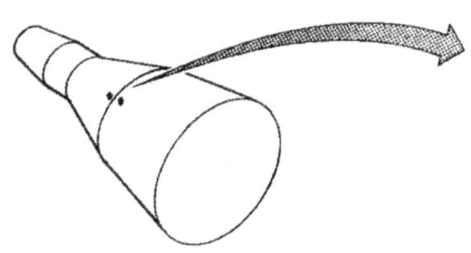
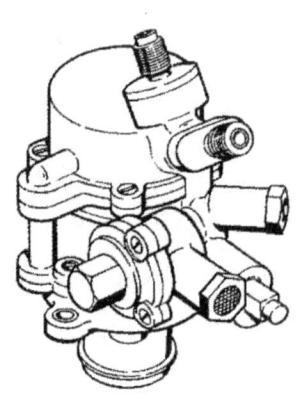
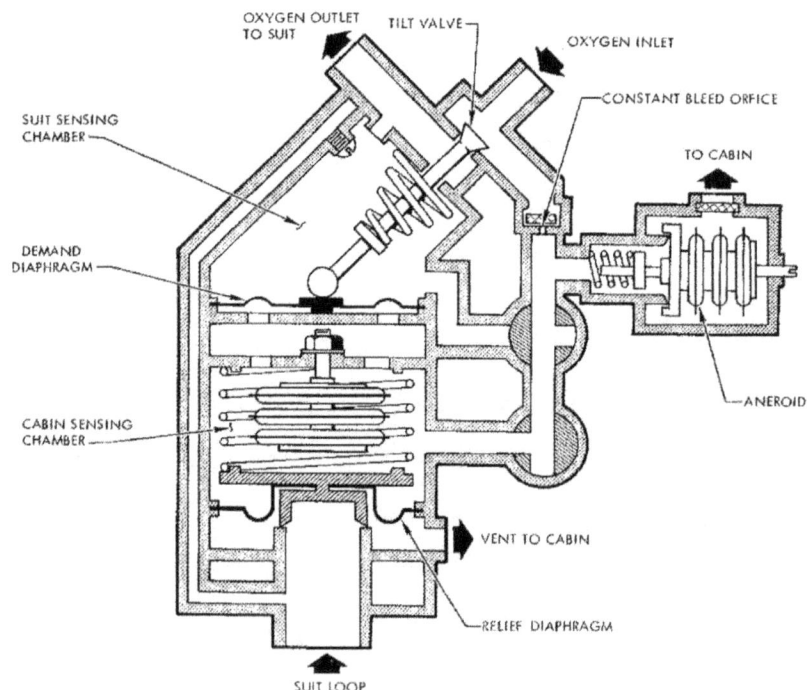

Figure 6-13 Suit Oxygen Demand Regulator

pressure. During descent, the suit demand regulator relieves the secondary oxygen rate flow through the relief portion of the valve, maintaining suit pressure 2 to 9 inches of water above cabin pressure.

A constant bleed and aneroid elements maintain the suit pressure at $3.5 ^{+0.4}_{-0}$ psia if cabin pressure decreases below this pressure. The bleed flow by-passes the tilt valve through a bleed orifice and is directed to the cabin pressure sensing side of the pressure demand diaphragm. A metering valve, controlled by an aneroid, regulates the reference pressure on the demand diaphragm. The regulator returns to normal operation when cabin pressure returns to $5.1 ^{+0.2}_{-0.1}$ psia. In the event that cabin decompression and a ruptured relief diaphragm in the regulator occur simultaneously, an aneroid over the relief diaphragm extends to control suit pressure at 3.9 psia maximum.

CABIN PRESSURE RELIEF VALVE (Figure 6-14)

The cabin pressure relief valve automatically controls the cabin-to-ambient differential pressure during launch, orbit and re-entry. Duplicate spring loaded poppet valves are controlled by servo elements within the valve.

The servo elements control spring loaded metering valves which determine the pressure within the diaphragm chamber behind the poppet, controlling the poppet position. A small inlet bleed orifice admits cabin pressure to the diaphragm chamber. When the poppet opens, a large orifice permits rapid change in pressure ensuring quick closure of the poppet.

During ascent the valve will relieve cabin pressure as ambient pressure decreases until cabin differential pressure is 5.5 to 6.0 psia. The valve closes main-

 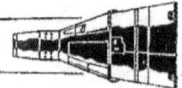

PROJECT GEMINI

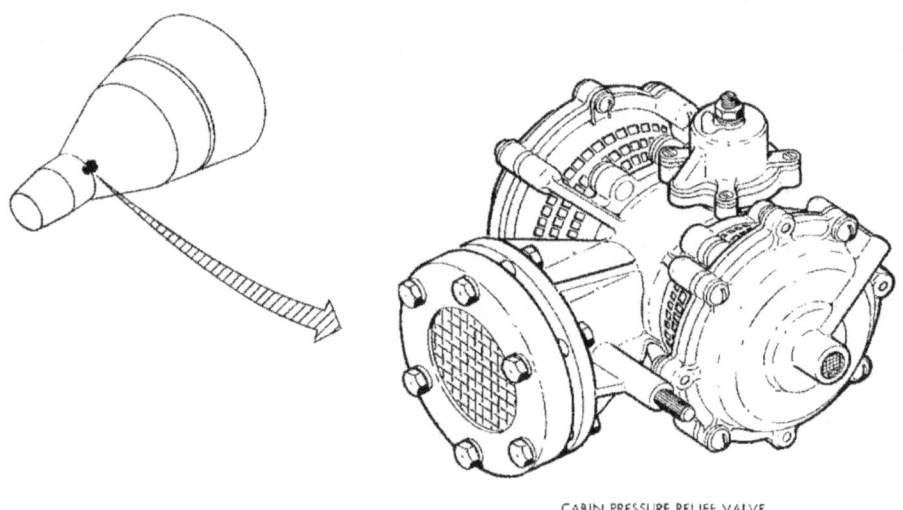

CABIN PRESSURE RELIEF VALVE

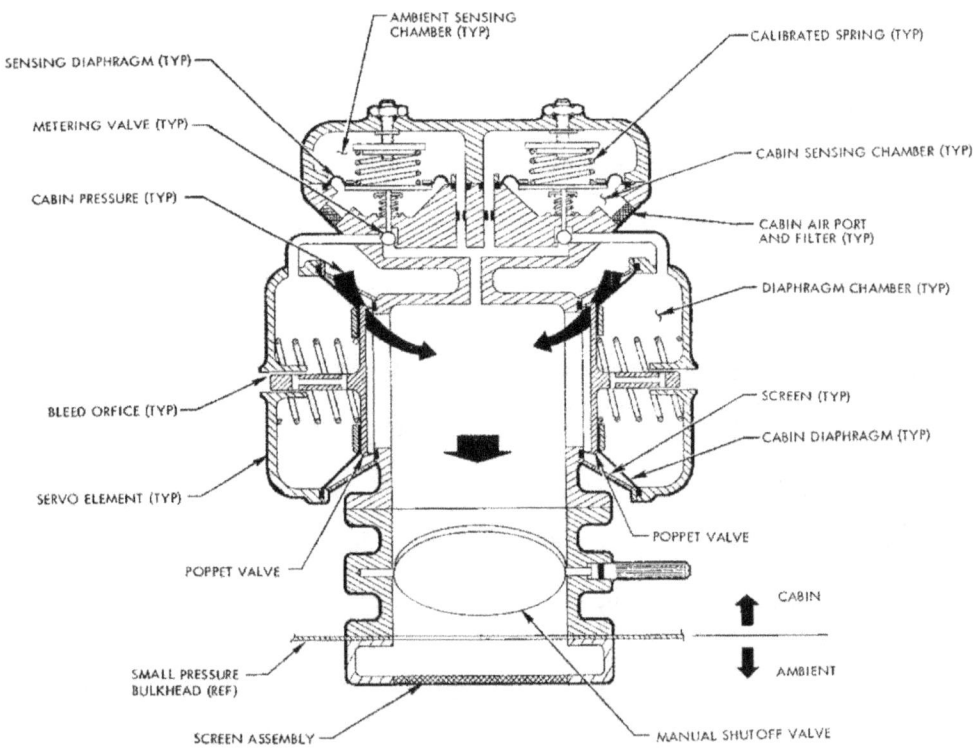

Figure 6-14 Cabin Pressure Relief Valve

taining differential pressure in this range. When cabin pressure decreases below 5.5 psia, the servo element closes the metering valve maintaining reference pressure within the diaphragm chamber at cabin pressure. The poppet is held closed by spring force and the zero differential between the diaphragm and the cabin prevents cabin pressure release. If cabin differential pressure exceeds 5.5 psia, the zero element retracts, opening the metering valves, allowing the diaphragm chamber to discharge to ambient. The discharge port, being larger than the inlet bleed orifice, permits the diaphragm chamber to approach external pressure. The cabin pressure reacting on the diaphragm overrides the poppet spring force, which opens permitting cabin pressure relief to ambient. During descent, as external pressure increases, ambient air is admitted to the cabin by the valve to reduce the differential pressure. As external pressure increases above the cabin pressure, the metering valves are held on their seats, preventing external pressure from entering the diaphragm chamber and retaining cabin pressure in the chamber. The poppet valve senses diaphragm chamber pressure versus ambient pressure. When the ambient pressure is 15 inches of water greater than cabin pressure, the poppet begins to open permitting ambient air to enter the cabin. The poppet opens fully when the differential pressure is 20 inches of water.

To preclude water entering the cabin during postlanding, a manual shutoff valve is provided.

SEDR 300

 **PROJECT GEMINI**

SUIT CIRCUIT COMPRESSOR (Figure 6-15)

Two electric motor driven, single stage compressors are incorporated in the suit circuit. One compressor is utilized for circulation of the gases within the suit circuit, supplying both suits. The other compressor remains in redundancy and is activated only by manual selection by the pilots. Either compressor can be manually selected by a switch on the center display panel, and both compressors can be selected simultaneously.

When secondary oxygen flow rate is selected, the compressor is automatically de-energized. Re-entry is made using the secondary rate. At an altitude of 25,600 feet or below, the manual inflow valve is opened which re-energizes the compressor. The suit compressor provides ventilation during landing and for a twelve hour postlanding period, or until the batteries fail.

SOLIDS TRAP (Figure 6-16)

A solids trap is located in the oxygen outlet duct of each suit. A cylindrical 40 micron filter strains the gaseous flow in the suit circuit removing the solid matter. In the event that the trap becomes choked with collected solids, an integral by-pass opens when the differential pressure across the screen exceeds 0.50 inches of water.

DUAL CABIN PRESSURE REGULATOR (Figure 6-17)

The cabin pressure regulator maintains cabin pressurization by providing makeup oxygen to the cabin on demand. The regulator contains two aneroid elements which individually sense cabin pressure. When cabin pressure decreases, the aneroids expand, forcing metering pins open and permitting oxygen flow into

SEDR 300
PROJECT GEMINI

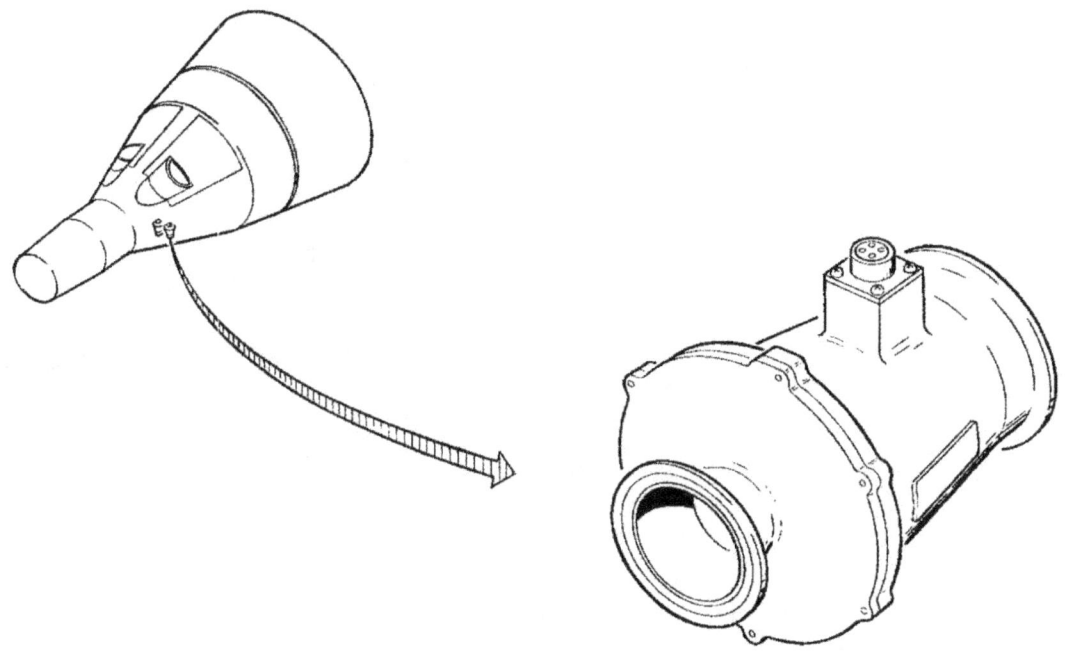

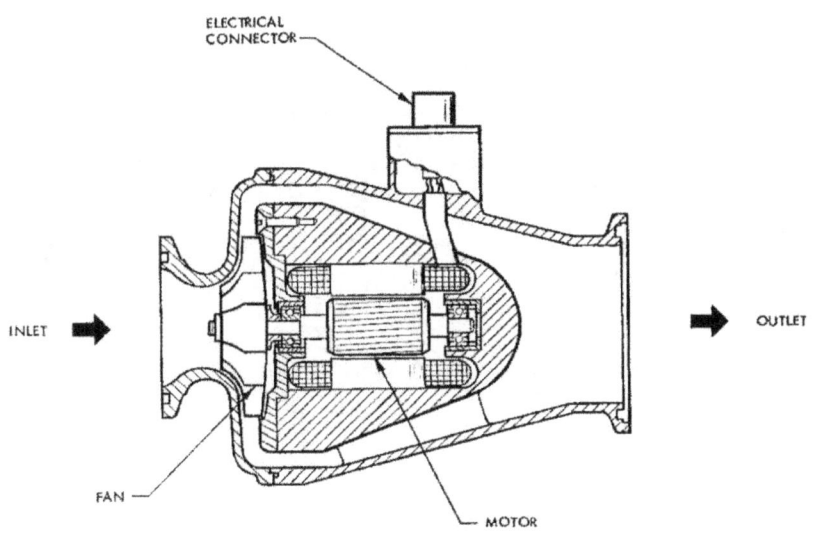

Figure 6-15 Suit Circuit Compressor

SEDR 300
PROJECT GEMINI

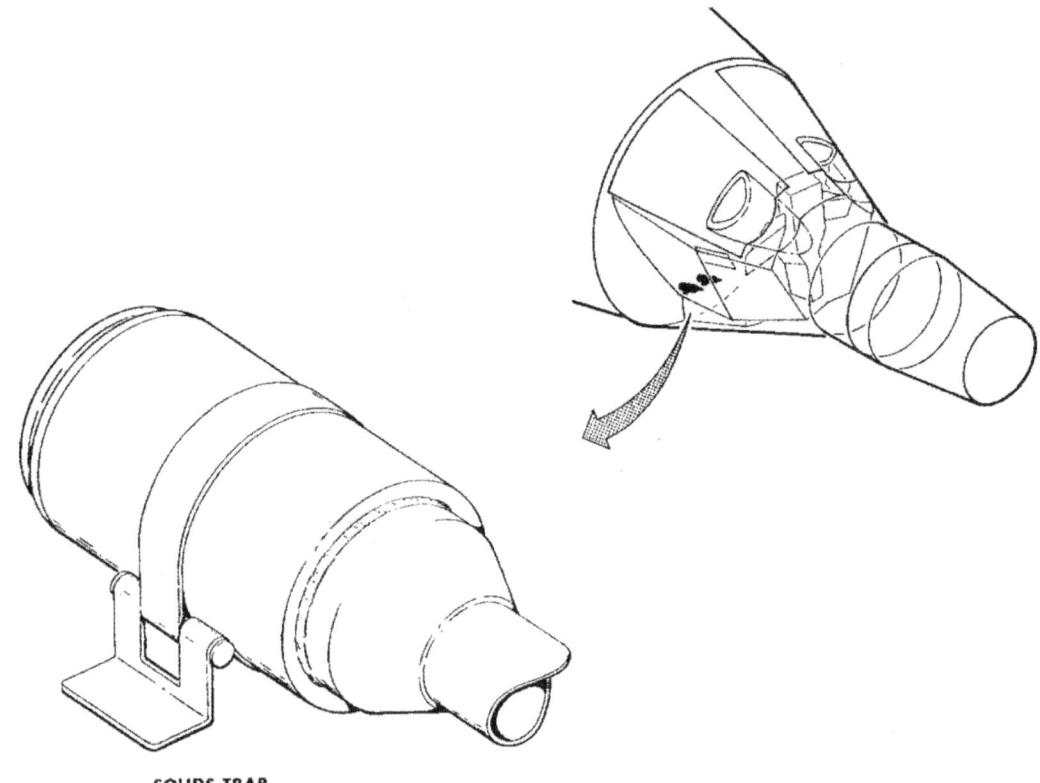

SOLIDS TRAP

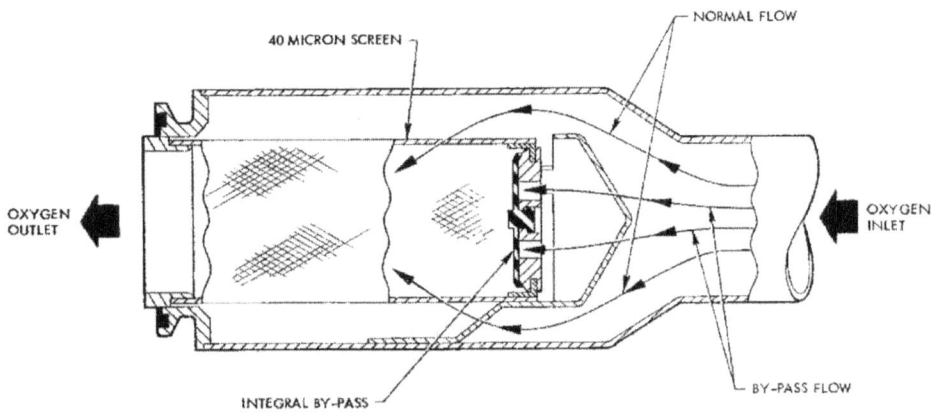

Figure 6-16 Suit Circuit Solids Trap

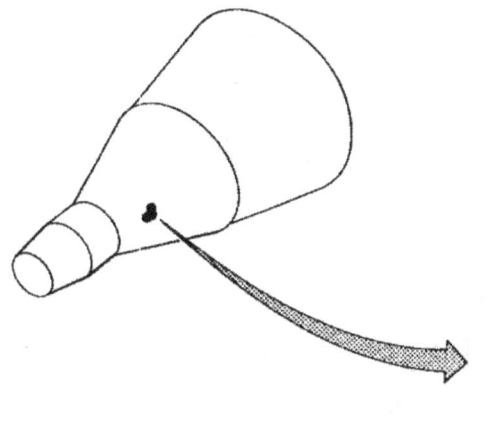
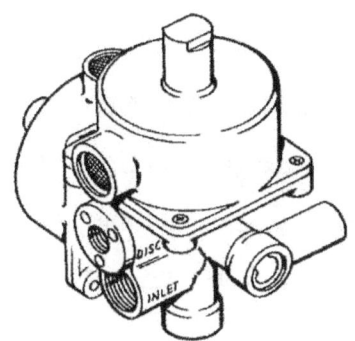
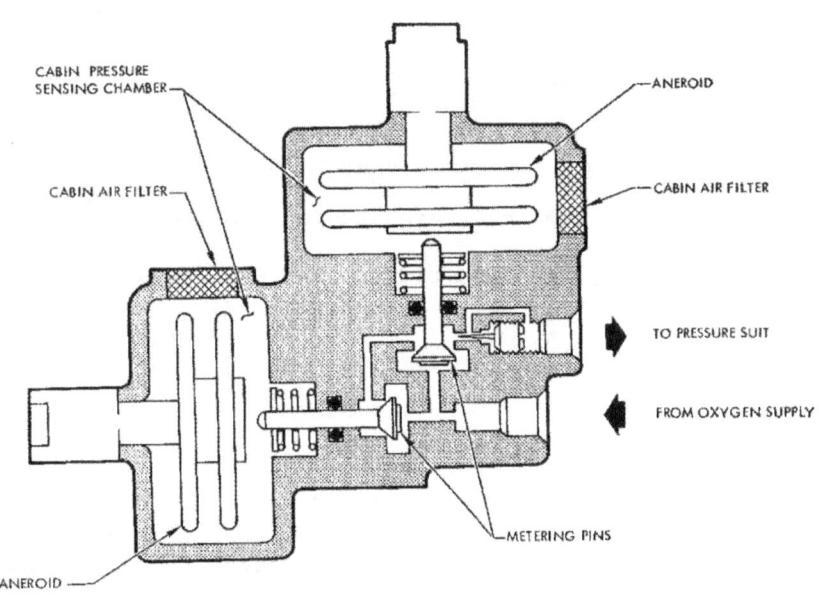

Figure 6-17 Dual Cabin Pressure Regulator

the cabin, maintaining cabin pressure at $5.1 \, ^{+0.2}_{-0.1}$ psia. If the cabin is punctured or develops leakage greater than the flow capacity of the valve $(4.79 \pm 0.48) \, 10^{-3}$ lb/min, oxygen flow to the cabin is stopped when the cabin pressure decreases to $4.1 \, ^{+0.2}_{-0.1}$ psia, by the aneroids expanding enough to cause the metering pins to close off the oxygen.

PRIMARY SUPERCRITICAL OXYGEN CONTAINER (Figure 6-18)

The primary oxygen container is a double walled tank. A dual concentric cylinder, quantity measuring devices, heaters and heat transfer spheres are internal to the container. The tank contains two heaters. The first is a 12.0 ± 2 watt heater which is activated either manually by a switch located on the center panel, or automatically by a pressure switch. The pressure switch controls the activation of the heating element in the tank to automatically maintain the cryogens in a supercritical state. The switch de-energizes the heater circuit when the pressure in the tank is between 875 to 910 psig, and closes the circuit 15 to 75 psig below the opening pressure. The second heater is a $325 \, ^{+50}_{-0}$ watt heater manually controlled by a switch located on the overhead switch/circuit breaker panel.

The pressure relief valve maintains the oxygen pressure within the container at $1000 \, ^{+0}_{-55}$ psig, and prevents overpressurization of the containers.

Provisions for servicing the primary oxygen container from a ground supply source of oxygen are provided.

SECONDARY OXYGEN CONTAINER (Figure 6-19)

The secondary oxygen container is a cylindrical shaped container, having a useful oxygen capacity of 6.5 pounds at an operating pressure of 5000 psig.

 **PROJECT GEMINI**

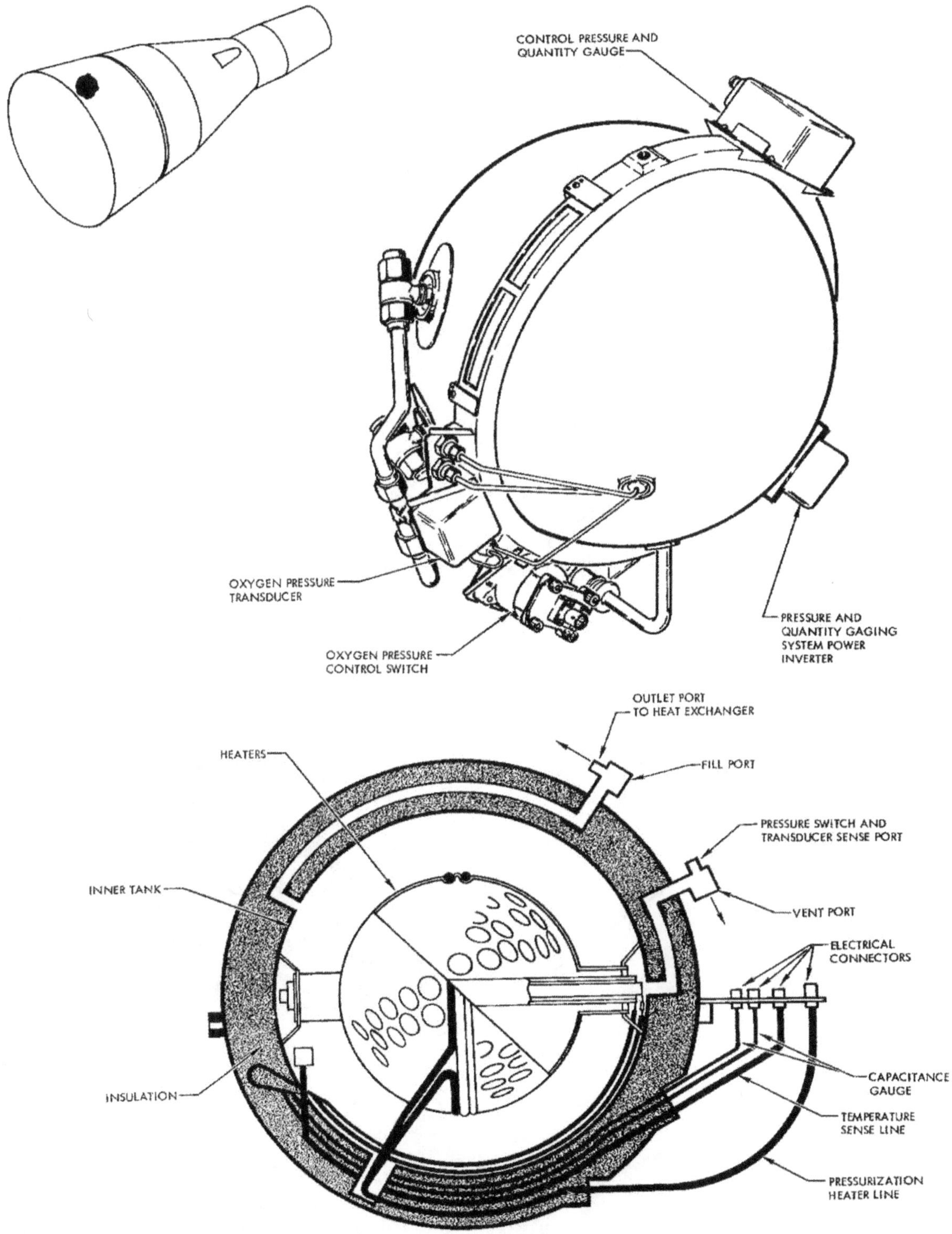

Figure 6-18 Supercritical Primary Oxygen Container

6-53

 **SEDR 300**

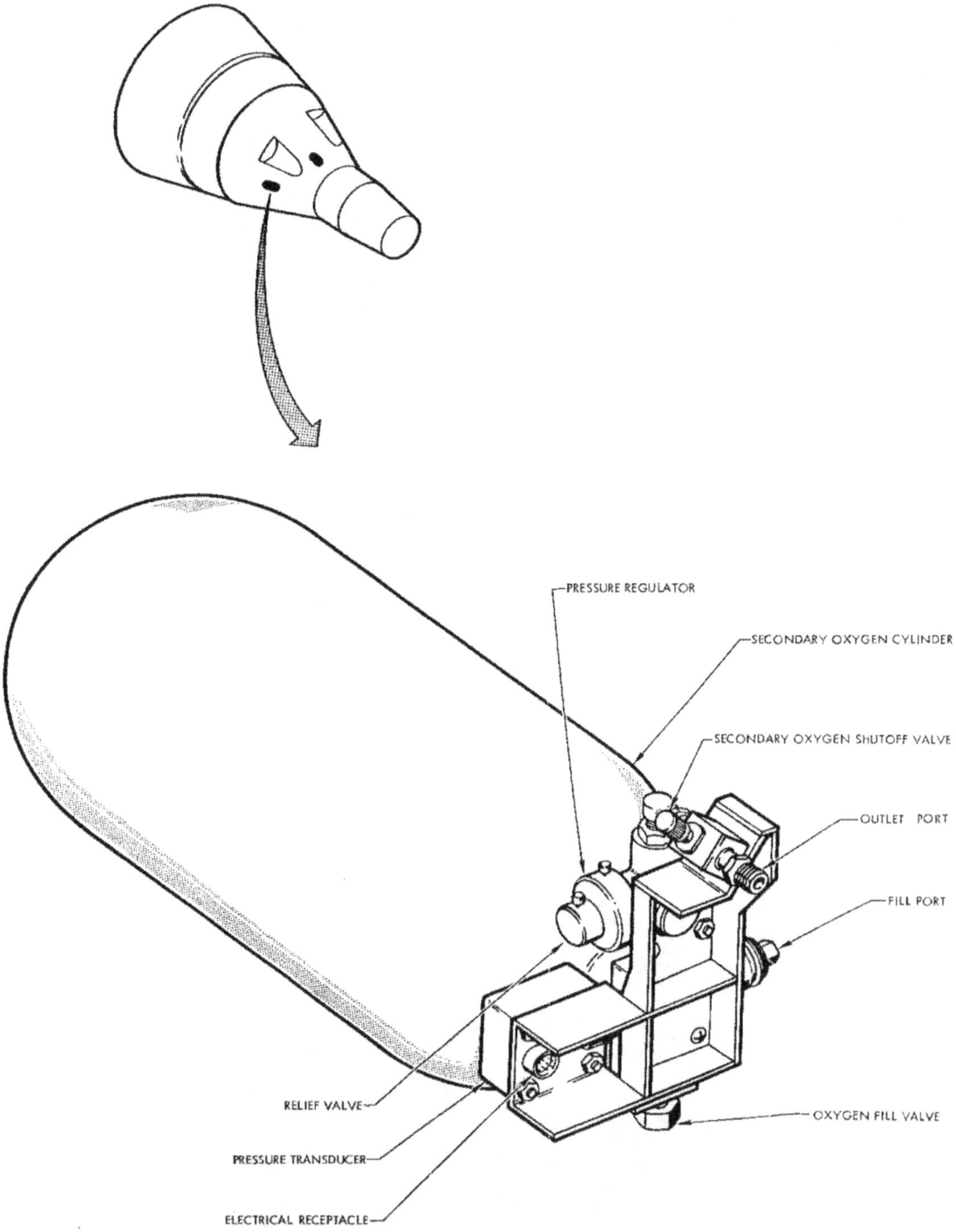

PROJECT GEMINI

Figure 6-19 Secondary Oxygen Tank

COOLING SYSTEM

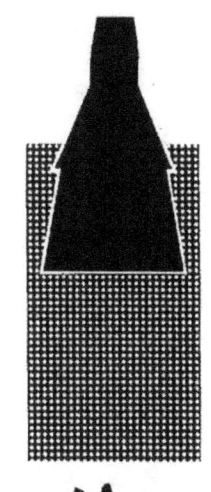

Section VII

TABLE OF CONTENTS

TITLE	PAGE
SYSTEM DESCRIPTION	7-3
SYSTEM DISPLAYS AND CONTROLS	7-4
SYSTEM OPERATION	7-5
SYSTEM UNITS	7-10

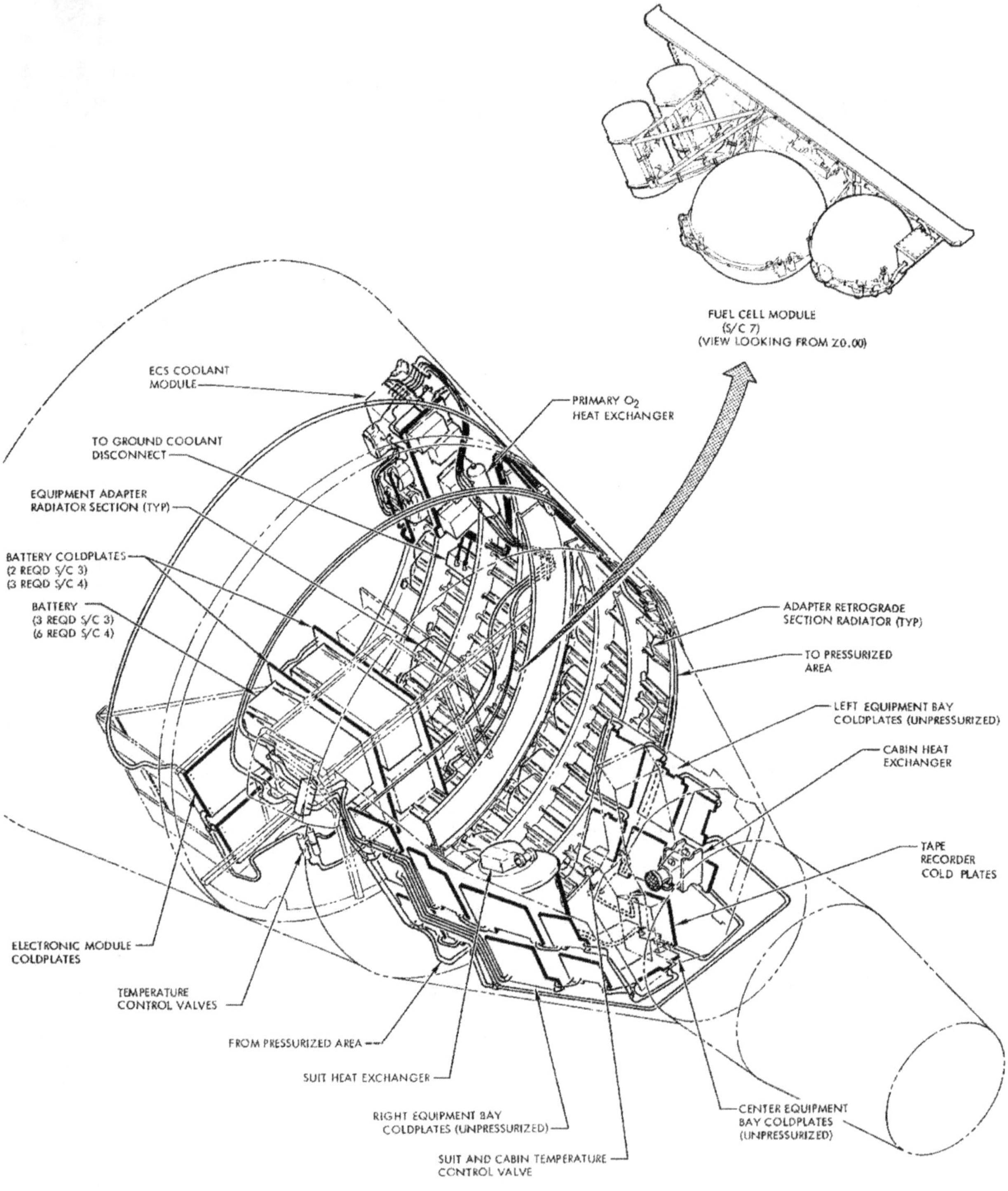

PROJECT GEMINI

SEDR 300

Figure 7-1 Spacecraft Coolant System

SECTION VII COOLING SYSTEM

SYSTEM DESCRIPTION (Figure 7-1)

The spacecraft cooling system consists basically of two identical temperature control circuits functioning independently of each other to provide the cooling requirements for the spacecraft. Each cooling circuit consists of a pump package, thermostatic and directional control valves, various type heat exchangers, radiators, filters, and the necessary plumbing required to provide a closed circuit. The cooling system may be operated in either the primary and/or secondary circuit, and is capable of carrying maximum heat loads in either circuit.

The equipment coldplates, cabin and suit heat exchangers are located in the re-entry module. The upper radiator panels are located in the retrograde section. The pump package, battery coldplates, filters, electronic equipment coldplates, ground launch cooling and regenerative heat exchangers and the lower radiator panels are located in the adapter equipment section. System manual controls are located on the pilots' pedestal console and the control switches, warning lights, and indicators are located on the center panel.

S/C 7 is provided with a fuel cell by-pass valve for ground operation of the cooling system. The valve design is such that fuel cell coolant can bleed but not flow. For fuel cell operation and prior to flight, the by-pass valve is placed in the normal position.

During orbital flight, Monsanto MCS-198 coolant is supplied throughout the cooling system and thermostatic control valves regulate the coolant temperature. Temperature sensors, located in the system, provide the necessary telemetering of system temperatures to ground stations.

SYSTEM DISPLAYS AND CONTROLS

The displays and controls for the coolant system are provided in the cabin and function as specified (Figure 3-6).

SUIT AND CABIN TEMPERATURE CONTROLS

Dual concentric knobs are mounted between the ejection seats for suit and cabin temperature control. These knobs control the operation of valves regulating the flow rate of primary and secondary coolant through the suit and cabin heat exchangers. Clockwise rotation results in increased temperatures.

CABIN AND SUIT TEMPERATURE INDICATOR

A dual indicator provides for monitoring temperatures in the suit and cabin circuits. Range markings are calibrated in degrees Fahrenheit.

PRIMARY AND SECONDARY PUMP SWITCHES

These switches are connected to the coolant pumps power supplies. One switch is provided for each power supply. Each switch has two positions: ON and OFF. These switches are located on the center panel. S/C 7 B pump switch in each loop changes the flow rate from 183 lb/hr to 140 lb/hr.

PRIMARY AND SECONDARY PUMP LIGHTS

Pump lights illuminate when the pumps are activated. They are located above their switches on the center panel. The RES LO lights illuminate when the coolant level in the reservoir is low.

EVAP PRESS INDICATOR

This light illuminates when pressure in the evaporator builds up to $4.0 \, ^{+0.0}_{-0.3}$ psig and extinguishes when the pressure falls within the range of $3.1 ^{+0.3}_{-0.0}$ psig.

EVAPORATOR PRESSURE HEATER SWITCH

This switch is connected to the evaporator heater and is used to heat the excessive water in the evaporator before dumping.

SYSTEM OPERATION

The cooling circuit, in which the cooling system operates, is dependent upon the temperature loads generated by the equipment, spacecraft phase of flight and the temperature within the spacecraft cabin. Cooling is provided throughout the mission up to pre-retrograde firing. At this time, the coolant pump packages are jettisoned with the adapter equipment section, terminating spacecraft cooling.

The primary circuit operates continuously providing the required cooling during low temperature loads. The secondary circuit is used, in conjunction with the primary circuit, during phases of high temperature loads; namely - launch, rendezvous, and pre-retrograde. Under normal heat loads, the A pump in the primary circuit provides the required cooling. Under peak heat loads, the A pump in the secondary circuit is used with the primary circuit A pump to provide maximum cooling. In the event of an A pump malfunction in either circuit, the B pump in that circuit is used. In the event of both pumps failing in one circuit, both pumps of the remaining circuit can be used to provide the required cooling.

The coolant pump inverters for pumps A and B are the same in S/C 3 and 4, and for the A pumps in S/C 7, which are capable of flowing 183 lb/hrs of coolant through each pump. The B pumps inverters for S/C 7 change the flow rate to 140 lb/hr.

PRE-LAUNCH (Figure 7-2)

During pre-launch an external supply of Monsanto MCS-198 coolant is circulated

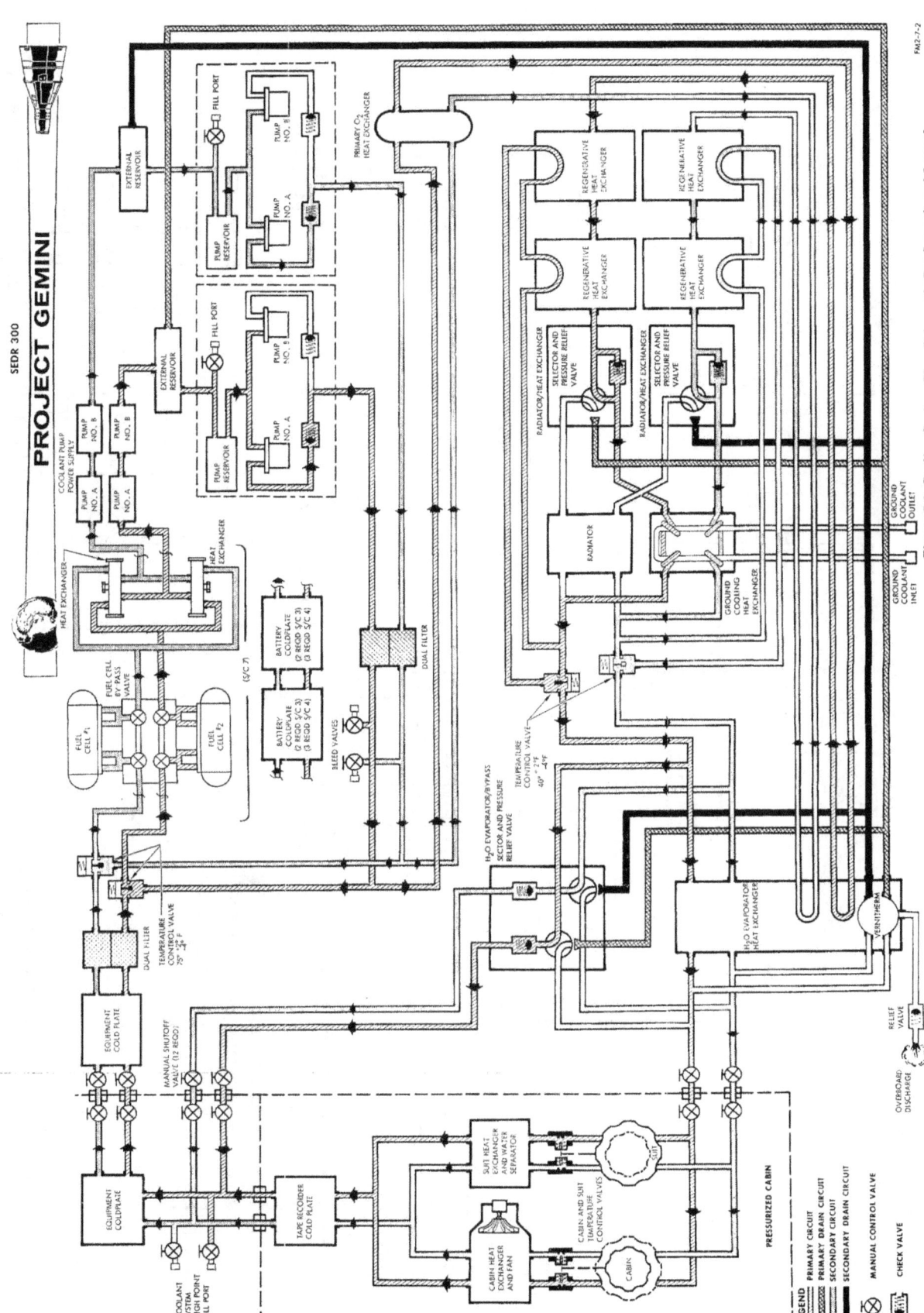

Figure 7-2 Cooling System Flow Schematic Pre-Launch and Launch

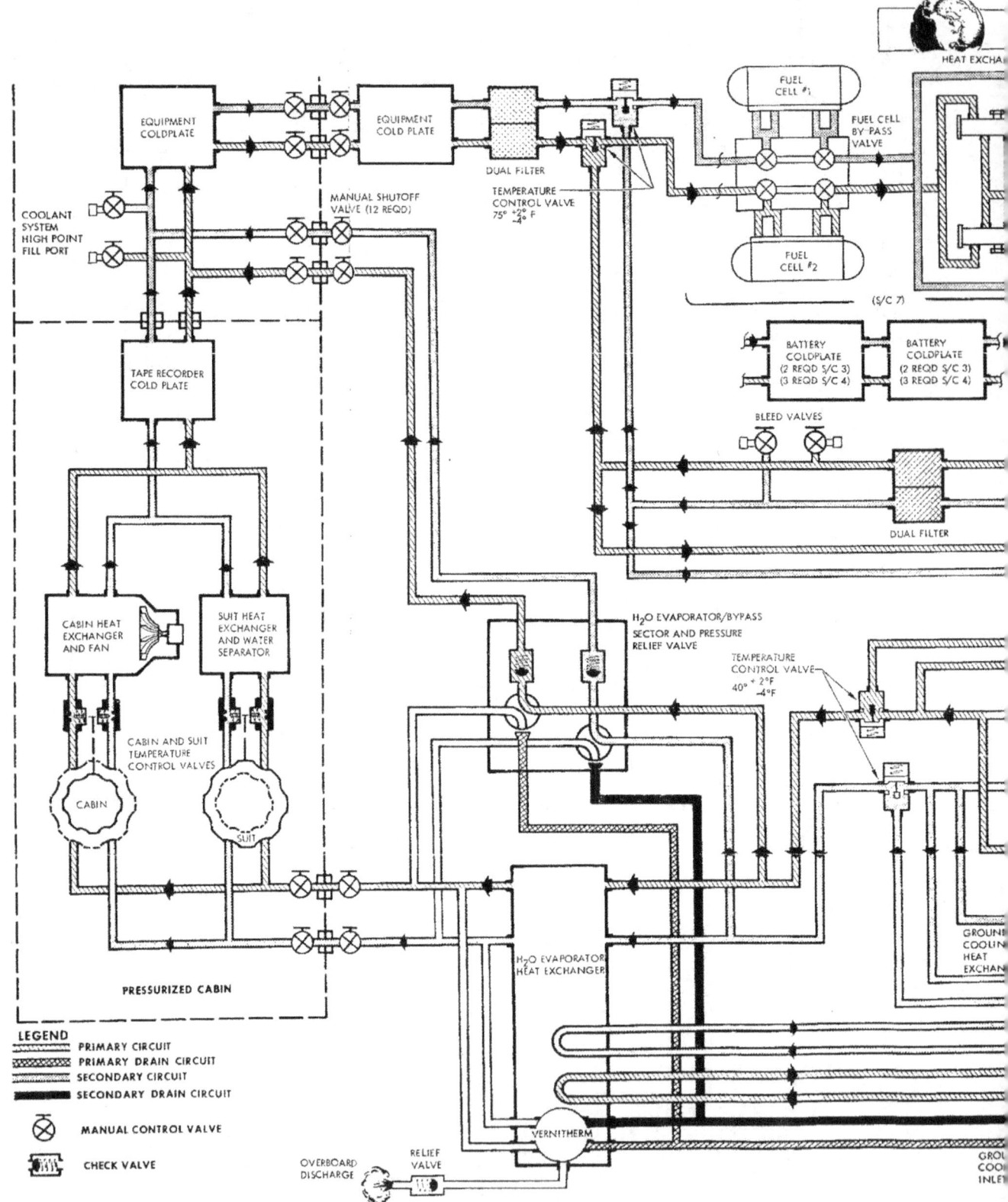

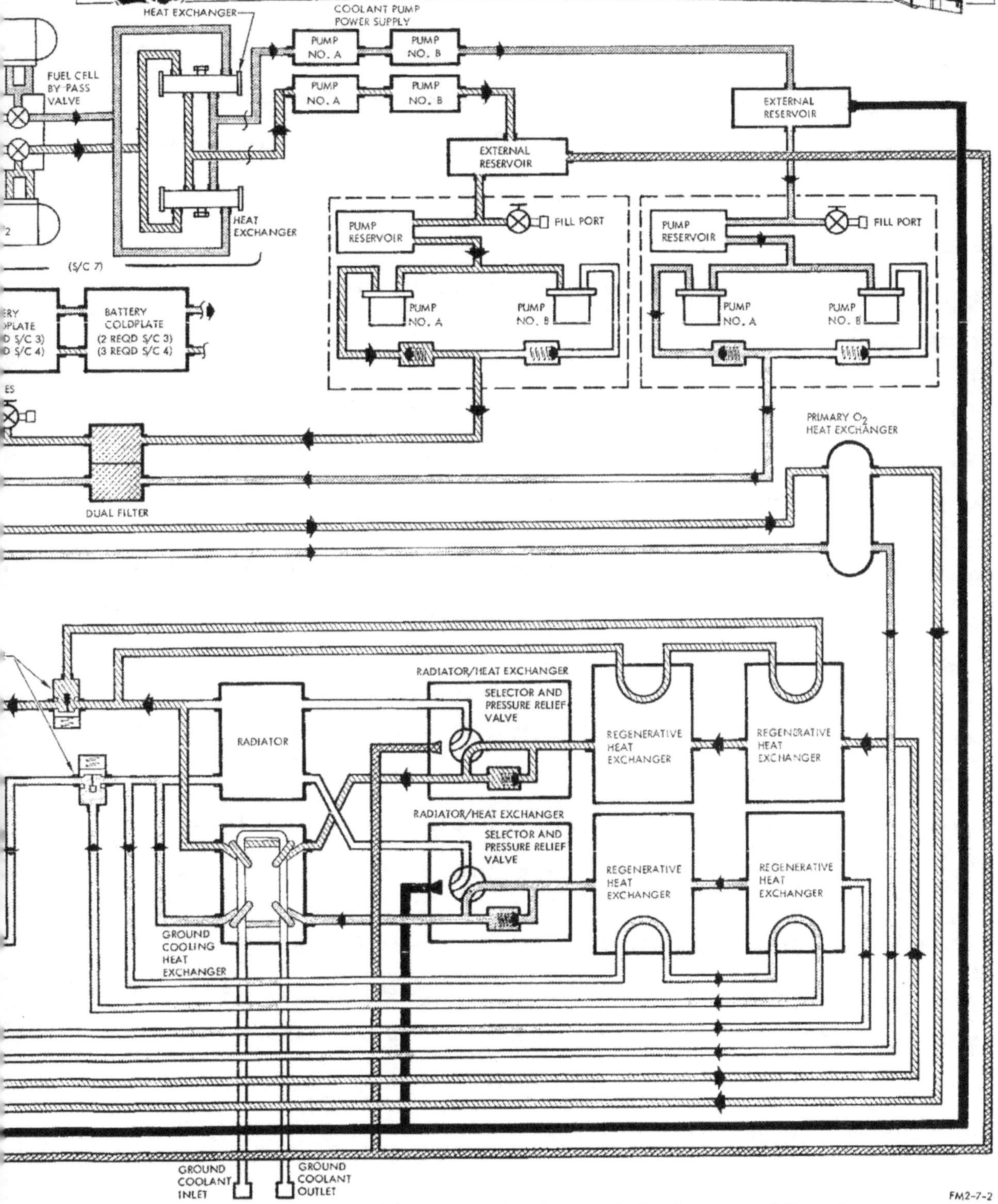

SEDR 300

PROJECT GEMINI

Figure 7-2 Cooling System Flow Schematic Pre-Launch and Launch

FM2-7-2

SEDR 300
PROJECT GEMINI

through the spacecraft ground cooling heat exchanger providing temperature control of the cooling system coolant. The A pumps of the primary and secondary cooling circuits are activated, using an external power source, to provide the required cooling for spacecraft equipment and cabin. The spacecraft radiator switch, located on the center panel, is placed in the BYPASS position so the cooling system coolant by-passes the radiators and is directed through the ground cooling heat exchanger.

Coolant is circulated through each coolant loop by a positive-displacement gear pump. S/C 3, 4, and 7 are provided with 2 pumps in each loop. Selection of loops and number of pumps is controlled manually.

The coolant is filtered, as it leaves the pump, and simultaneously flows to the inlet of the battery coldplate or fuel cell temperature control valve and primary oxygen heat exchanger.

The temperature control valve maintains the cooling temperature at the fuel cell or battery coldplate inlet at $75° {}^{+2°}_{-4°}$ F. Temperature increasing above setting will reduce by-pass flow. Coolant temperature from by-pass line varies from $80°$F to $165°$F. Coolant temperature from equipment lines varies from $60°$F to $125°$F.

Coolant enters the primary oxygen heat exchanger and then is routed around the steam discharge lines in the water boiler before it passes through the regenerative heat exchanger. It then passes through the selector and pressure relief valve. This selector valve is electrically actuated and when in the radiator by-pass position allows the coolant to pass through the ground cooling heat exchanger where the external supply of coolant flowing through the ground cooling heat exchanger absorbs the heat from the spacecraft's coolant system.

The ground coolant heat exchanger has an airborne flow capacity of 336 lb/hr, per coolant loop, at $125°F$. It has a ground coolant flow capacity of 425 lb/hr at $40°F$.

The coolant is now ready to pass through the temperature control valve. This valve maintains the outlet temperature at $40°{}^{+2°}_{-4°}F$. If the coolant entering the valve from the ground heat exchanger is below this range, a portion of the coolant is directed through the regenerative heat exchanger and then mixed at the valve.

The coolant then flows through the water evaporator to the cabin and suit manual temperature control valves. These valves meter the coolant flow through the cabin and suit heat exchangers. The evaporator selector valve relief portion allows part of the coolant to by-pass the cabin and suit heat exchangers depending on the setting of the manual control valves. The selector portion of this valve allows the by-pass fluid to come from either downstream or upstream of the evaporator. The coolant continues through the various coldplates until it reaches the battery coldplates for S/C 3 and 4 or through the fuel cells on S/C 7. The coolant has now returned to the reservoir where the cycle is ready to be repeated.

Shortly before launch, the external cooling and electrical power are disconnected.

LAUNCH (Figure 7-2)

During launch, the launch cooling heat exchanger goes into operation in the following sequences. The heat transfer characteristics and capabilities of the ground cooling heat exchanger no longer exist. The MCS 198 coolant fluid now with no place to dissipate its internal heat, which is constantly being generated by and absorbed from the loop components, circulates about the vernatherm section of the heat exchanger. When the coolant temperature exceeds $46 ^{+4°}_{-2°}$ F, the vernatherm pilot valve opens to pressurize a doughnut shaped bellows which unseats the poppet valve exposing the water in the heat exchanger core to reduce pressure as altitude increases during launch.

When S/C altitude exceeds 100,000 feet, water in the heat exchanger will boil absorbing heat from the coolant. This absorbed heat is then expelled overboard in the form of steam.

When the coolant reaches a temperature of $46°$F, the vernatherm section repositions to relieve pressure to the doughnut shaped bellows holding the poppet open. As this pressure diminishes, a spring behind the poppet will reposition it to the closed position. The evaporator selector valve is positioned to allow all flow to go through the evaporator.

The water boiler water reservoir is constantly replenished from the suit heat exchanger water separator, and if the need arises, from the drinking water supply tank.

 SEDR 300
PROJECT GEMINI

ORBIT (Figure 7-3)

After orbiting for approximately 30 minutes, to allow the radiator to cool after being subject to launch heating, the coolant flow is directed through the space radiators by manual selection of the radiator switch located on the center panel. This by-passes the ground cooling heat exchanger. The evaporator selector valve is also positioned so that only the flow to the suit and cabin heat exchangers pass through the evaporator.

Prior to retrograde firing, the coolant pump packages, radiators, batteries and various heat exchangers are jettisoned with the adapter equipment section. Prior to adapter jettisoning and retrograde firing the A coolant pumps for both the primary and secondary cooling circuits are activated. The suit, cabin, and equipment bays are cooled to as low a temperature as possible, before the adapter equipment section is jettisoned.

SYSTEM UNITS

PUMP PACKAGE (Figure 7-4)

The pump package for each coolant circuit incorporates two constant displacement electrical pumps, two pump inverters, an external reservoir, filters, relief and check valves. The pump package is located in the adapter equipment section. Pump selection is provided by switches on the pilots' center panel. A pump failure warning light is provided on the center panel. When a pump is activated the coolant flows from the reservoir to the pump, which circulates the coolant through the cooling circuit. The coolant returns to an external reservoir that compensates for thermal expansion, contraction, and leakage of the coolant. A

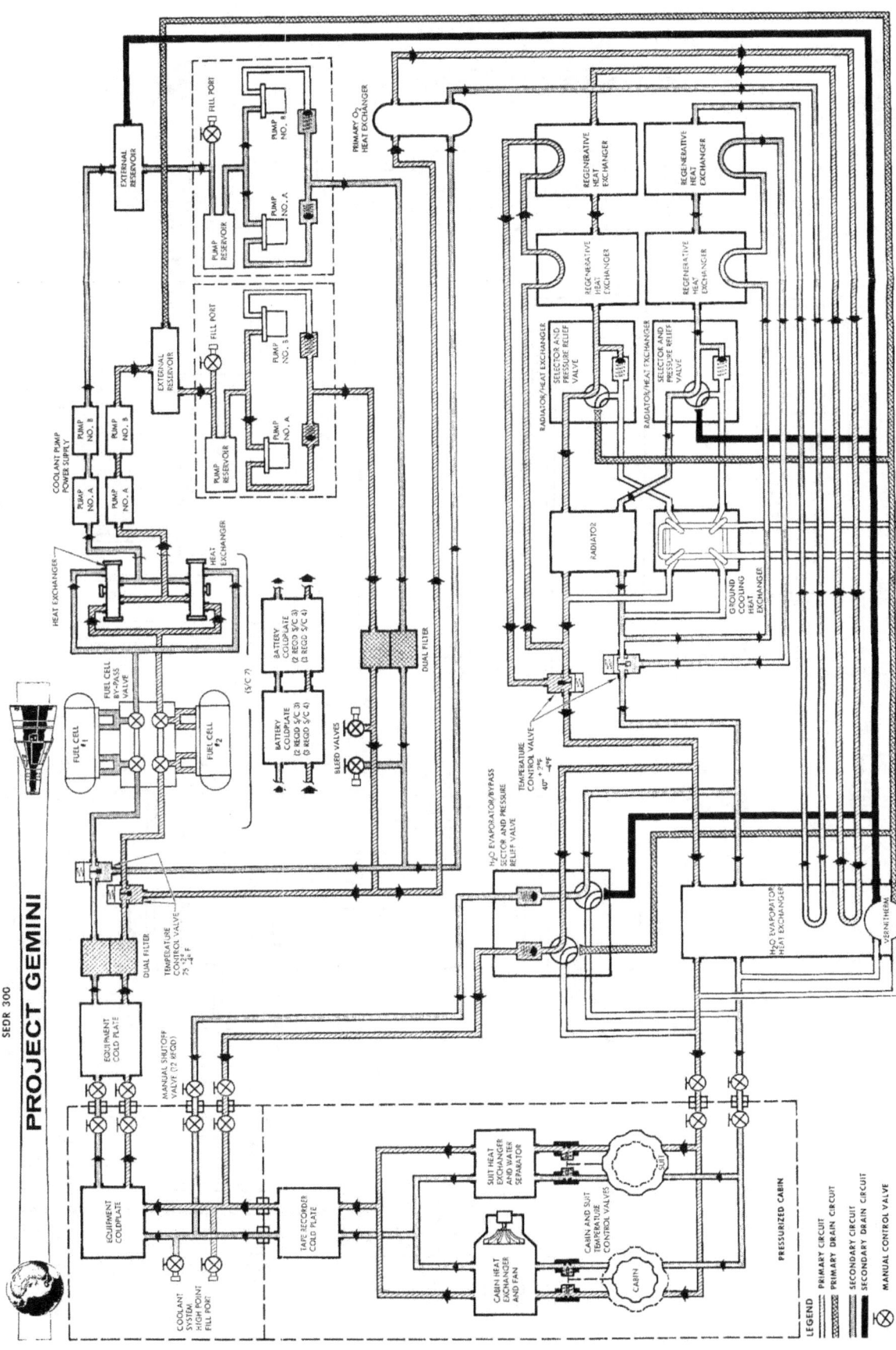

Figure 7-3 Cooling System Flow Schematic-Orbit

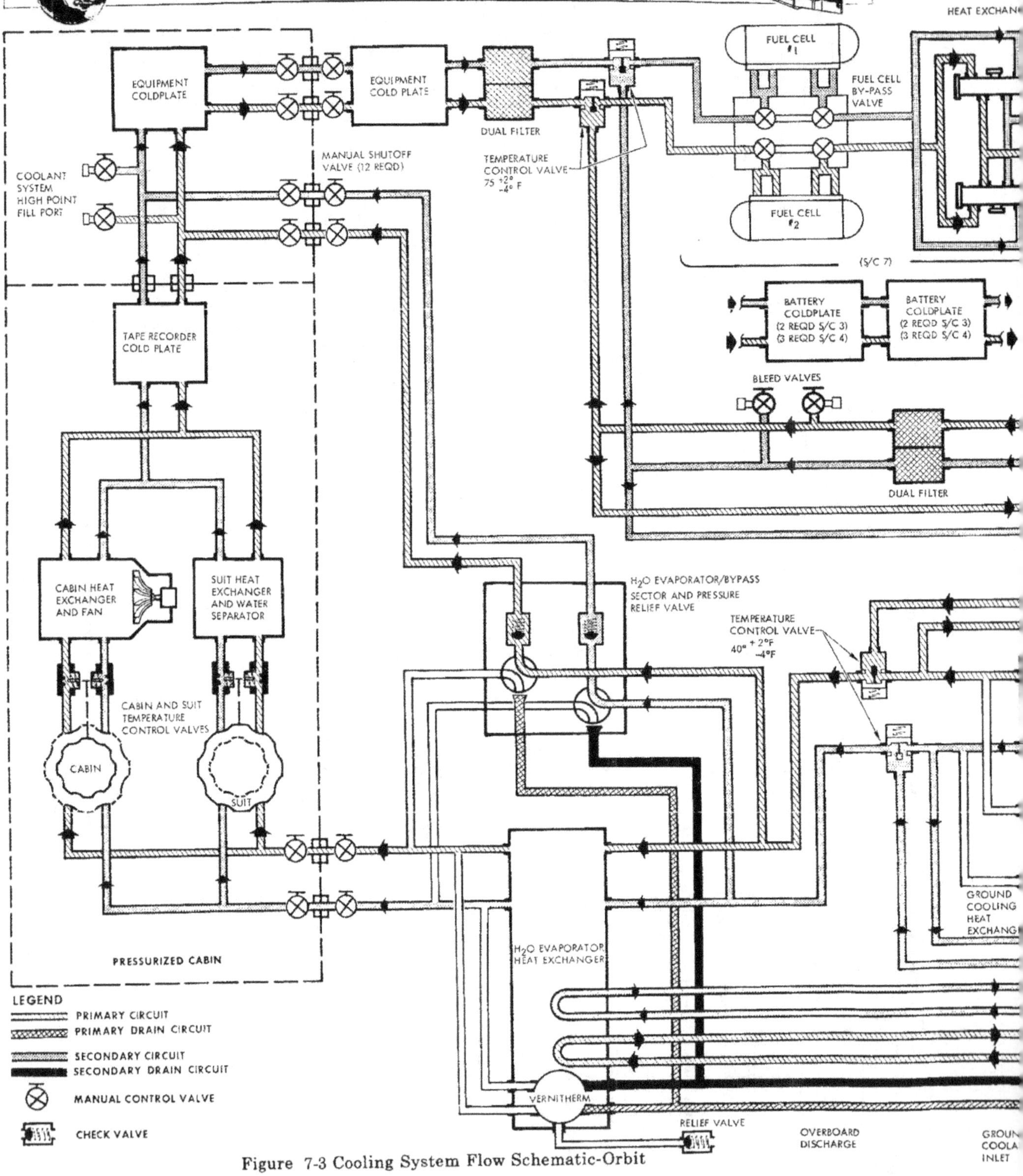

Figure 7-3 Cooling System Flow Schematic-Orbit

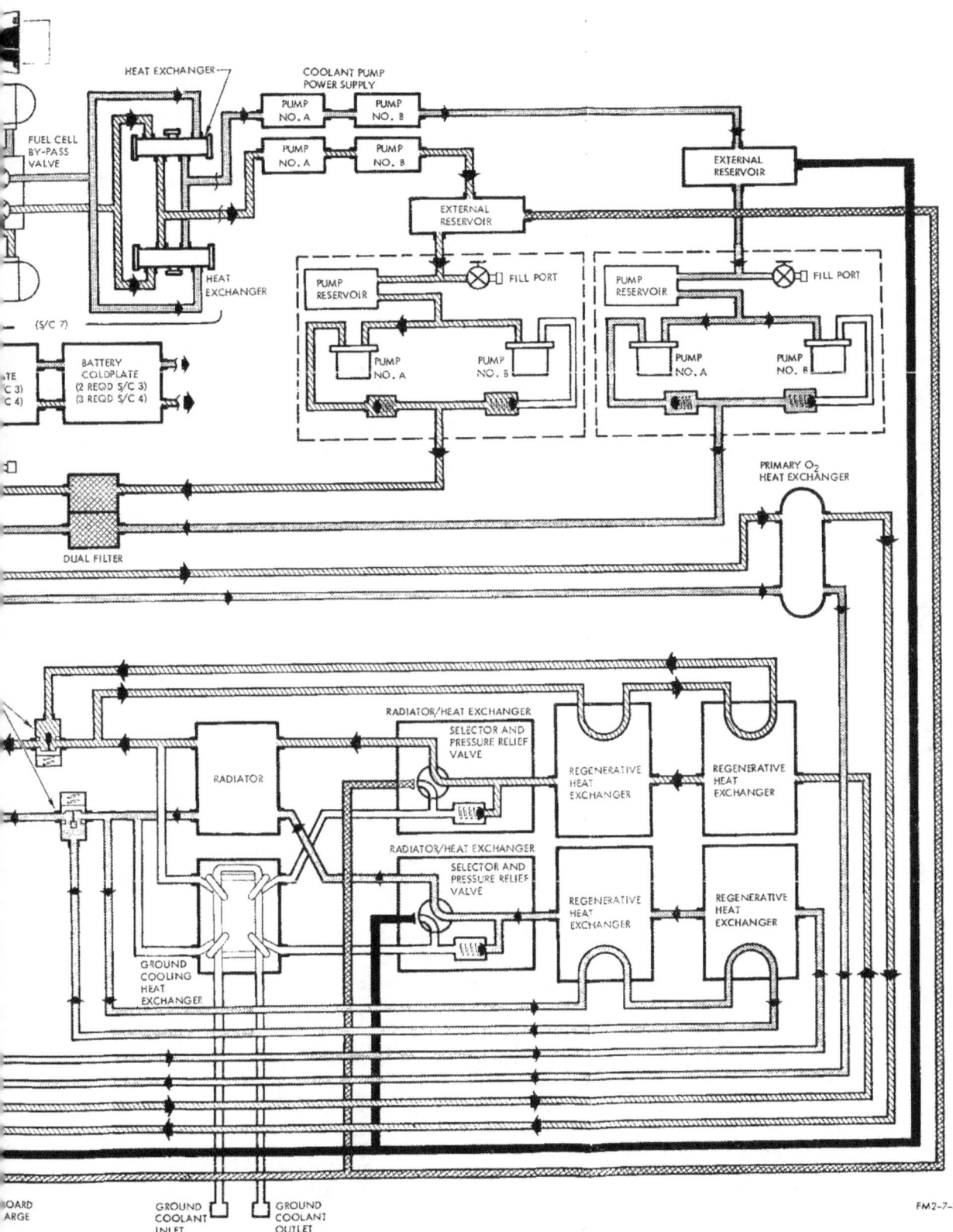

 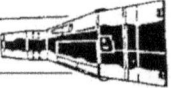

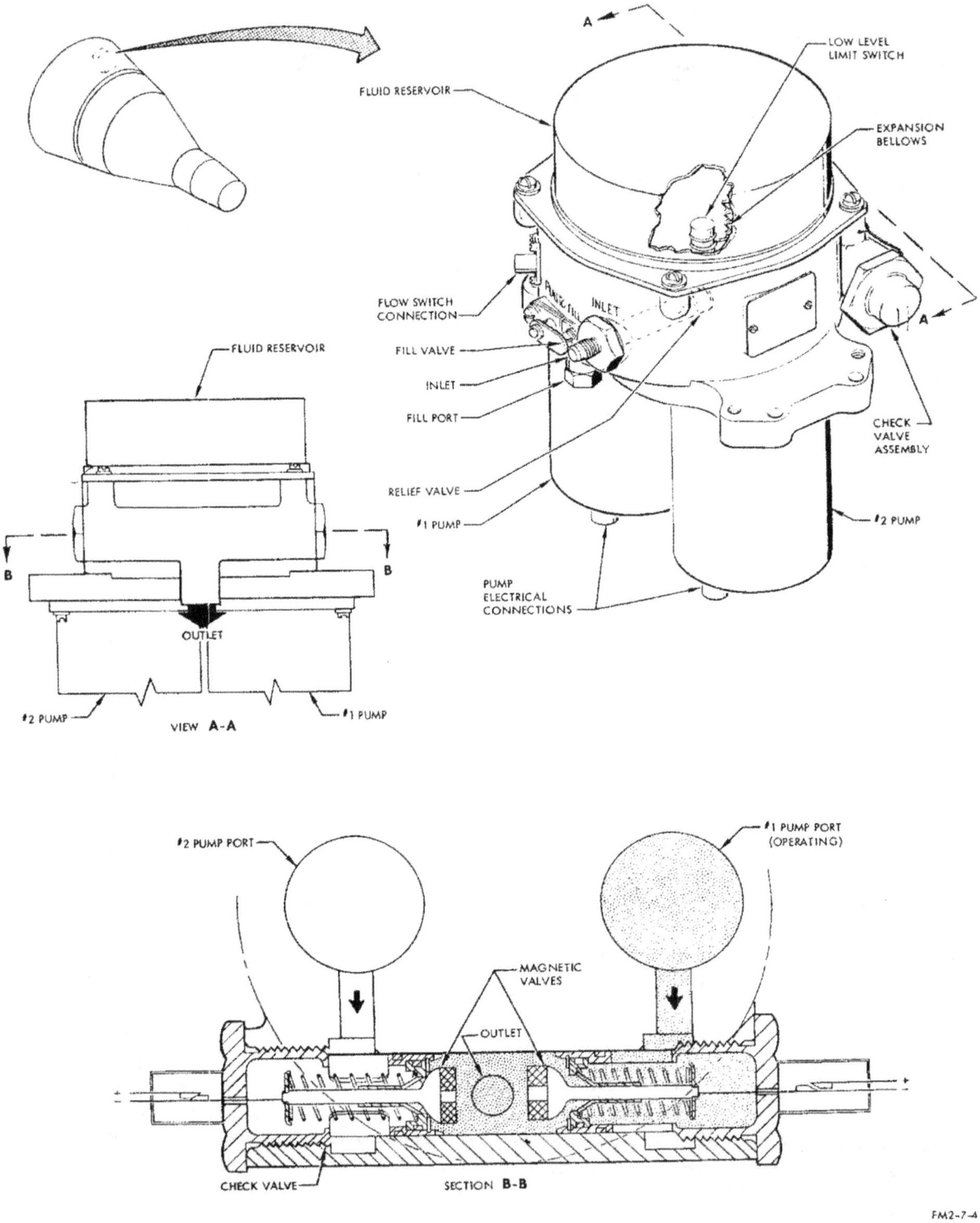

SEDR 300

PROJECT GEMINI

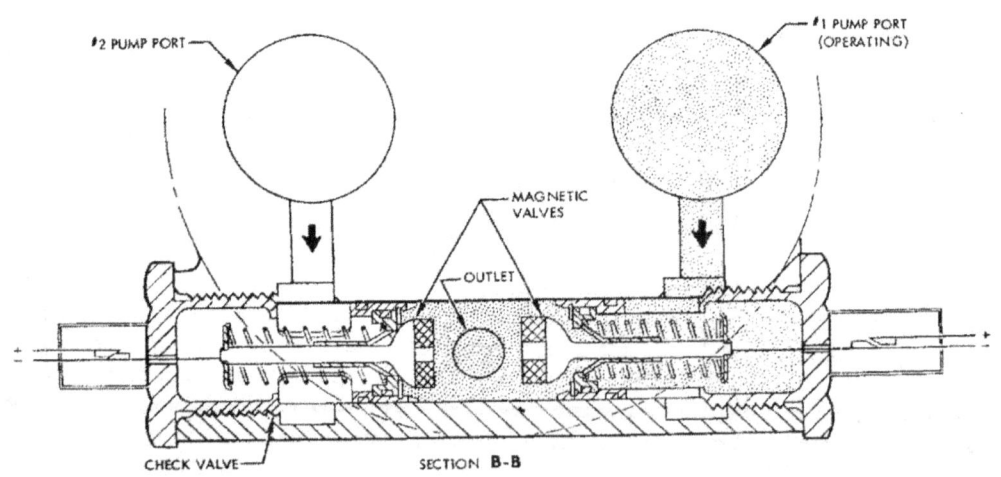

Figure 7-4 Coolant Pump Package

100 micron filter downstream of the pump prevents contamination of the cooling system. Check valves in the pump package prevent the operating pump from pumping coolant into the redundant pump. Flow sensing switches illuminate a pump failure lamp on the pilots' center panel in the event of pump failure.

RADIATOR (Figure 7-5)

The spacecraft radiator consists of two circumferential radiator panels made of 0.25 inch diameter cooling tubes. There are four sections of tubing to each radiator panel. The tubing is manufactured as part of the spacecraft structure. Each panel incorporates two parallel cooling circuits, one for the primary cooling circuit and the other for the secondary circuit. During orbit/, the cooling system coolant is circulated through the radiator. The heat of the coolant radiates into space, lowering the temperature of the coolant.

COLDPLATES (Figure 7-6)

The coldplates, other than the battery coldplates, are plate fin constructed units incorporating parallel coolant system passages. Coldplates are fabricated from aluminum. Battery, electrical, electronic and other heat generating components are mounted on coldplates. The coolant flowing through the coldplates absorbs the heat generated by the components, preventing overheating of the operating equipment.

HEAT EXCHANGERS (Figure 7-7, 7-8)

Two types of heat exchangers are used in the spacecraft; namely, plate fin constructed and shell and tube constructed heat exchangers. The suit, cabin, water evaporator, ground cooling and regenerative heat exchangers are of plate

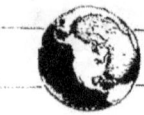

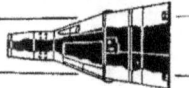

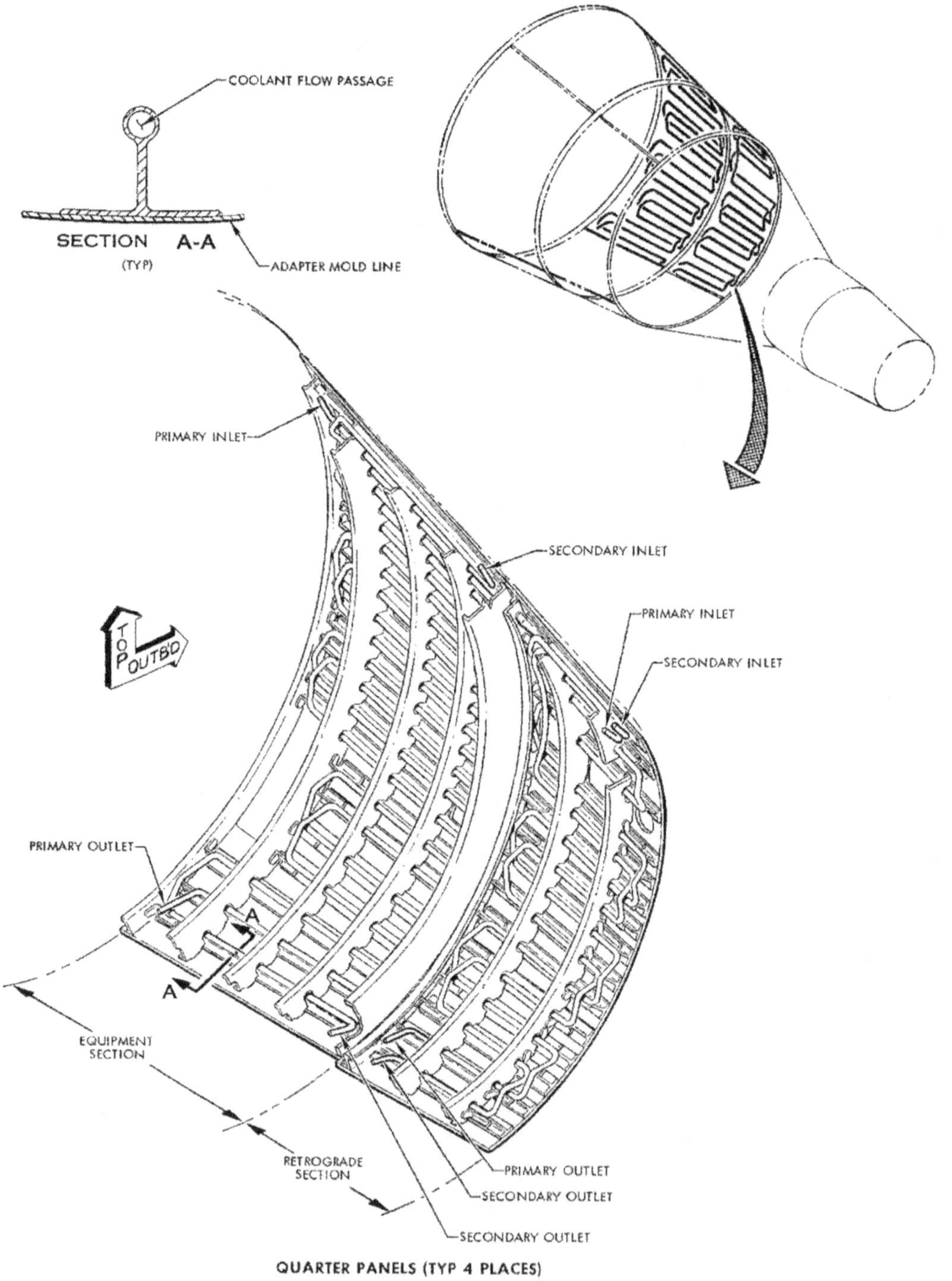

Figure 7-5 Radiator Stringer Assembly

FM2-7-5

 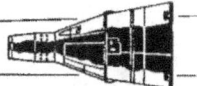

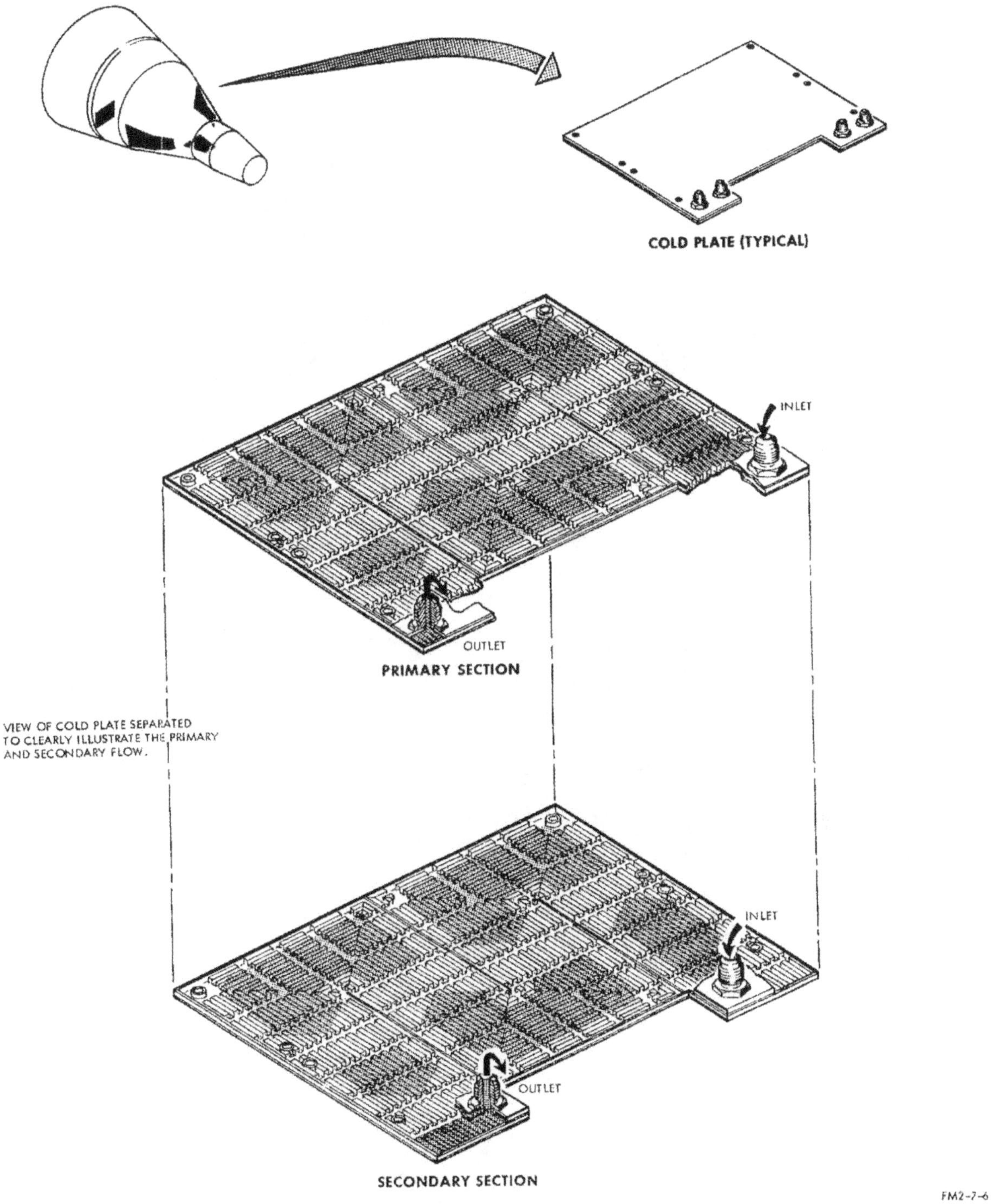

SEDR 300
PROJECT GEMINI

Figure 7-6 Cold Plate

PROJECT GEMINI

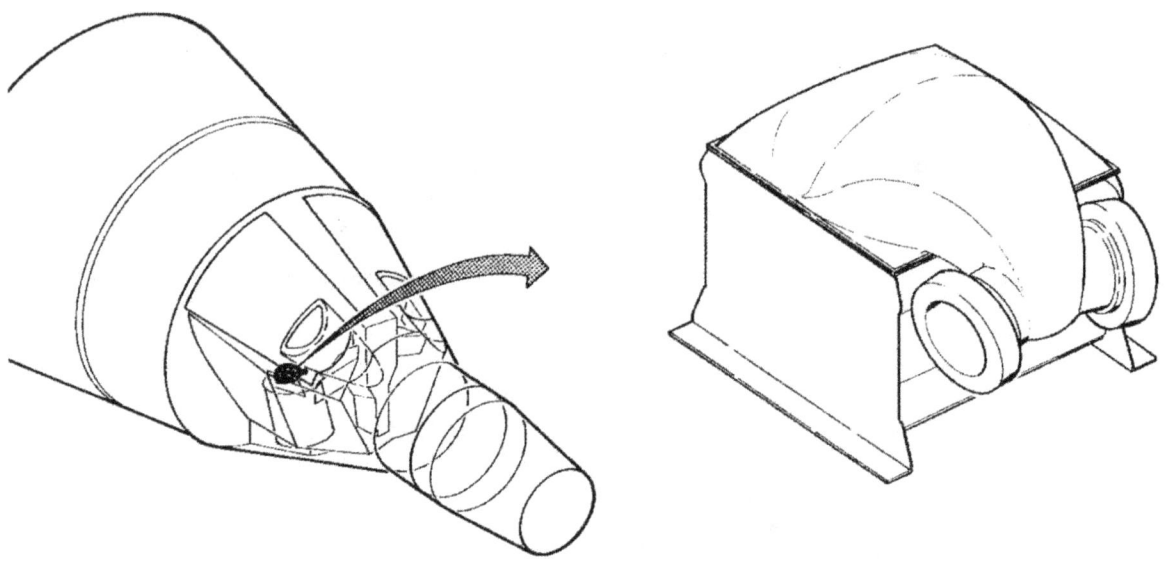

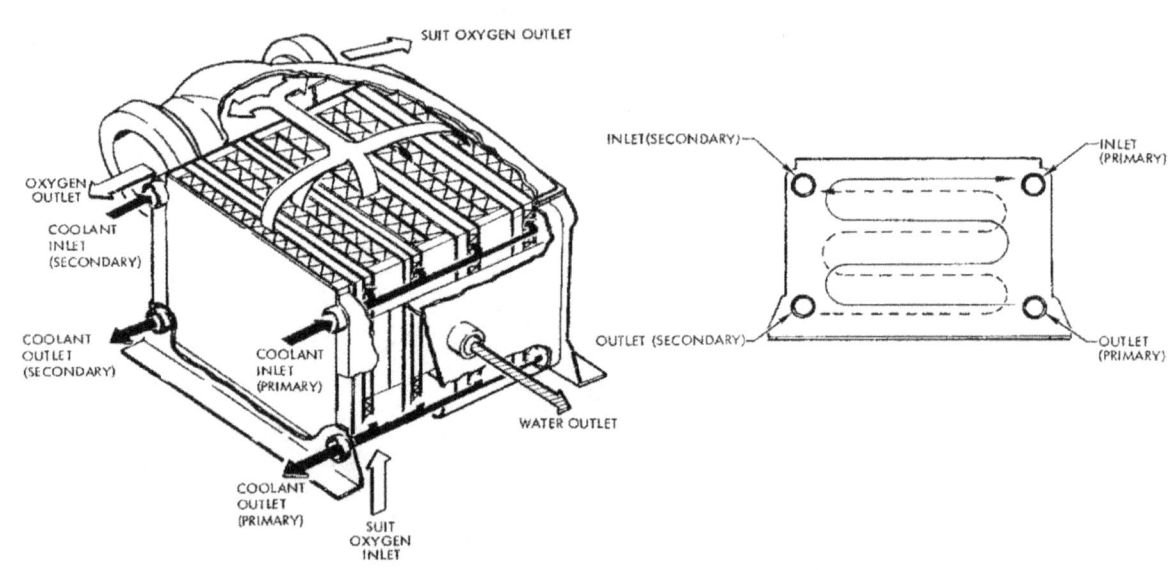

Figure 7-7 Heat Exchanger-Suit

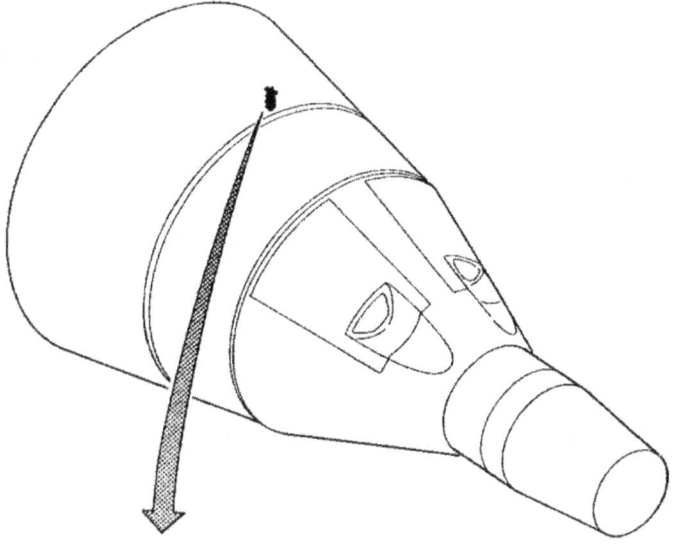

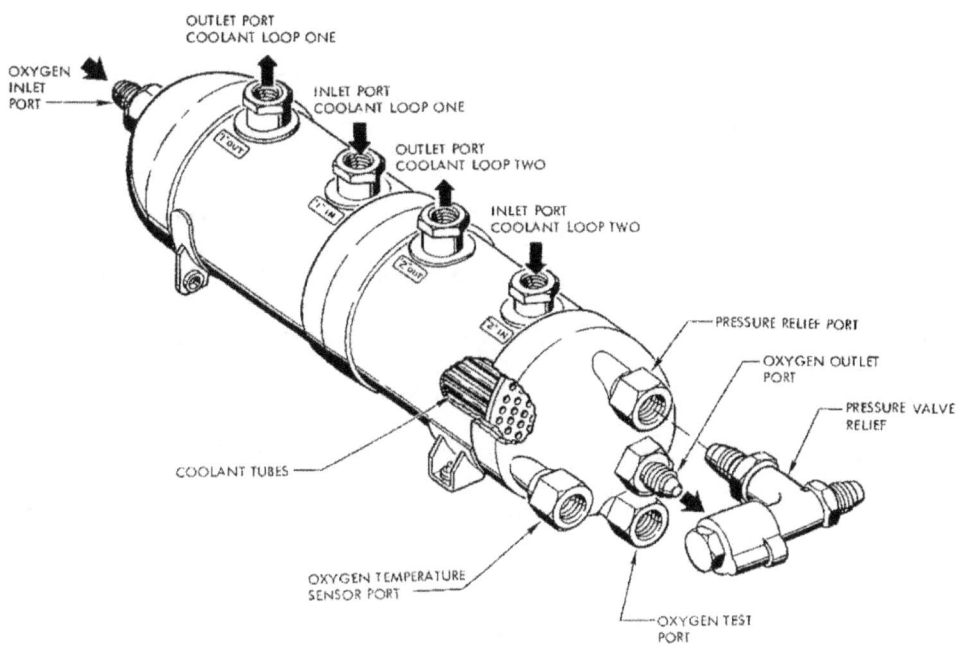

Figure 7-8 Coolant Tube Type Heat Exchanger

 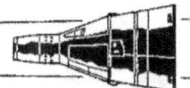

SEDR 300

PROJECT GEMINI

construction. The primary oxygen heat exchanger is of shell and tube construction. The coolant absorbs heat from the cabin, suit and regenerative heat exchangers. The ground cooling and water evaporator heat exchangers permit heat transfer to cool the coolant. The primary oxygen heat exchanger is designed so heat transfer will heat the primary oxygen to a desired temperature.

TEMPERATURE CONTROL VALVE (Figure 7-9)

Temperature control valves are provided in both the primary and secondary cooling circuits. These valves are located at the radiator outlets and at the inlets to the battery coldplates or fuel cells.

The temperature control valve, located in the coolant system radiator outlet, automatically maintains the coolant outlet temperature at $40 \, {}^{+2°}_{-4°}$ F as long as the radiator capacity has not been exceeded.

The temperature control valve, located in the battery coldplate inlet, automatically maintains the coolant inlet temperature at $75 \, {}^{+3°}_{-4°}$ F or above.

The temperature control valve contains a piston that regulates the inlet flow to the valve. The piston is spring loaded on one side. A thermostatic actuator on the opposite side of the piston determines piston movement, which in turn regulates the coolant flow through the valve. The thermostatic actuator, which is located to accurately sense mixing temperature, consists of an encapsulated wax pellet that expands or contracts as temperature varies. As temperature around the pellet increases, the wax expands exerting pressure on the diaphragm. The diaphragm moves a piston, which in turn controls the inlet flow to the valve. Temperature reduction around the wax decreases the pressure in the pellet cup

PROJECT GEMINI

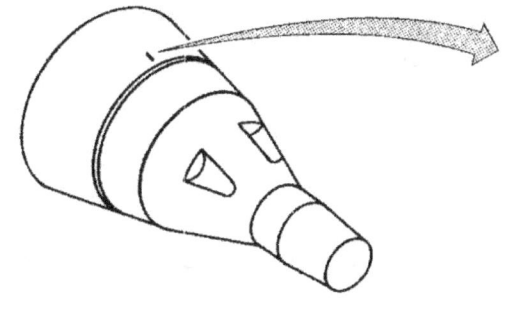
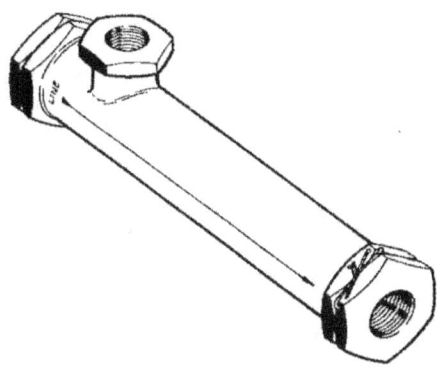
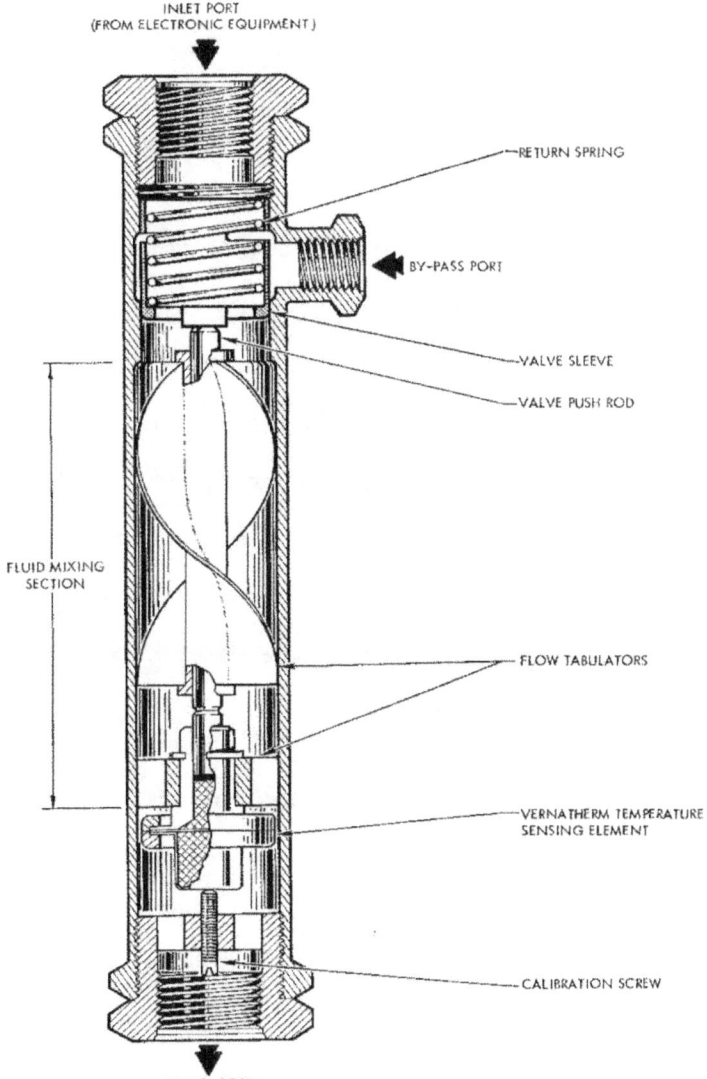

Figure 7-9 Coolant Temperature Control Valve

allowing the spring to reposition the piston regulating the flow of coolant through the valve.

LAUNCH COOLING HEAT EXCHANGER (Figure 7-10, 7-11)

The launch cooling heat exchanger is located in the adapter section. Via its relief valve, it can dump liquids overboard; or if the temperature control valve senses temperatures greater than $50°F$, it can control the outlet temperature of the primary and secondary coolants to $46^{+4}_{-2}°F$. In addition, it serves as a water reservoir, storing water until it is needed for cooling.

This evaporator consists of a wicking type heat exchanger and is capable of storing seven pounds of water. A temperature control valve has been set to control the outlet coolant temperature to $46^{+4}_{-2}°F$. A relief valve opens and allows excess water to be dumped overboard at 2.75 ± 0.25 psid and reseats at 2.0 psid minimum. An electrical heater is provided in the poppet to prevent ice formation. Coolant flow capacity is 366 lb/hr at $40°F$. Water flow capacity is 3 lb/min when cooling is not required from the evaporator. Maximum operating pressure in the fluid heater coolant circuits is 230 psig, and 100 psig in the core circuits. Maximum operating pressure in the water circuit is 20 psig with exit port relief valve in normal operation.

The steam exit duct is continuously heated by coolant coming from the primary oxygen heat exchanger to prevent ice formation.

A loss of pressure in either coolant loop will not affect the operation of the valve.

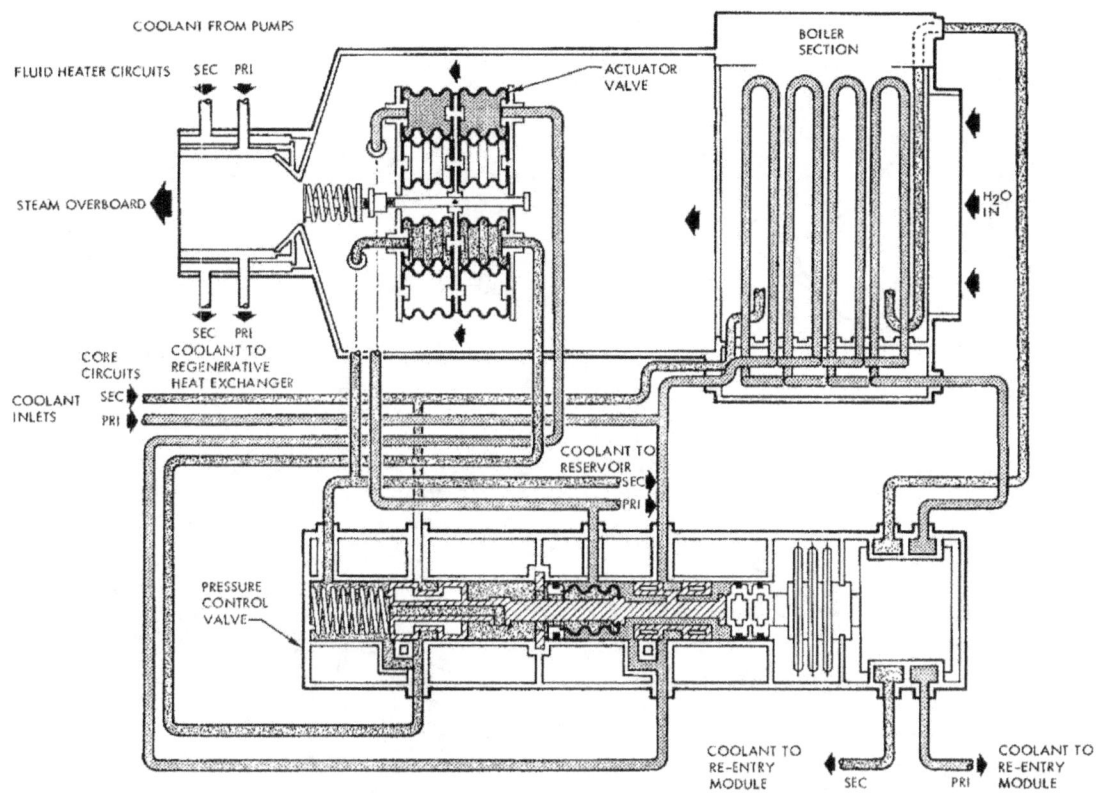

Figure 7-10 Launch Cooling Heat Exchanger Schematic

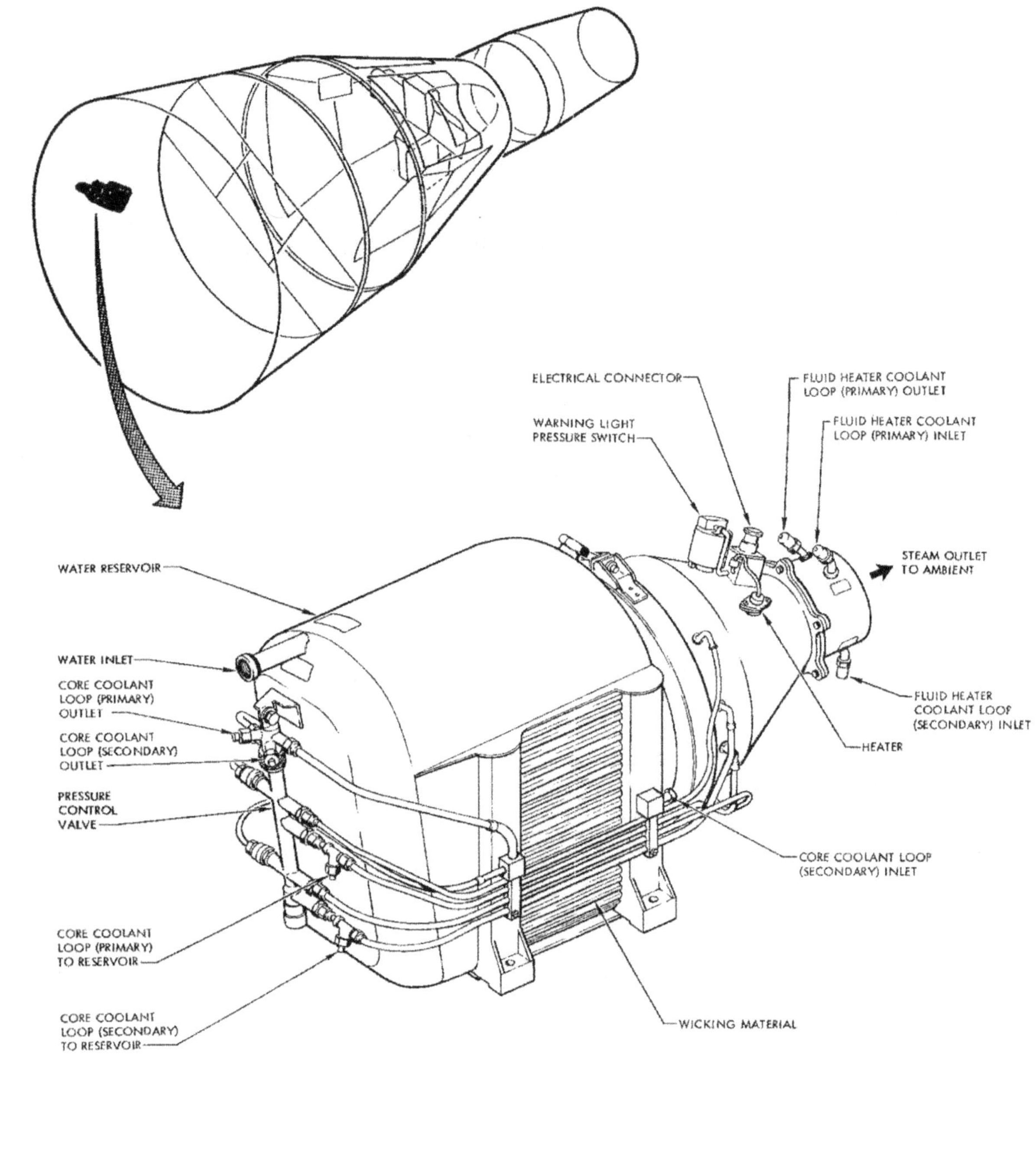

PROJECT GEMINI

SEDR 300

Figure 7-11 Launch Cooling Heat Exchanger

GUIDANCE and CONTROL SYSTEM

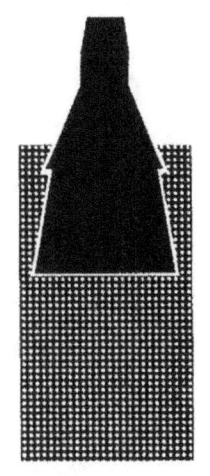

Section VIII

REFER TO THE SEDR 300 CONFIDENTIAL SUPPLEMENT FOR INFORMATION CONCERNING THE GEMINI GUIDANCE AND CONTROL SYSTEM.

COMMUNICATION SYSTEM

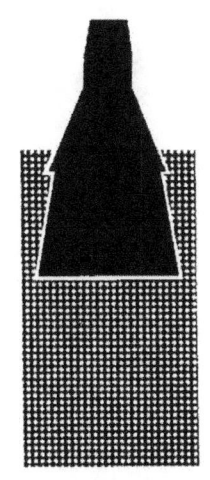

Section IX

TABLE OF CONTENTS

TITLE	PAGE
SYSTEM DESCRIPTION	9-3
ANTENNAS	9-3
BEACONS	9-5
VOICE COMMUNICATIONS	9-7
TELEMETRY TRANSMITTERS	9-7
FLASHING RECOVERY LIGHT	9-8
DIGITAL COMMAND SYSTEM (DCS)	9-8
SYSTEM OPERATION	9-8
VOICE TAPE RECORDER	9-8
PRE-LAUNCH	9-9
SPACECRAFT/LAUNCH VEHICLE SEPARATION	9-13
ORBIT	9-14
ADAPTER SEPARATION	9-20
RE-ENTRY	9-21
LANDING THROUGH RECOVERY	9-22
SYSTEM UNITS	9-23
ANTENNAS	9-23
BEACONS	9-44
VOICE COMMUNICATION	9-55
TELEMETRY TRANSMITTERS	9-70
FLASHING RECOVERY LIGHT AND POWER SUPPLY	9-72
DIGITAL COMMAND SYSTEM	9-74

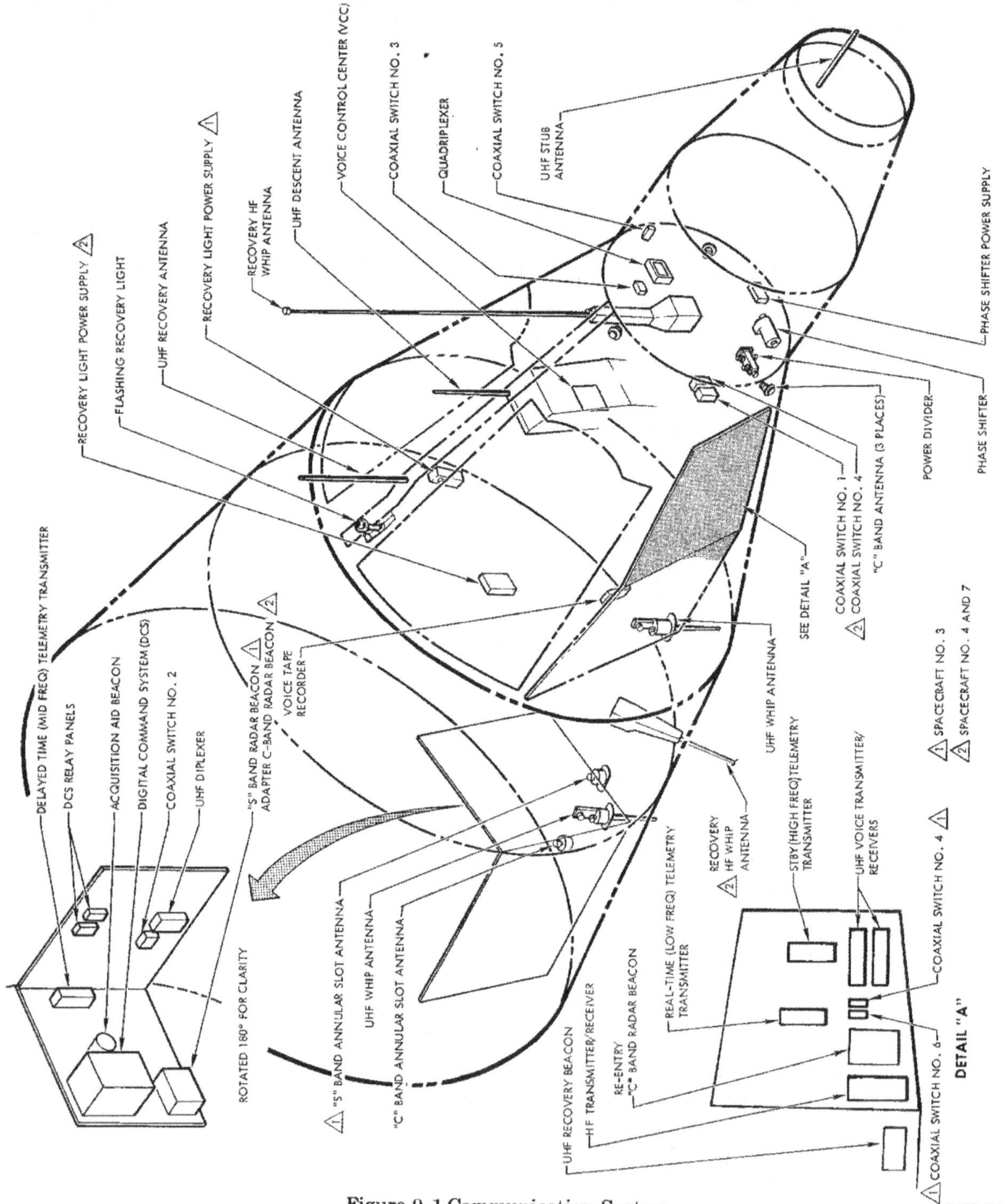

SEDR 300

PROJECT GEMINI

Figure 9-1 Communication System

FMG2-125

SEDR 300
PROJECT GEMINI

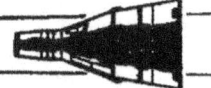

SECTION IX COMMUNICATION SYSTEM

SYSTEM DESCRIPTION

The communication system provides the only communication link between the ground and the Gemini Spacecraft. The system provides for the following capabilities: radar tracking of the spacecraft; two-way voice communications between the ground and the spacecraft, and between the Flight Crew; ground command to the spacecraft; instrumentation system data transmission, and post landing and recovery aid data transmission. To make possible these various capabilities, the communication system contains components that may be divided into the following categories: antennas, including multiplexers and coaxial switches; beacons; voice communications; telemetry transmitters; flashing recovery light; and digital command system. The flashing recovery light and the UHF recovery beacon are grouped together in a category called the electronic recovery aids (ERA).

The communication system components are located throughout the spacecraft with the largest concentration being in the re-entry module right equipment bay and the electronic module of the equipment adapter as illustrated in Figure 9-1.

ANTENNAS

Eight antennas and one antenna system provide transmission and/or reception capabilities for the various communication system components. The spacecraft communication system (Figure 9-2) contains the following antennas: UHF recovery, UHF stub, UHF descent, two UHF whips, two HF whips on S/C 4 and 7, and one on S/C 3, S-band annular slot (S/C 3 only), C-band annular slot, and C-band antenna system consisting of a power divider, a phase shifter, phase

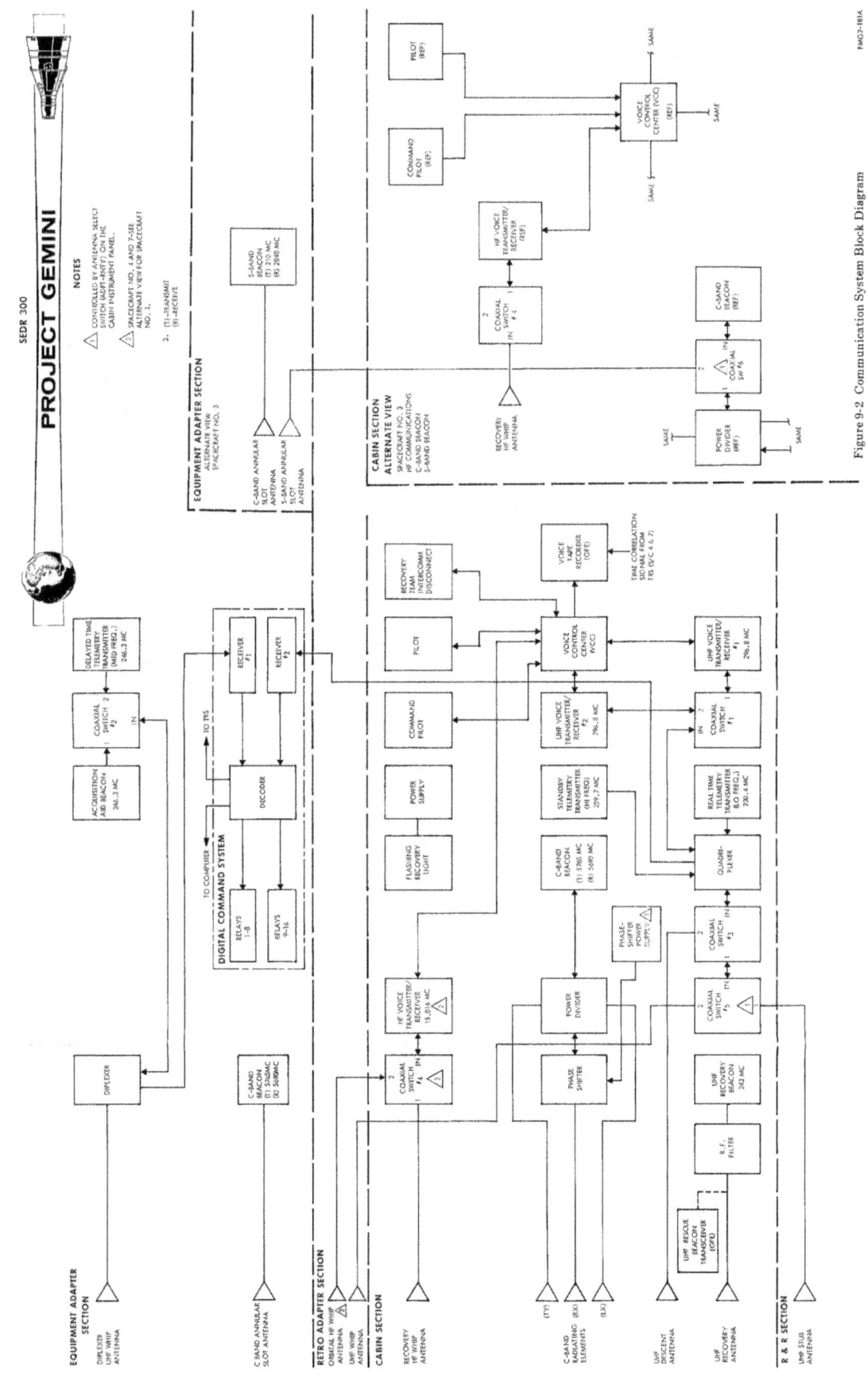

Figure 9-2 Communication System Block Diagram

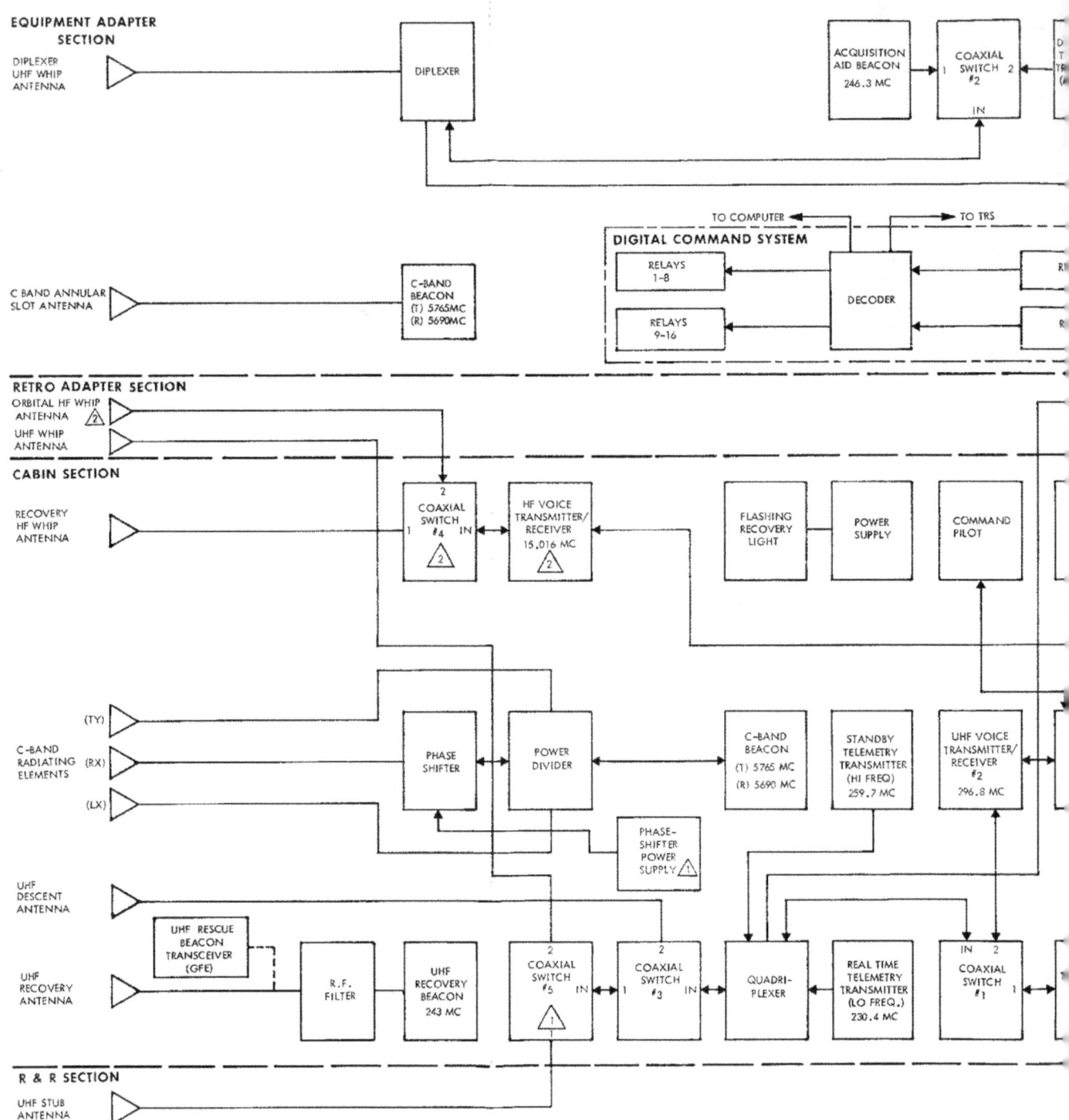

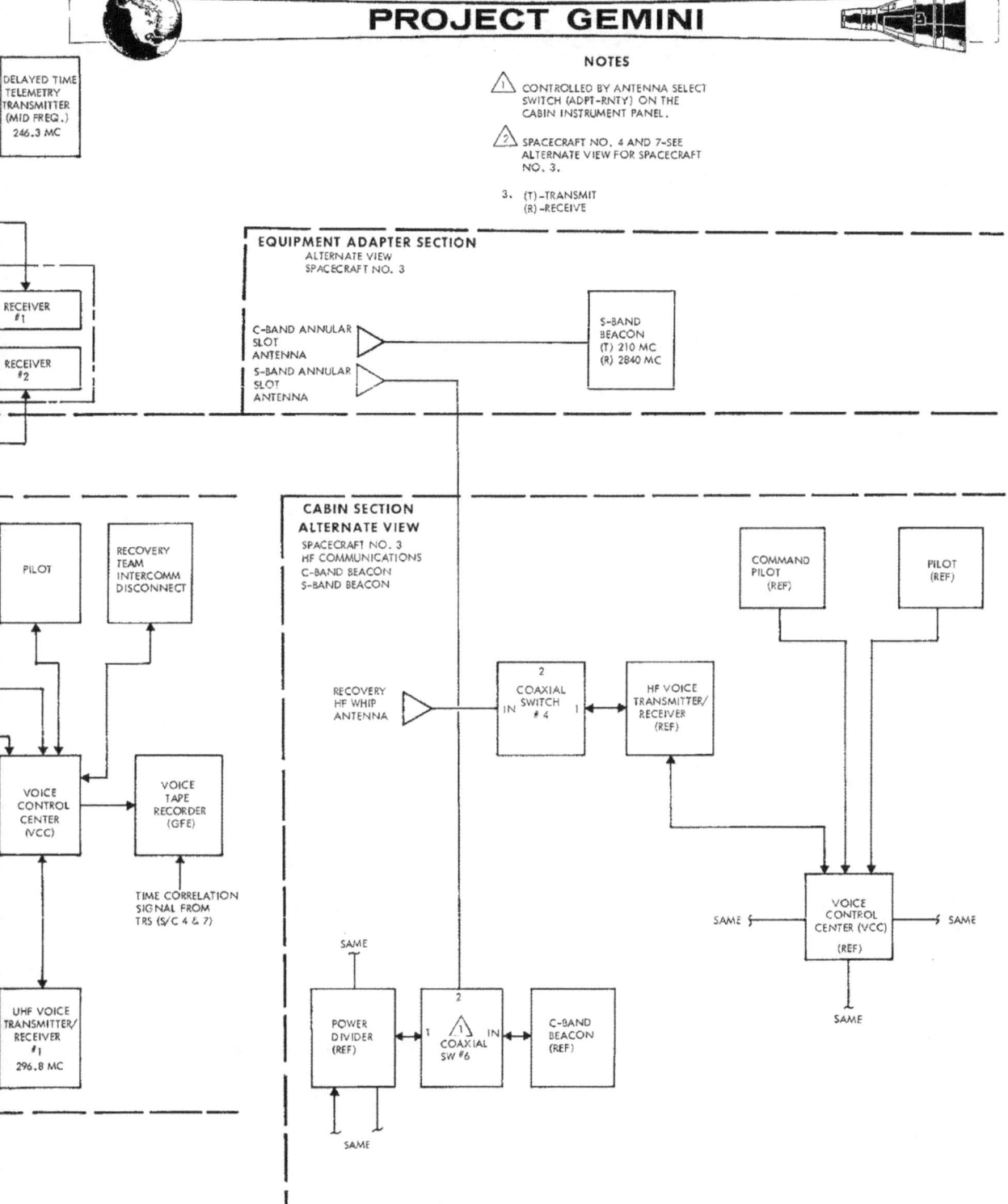

Figure 9-2 Communication System Block Diagram

shifter power supply, and three radiating elements. Antenna usage is illustrated in Figure 9-3 and described under system units.

To provide most efficient antenna usage, a diplexer and a quadriplexer are utilized in conjunction with the UHF whip and the UHF stub antennas. The multiplexers make it possible to use more than one transmitter and/or receiver with a single antenna.

Five coaxial switches on S/C 4 and 7, and six on S/C 3 allow antenna and transmitter/receiver switching for best communication coverage during the various phases of the mission (launch, orbit, re-entry and recovery).

BEACONS

Four beacons provided by the communication system establish the capability of tracking and locating the spacecraft during the entire spacecraft mission. The four beacons are: an acquisition aid beacon; a recovery beacon; two C-band beacons on S/C 4 and 7, and a C-band and an S-band beacon on S/C 3. The acquisition aid beacon, operating on a fixed frequency, is used to determine when the spacecraft is within the range of a ground tracking station, and provides information for orientating the ground station antennas during the orbital phase of the mission. The recovery beacon is also merely a transmitter that operates on the international distress frequency and is used by the recovery forces to determine spacecraft location.

The C-band and S-band beacons are transponders which, when properly interrogated by a ground station, provide signals for accurate spacecraft tracking.

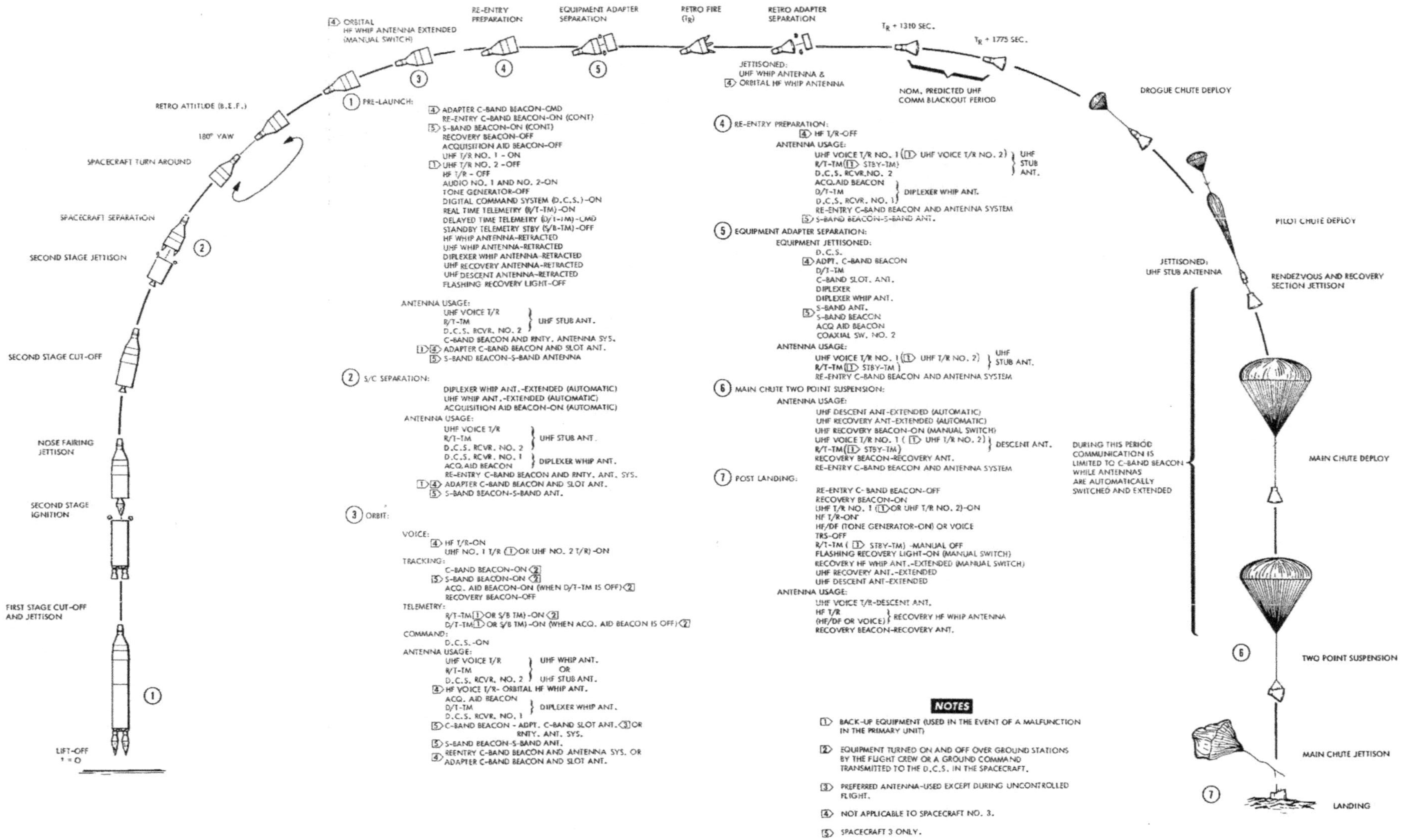

Figure 9-3 Communication System Sequential Diagram

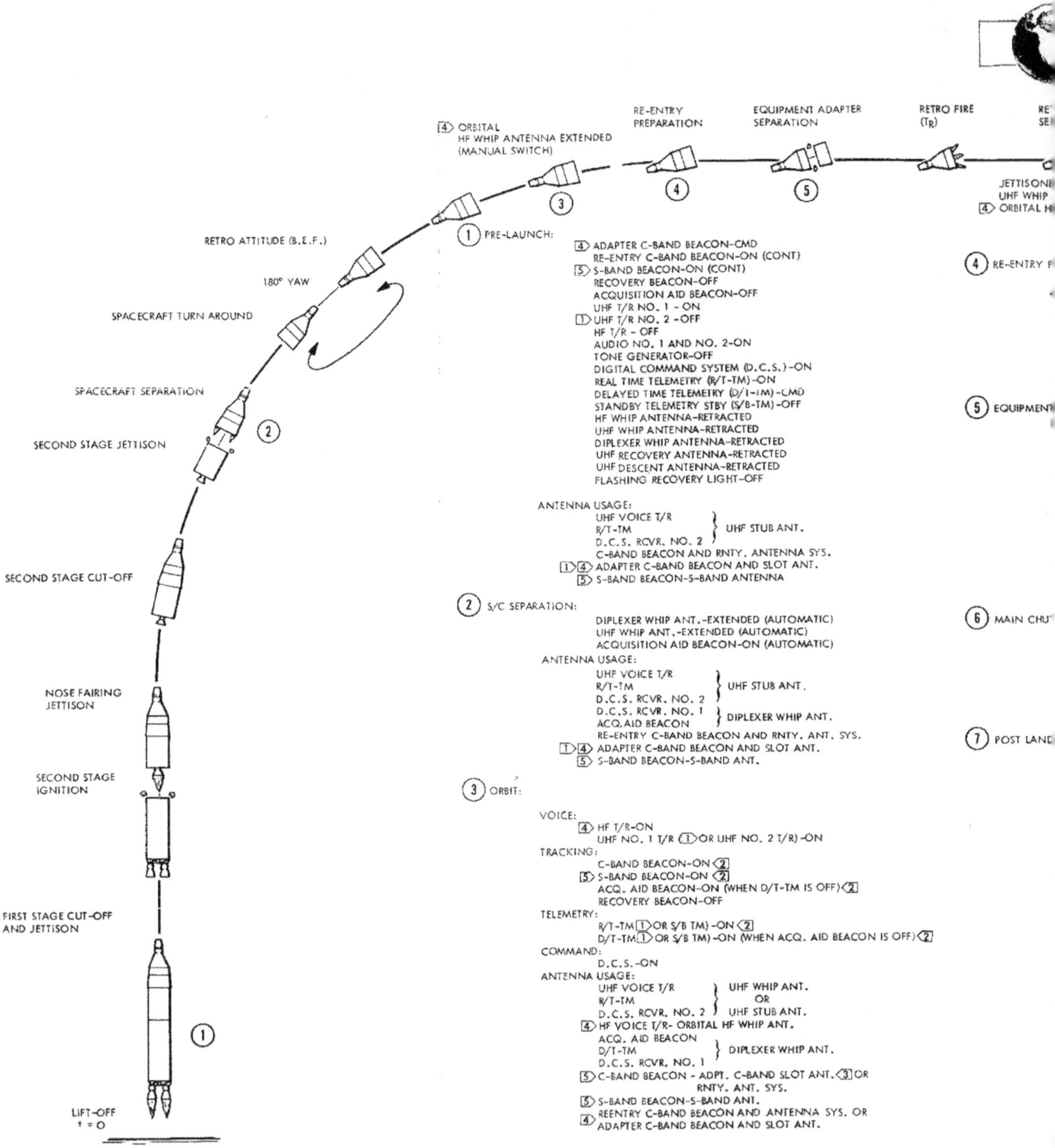

RE-ENTRY PREPARATION

EQUIPMENT ADAPTER SEPARATION

RETRO FIRE (T_R)

[4] ORBITAL HF WHIP ANTENNA EXTENDED (MANUAL SWITCH)

JETTISONE UHF WHIP
[4] ORBITAL H

RETRO ATTITUDE (B.E.F.)

180° YAW

SPACECRAFT TURN AROUND

SPACECRAFT SEPARATION

SECOND STAGE JETTISON

SECOND STAGE CUT-OFF

NOSE FAIRING JETTISON

SECOND STAGE IGNITION

FIRST STAGE CUT-OFF AND JETTISON

LIFT-OFF
t = 0

(1) PRE-LAUNCH:
    [4] ADAPTER C-BAND BEACON-CMD
    RE-ENTRY C-BAND BEACON-ON (CONT)
    [5] S-BAND BEACON-ON (CONT)
    RECOVERY BEACON-OFF
    ACQUISITION AID BEACON-OFF
    UHF T/R NO. 1 - ON
    [1] UHF T/R NO. 2 -OFF
    HF T/R - OFF
    AUDIO NO. 1 AND NO. 2-ON
    TONE GENERATOR-OFF
    DIGITAL COMMAND SYSTEM (D.C.S.)-ON
    REAL TIME TELEMETRY (R/T-TM)-ON
    DELAYED TIME TELEMETRY (D/T-TM)-CMD
    STANDBY TELEMETRY STBY (S/B-TM)-OFF
    HF WHIP ANTENNA-RETRACTED
    UHF WHIP ANTENNA-RETRACTED
    DIPLEXER WHIP ANTENNA-RETRACTED
    UHF RECOVERY ANTENNA-RETRACTED
    UHF DESCENT ANTENNA-RETRACTED
    FLASHING RECOVERY LIGHT-OFF

ANTENNA USAGE:
    UHF VOICE T/R
    R/T-TM   } UHF STUB ANT.
    D.C.S. RCVR. NO. 2
    C-BAND BEACON AND RNTY. ANTENNA SYS.
    [1][4] ADAPTER C-BAND BEACON AND SLOT ANT.
    [5] S-BAND BEACON-S-BAND ANTENNA

(2) S/C SEPARATION:
    DIPLEXER WHIP ANT.-EXTENDED (AUTOMATIC)
    UHF WHIP ANT.-EXTENDED (AUTOMATIC)
    ACQUISITION AID BEACON-ON (AUTOMATIC)
ANTENNA USAGE:
    UHF VOICE T/R
    R/T-TM   } UHF STUB ANT.
    D.C.S. RCVR. NO. 2
    D.C.S. RCVR. NO. 1   } DIPLEXER WHIP ANT.
    ACQ. AID BEACON
    RE-ENTRY C-BAND BEACON AND RNTY. ANT. SYS.
    [1][4] ADAPTER C-BAND BEACON AND SLOT ANT.
    [5] S-BAND BEACON-S-BAND ANT.

(3) ORBIT:
VOICE:
    [4] HF T/R-ON
    UHF NO. 1 T/R [1] OR UHF NO. 2 T/R)-ON
TRACKING:
    C-BAND BEACON-ON [2]
    [5] S-BAND BEACON-ON [2]
    ACQ. AID BEACON-ON (WHEN D/T-TM IS OFF) [2]
    RECOVERY BEACON-OFF
TELEMETRY:
    R/T-TM [1] OR S/B TM) -ON [2]
    D/T-TM [1] OR S/B TM) -ON (WHEN ACQ. AID BEACON IS OFF) [2]
COMMAND:
    D.C.S.-ON
ANTENNA USAGE:
    UHF VOICE T/R   UHF WHIP ANT.
    R/T-TM   } OR
    D.C.S. RCVR. NO. 2   UHF STUB ANT.
    [4] HF VOICE T/R- ORBITAL HF WHIP ANT.
    ACQ. AID BEACON
    D/T-TM   } DIPLEXER WHIP ANT.
    D.C.S. RCVR. NO. 1
    [5] C-BAND BEACON - ADPT. C-BAND SLOT ANT. [3] OR
        RNTY. ANT. SYS.
    [5] S-BAND BEACON-S-BAND ANT.
    REENTRY C-BAND BEACON AND ANTENNA SYS. OR
    [4] ADAPTER C-BAND BEACON AND SLOT ANT.

(4) RE-ENTRY F

(5) EQUIPMENT

(6) MAIN CHU

(7) POST LAND

 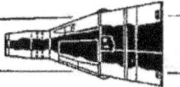

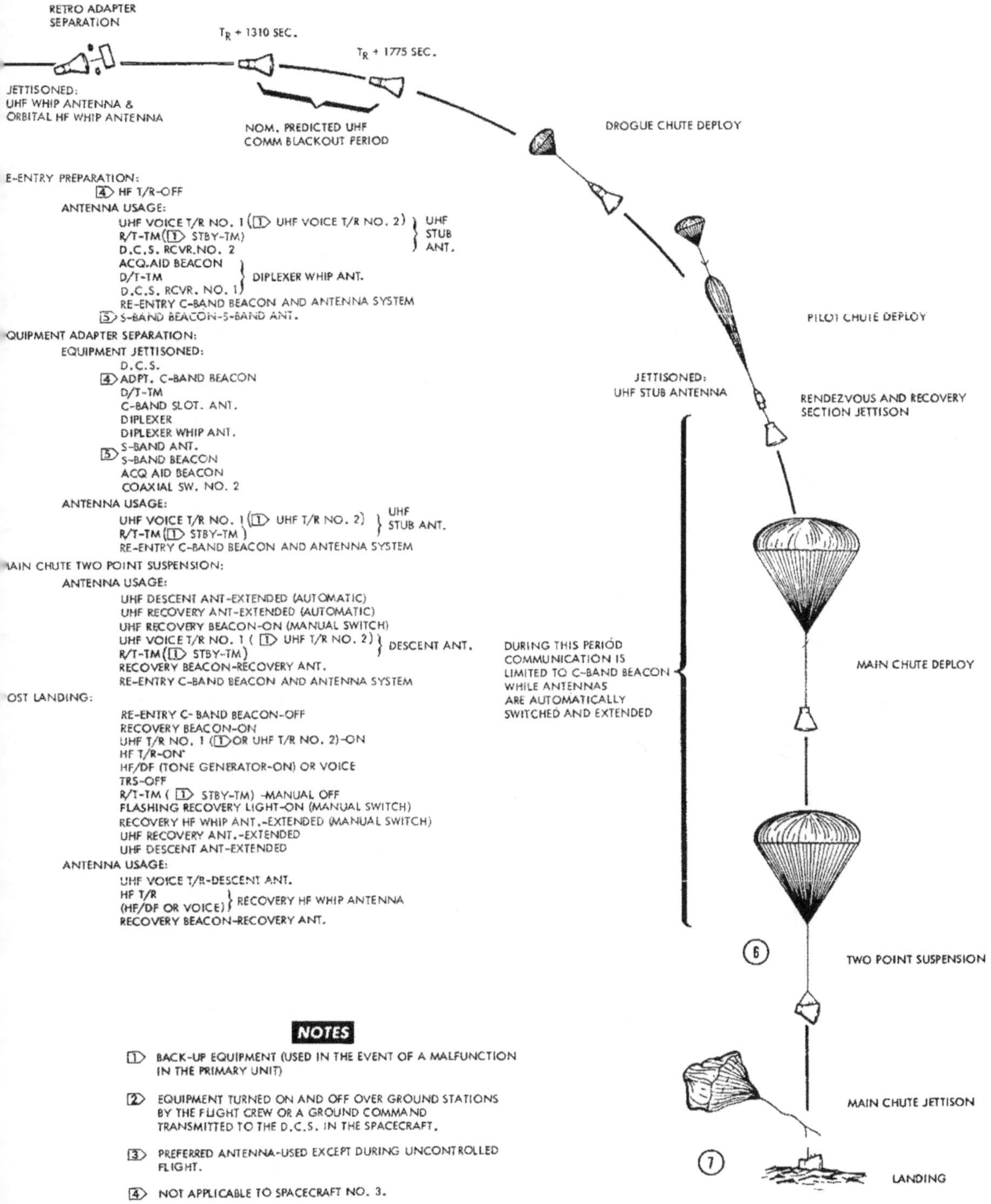

PROJECT GEMINI

SEDR 300

RETRO ADAPTER SEPARATION

JETTISONED:
UHF WHIP ANTENNA &
ORBITAL HF WHIP ANTENNA

T_R + 1310 SEC.

T_R + 1775 SEC.

NOM. PREDICTED UHF
COMM BLACKOUT PERIOD

DROGUE CHUTE DEPLOY

RE-ENTRY PREPARATION:
    [4] HF T/R-OFF
    ANTENNA USAGE:
        UHF VOICE T/R NO. 1 ([1] UHF VOICE T/R NO. 2) } UHF
        R/T-TM ([1] STBY-TM) } STUB
        D.C.S. RCVR. NO. 2 } ANT.
        ACQ. AID BEACON
        D/T-TM } DIPLEXER WHIP ANT.
        D.C.S. RCVR. NO. 1
        RE-ENTRY C-BAND BEACON AND ANTENNA SYSTEM
        [5] S-BAND BEACON-S-BAND ANT.

EQUIPMENT ADAPTER SEPARATION:
    EQUIPMENT JETTISONED:
        D.C.S.
        [4] ADPT. C-BAND BEACON
        D/T-TM
        C-BAND SLOT. ANT.
        DIPLEXER
        DIPLEXER WHIP ANT.
        [5] S-BAND ANT.
        S-BAND BEACON
        ACQ AID BEACON
        COAXIAL SW. NO. 2
    ANTENNA USAGE:
        UHF VOICE T/R NO. 1 ([1] UHF T/R NO. 2) } UHF
        R/T-TM ([1] STBY-TM) } STUB ANT.
        RE-ENTRY C-BAND BEACON AND ANTENNA SYSTEM

PILOT CHUTE DEPLOY

JETTISONED:
UHF STUB ANTENNA

RENDEZVOUS AND RECOVERY
SECTION JETTISON

MAIN CHUTE TWO POINT SUSPENSION:
    ANTENNA USAGE:
        UHF DESCENT ANT-EXTENDED (AUTOMATIC)
        UHF RECOVERY ANT-EXTENDED (AUTOMATIC)
        UHF RECOVERY BEACON-ON (MANUAL SWITCH)
        UHF VOICE T/R NO. 1 ([1] UHF T/R NO. 2) } DESCENT ANT.
        R/T-TM ([1] STBY-TM)
        RECOVERY BEACON-RECOVERY ANT.
        RE-ENTRY C-BAND BEACON AND ANTENNA SYSTEM

DURING THIS PERIOD
COMMUNICATION IS
LIMITED TO C-BAND BEACON
WHILE ANTENNAS
ARE AUTOMATICALLY
SWITCHED AND EXTENDED

MAIN CHUTE DEPLOY

POST LANDING:
    RE-ENTRY C-BAND BEACON-OFF
    RECOVERY BEACON-ON
    UHF T/R NO. 1 ([1] OR UHF T/R NO. 2)-ON
    HF T/R-ON
    HF/DF (TONE GENERATOR-ON) OR VOICE
    TRS-OFF
    R/T-TM ([1] STBY-TM) -MANUAL OFF
    FLASHING RECOVERY LIGHT-ON (MANUAL SWITCH)
    RECOVERY HF WHIP ANT.-EXTENDED (MANUAL SWITCH)
    UHF RECOVERY ANT.-EXTENDED
    UHF DESCENT ANT-EXTENDED
    ANTENNA USAGE:
        UHF VOICE T/R-DESCENT ANT.
        HF T/R
        (HF/DF OR VOICE) } RECOVERY HF WHIP ANTENNA
        RECOVERY BEACON-RECOVERY ANT.

(6) TWO POINT SUSPENSION

MAIN CHUTE JETTISON

(7) LANDING

NOTES

[1] BACK-UP EQUIPMENT (USED IN THE EVENT OF A MALFUNCTION IN THE PRIMARY UNIT)

[2] EQUIPMENT TURNED ON AND OFF OVER GROUND STATIONS BY THE FLIGHT CREW OR A GROUND COMMAND TRANSMITTED TO THE D.C.S. IN THE SPACECRAFT.

[3] PREFERRED ANTENNA-USED EXCEPT DURING UNCONTROLLED FLIGHT.

[4] NOT APPLICABLE TO SPACECRAFT NO. 3.

[5] SPACECRAFT 3 ONLY.

Figure 9-3 Communication System Sequential Diagram

FM2-9-3

During the recovery phase of the mission, the UHF rescue beacon transceiver, which is Government Furnished Equipment (GFE), may be connected to the UHF recovery antenna.

VOICE COMMUNICATIONS

Voice communications is provided by one HF and two UHF transmitter/receivers and the voice control center (VCC). The VCC contains all necessary controls and switches required for various keying modes, transmitter/receiver selection, squelch, volume control, and voice recording. The HF voice transmitter/receiver may also be used for direction finding (DF) purposes during the post-landing phase of the mission.

On S/C 7, an intercomm connector is provided to enable communications between the Flight Crew and Frogmen prior to opening the spacecraft hatches during the recovery phase of the mission. Light weight headsets are also supplied for use when the spacesuit helmets are removed during orbit, or during post-landing if the helmets or entire spacesuit is removed prior to recovery.

TELEMETRY TRANSMITTERS

Receiving inputs from the pulse code modulated (PCM) programmer and the on-board tape recorder, three telemetry transmitters transmit vital spacecraft systems parameters to the ground stations. The three transmitters operate on different frequencies and are identified as real-time, delayed-time, and stand-by transmitters. The stand-by transmitter is only used in case of real-time, or delayed-time transmitter failure.

PROJECT GEMINI
SEDR 300

FLASHING RECOVERY LIGHT

The flashing recovery light, used during the post-landing phase of the mission, contains its own power supply and improves visual spacecraft location.

DIGITAL COMMAND SYSTEM (DCS)

The digital command system (DCS) provides the command capability between the ground and the spacecraft. The DCS consists of two UHF receivers, a decoder, and two relay packages and is operational from pre-launch until equipment adapter section separation. Basically, the DCS receives and decodes two types of commands: a real-time command (RTC) for spacecraft equipment utilization, or stored program commands (SPC) that provide digital information for various spacecraft systems. Real-time commands operate DCS relays that control power directly or energize relays in the spacecraft electrical system that determine equipment usage. Stored program commands are received and decoded for usage by either the Time Reference System (TRS) or the spacecraft computer.

SYSTEM OPERATION

The communication system is semi-automatic in operation. The sequence and theory of operation of the communication system is described in the following paragraphs and illustrated in Figures 9-2 and 9-3. Individual components are described in greater detail under Systems Units.

VOICE TAPE RECORDER

Voice tape recordings may be made during the mission, in accordance with the applicable flight plan, by placing either MODE switch on the VCC to the RCD position on S/C 3 and 4 or by placing the RECORD switch on the VCC to the MOM or CONT position on S/C 7. The TONE VOX, AUDIO & UHF T/R 1 and 2 circuit

SEDR 300
PROJECT GEMINI

breakers must be in the ON position. Each tape cartridge allows approximately one hour of recording time and is easily changed by the pilot. An end-of tape light on the voice recorder illuminates for two seconds when two minutes of recording time remains on the tape. The end-of tape light remains lighted when the end of the tape is reached.

On S/C 4 and 7 a digital timing signal is applied to one channel of the tape for time correlation of the voice recording. On S/C 3 the time correlation information is obtained by recording the GMT or Event Timer reading on the voice channel. More detailed information on the voice tape recorder and other components of the communication system is provided under systems units which describe individual component operation.

PRE-LAUNCH

C-Band Radar Beacons - Spacecrafts No. 4 and 7

During pre-launch, the BEACONS - C circuit breaker on S/C 7 and the BEACONS - C RNTY and C ADPT circuit breakers on S/C 4 are placed to the ON position to arm the C-RNTY and C-ADPT BEACON CONTROL switches. The C-RNTY switch is placed in the CONT position during pre-launch to enable the re-entry C-band beacon to transpond when properly interrogated by a ground station. The C-ADPT switch is placed in the CMD position during pre-launch. The CMD position, enables the ground station, during launch, to activate the adapter C-band beacon via a DCS channel if the need arises. After the adapter C-band beacon is activated, it will transpond when properly interrogated by a ground station. The C-band antenna system, used with the re-entry C-band beacon, is operational when the

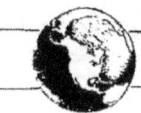

PROJECT GEMINI

SEDR 300

ANT SEL switch is placed in the RNTY position. The ANT SEL switch is armed when the COAX CNTL circuit breaker is positioned to ON. The ANT SEL switch controls application of power to the phase shifter power supply in the C-band antenna system.

C-Band Radar Beacon - Spacecraft No. 3

The C-band radar beacon is energized during pre-launch by placing the BEACONS-C circuit breaker on the overhead switch/circuit breaker panel to the ON position to arm the C-BAND BEACON CONTROL switch. The switch contains two positions, CONT and CMD. The CONT position is utilized during pre-launch to enable the C-band beacon to transpond when properly interrogated by a ground station. The CMD position will be utilized and described during the orbital phase of the mission.

The C-band antenna system is used with the C-band beacon during pre-launch. Proper antenna selection is made when the ANT SEL switch is placed in the RNTY position. The ANT SEL switch is armed from the common control bus through the COAX CNTL circuit breaker of the left switch/circuit breaker panel. The RNTY position of the ANT SEL switch places coaxial switch no. 6 to position no. 1; thus connecting the C-band beacon to the C-band antenna system.

S-Band Radar Beacon - Spacecraft No. 3

The S-band radar beacon is utilized during pre-launch as a back up tracking system for the C-band beacon. Actuation of the S-band beacon is similar to that of the C-band beacon. The BEACONS-S circuit breaker is placed in the ON position and the S-BAND BEACON CONTROL switch is placed in the CONT position to enable the S-band beacon to transpond when properly interrogated by a ground station. The CMD position of the S-BAND BEACON CONTROL switch will be discussed during the orbital phase of the mission. The S-band radar beacon uses the S-band annular slot antenna which is the only antenna provided for S-band operation.

UHF Transmitter/Receiver

The no. 1 UHF voice transmitter/receiver will be utilized during pre-launch unless some malfunction occurs in which case the no. 2 transmitter/receiver can be selected. For operation of the no. 1 or no. 2 UHF voice transmitter/receiver, stand-by power is applied through the AUDIO & UHF T/R 1 and 2 circuit breakers which must be in the ON position. The selected transmitter/receiver will be powered by placing the UHF select switch to either the no. 1 or no. 2 position and the MODE switch of no. 1 or no. 2 AUDIO to the UHF position. The UHF select switch also operates coaxial switch no. 1 to connect the selected transmitter/receiver to the quadriplexer. Coaxial switch power is obtained from the common control bus through the ON position of the UHF RELAY circuit breaker. The method of keying the UHF transmitter/receiver is selected by positioning the KEYING switch on the VCC to voice operated relay (VOX), push-to-talk (PTT), or continuous intercomm/push-to-talk transmitter keying (CONT INT/PTT).

The desired antenna usage is determined by placing the ANT CNTL circuit breaker to the ON position. This places coaxial switch no. 3 to position no. 1; thus connecting the quadriplexer to the IN position of the coaxial switch no. 5. Coaxial switch no. 5 was placed in position no. 1 when the ANT SEL switch was placed in the RNTY position during the C-band beacon operation. With coaxial switch no. 5 in position no. 1 the UHF stub antenna is available for UHF voice transmission and reception. Prior to umbilical release, voice communication is accomplished between the spacecraft and the ground complex through a hardline utilizing the headset and microphone amplifiers of the VCC. After umbilical release, voice transmission to the ground complex is accomplished by means of the UHF voice transmitter/receiver.

Real-Time Telemetry Transmitter

The real-time telemetry transmitter will be operating during the pre-launch phase of the mission. The real-time telemetry transmitter is powered by placing the RT XMTR circuit breaker in the ON position and the TM CONTROL switch to the R/T & ACQ position.

The real-time telemetry transmitter utilizes the UHF stub antenna via the quadriplexer and coaxial switches no. 3 and no. 5 the same as the UHF transmitter/receiver.

In case of failure of the real-time telemetry transmitter, the stand-by telemetry transmitter may be used for real-time transmission. To operate the stand-by transmitter, the STBY XMTR CNTL and PWR circuit breakers must be in the ON position. Selection can then be made by a ground command via a DCS channel, if the STBY TM CONTROL is in the OFF position. If the Flight Crew makes the selection, the STBY TM CONTROL switch will be placed in the R/T position. The stand-by, operating as the real-time telemetry transmitter, utilizes the UHF stub antenna for transmission.

Non-Operational Components

All components other than those described previously, will be non-operational,

SEDR 300

PROJECT GEMINI

except for the DCS, during the pre-launch phase of the mission. To assure the off condition of those components, the following switches should be in the position specified below:

    HF select (on VCC)   - OFF
    BEACON CONTROL - RESC   - OFF
    HF ANT   - OFF

To assure proper sequential actuation of the various communication components, the following circuit breakers (in addition to those previously described) must be placed into the ON position prior to launch:

    WHIP ANTENNAS - HF
    WHIP ANTENNAS - UHF
    WHIP ANTENNAS - DIPLEX
    HF XMTR-RCVR - RE-ENTRY (ON S/C 3)
                        HF T/R (ON S/C 4 AND 7)
    BEACONS - ACQ
    BEACONS - RESC
    XMTRS - DT
    TAPE RCDR - CNTL

SPACECRAFT/LAUNCH VEHICLE SEPARATION

Equipment usage, after spacecraft/launch vehicle separation, is identical to that described under pre-launch except for the following: Upon closure of any two of the three spacecraft separation sensors, the acquisition aid beacon will be energized. The UHF whip antenna solenoid actuators will be powered and initiate the release mechanism of the UHF whip antennas, on the retro and

SEDR 300
PROJECT GEMINI

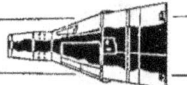

equipment adapter sections, allowing them to self extend.

The acquisition aid beacon transmits via the diplexer and UHF whip antenna on the equipment adapter section. Placing the TAPE RCDR - CNTL circuit breaker to ON and the TM CONTROL switch to R/T & ACQ, during pre-launch, placed coaxial switch no. 2 in position no. 1 to connect the acquisition aid beacon to the diplexer.

ORBIT

During orbit, operation of the telemetry transmitters and beacons will noramlly be controlled by ground commands via DCS channels. To operate from ground commands, the following switches are placed in the CMD position: on S/C 4 and 7 the C-ADPT, C-RNTY and T/M CONTROL; on S/C 3 the C-BAND, S-BAND and the T/M CONTROL.

On S/C 7, to allow uninterrupted sleep of the Flight Crew during extended missions, a SILENCE switch has been added to the VCC. The SILENCE switch, in the NORM position, applies power to both headsets of the Flight Crew. In the NO. 1 or NO. 2 position, the corresponding pilot headset amplifier will not be powered to allow uninterrupted sleep.

HF Voice Transmitter/Receiver

On S/C 3, HF communications is not used during orbit, to avoid extending the recovery HF whip antenna on the re-entry module prior to landing. On S/C 4 and 7, HF communications, during orbit, is via the orbital HF whip on the retro adapter section. At orbital insertion, the Flight Crew will extend the orbital HF whip by placing the HF ANT switch to the EXT position. This also places

9-14

PROJECT GEMINI

coaxial switch no. 4 in position no. 2 to allow HF voice transmission and reception via the orbital HF whip. On S/C 4, power is also applied to the retract mechanism of the recovery HF whip to assure it remains in the retracted position. After extension of the orbital HF whip (approximately one minute), the HF ANT switch is returned to the OFF position.

Stand-by power is applied to the HF transmitter/receiver by the HF T/R circuit breaker which was positioned to ON during pre-launch. The HF transmitter/receiver is powered by positioning the HF select switch (on the VCC) to RNTY and the MODE switch of No. 1 or No. 2 AUDIO to HF. The method of keying for the HF transmitter/receiver is selected by positioning the KEYING switch on the VCC to voice operated relay (VOX), push-to-talk (PTT) or continuous intercom/push-to-talk transmitter keying (CONT INT/PTT). During orbit, either of the three keying modes may be selected by the Flight Crew.

UHF Voice Transmitter/Receiver

UHF voice transmitter/receiver operation is identical to that described under pre-launch with the following exception. Preferred antenna usage during orbit for UHF transmission and reception is via the retro adapter UHF whip antenna. The retro adapter UHF whip antenna is selected by placing the ANT SEL switch to the ADPT position which places coaxial switch no. 5 to position no. 2. Although preferred UHF transmission and reception is via the UHF whip antenna, the UHF stub antenna may be utilized during orbit by placing the ANT SEL switch to the RNTY position. On S/C 3, the ANT SEL switch also controls coaxial switch no. 6 and the C-band radar beacon will be switched from the C-band slot antenna on the equipment adapter section to the C-band antenna system on the re-entry

9-15

module when the ANT SEL switch is placed to the RNTY position.

Delayed Time Telemetry Transmitter

The acquisition aid beacon will be operating continuously throughout the orbital phase of the mission except when the delayed-time telemetry transmitter is operating. At the time the ground station receives the acquisition aid beacon, the ground station will initiate the delayed-time telemetry transmitter to transmit data stored on the on-board recorder while the spacecraft was between ground stations. Delayed-time transmission may also be initiated by the Flight Crew by placing the T/M CONTROL switch to the R/T - D/T position which initiates both real-time and delayed-time telemetry transmission.

Real-time and delayed-time transmission will normally be initiated from the ground station via DCS channels. At the time the delayed-time telemetry transmitter is initiated, the acquisition aid beacon will be turned off and coaxial switch no. 2 will be placed in position no. 2, allowing telemetry transmission via the diplexer and UHF whip antenna on the equipment adapter section.

At the time the spacecraft goes out of range of the ground station, delayed-time telemetry transmission will cease and the acquisition aid beacon will resume transmission. This function will normally be performed by the ground station but may be performed by the Flight Crew. If this function is performed by the Flight Crew, the T/M CONTROL switch may be placed to either the CMD or the R/T & ACQ position. If the R/T & ACQ position is selected, the delayed-time transmitter will be turned off and the real-time transmitter and the acquisition aid beacon will be transmitting. If the CMD position is selected, only the acquisition aid beacon will be operating; however, the ground station has the capa-

SEDR 300
PROJECT GEMINI

bility of energizing the real-time telemetry transmitter via a DCS command.

Either of the three previously described methods of disabling the delayed-time telemetry transmitter will also operate coaxial switch no. 2, placing it into position no. 1 to allow acquisition aid beacon transmission via the diplexer and UHF whip antenna.

In case of failure of the delayed-time telemetry transmitter, the stand-by telemetry transmitter may be used for delayed-time transmission. With the STBY TM CONTROL switch in the OFF position, the stand-by transmitter is switched to delayed-time transmission by a ground command via a DCS command. Selection by the Flight Crew is accomplished by placing the STBY TM CONTROL switch to D/T position. Delayed-time transmission via the stand-by telemetry transmitter utilizes either the UHF stub or the UHF whip antenna on the retro adapter, depending upon the setting of the ANT SEL switch.

Real-Time Telemetry Transmitter

Orbital operation of the real-time telemetry transmitter is similar to that of the delayed-time telemetry transmitter in that the real-time telemetry transmitter will only be operational while the spacecraft is within range of a ground station. The real-time telemetry transmitter is actuated by a DCS command from the ground station or by the Flight Crew as described under delayed-time telemetry transmitter operation. Real-time transmission is accomplished by means of either the UHF stub or the retro section UHF whip antenna, depending upon the position of the ANT SEL switch. In case of failure of the real-time telemetry transmitter, the stand-by transmitter may be used for real-time

 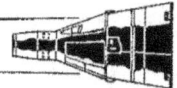

transmission. With the STBY TM CONTROL switch in the OFF position, selection is performed by the ground station via a DCS command. Selection by the Flight Crew is accomplished by placing the STBY TM CONTROL switch to the R/T position. The stand-by telemetry transmitter transmits via the UHF stub or the retro section UHF whip antenna, depending upon the position of the ANT SEL switch.

It should be noted that the stand-by telemetry transmitter may only be used for delayed-time, or real-time transmission but may not be used simultaneously for both. In the event that both the real-time and delayed-time transmitters fail, it is up to the ground station to determine the purpose for which the stand-by transmitter will be used.

C-Band Radar Beacons - Spacecrafts No. 4 and 7

During orbit, the C-band beacons will be operational only while the spacecraft is within range of a ground station. Normally, the adapter C-band beacon will be used during stabilized orbital flight and the re-entry C-band beacon used during orbital roll maneuvers. Operation of the beacons is similar to that described under pre-launch. The C-RNTY and C-ADPT BEACON CONTROL switches are normally kept in the CMD position. When the spacecraft comes within range of a ground station, as determined from the acquisition aid beacon signal, power to the desired C-band beacon will be actuated by ground command via a DCS command. The Flight Crew may actuate the desired beacon by placing the C-ADPT or C-RNTY BEACON CONTROL switch to the CONT position. After beacon power is actuated, the selected C-band beacon will transpond when properly interrogated by a ground station. When the re-entry C-band beacon is selected, the ANT SEL switch should be placed in the RNTY position to energize the phase shifter and provide optimum radiation coverage.

PROJECT GEMINI

SEDR 300

C-Band Radar Beacon - Spacecraft No. 3

During orbit, the C-band beacon will be operational only while the spacecraft is within the range of a ground station. Basically, operation of the beacon is similar to that described under pre-launch. The C-BAND BEACON CONTROL switch is normally kept in the CMD position. When the spacecraft comes within range of a ground station, as determined from the acquisition aid beacon signal, the C-band beacon power will be actuated by ground command via a DCS command. The Flight Crew may actuate the C-band beacon by placing the C-BAND BEACON CONTROL switch to the CONT position. After power is actuated, the C-band beacon will transpond when properly interrogated by a ground station.

Preferred antenna usage for the C-band radar beacon during orbit is the C-band annular slot antenna of the adapter. Antenna usage is dependent upon the position of the ANT SEL switch. With the ANT SEL switch in the ADPT position, coaxial switch no. 6 is placed to position no. 2 connecting the C-band beacon to the C-band annular slot antenna. If the ANT SEL switch is in the RNTY position, coaxial switch no. 6 is in position no. 1 making the C-band antenna system available for C-band beacon transmission and reception.

S-Band Radar Beacon - Spacecraft No. 3

The S-band beacon, used as back-up tracking for the C-band beacon, will be operational during the same intervals as the C-band beacon. The S-BAND BEACON CONTROL switch is normally kept in the CMD position. When the spacecraft comes within range of a ground station, as determined from the acquisition aid beacon signal, the S-band beacon power will be actuated by ground command via a DCS command. The Flight Crew may activate the beacon by placing the S-BAND

SEDR 300

PROJECT GEMINI

BEACON CONTROL switch to the CONT position. After the S-band beacon power is actuated, the beacon will transpond when properly interrogated by a ground station. The S-band beacon transmission and reception is via the S-band annular slot antenna on the equipment adapter section.

ADAPTER SEPARATION

Shortly prior to equipment section separation, the Flight Crew will, if not already using the re-entry module antennas, place the ANT SEL switch to RNTY and the T/M CONTROL switch to R/T & ACQ. On S/C 4 and 7 the Flight Crew will place the C-RNTY BEACON CONTROL switch to CONT. On S/C 3 the Flight Crew will place the C-BAND BEACON CONTROL switch to CONT. The C-band antenna system and UHF stub antenna will then be used for the required transmission and reception.

On S/C 4 and 7, the Flight Crew will disable HF voice communications by placing the HF select switch to the OFF position. On S/C 4, the orbital HF whip on the retro adapter section will remain extended. On S/C 7, the orbital HF whip may be retracted by holding the HF ANT switch in the RET position for approximately 1.5 minutes for complete retraction.

At equipment section separation, the following communications components will be jettisoned with the equipment adapter section:

        digital command system (DCS)
        delayed-time telemetry transmitter
        diplexer
        C-band annular slot antenna
        adapter C-band radar beacon (on S/C 4 and 7)

PROJECT GEMINI
SEDR 300

        S-band beacon and annular slot antenna (on S/C 3)

        diplexer UHF whip antenna

        acquisition aid beacon

        coaxial switch no. 2

This limits telemetry data transmission to real-time, voice communication to UHF, and tracking data to the re-entry C-band beacon.

Following equipment section separation and retro firing, retro section separation will occur at which time the retro UHF whip antenna and orbital HF whip antenna (on S/C 4 and 7) will be jettisoned.

RE-ENTRY

During the re-entry phase of the mission, two short duration communication blackout periods exist. The first period, from approximately $T_R + 1310$ seconds to $T_R + 1775$ seconds, is caused by an ionization shield around the spacecraft. This ionization is because of the extremely high temperatures created upon re-entry into the earth's atmosphere. The second blackout period occurs at rendezvous and recovery (R & R) section separation when the UHF stub antenna is jettisoned. This period is terminated at two-point suspension which occurs shortly after main parachute deployment.

At R & R separation, energized chute deploy time delay relays initiate coaxial switch no. 3, placing it to position no. 2. This makes the UHF descent antenna available for real-time telemetry transmission and UHF voice communications.

At two-point suspension, the UHF recovery and UHF descent antennas are automatically extended. The Flight Crew will initiate the UHF recovery beacon by

 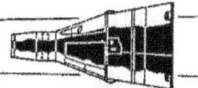

placing the RESC BEACON CONTROL switch to the ON position on S/C 3 and to the W/O LT position on S/C 4 and 7.

Antenna usage during re-entry will be as follows: prior to R & R separation, real time telemetry transmission and UHF voice communication will be via the UHF stub antenna. After two-point suspension, the UHF descent antenna will be used instead of the UHF stub. The re-entry C-band beacon and C-band antenna system will be used for tracking and the UHF recovery beacon will use the UHF recovery antenna.

LANDING THROUGH RECOVERY

Upon impact, the Flight Crew will jettison the main parachute by actuating the PARA JETT switch. This will also extend the flashing recovery light. On S/C 3, the recovery light is energized automatically at extension. On S/C 4 and 7 the recovery light is energized by changing the RESC BEACON CONTROL switch from the W/O LT position to the ON position.

The C-band beacon will be turned off by the Flight Crew placing the C-RNTY switch on S/C 4 and 7 or the C-BAND switch on S/C 3 to the CMD position. The real-time telemetry transmitter will be turned off by placing the T/M CONTROL switch to the CMD position. If the Flight Crew selected the stand-by telemetry transmitter for real-time transmission, the stand-by transmitter will be turned off by placing the STBY TM CONTROL switch to the OFF position.

The Flight Crew will extend the recovery HF whip antenna by using the HF ANT switch as follows: on S/C 3, by placing the switch to EXT; on S/C 4, by placing the switch to PST LDG; on S/C 7, by holding the switch in the EXT position for approximately one minute. HF voice communication is then possible by placing

SEDR 300
PROJECT GEMINI

the HF select switch to the RNTY position and either MODE switch to HF. The HF transmitter/receiver can also be used to transmit a direction finding signal by placing either MODE switch to HF/DF.

During the recovery phase of the mission, the UHF rescue beacon transceiver may be connected to the UHF recovery antenna. The UHF recovery beacon can be turned-off by positioning the RESC BEACON CONTROL switch to OFF. On S/C 4 and 7 lightweight headsets are provided to replace the Spacesuit Helmets if the helmets or spacesuits are removed and the Flight Crew remains inside the Spacecraft. On S/C 7 a recovery team disconnect is provided for intercomm conversation between the Flight Crew and Frogmen prior to opening the Spacecraft hatches.

SYSTEM UNITS

ANTENNAS

UHF Descent and UHF Recovery Antennas

Purpose: The UHF descent antenna provides for simultaneous transmission for the real-time and stand-by telemetry transmitters and transmission and reception for the UHF voice transmitter/receiver. The UHF recovery antenna provides transmission capability for the UHF recovery beacon. The two antennas are only used from two-point suspension of the main parachute through final recovery of the spacecraft.

Physical Characteristics: The two antennas, being similar in physical appearance, are shown in Figure 9-4. Both antennas are mounted in the parachute cable trough where they are stowed until main parachute two point suspension during

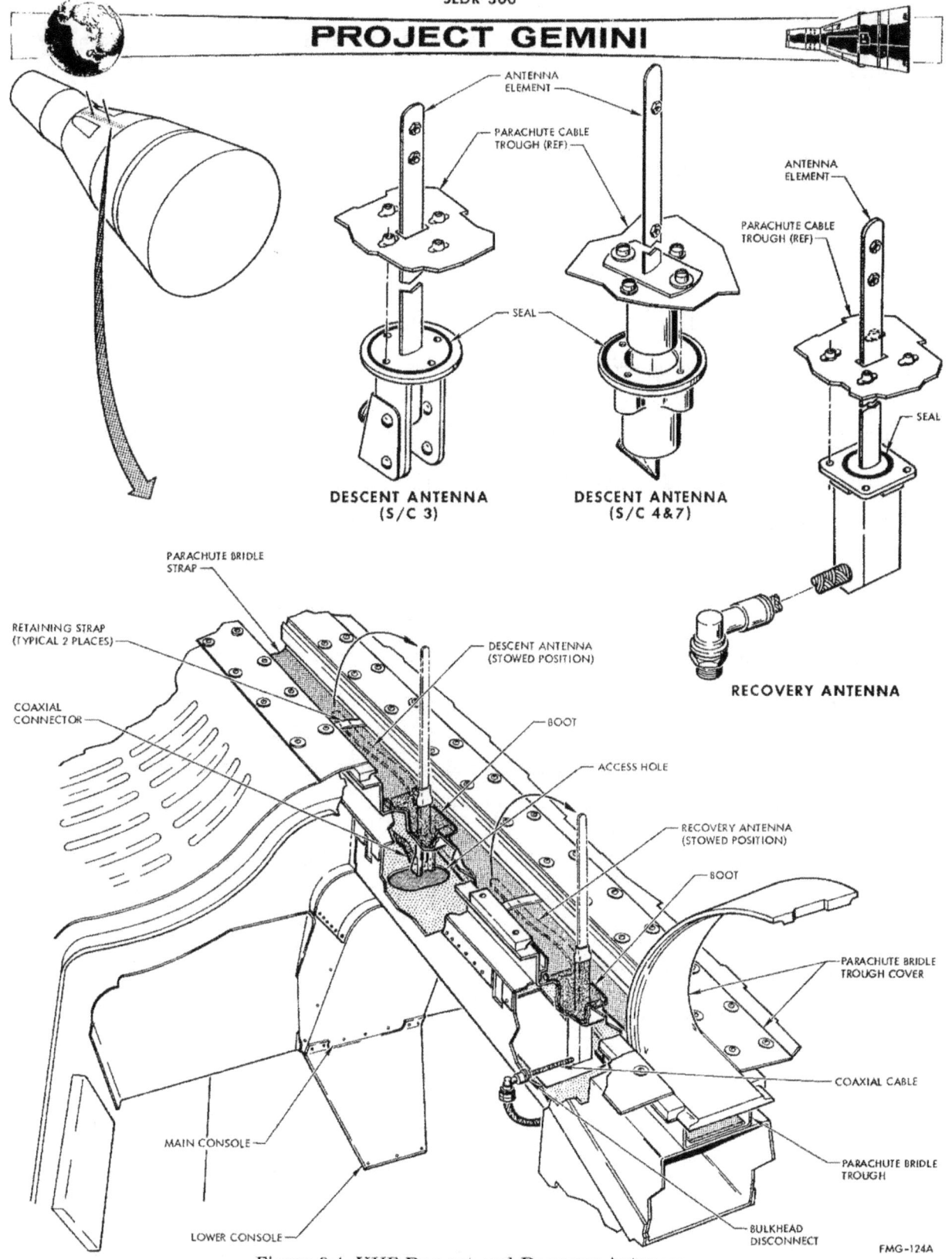

Figure 9-4 UHF Descent and Recovery Antennas

the landing phase of the mission.

The element of each antenna consists of two one-half inch wide gold plated steel blades bolted together at two places. The UHF descent antenna is approximately 17.28 inches long on S/C 3, and 16 inches long on S/C 4 and 7. The UHF recovery antenna is approximately 18 inches long.

Mechanical Characteristics: For rigidity, the antenna element is shaped in a 0.5 inch wide arc having a radius of 1.5 inches. The two laminations of steel blades, compounding a single antenna element, are rigidly secured at the lower half of the antenna. To allow a slight displacement of the two laminations with respect to each other during stowage and deployment, two nuts and bolts placed through elongated holes secure the two laminations together at the upper half of the antenna element.

The antennas are bent towards the small end of the spacecraft for stowage and are held in place by a retaining strap. The strap is broken when the landing system shifts from single point to two point suspension, allowing the antennas to self extend.

Each of the two antennas provide a radiation pattern which is identical to that of a quarter wave stub.

UHF Stub Antenna

Purpose: The UHF stub antenna (Figure 9-5) provides for simultaneous transmission of the real-time and delayed-time telemetry transmitters, transmission and reception for the UHF voice transmitter/receivers, and reception for DCS receiver no. 2. The antenna may be used from pre-launch until separation of

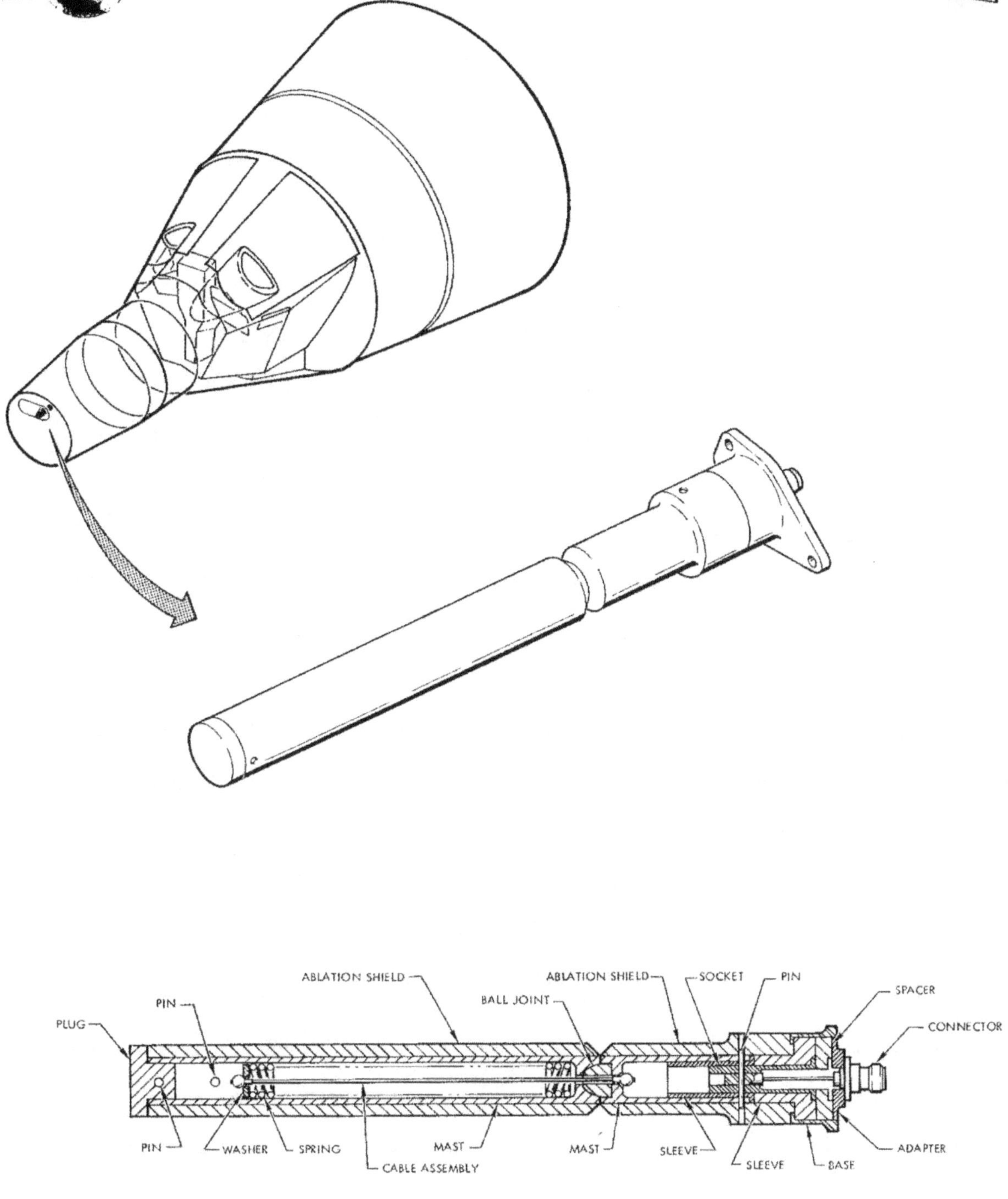

SEDR 300
PROJECT GEMINI

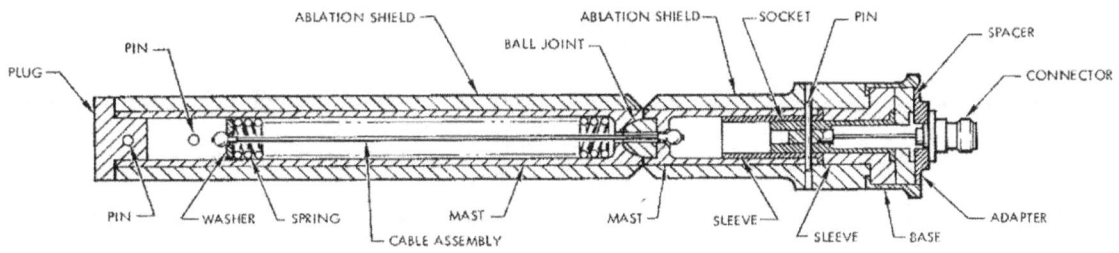

Figure 9-5 UHF Stub Antenna

the R & R section during re-entry but is normally used from pre-launch to orbital insertion and from re-entry preparation to R & R section separation.

Physical Characteristics: The UHF stub antenna, physically constructed as illustrated in Figure 9-5, is mounted in the nose of the R & R section. The antenna protrudes forward from the R & R section and is covered by the nose fairing during the boost phase of the mission. The antenna consists of a mast and base and weighs approximately 1.1 pounds. The mast is constructed of 3/4 inch cobalt steel, machined to tubular form, and covered by a teflon ablation shield for protection during re-entry. The antenna is approximately 13.5 inches long including the connector and 1.25 inches in diameter over the ablation material. The mast consists of two sections. The front section is mounted on a cobalt steel ball joint and retained to the rear section by a spring loaded cable. Electrical contact between the mast sections is made through the ball joint and the spring loaded cable assembly. The ball joint allows the front section of mast to be deflected to approximately 90 degrees in any direction around the antenna axis. The spring of the cable assembly is pre-loaded to approximately 45 pounds to cause the front section, when deflected, to return to the erected position.

The RF connector is press fitted into a socket and makes contact to the mast through the socket and sleeve which are the same material as the mast. The shell of the RF connector is mounted to the base which is isolated from the mast by a teflon spacer and sleeve.

Mechanical Characteristics: The UHF stub is a quarter wave length antenna. The radiating length of the antenna, mounted in the R & R section, is approximately 11.2 inches.

UHF Whip Antennas

Purpose: Two identical UHF whip antennas (Figure 9-6) provide the required UHF transmission and reception facilities during orbit. One of the UHF antennas is located on the equipment adapter section and serves the DCS receiver no. 1, the acquisition aid beacon or delayed-time telemetry transmitter. The second UHF antenna, mounted on the retro adapter section, serves the real-time and stand-by telemetry transmitters, the UHF voice transmitter/receiver, and DCS receiver no. 2.

Physical Characteristics: The UHF whip antenna is self extendable and requires no power other than that required for initial release. The antenna element is a tubular device made from a 2 inch wide beryllium copper strip processed in the form of a tube. The antenna, when fully extended, forms an element that is approximately 12 inches long and 1/2 inch in diameter. During stowage, the tube is opened flat, wound on the inside of a retaining drum, and latched in position. Upon release of the latch by a solenoid, the extension of the antenna depends entirely on the energy stored in the rolled strip material. This energy is sufficient to erect the antenna at a rate of 5 feet/second into its tubular form. In the stored condition, the antenna is flush with the outer skin of the spacecraft.

Mechanical Characteristics: The antenna element is retained inside the housing by a metal lid. A metal post is attached to the lid and passes through the center of the coiled antenna. The bottom of the post is grooved to accept a forked latch which holds the catch post assembly firmly in position prior to

SEDR 300
PROJECT GEMINI

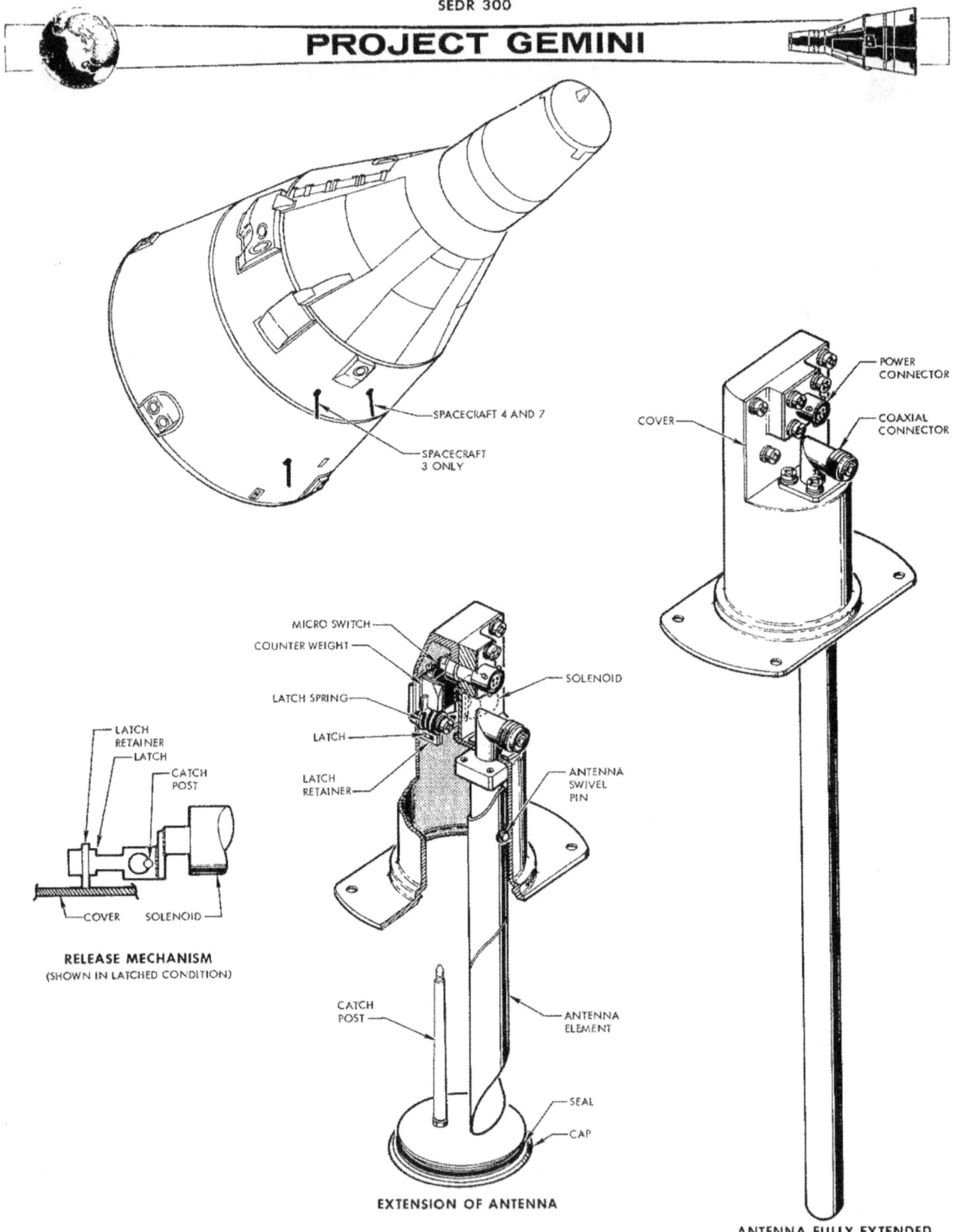

Figure 9-6 UHF Whip Antenna

PROJECT GEMINI

initiation. The forked latch is attached to a miniature pull-solenoid which is spring loaded in the extended position to ensure that launch shock and vibration loads will not cause inadvertent antenna extension. When a voltage from the sequence system is applied to the antenna solenoid at approximately spacecraft separation, the latch will be withdrawn allowing the antenna cap to eject and the antenna to extend. As the catch post assembly is ejected, a microswitch in series with the solenoid coil, opens the circuit to the coil thus preventing further current drain from the power source.

The two antennas are jettisoned with the corresponding adapter section.

HF Whip Antennas

Purpose: The HF whip antennas (one on S/C 3, two on S/C 4 and 7) provide for transmission and reception of the HF voice transmitter/receiver. On S/C 3, the antenna is utilized during the post-landing phase of the mission. On S/C 4 and 7, the antennas are utilized during the orbital and post-landing phases of the mission.

Physical Characteristics: The HF whip antennas are physically constructed as illustrated in Figure 9-7. The recovery HF whip antenna located on the re-entry module is mounted on the small pressure bulkhead, outside the pressurized area of the spacecraft. The orbital HF whip antenna (on S/C 4 and 7) is located on the retro adapter section. The antenna mechanism housing, approximately 6.25 inches wide and 22.4 inches high, completely encloses all parts of the antenna, including storage space for the antenna elements.

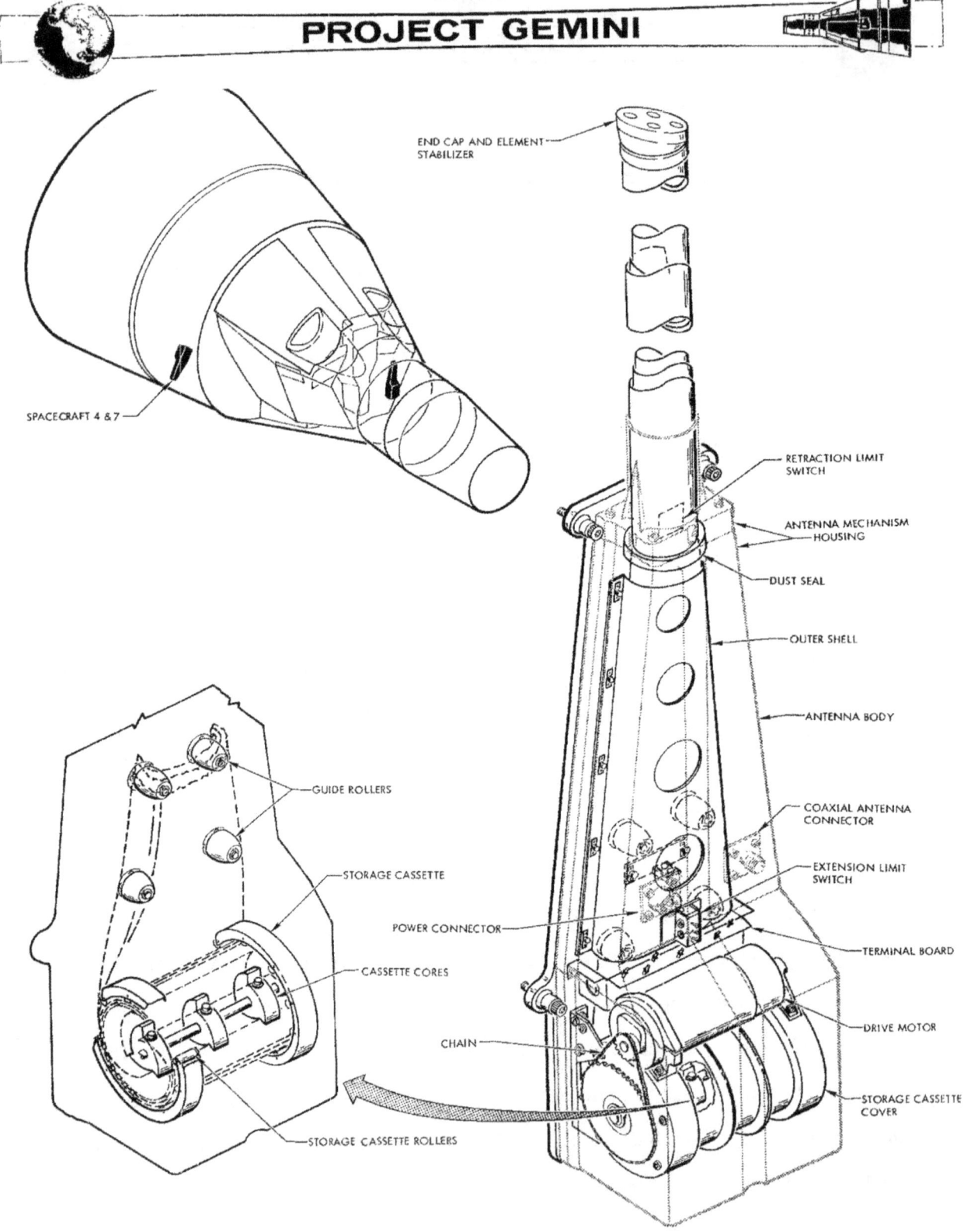

Figure 9-7 HF Whip Antennas

The recovery HF whip antenna contains six elements which, when fully extended, comprise a single antenna mast approximately 13 feet 3 inches long. The orbital HF whip antenna contains three elements which, when fully extended, comprise a single antenna mast approximately 16 feet long on S/C 4, and approximately 13 feet long on S/C 7. The mast is one inch in diameter on all spacecrafts.

Two connectors, supported by the antenna body, provide a means of applying power and connecting the antenna to the RF connector on the HF voice transmitter/receiver. The total weight of the recovery HF whip antenna is approximately 9.0 pounds. The 16 foot version of the orbital HF whip antenna weighs approximately 7.5 pounds and the 13 foot version 6.0 pounds. The main supporting structure of the antenna mechanism housing is the antenna body consisting of a thin fiberglass shell. The outer shell is made in two sections which mate together and form a completely sealed envelope around all moving parts. The antenna mast elements are heat treated stainless steel strips and are stored in a DC motor driven cassette.

Mechanical Characteristics: The strip material comprising the antenna elements is heat treated into a material circular section in such a manner that the edges of the material overlap approximately 180°. When the antenna is retracted, the tubular elements are continuously transformed by guide rollers into a flattened condition and stored in a strained manner in a cassette. Extension and retraction of the antenna is accomplished by a motor which, by means of a chain, drives the storage cassette core. Because of the natural physical shape of the antenna elements, the antenna has a tendency to self-extend; thus providing an extension time of approximately 25 seconds. The retraction time is approxi-

SEDR 300

PROJECT GEMINI

mately 40 seconds. The antenna is stopped within its desired limits by two micro switches, one for extension and one for retraction, which automatically cut the power applied to the motor at the time the extreme limits of the antenna are reached.

The RF connection to the antenna is obtained by a wiper arm sliding on the cassett core drive shaft.

Spacecraft 3 contains no orbital HF whip antenna. After landing, the recovery HF whip is extended or retracted by positioning the HF ANT switch to EXT or RET respectively.

On S/C 4, the HF whip antennas are operated as follows: spacecraft control bus voltage is supplied through the WHIP ANTENNAS-HF circuit breaker to the HF ANT switch. The orbital HF whip antenna is extended during orbit by positioning the HF ANT switch to EXT. The orbital HF antenna is not retracted during orbit, but is jettisoned in the extended position with the retro adapter section. After landing, the recovery HF whip antenna is extended by positioning the HF ANT switch to PST LDG, and is retracted by positioning the HF ANT switch to EXT.

On S/C 7, extension of the HF whip antennas is controlled through the HF ANT switch and LANDING switch. The HF antennas are operated as follows: Spacecraft control bus voltage is supplied through the WHIP ANTENNAS-HF circuit breaker to the HF ANT switch, which has momentary type contacts. During orbit, the LANDING switch is in the SAFE position and the orbital HF whip antenna can be extended or retracted by holding the HF ANT switch in the EXT or RET position respectively. During re-entry, the LANDING switch is placed in the ARM

9-33

position. After landing, the recovery HF whip antenna can be extended or retracted by holding the HF ANT switch in the EXT or RET position respectively. The HF ANT switch should be held in the EXT position for approximately one minute for full extension of the antennas, and in the RET position for approximately 1.5 minutes for full retraction.

S-Band Annular Slot Antenna - Spacecraft No. 3

Purpose: The S-band annular slot antenna (Figure 9-8) serves the S-band radar beacon.

Physical Characteristics: The antenna is mounted on the equipment section of the adapter such that the antenna is flush with the outer skin of the spacecraft. The S-band antenna is approximately 2.5 inches in diameter, 2.14 inches long, and weighs 14 ounces maximum. The antenna contains one coaxial connector for attachment through coaxial cable to the S-band beacon.

Mechanical Characteristics: The antenna provides a radiation pattern identical to that of a quarter wave stub on a ground plane. The antenna is used for both reception and transmission of the S-band beacon and is jettisoned with the equipment adapter section.

C-Band Annular Slot Antenna

Purpose: The C-band annular slot antenna (Figure 9-8) serves the adapter C-band radar beacon on S/C 4 and 7. On S/C 3, which contains only one C-band radar beacon, the C-band annular slot antenna and C-band antenna system are connected to a coaxial switch for alternate use with the C-band beacon. The C-band annular slot antenna is normally used during the stabilized orbital

 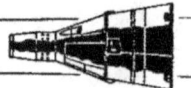

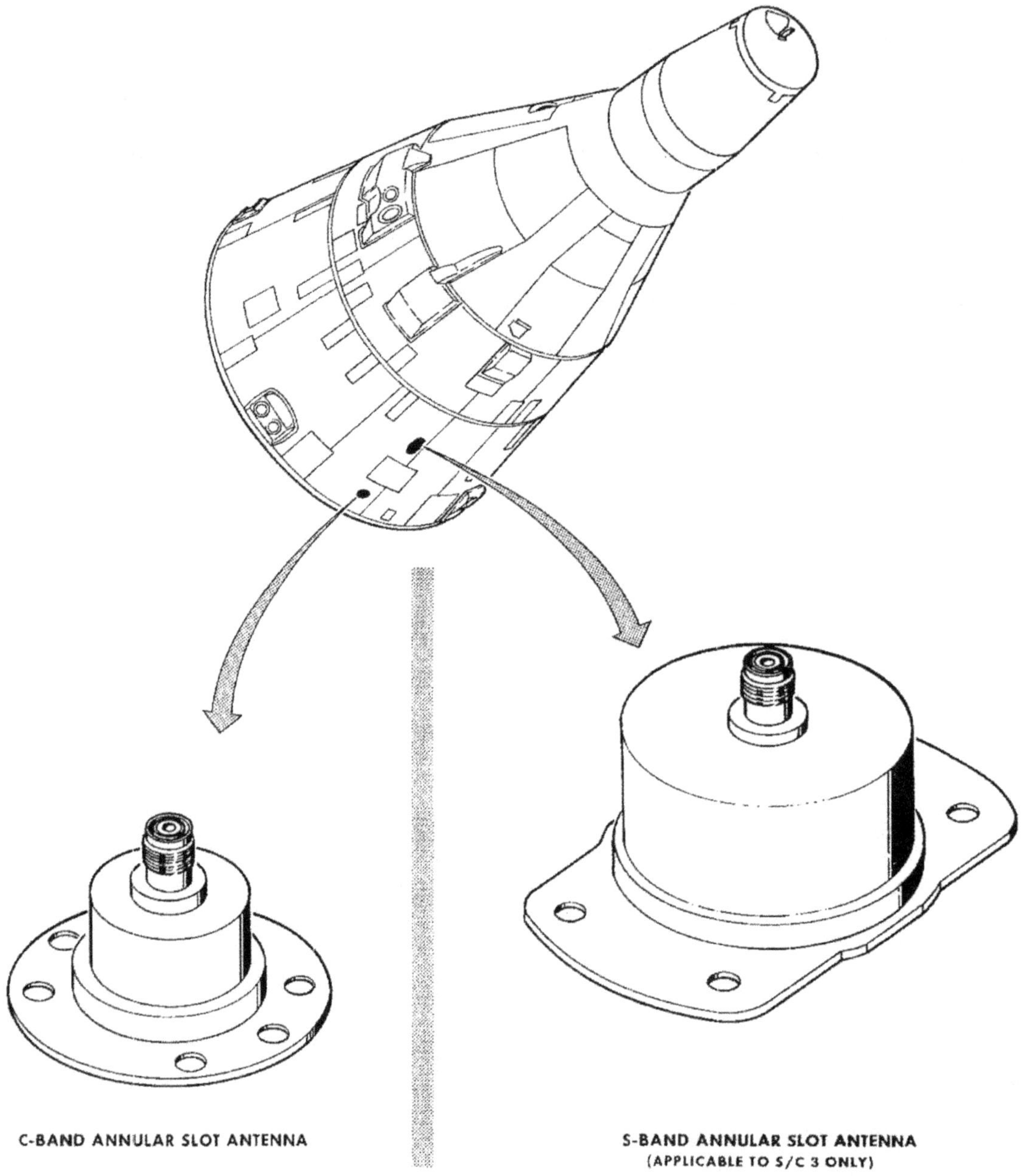

Figure 9-8 C- and S-Band Annular Slot Antennas

phase of the mission.

Physical Characteristics: The C-band annular slot antenna is mounted on the equipment section of the adapter such that the antenna is flush with the spacecraft outer skin. The antenna is approximately 1.4 inches in diameter, 1.34 inches long, and weighs 8 ounces maximum. The antenna contains one coaxial connector to provide a means of establishing an RF link to the C-band beacon.

Mechanical Characteristics: The antenna radiation pattern is identical to that of a quarter wave stub on a ground plane. The antenna is used for both reception and transmission of the C-band beacon and is jettisoned with the equipment adapter section.

C-Band Antenna System

Purpose: The C-band antenna system consisting of a power divider, a phase shifter, and three radiating elements, provides transmission and reception capability for the re-entry module C-band radar beacon. The power divider provides equal transmission power to the three radiating elements. A phase shifter is in series with one of the antennas to compensate for areas of low or no radiation coverage between lobes of the three individual radiation patterns. A phase shifter power supply provides the phase shifter with 26 VAC 453 CPS power. The antenna system provides the circular radiation pattern around the spacecrafts longitudinal axis required for ascent, descent and roll spacecraft attitudes.

Physical Characteristics: The power divider, phase shifter, phase shifter power supply, and radiating elements are shown in Figure 9-9. The power divider, phase shifter, and phase shifter power supply are mounted on the small pressure

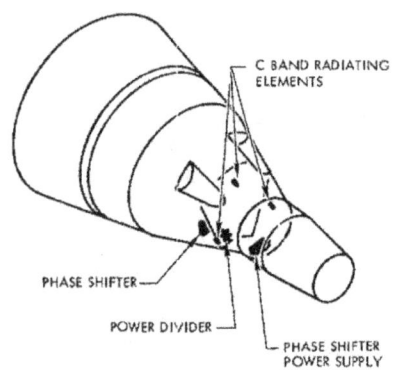
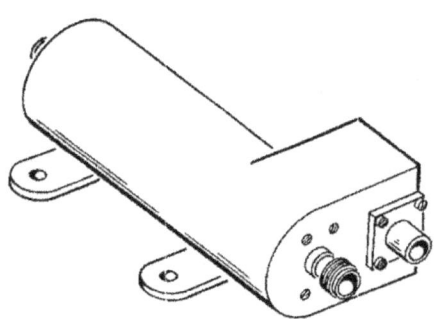
PHASE SHIFTER

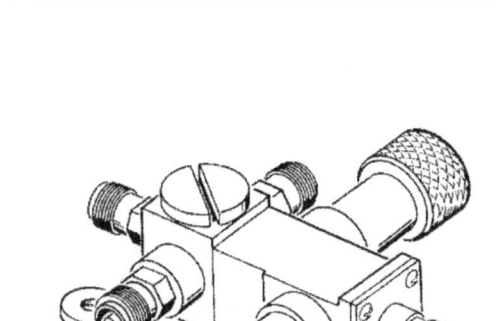
POWER DIVIDER

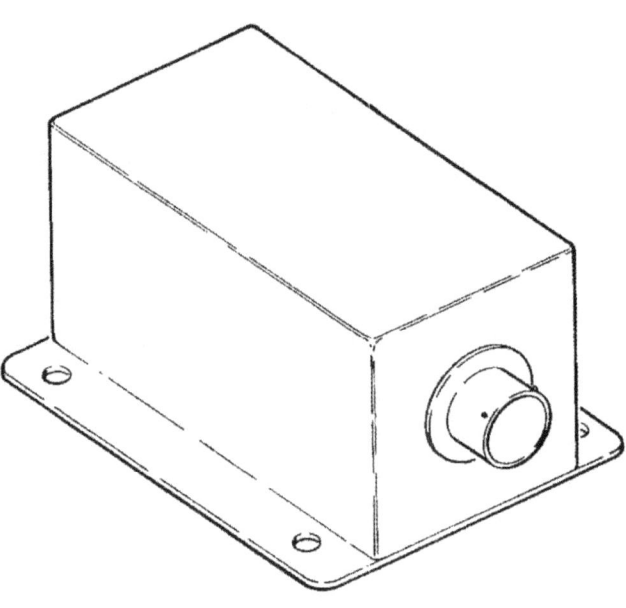
PHASE SHIFTER POWER SUPPLY

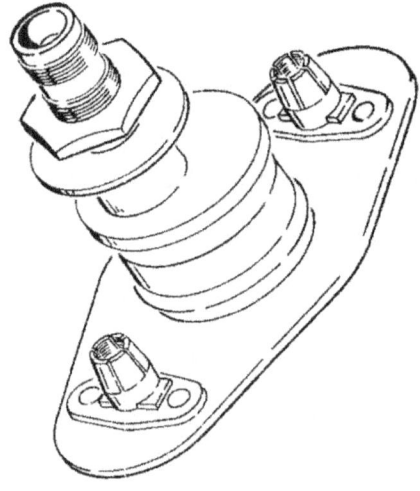
C BAND RADIATING ELEMENTS

Figure 9-9 C Band Antenna System

SEDR 300
PROJECT GEMINI

bulkhead, outside the pressurized area of the spacecraft. The power divider measures approximately 3.86 inches over the connectors, 4.0 inches over the tuning knobs, and weighs approximately 6.5 ounces. The phase shifter is approximately 5.8 inches long, 2.84 inches wide at the large end, 1.4 inches high, has a diameter at the small end of about 1.5 inches, and weighs approximately 12 ounces. The phase shifter power supply measures approximately 1.5 inches wide, 1.75 inches high, 3.5 inches long over the connector, and weighs approximately 8 ounces. The three C-Band radiating elements are mounted flush with the outside skin of the spacecraft and spaced approximately 120° apart. Each antenna unit is approximately 3.4 inches long, 1.8 inches wide, has a depth of 2.21 inches over the connector, and weighs approximately 3.5 ounces.

Electrical Characteristics: The power divider, phase shifter and radiating elements comprise an antenna system that satisfies the transmission and reception requirements for the C-band radar beacon during the launch and re-entry phases of the mission.

The power divider is basically a cavity type power splitter. During beacon transmission, power is delivered to the power divider where it is divided equally among the C-band radiating elements. The power divider compensates for the loss of power due to the phase shifter in series with the right antenna. The power divider also contains a double stub tuner to compensate for mismatch between the C-band beacon, the C-band radiating elements, and the phase shifter. Tuning is accomplished by means of a self-locking tuning shell located underneath each tuning stub cap.

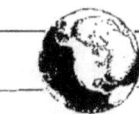 

SEDR 300

PROJECT GEMINI

The phase shifter is an AC operated device. A 26 VAC, 453 CPS input to the phase shifter, is half wave rectified and applied across a coil wound around a ferrite material. Due to the characteristics of the ferrite material, the RF signal from the power divider is delayed 0 to 180° ± 20° at the rate of 453 cycles per second. The changing phase shift of the RF power on one of the C-band radiating elements with respect to the other two, shifts the lobe of that antenna by approximately ± 45 degrees; thus, giving the effect of an almost ideal circular radiation pattern around the longitudinal axis of the spacecraft. The combination of the three antenna elements gives a radiation pattern which extends in all directions except forward and aft of the spacecraft.

The phase shifter power supply is a DC-AC Inverter which supplies a nominal 26 VAC, 453 CPS power to operate the phase shifter. The power supply is a hermetically sealed solid-state unit consisting of a voltage regulator, single-stage oscillator, buffer stage and a push-pull output stage which has a transformer coupled output. The power supply provides a minimum output of 21 volts RMS at 453 ± 17 CPS with an input voltage range from 20 to 30 VDC. Input voltage is applied from the spacecraft main bus via the BEACON-C (BEACON C-RNTY on S/C 4 and 7) circuit breaker, BEACON CONTROL -C RNTY (-C BAND on S/C 4 and 7) switch and the RNTY position of the ANT SEL switch. Maximum input current is 370 milliamperes.

Multiplexers (UHF Diplexer and UHF Quadriplexer)

Purpose: The UHF diplexer provides isolation between DCS receiver no. 1 and the acquisition aid beacon or the delayed-time telemetry transmitter operating into a common antenna. The UHF quadriplexer provides isolation between the stand-by telemetry transmitter, the real-time telemetry transmitter, a UHF voice transmitter/receiver and DCS receiver no. 2 operating into a common antenna via coaxial switches.

SEDR 300

PROJECT GEMINI

Physical Characteristics: The physical representation and approximate location of the UHF diplexer and the UHF quadriplexer is shown in Figure 9-10. The diplexer is located on the electronic module of the equipment adapter section. The quadriplexer is located forward of the small pressurized bulkhead outside the pressurized area of the cabin.

The diplexer is approximately 4.5 inches wide, 4 inches high, and 2.7 inches deep; contains two input and one output connectors, and weighs approximately 1.25 pounds. The UHF quadriplexer is approximately 5.75 inches wide, 5.5 inches deep, and 4.1 inches high; weighs approximately 2.75 pounds, and has four input and one output connectors.

Electrical Characteristics: Figure 9-10 shows the schematic of the UHF diplexer and the UHF quadriplexer. Each channel consists of a high Q cavity tuned to the corresponding operating frequency. All channels are isolated from each other without appreciably attenuating the RF signals passing through it. Each channel can be re-tuned if the signal operating frequency is changed.

The diplexer isolates DCS receiver no. 1 and the acquisition aid beacon or the delayed-time telemetry transmitter, depending upon the position of coaxial switch no. 2. The diplexer operates into the UHF whip antenna on the equipment adapter section.

The UHF quadriplexer isolates the real-time telemetry transmitter, the stand-by telemetry transmitter, one of the two UHF voice transmitter/receivers, and DCS receiver no. 2. The quadriplexer operates into one of the following three antennas, depending on the position of the coaxial switches in series with the antennas: UHF stub antenna, UHF descent antenna, or the UHF whip antenna on the retro

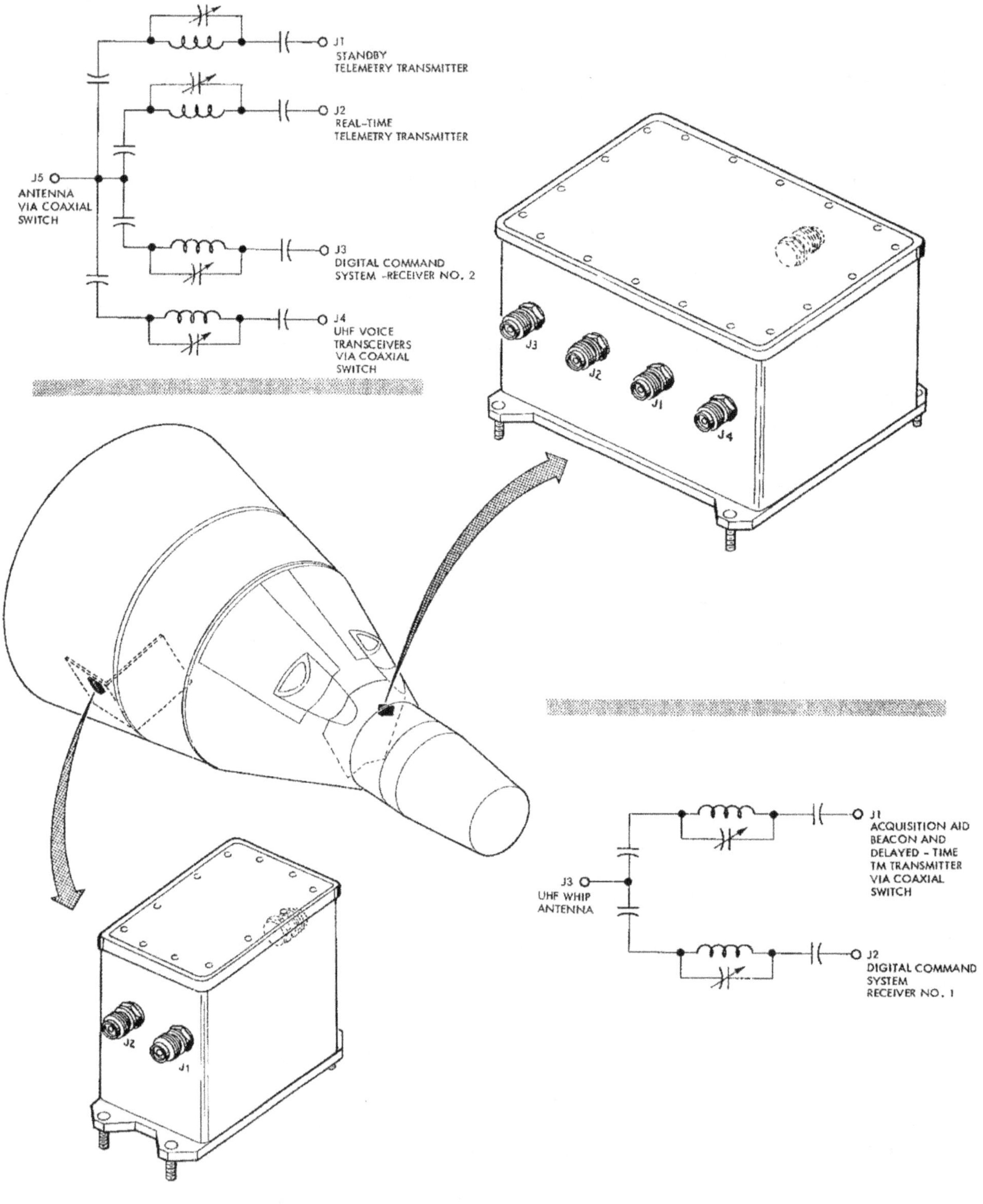

PROJECT GEMINI

SEDR 300

Figure 9-10 UHF Diplexer/UHF Quadriplexer

9-41

adapter section.

Coaxial Switches

Purpose: Five coaxial switches on S/C 4 and 7 (six on S/C 3) are provided to perform the following functions: (1) select either the acquisition aid beacon or the delay-time telemetry transmitter output as the input to the diplexer; (2) select one of the two UHF voice transmitter/receiver outputs as the input to the quadriplexer; (3) connect the HF voice transmitter/receiver to the recovery whip antenna on the re-entry module, or (on S/C 4 and 7) the orbital HF whip antenna on the adapter; (4) connect the output of the quadriplexer to either the UHF descent antenna, or through coaxial switch no. 5 to either the UHF stub or the retro adapter UHF whip antenna; (5) on S/C 3 only, connect either the C-band antenna system or the C-band annular slot antenna to the C-band radar beacon.

Physical Characteristics: The physical construction and approximate location of the coaxial switches is shown in Figure 9-11. The location of the switches is as follows:

Coaxial switch no. 1; approximately five inches from the small end of the cabin in the fourth quadrant.

Coaxial switch no. 2; approximately 10 inches from the forward (small) end of the adapter equipment section in the third quadrant.

Coaxial switch no. 3; approximately 10 inches from the small end of the cabin in the third quadrant.

Coaxial switch no. 4; on S/C 3, located at center of forward right equipment bay in the cabin; on S/C 4 and 7, located adjacent to coaxial switch no. 1.

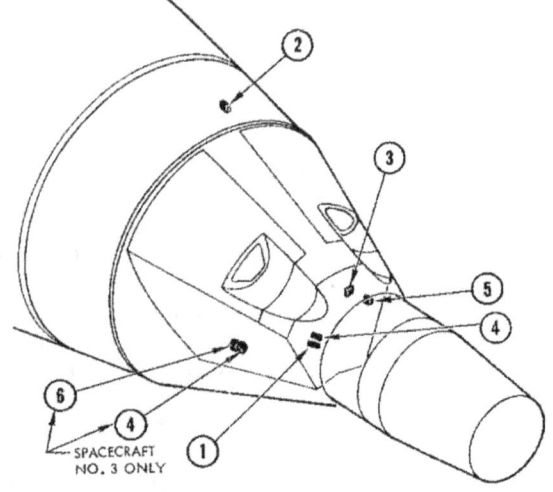
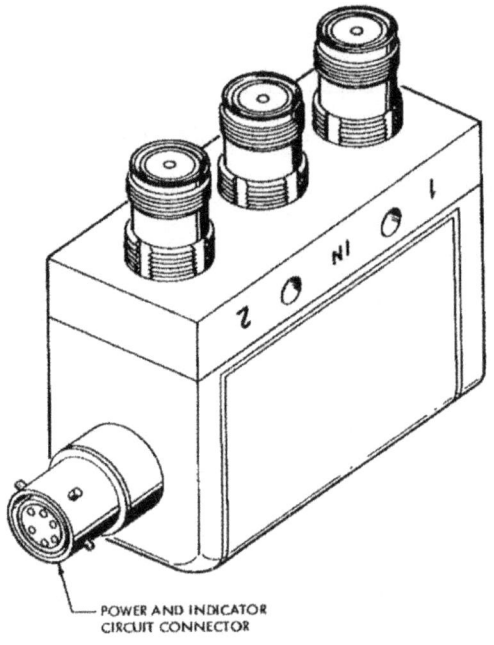

ITEM	FUNCTION
1	UHF TRANCEIVERS
2	TELEMETRY TRANSMITTER/ACQUISITION AID BEACON
3	DESCENT ANTENNA
4	HF TRANCEIVERS
5	UHF WHIP/UHF STUB
6	C-BAND ANTENNAS (S/C 3 ONLY)

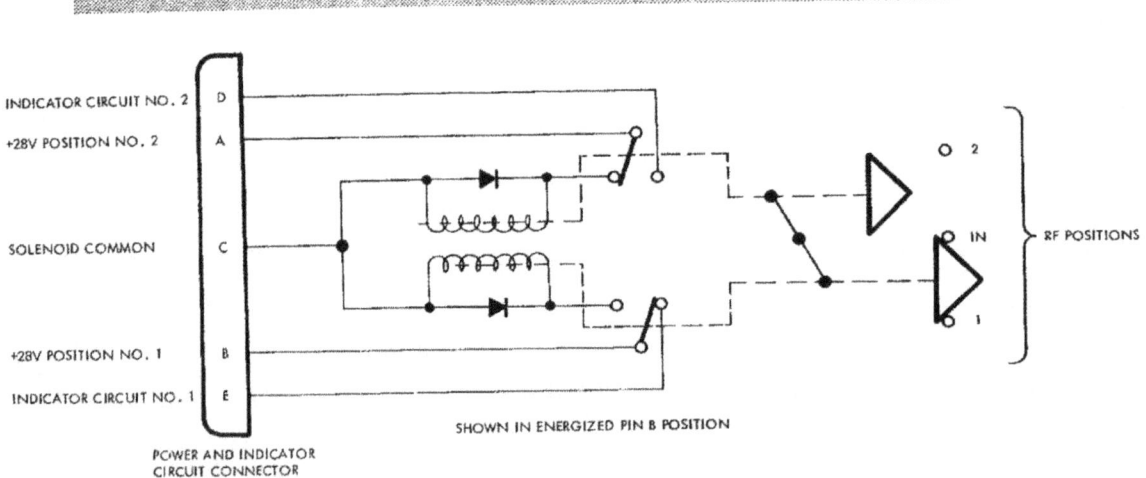

Figure 9-11 RF Coaxial Switches

Coaxial switch no. 6; at center of forward right equipment bay in the cabin. This switch used on S/C 3 only.

Each switch contains a power connector, an input connector, and two output connectors, and weighs approximately 0.5 pounds. The dimensions of each switch are approximately 2.65 inches long, 1.82 inches high and 1 inch wide.

Electrical Characteristics: The coaxial switches are identical and may be used interchangeably. Basically, the coaxial switches provide single pole, double throw switching action as illustrated in Figure 9-11. The switch, having a 20 millisecond maximum operation time, operates on 3 amperes at 28 VDC employing a latching solenoid break-before-make switching action. The coaxial switches are designed to work in the UHF and C-band frequency range. Pins D and E of each switch are brought out to AGE test points to permit monitoring of the switch positions prior to lift off. Pins A and B of each switch are utilized to accomplish the switching action.

BEACONS

Re-entry C-Band Radar Beacon

Purpose: The re-entry C-band radar beacon provides tracking capability of the spacecraft during its entire mission on S/C 3. On S/C 4 and 7 the re-entry C-band beacon is used from lift-off to orbital insertion, from pre-retro to landing, during orbital roll maneuvers, and in the event of adapter C-band beacon failure.

Physical Characteristics: The re-entry C-band radar beacon is a sealed unit which measures approximately 7.64 x 6.14 x 3.02 inches, and weighs approximately 8.3 pounds. As shown in Figure 9-12, the beacon has power, antenna, and test connec-

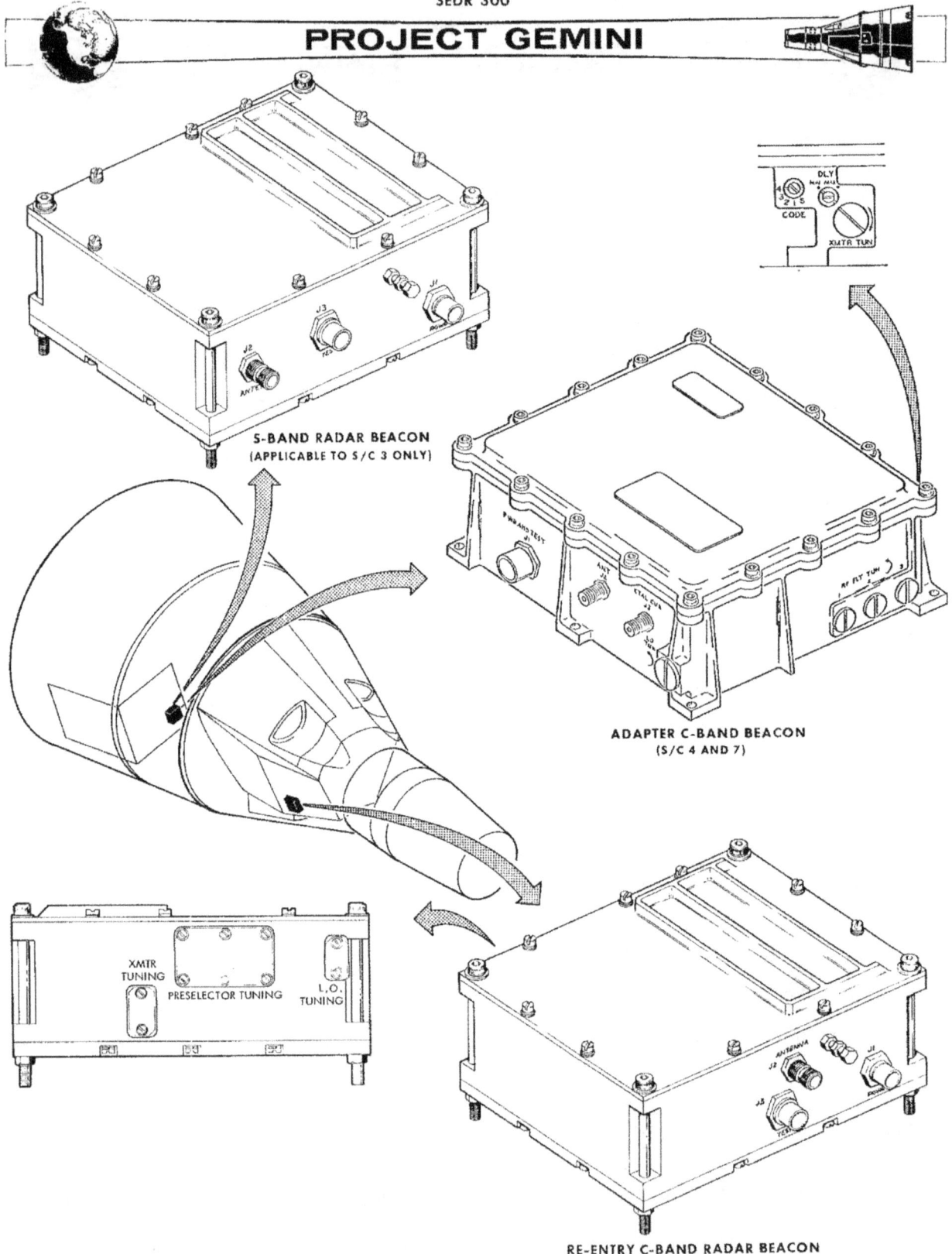

Figure 9-12 C-Band and S-Band Radar Beacons

tors. Located on the rear of the beacon are various adjustment for transmitter, pre-selector, and local oscillator tuning. Solid state modular circuitry is used throughout the beacon with the exception of the transmitter magnetron and the local oscillator cavity. The beacon is mounted on the right forward equipment bay of the re-entry module.

Electrical Characteristics: The re-entry C-band radar beacon is a transponder which upon reception of a properly coded interrogation signal from a ground radar tracking station, transmits a pulse modulated signal back to the tracking station. By measuring the elapsed time between transmission and reception at the tracking station, and compensating for the time delay of the beacon, the position of the spacecraft can be determined. The block diagram of the beacon is shown in Figure 9-13. The signal arriving at the antenna is routed via the directional coupler to one half of a dual ferrite circulator. The ferrite circulator isolates the transmitter from the receiver; thus, providing the capability of receiving and transmitting on the same antenna. The beacon utilizes a superheterodyne receiver which is tunable, by means of a three stage pre-selector, over a range of 5600 MC to 5800 MC. The assigned receiver center frequency is 5690 MC. The output of the pre-selector is combined with the local oscillator frequency in the crystal mixer to produce an output IF frequency of 80 MC. The local oscillator is of the metal-ceramic triode cavity type. The mixer contains a ferrite circulator for isolation between the local oscillator, mixer, and pre-selector. The output of the mixer is amplified by three tuned IF amplifier stages followed by a video detector and a video pre-amplifier. Additional amplification is obtained by a pulse amplifier whose output is supplied to the decoder. The purpose of the decoder is to initiate triggering of the transmitter after a correctly coded signal has been received.

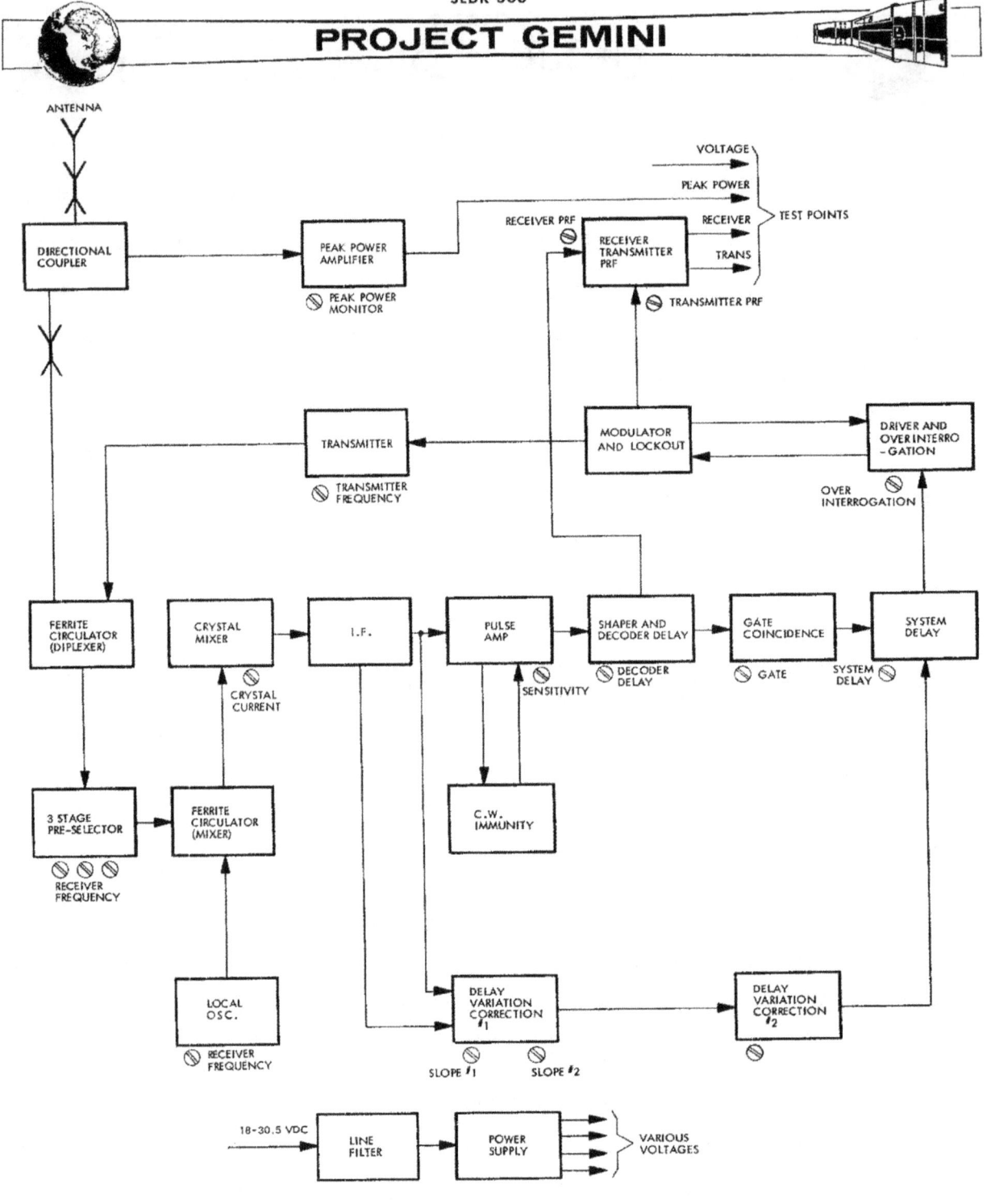

Figure 9-13 Re-Entry C-Band Radar Beacon Block Diagram

The system delay in conjunction with the delay variation correction circuitry, provides for a constant fixed delay used in determining the exact position of the spacecraft. The beacon incorporates a CW immunity circuit that prevents the transmitter from being triggered by random noise. The noise level is reduced below the triggering level of the transmitter by controlling the gain of the pulse amplifier. The transmitter utilizes a magnetron and provides a one kilowatt peak pulse modulated signal at a frequency of 5765 MC to the power divider. The beacon is powered by a DC-DC converter employing a magnetic amplifier and silicon controlled rectifiers. The converter provides voltage regulation for input voltage variations between 18 and 32.5 VDC. The input to the converter is filtered by a pi-type filter to minimize any line voltage disturbances.

Adapter C-Band Radar Beacon - Spacecrafts No. 4 and 7

Purpose: The adapter C-band beacon provides tracking capability of the spacecraft during the orbital phase of the mission and is jettisoned with the equipment adapter section.

Physical Characteristics: The adapter C-band beacon is a sealed unit and measures approximately 9.34 x 8.03 x 3.26 inches. As shown in Figure 9-12, the adapter beacon has a power and test connector, an antenna connector, and a crystal current test point connector. The beacon contains external adjustments for local oscillator, pre-selector (RF filter), and transmitter tuning; switches for selecting the desired interrogation code, and one of two pre-set transponder fixed delay times. These adjustments and switches are accessible by removing pressure sealing screws. The beacon employs solid state circuitry, except for the transmitter magnetron and receiver local oscillator. The adapter beacon is located on the electronic module of equipment adapter section and uses the C-band annular slot antenna for reception and transmission.

Electrical Characteristics: The adapter C-band radar beacon is a transponder, which employs the same basic operating principles as the re-entry C-band beacon to provide spacecraft location data upon receipt of a properly coded interrogation signal. A block diagram of the adapter C-band beacon is shown in Figure 9-14. The interrogation signal is fed from the antenna to the diplexer. The diplexer is a ferrite circulator which couples the received signal to the RF filter (pre-selector) and also isolates the receiver from the transmitter to permit use of a common antenna for reception and transmission. The superheterodyne receiver frequency is tunable from 5395 MC to 5905 MC. The assigned operating center frequency is 5690 MC and is selected by adjustment of the RF filter.

The RF filter is a three-stage preselector, employing three separately tuned coaxial resonator cavities to provide adequate RF selectivity, and to protect the mixer crystal from damage due to transmitter power reflected by the antenna.

The output of the pre-selector is combined with the local oscillator output in the mixer stage to provide a 60 MC output to the IF amplifier. The mixer consists of a coaxial directional coupler and a mixer crystal. The directional coupler isolates the local oscillator output from the antenna and directs it to the mixer crystal. The local oscillator is a re-entrant cavity type employing a planar triode to generate the CW signal required to operate the mixer.

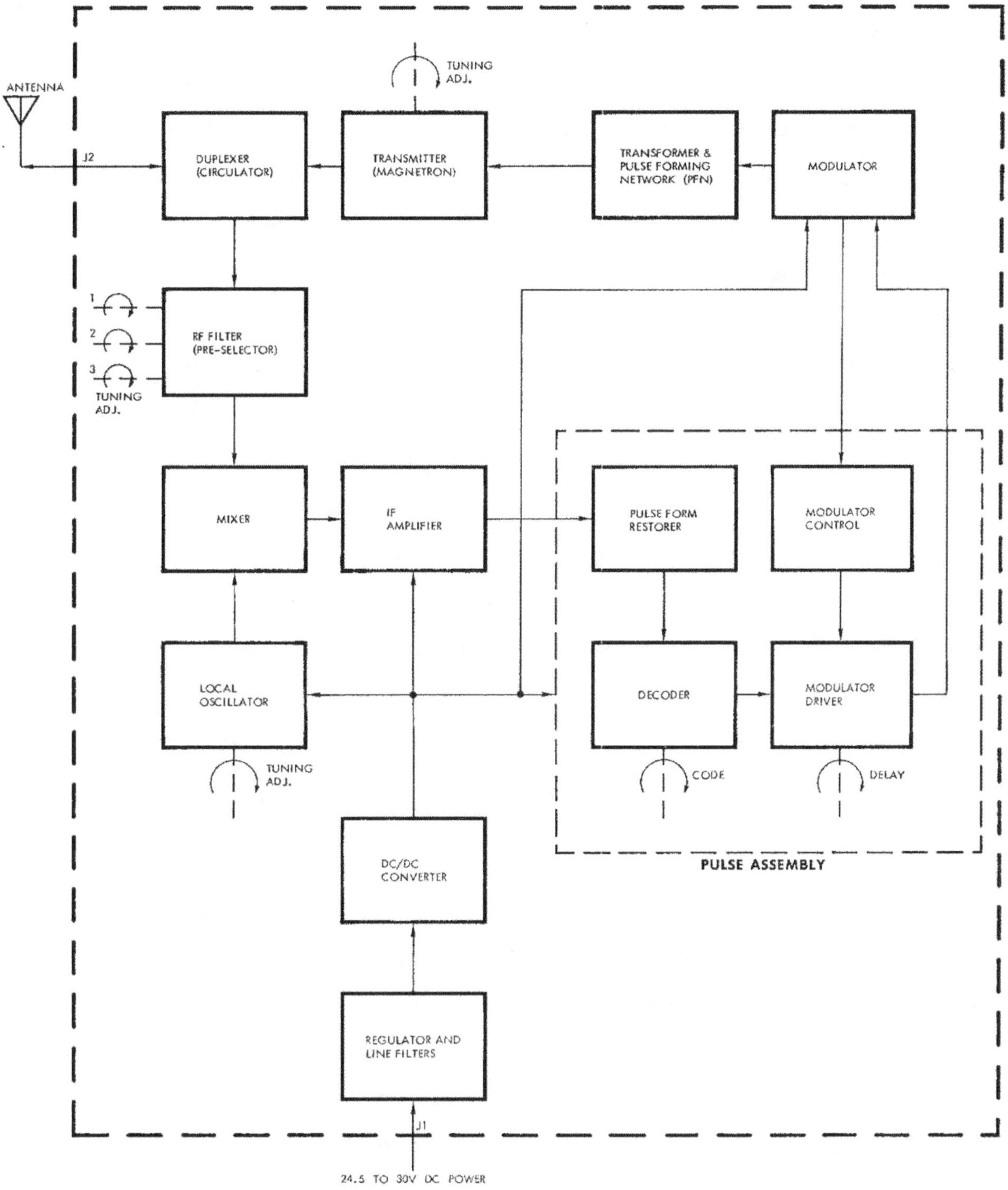

SEDR 300

PROJECT GEMINI

Figure 9-14 Adapter C-Band Radar Beacon Block Diagram

PROJECT GEMINI

The IF amplifier is a linear-logarithmic, high gain amplifier composed of an input stage, five amplifier stages, a summing line and a video amplifier. The amplified video output is fed to the pulse form restorer circuits which prevent a ranging error due to variations in receiver input signal levels, and also provides a standard amplitude pulse to the decoder for each input signal exceeding its triggering threshold. The decoder determines when a correctly coded signal is received and supplies an output to the modulator driver. The type code to be accepted is selected by the CODE switch. Single pulse, two pulse or three pulse codes may be selected. The modulator driver and control circuits initiate and control triggering of the transmitter modulator. The modulator driver supplies two fixed values of overall system delay. The desired delay is selected by the position of the DLY switch. An alternate value of maximum delay is available by removing any internal jumper lead. The modulator control furnishes the trigger and turn-off pulse for the modulator and limits modulator triggers to prevent the magnetron duty cycle from being exceeded, regardless of the interrogating signal PRF. The modulator circuit employs silicon controlled rectifiers which function similar to a thyratron, but require a much shorter recovery time.

The associated modulator pulse forming network (PFN) and transformer provide the necessary pulse to drive the transmitter magnetron. The desired pulse width is selected by the internal connections made to the PFN. The transmitter magnetron frequency is tunable from 5400 MC to 5900 MC. The assigned transmitter center frequency is 5765 MC. A minimum of 500 watts peak pulse power is supplied to the antenna under all conditions of rated operation.

The transponder power supply consists of input line filters, a series regulator, and a DC-DC converter. The power supply furnishes the required regulated output voltages with the unregulated input voltage between 21 and 30 VDC. The converter employs a conventional RL coupled multivibrator and full wave rectifier circuits.

S-Band Radar Beacon - Spacecraft No. 3

Purpose: The S-band radar beacon provides back-up tracking capability of the spacecraft for the C-band radar beacon from lift-off through equipment adapter separation at which time the beacon will be jettisoned.

Physical Characteristics: The S-band radar beacon is a sealed unit and is approximately 7.77 inches long, 6.27 inches wide, and 3.52 inches high. The S-band beacon contains a power connector, a test connector, and an antenna connector, as illustrated in Figure 9-12.

Electrical Characteristics: The operation of the S-band radar beacon is identical to that of the re-entry module C-band radar beacon except the operating frequency which is 2840 MC for the receiver and 2910 MC for the transmitter.

Acquisition Aid Beacon

Purpose: Unlike the C-band and S-band beacons, that provide accurate tracking data, the acquisition aid beacon is merely a transmitter used to determine when the spacecraft comes within range of a ground tracking station. When the spacecraft comes within the range of a ground tracking station, the acquisition aid beacon is disabled and remains off until the spacecraft is again out of range.

Physical Characteristics: The acquisition aid beacon, shown in Figure 9-15, is cylindrical, having a diameter of approximately 2.6 inches, and a height of approximately 3.5 inches. The acquisition aid beacon located as shown in Figure 9-15 has a power connector, a coaxial antenna connector and weighs approximately 17 ounces.

Electrical Characteristics: The acquisition aid beacon consists of a transmitter, DC-DC voltage regulator, and a low pass output filter.

The transmitter is an all transistorized unit, containing a push-pull output stage to obtain a minimum output of 250 milliwatts at a frequency of 246.3 MC. The transmitter frequency is derived by taking the basic frequency of an oscillator and multiplying it through a series of tripler and doubler stages.

The transmitter is powered by a DC-DC voltage regulator. The regulator is completely transistorized and provides a regulated output voltage of 28 VDC. To reduce the probability of obtaining a spurious output signal, a band pass filter is placed in the output circuit.

UHF Recovery Beacon

Purpose: The UHF recovery beacon, operating on the international distress frequency of 243 mc, serves as a recovery aid by providing information regarding location of the spacecraft to the recovery crew.

Physical Characteristics: The UHF recovery beacon and its approximate location is shown in Figure 9-15. The beacon is mounted on the aft right equipment bay of the re-entry module. The recovery beacon is approximately 9.0 inches long, 4.0 inches wide, 2.5 inches high, and weighs 3.9 pounds maximum. The beacon

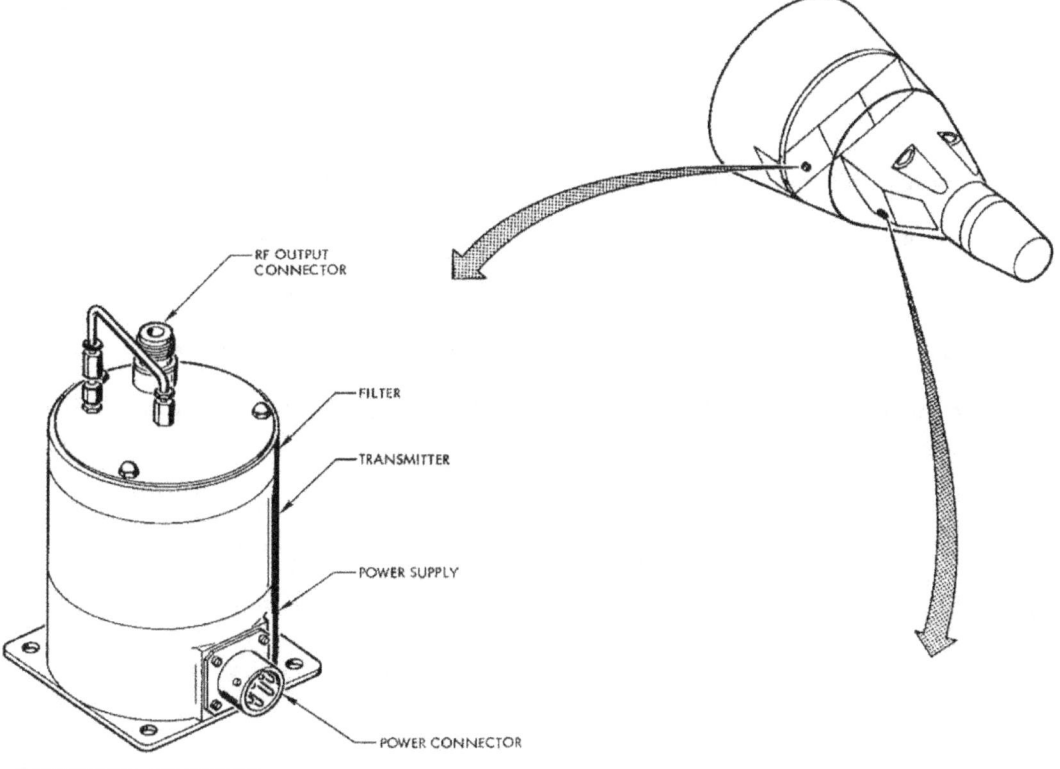

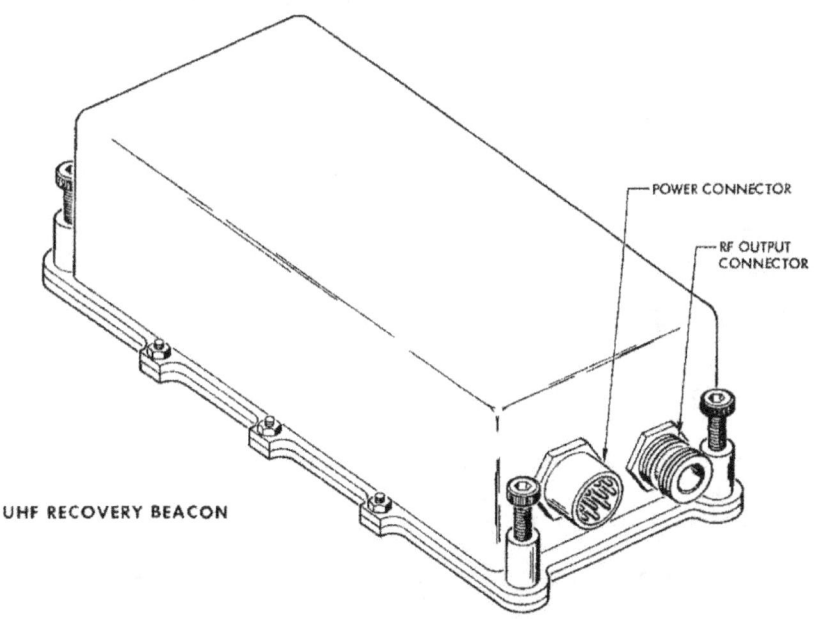

Figure 9-15 Acquisition Aid and UHF Recovery Beacons

contains one multipin power connector, and one coaxial type RF connector.

Electrical Characteristics: The UHF recovery beacon consists of a spike eliminator, a regulator, a DC-DC converter, a pulse coder, a modulator, and a transmitter. Spacecraft main bus voltage is fed to the switching type regulator through the spike eliminator filter. The voltage regulator provides a DC regulated output voltage of 12 VDC to the DC-DC converter, the transmitter tube filaments, and the pulse coder.

The DC-DC converter is a solid state device providing two high voltage outputs to the transmitter and modulator. The pulse coder, a solid state device, operates with the modulator to apply correctly coded high voltage pulses to the transmitter for plate modulation of the power amplifiers.

The transmitter consists of an oscillator stage, a doubler stage, and a power amplifier. The transmitter power amplifier provides a UHF pulse coded output having a peak power of at least 50 watts to the UHF recovery antenna. An external RF band-pass filter is installed between the transmitter output and the antenna to reduce spurious RF radiations, especially at the UHF voice transmitter frequencies.

VOICE COMMUNICATION

Voice Control Center

Purpose: The Voice Control Center (VCC) contains all switches and controls for selecting the type of voice communication and the desired operating mode. The VCC also contains microphone and headset amplifiers, an alarm tone generator, and voice actuated transmitter keying circuitry.

SEDR 300
PROJECT GEMINI

Physical Characteristics: The VCC and its approximate location is shown in Figure 9-16. The VCC is mounted in the center panel of the spacecraft cabin such that the controls and switches comprise a part of the center panel. The VCC is modular constructed, approximately 6.4 inches wide, 6.4 inches high, 5.5 inches deep, and weighs approximately 6.3 pounds.

Five connectors located on the rear of the unit provide a means of connecting to the other voice communication system components and test connectors. The function of each connector is listed on Figure 9-16. The VCC front panel consists of three groups of switches and controls. The NO. 1 and NO. 2 AUDIO groups on S/C 3 and 4 each consist of a MODE switch for selection of voice recording (RCD), UHF, intercommunication (INT), HF, or HF/DF transmission, and three thumb-wheel-type multi-detent volume controls, one for each of the above mentioned modes.

On S/C 7 the RCD position has been removed from the MODE switch, and a separate RECORD switch has been added to allow recordings to be made in any mode of operation. To allow uninterrupted sleep during extended spacecraft missions. A SILENCE switch has also been added on S/C 7. The SILENCE switch in the NORM position allows reception for both pilots. The NO. 1 position, allows for uninterrupted sleep of the command pilot by removing power from the command pilot's headset amplifiers. The NO. 2 position allows uninterrupted sleep for the pilot by removing power from the pilot's headset amplifier; thus, making reception impossible.

PROJECT GEMINI

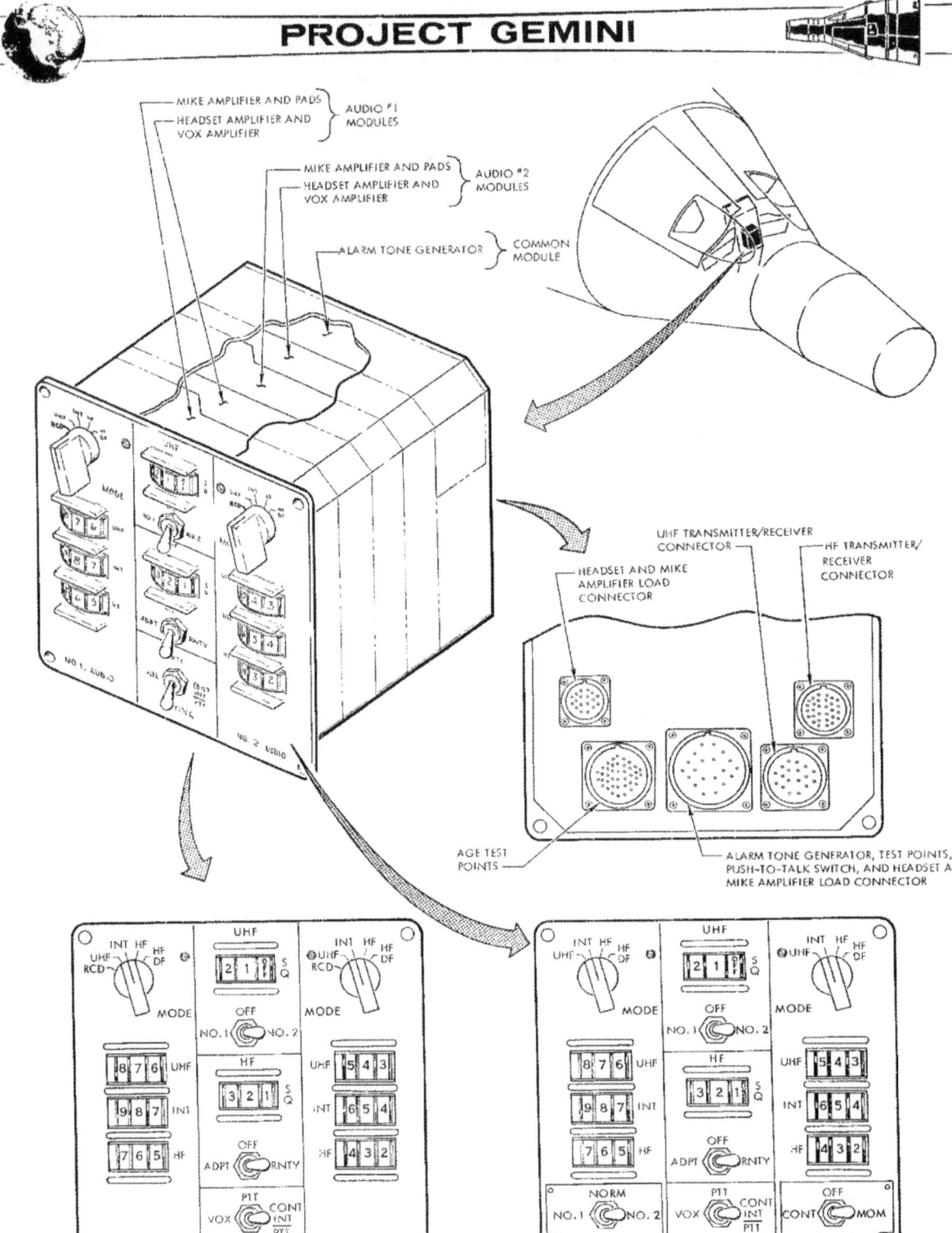

Figure 9-16 Voice Control Center

9-57

SEDR 300
PROJECT GEMINI

The center group consists of a KEYING switch, a HF select switch, a UHF select switch, and thumb-wheel-type squelch controls for UHF and HF circuitry. The KEYING switch provides for selection of push-to-talk (PTT), voice operated relay (VOX), or continuous intercommunication with push-to-talk keying (CONT INT/PTT) for the voice transmitters. The UHF and HF select switches provide capability of selecting the desired transmitter/receiver. The ADPT position of the HF select switch is not used.

Electrical Characteristics: The VCC contains two headset and two microphone amplifiers for each of the two audio channels.

Figure 9-17 shows a functional block diagram of the VCC. An audio signal from the microphone in the Flight Crews helmets or light weight headsets, is amplified by two microphone amplifiers and then applied to the MODE switch. With the MODE switch in the HF position, the output of the microphone amplifiers (MA) is applied to the HF transmitter. When the MODE switch is in the INT position, the output of the microphone is applied to the four headset amplifiers via the two INT volume controls. The outputs of the headset amplifiers are applied to the headsets of the Flight Crew. With the MODE switch in the UHF position, the output of the microphone amplifiers is applied to the UHF transmitters. The UHF select switch selects either UHF transmitter NO. 1 or NO. 2 and also operates coaxial switch no. 1 to connect the selected transmitter output to the UHF quadriplexer. The desired keying mode is selected by a common KEYING switch. Three methods may be selected to key the voice transmitters. The VOX position enables keying of the selected transmitter at the instant the microphone provides an output signal. The PTT position enables keying of the

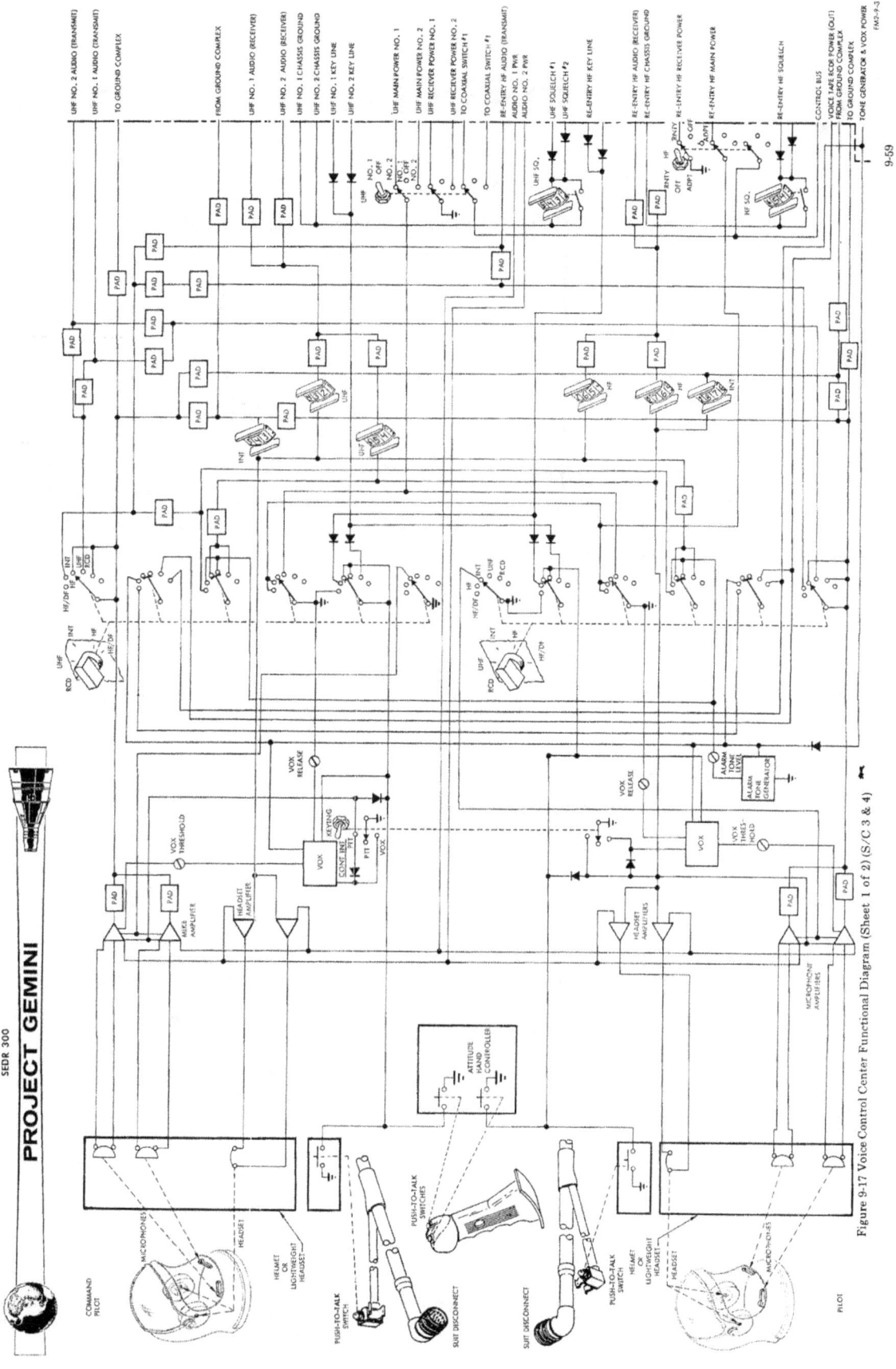

Figure 9-17 Voice Control Center Functional Diagram (Sheet 1 of 2) (S/C 3 & 4)

 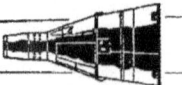

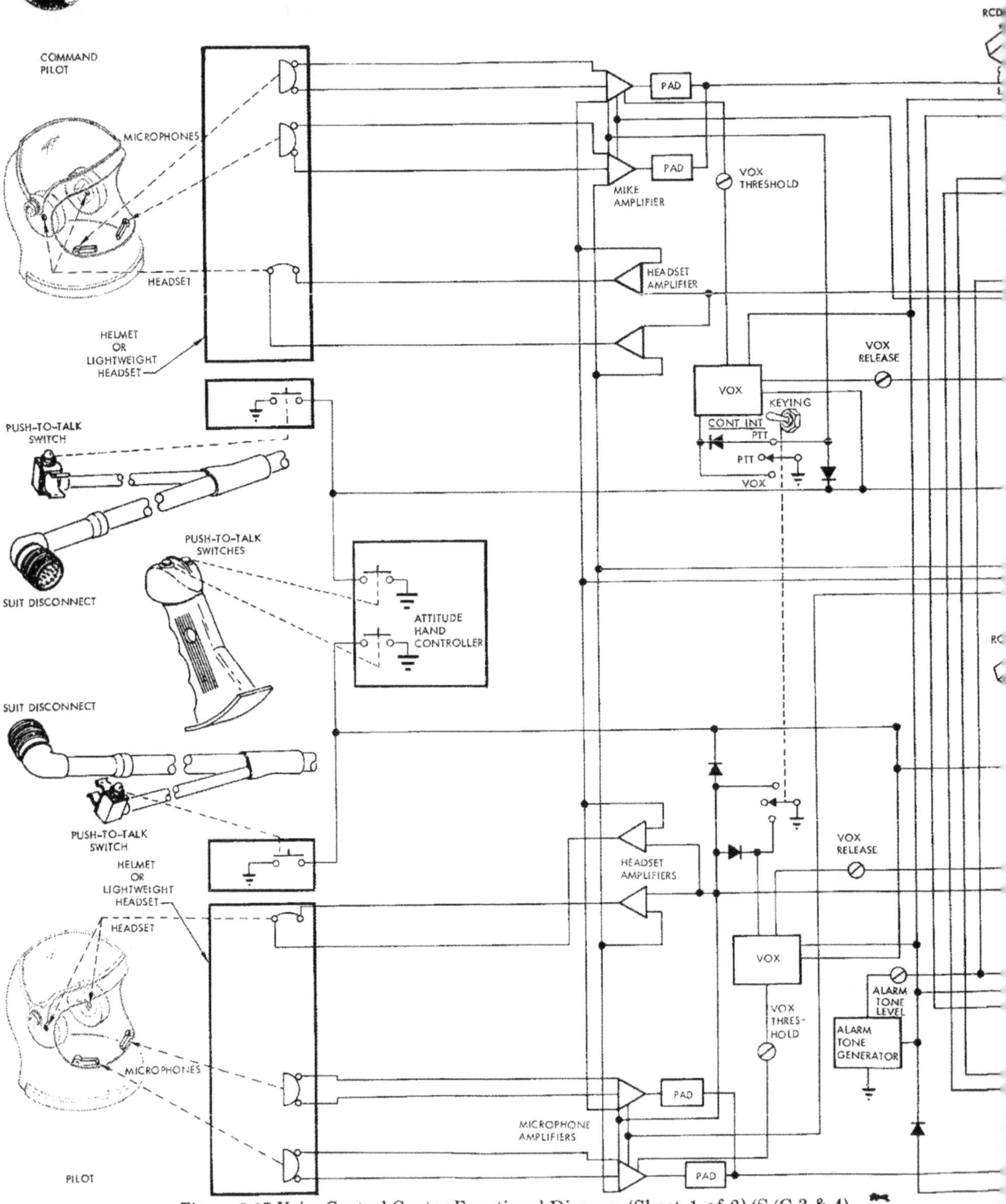

PROJECT GEMINI

SEDR 300

Figure 9-17 Voice Control Center Functional Diagram (Sheet 1 of 2) (S/C 3 & 4)

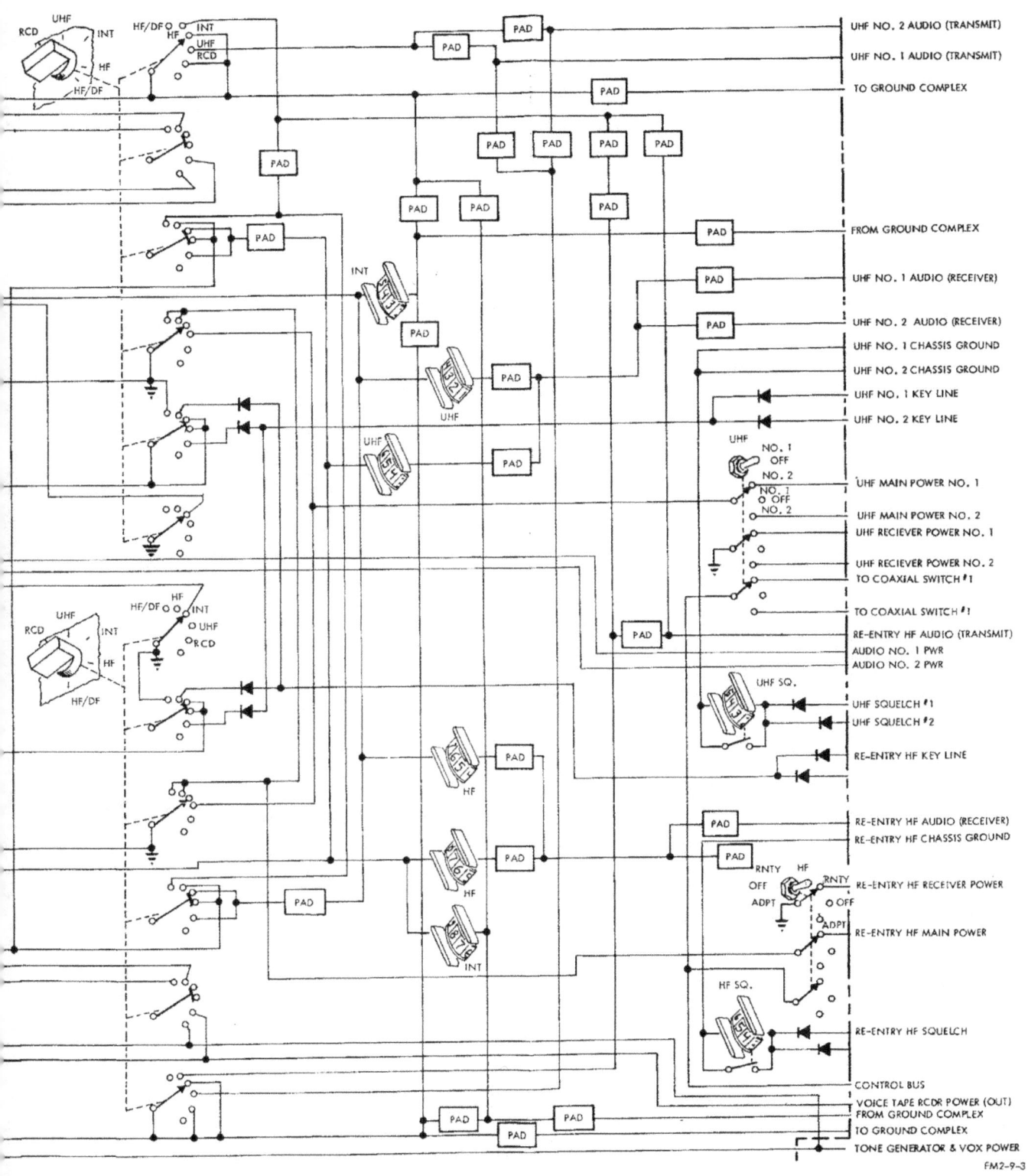

FM2-9-3

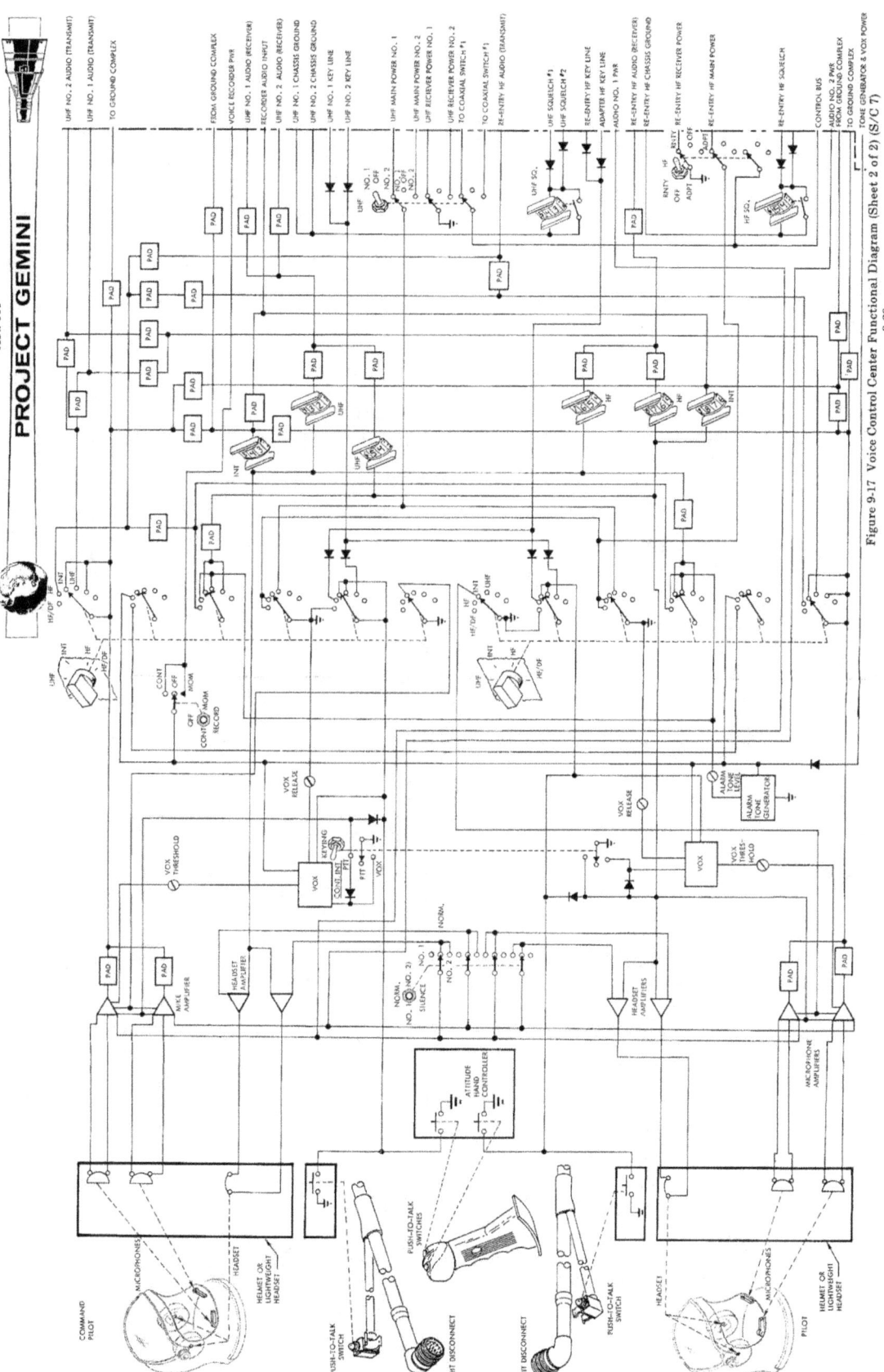

Figure 9-17 Voice Control Center Functional Diagram (Sheet 2 of 2) (S/C 7)

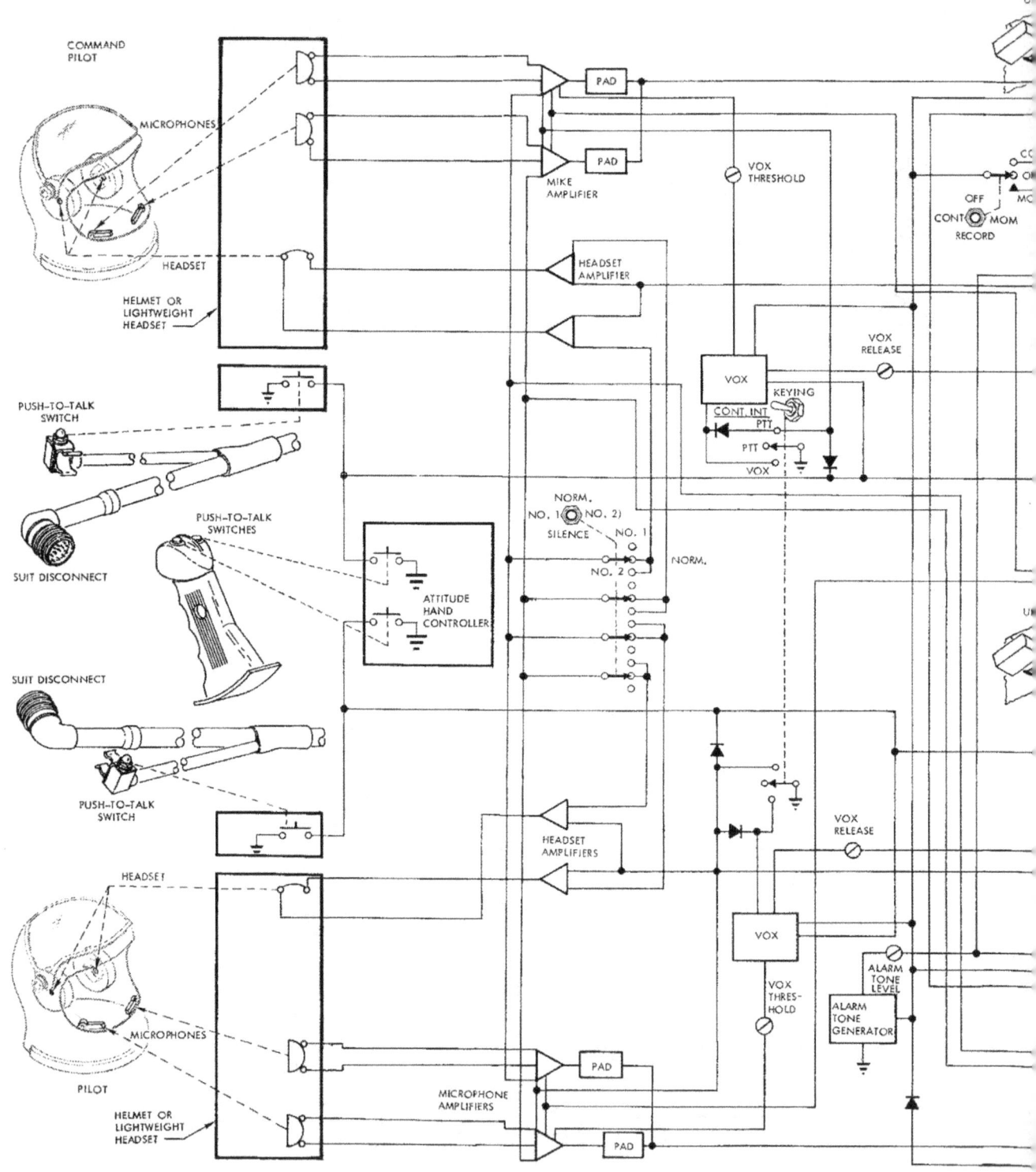

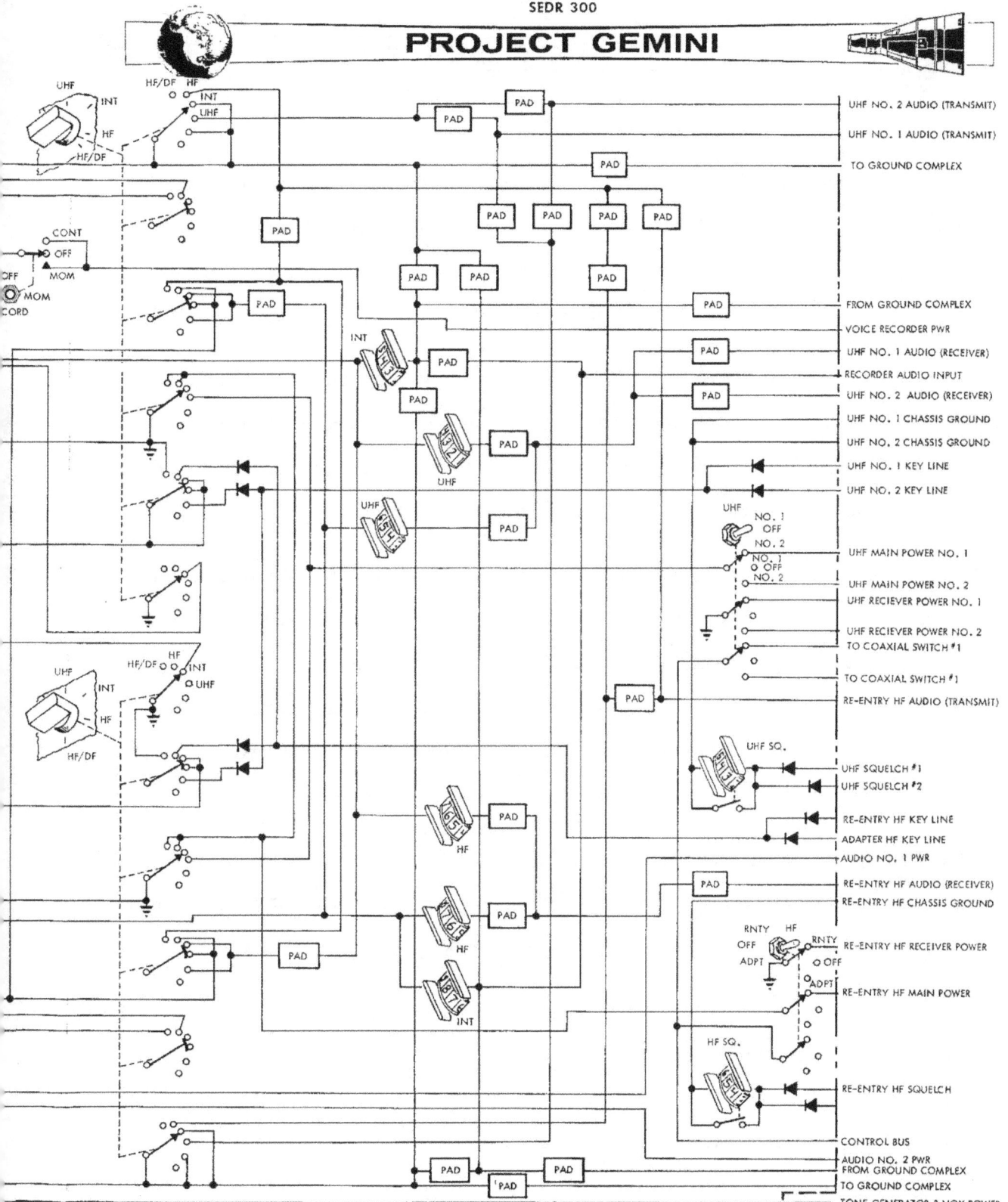

Figure 9-17 Voice Control Center Functional Diagram (Sheet 2 of 2) (S/C 7)

transmitter when the push-to-talk switch, located on the suit disconnect cables or on the attitude control handle, is depressed. The CONT INT/PTT position provides continuous intercommunication between the Flight Crew, and push-to-talk keying for transmission from the spacecraft to the ground station.

The VCC also controls the power supplies of the transmitter/receivers by means of ground switching. With the MODE switch in a position other than HF and the HF select switch in the RNTY position, a ground is supplied to the HF transmitter/receiver auxiliary power supply to power the HF receiver.

With the HF select switch in RNTY and the MODE switch in the HF position, a ground is supplied to the HF transmitter/receiver main power supply to power the HF receiver and transmitter. The UHF circuitry operates on the same principle as the HF. The UHF select switch supplies power ground for the selected receiver. The MODE switch (UHF position) together with the UHF select switch, supplies a power return for the UHF transmitter and receiver.

The HF/DF position of the MODE switch is utilized for direction finding purposes. With the MODE switch in HF/DF and the HF select switch in the RNTY position, the HF transmitter is modulated by a 1,000 CPS tone which is utilized to determine spacecraft location.

UHF Voice Transmitter/Receivers

Purpose: Two UHF voice transmitter/receivers are provided for redundant line-of-sight voice communication between the spacecraft and the ground.

Physical Characteristics: The UHF voice transmitter/receiver and their approximate location are shown in Figure 9-18. Both transmitter/receivers are identical

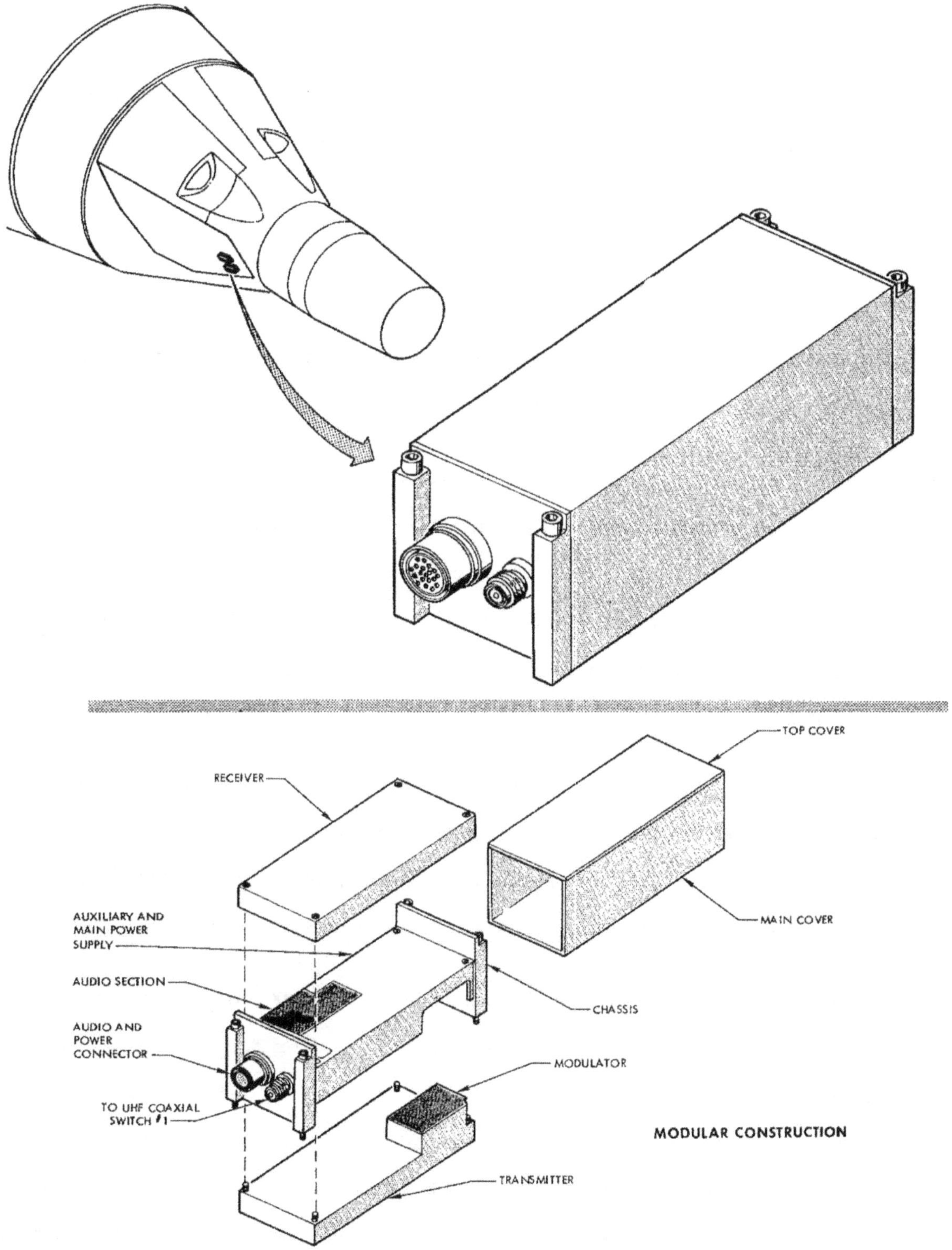

SEDR 300
PROJECT GEMINI

Figure 9-18 UHF Voice Transmitter/Receiver

9-62

and are mounted side by side in the forward right equipment bay of the re-entry module. Each transmitter/receiver is a hermetically sealed modular constructed unit, approximately 7.7 inches long, 2.8 inches wide, 2.4 inches deep, and weighs approximately 3.0 pounds. Each unit has a multipin audio and power connector, and a coaxial type RF connector.

Electrical Characteristics: The UHF voice transmitter/receiver consists of a transmitter, receiver, and power supply.

The transmitter consists of a crystal controlled oscillator, two RF amplifiers, a driver, and a push-pull power amplifier. All stages except the driver and power amplifier are transistorized. The transmitter is fixed-tuned at 296.8 MC and is capable of producing an RF power output of 3.0 watts into a 50 ohm resistive load. The transmitter is AM voice modulated by a transistorized modulator stage.

The AM superheterodyne receiver is fully transistorized, is fixed-tuned at a frequency of 296.8 MC and contains a squelch circuit for noise limiting. The squelch threshold is manually controlled. An automatic volume control stage is also incorporated to provide a constant audio output with input signal variations.

The UHF voice transmitter/receiver is powered by two DC-DC converters comprising an auxiliary and a main power supply. DC operating power for the two power supplies is limited by two circuit breakers located on the left switch/circuit breaker panel. One circuit breaker is provided for each unit. Actuation of the power supplies is accomplished by ground return switching through the voice control center. If the UHF select switch is in either the NO. 1 or NO. 2 position and the MODE switch is in a position other than UHF, a ground is supplied to the auxiliary power supply only, placing the transmitter/receiver into a receive condition. With the MODE switch in the UHF position, a ground is supplied to the main power supply, placing the selected UHF voice transmitter/receiver into a receive and transmit condition.

It should be noted that when the UHF transmitter is keyed, the UHF receiver is disabled and the Flight Crew cannot receive UHF voice transmissions from the ground station.

HF Voice Transmitter/Receiver

Purpose: The HF voice transmitter/receiver is provided to enable beyond the line-of-sight voice communication between the spacecraft and the ground.

Physical Characteristics: Figure 9-19 shows the modular construction and approximate location of the HF voice transmitter/receiver in the forward right equipment bay of the re-entry module. The unit weighs approximately 62 ounces, is approximately 8.5 inches long, 3.3 inches wide, and 2.9 inches deep. One multi-pin audio connector and one RF coaxial type connector are provided.

Electrical Characteristics: Basically, the HF voice transmitter/receiver is electrically identical to the UHF transmitter/receiver except for the operating frequency and power output. The HF transmitter and receiver are fixed tuned to a frequency of 15.016 MC and the HF transmitter provides an RF power output of 5 watts.

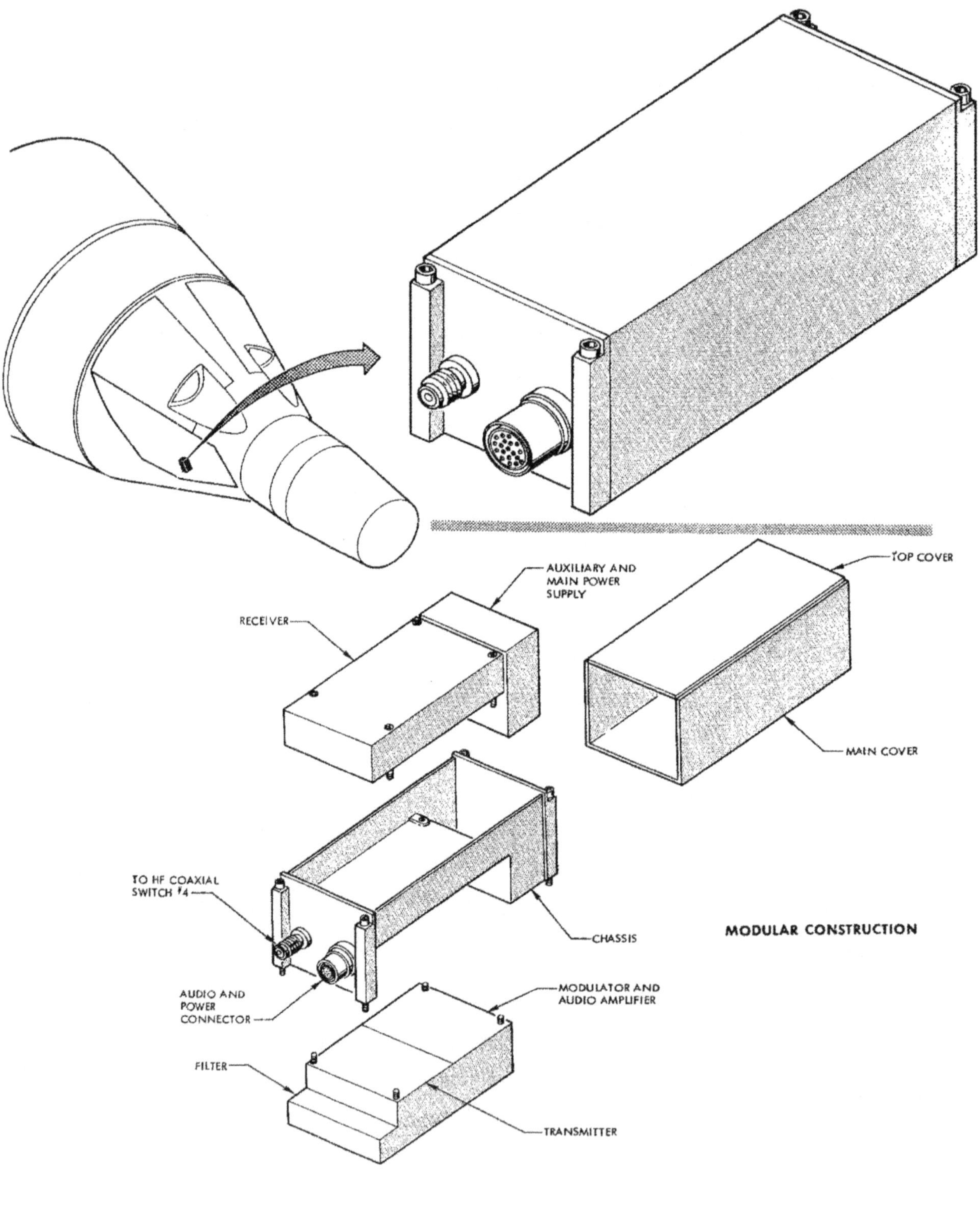

PROJECT GEMINI

SEDR 300

Figure 9-19 HF Voice Transmitter/Receiver

9-65

Actuation of the HF receiver and transmitter is accomplished through the VCC. If the HF Select switch is in RNTY and the MODE switch is in a position other than HF, the HF transmitter/receiver is in a receive condition. With the MODE switch in the HF position, the HF transmitter/receiver is placed in a receive and transmit condition.

It should be noted that when the HF transmitter is keyed, the HF receiver is disabled and the Flight Crew cannot receive HF voice transmissions from the ground station.

Voice Tape Recorder

Purpose: The voice tape recorder is provided so the Flight Crew can make voice recordings during the spacecraft mission in accordance with the applicable flight plan.

Physical Characteristics: The physical construction and approximate location of the voice tape recorder is shown in Figure 9-20. The voice tape recorder is located inside the cabin in a vertical position between the pilots seat and the right-hand side wall. The voice tape recorder assembly consists of the recorder, tape cartridge, and shock absorber mounting plate and is supplied as GFE equipment. The recorder is approximately 6.25 inches long, 2.87 inches wide, one inch thick, and weighs 30 ounces maximum without the tape cartridge. The shock absorber mounting plate is approximately 6.3 inches long, three inches wide, and weighs 20 ounces maximum. The tape cartridge is approximately 2.25 inches square, 3/8 inch thick, and weighs two ounces.

SEDR 300

PROJECT GEMINI

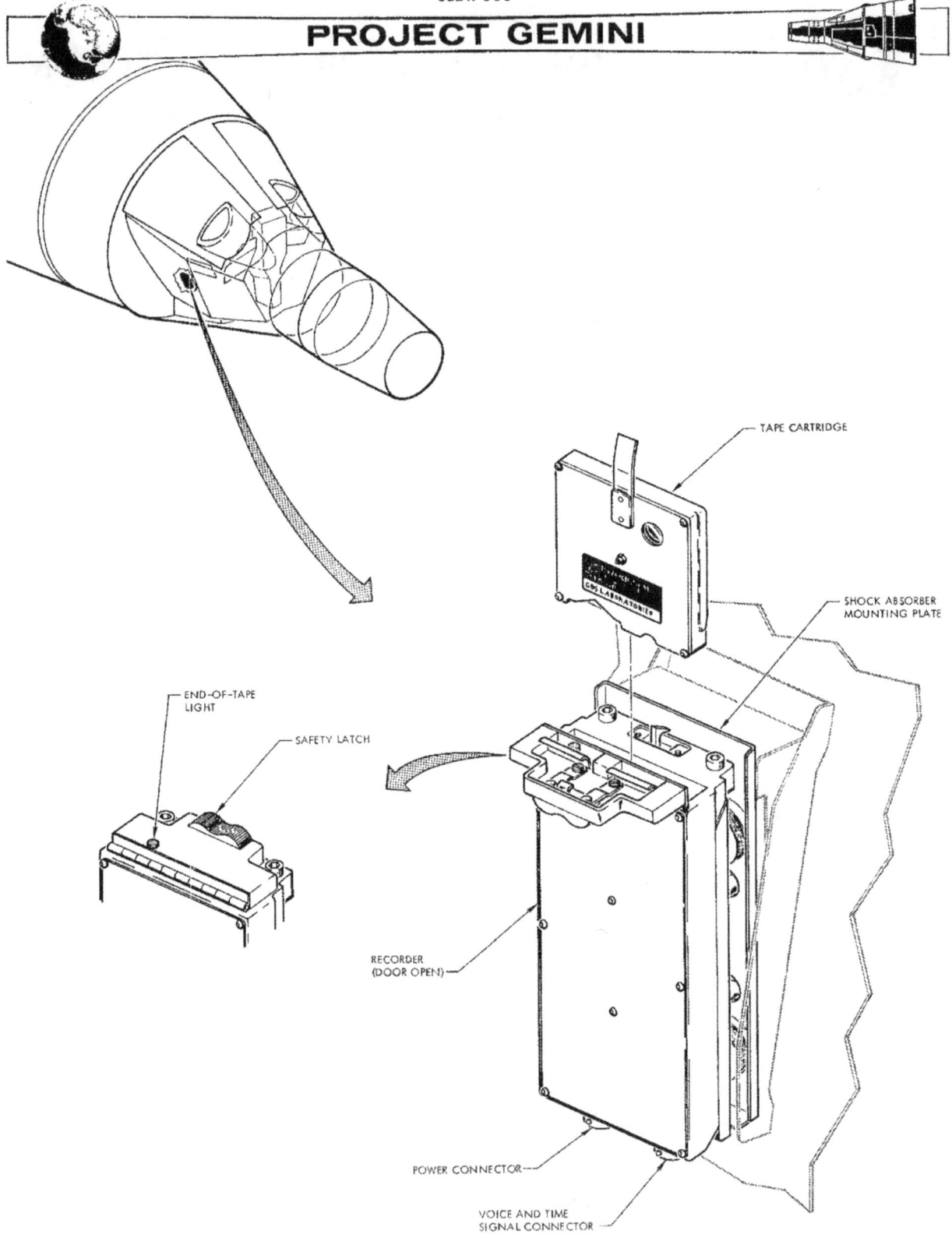

Figure 9-20 Voice Tape Recorder

The recorder contains a power connector and a signal connector located on the end as shown in Figure 9-20. The recorder is retained in the shock mount by guides and two allen-head bolts for easy removal. The door contains a red plastic lens so that light from the end-of-tape bulb is visible to the pilot. A safety latch prevents accidental opening of the door. The door is opened by pressing down on the latch and sliding it sideways. When the latch is released the spring loaded hinge causes the door to open, exposing the cartridge tab. Flat pressure springs on the door hold the inserted cartridge in place and maintains tape contact with the recorder head and end-of-tape contact.

The tape cartridge is guided into the recorder by step rails on each side of the cartridge. One step rail is slightly larger to insure correct insertion of the cartridge. When the recorder door is opened, a heavy tab on the cartridge springs up to provide easy removal. The cartridge contains approximately 180 feet of magnetic tape, a supply reel, a take-up reel, and associated gears and clutches.

Electrical Specifications: The recorder is a two-channel transistorized unit consisting of the cartridge hold-down mechanism, voltage regulator, voice amplifier, time signal amplifier, bias oscillator, motor drive circuit, synchronous drive motor, speed reduction unit, capstan, magnetic record head, and end-of-tape circuit.

When the tape cartridge is inserted and secured in the tape recorder, the pressure roller in the cartridge contacts the capstan and the tape is pressed against the record head and the end-of-tape contact.

The voice tape recorder is energized by spacecraft main bus power applied through the TONE VOX circuit breaker and on S/C 3 and 4 the RCD position of either MODE switch on S/C 7 the CONT or MOM position of the RECORD switch on the VCC. The voltage regulator supplies 15 VDC to the motor drive circuits, bias oscillator and amplifiers. With the VCC and recorder energized, voice signals from the Flight Crews microphone are applied through microphone amplifiers in the VCC to the recorder voice amplifier. The voice signal is amplified and applied to the lower record head for recording on the magnetic tape. The time channel is not utilized on S/C 3. On S/C 4 and 7 a digital timing signal is applied from a time correlation buffer, in the time reference system (TRS), to the recorder time signal amplifier. The timing signal is amplified and applied to the upper record head for recording on the magnetic tape.

Simultaneously with the voice or timing signal, a 20 KC bias current from the bias oscillator is applied to the recorder heads to make a linear recording.

The motor drive circuit consists of a 133 CPS oscillator, a driver and push-pull output stage used to drive the synchronous motor. Phase-shift capacitors are connected to one motor winding for self-starting. The motor speed of 8000 RPM is reduced through the speed reduction unit to a capstan speed of 122 RPM.

The end-of-tape circuit is energized by conductive foil on the tape contacting the recorder head and end-of-tape contact causing the end-of-tape light to illuminate. The end-of-tape light will illuminate for two seconds when two minutes of recording time remains on the tape. The light will remain illuminated when the end-of-tape is reached. Recordings cannot be made when the light

 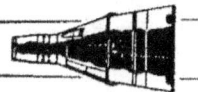

SEDR 300

PROJECT GEMINI

is illuminated. The pilot may remove the used tape cartridge, insert another cartridge and continue recording. Each cartridge provides approximately one hour of recording. The tape speed is approximately 0.6 inches per second.

TELEMETRY TRANSMITTERS

Purpose: The three telemetry transmitters provide a radio frequency (RF) link from the spacecraft to ground communication facilities for transmission of various data obtained by the instrumentation subsystem.

Physical Characteristics: The three telemetry transmitters are identical except for the operating frequency. The physical construction and approximate location of the transmitters in the spacecraft is shown in Figure 9-21. The transmitters are approximately 2.75 inches high, 2.25 inches wide, and 6.5 inches long and weighs approximately 41 ounces. Each transmitter contains a DC power connector, and RF output power connector, and a video input connector.

Two of the transmitters are located in the right forward equipment bay of the re-entry module, the third is located on the electronic module in the adapter equipment section.

Electrical Characteristics: The three telemetry transmitters are classified either by operating frequency (high, low, or mid) or by their function (real-time, delayed-time, or stand-by). The stand-by transmitter may be used either for real-time or delayed-time transmission depending upon the ground station command via the DCS or the setting of the STBY TM switch. The real-time telemetry transmitter, transmitting data directly to the ground station, is actuated

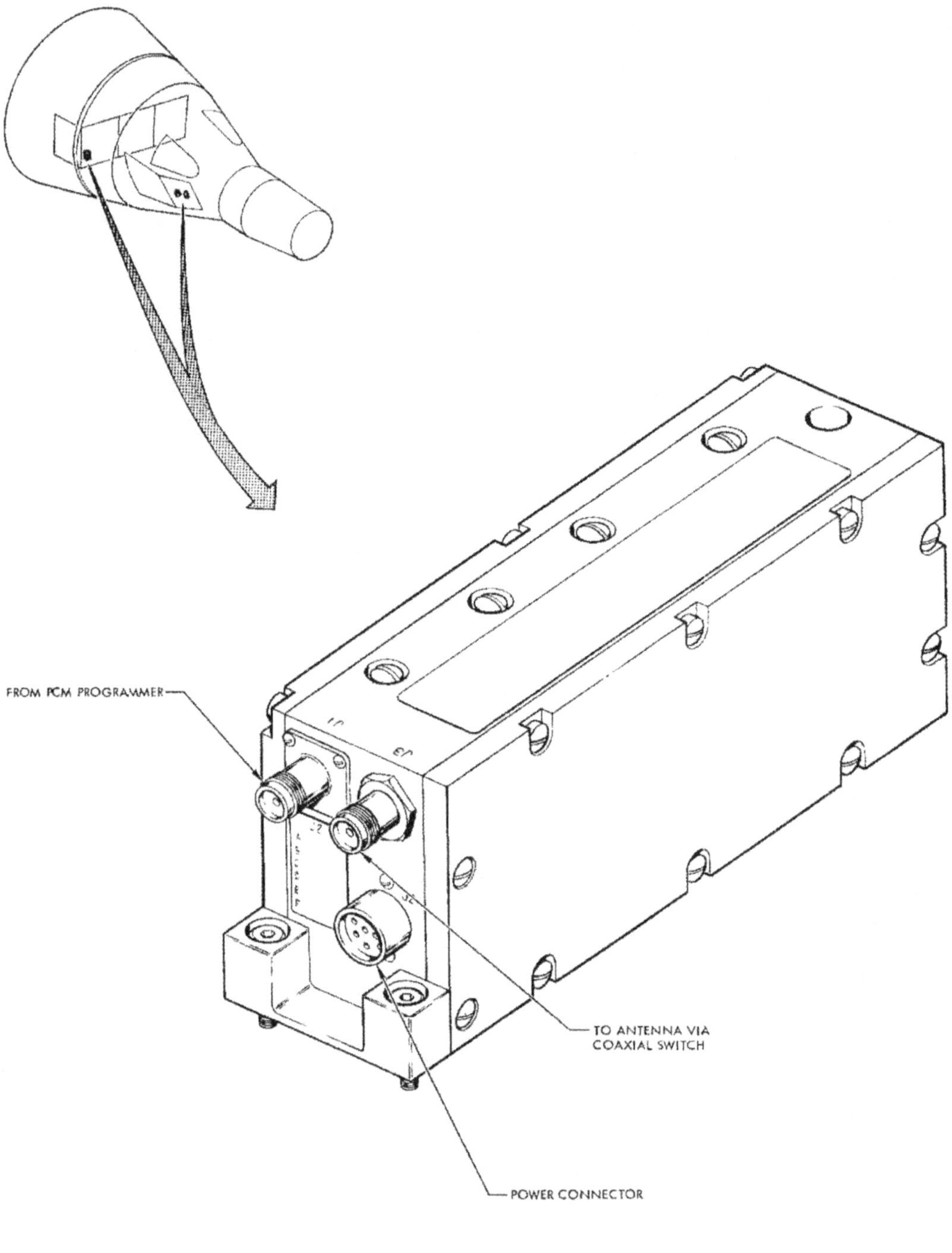

SEDR 300

 **PROJECT GEMINI**

Figure 9-21 Telemetry Transmitters

FM 2-9-20

either by the Flight Crew or by ground command. The delayed time telemetry transmitter transmits data that has been stored previously on an on-board recorder. The delayed time telemetry transmitter is actuated either by ground command or by the Flight Crew. The TM control switch allows actuation of either the real-time telemetry transmitter (R/T & ACQ), the real-time and delayed-time telemetry transmitters (R/T-D/T), or selection of the command (CMD) position. In the CMD position, either the real-time, delayed-time, or both telemetry transmitters, may be actuated by a DCS command.

The three frequency modulated (FM) transmitters provide minimum output power of two watts. The real-time low frequency telemetry transmitter operates at 230.4 MC. The delayed-time (mid-frequency) telemetry transmitter, receiving its input from an on-board tape recorder, operates at a frequency of 246.3 MC. The stand-by (high frequency) transmitter, operating at 259.7 MC, may be used either for real-time or delayed-time transmission in case one of the transmitters fails.

The three telemetry transmitters receive their pulse code modulated (PCM) inputs from the spacecraft instrumentation system.

FLASHING RECOVERY LIGHT AND POWER SUPPLY

Purpose: The flashing recovery light and power supply provide visual spacecraft location information.

Physical Characteristics: Figure 9-22 shows the physical representation and approximate location of the flashing recovery light and its power supply. The light is self-extended by a torsion spring. The plug applying power to

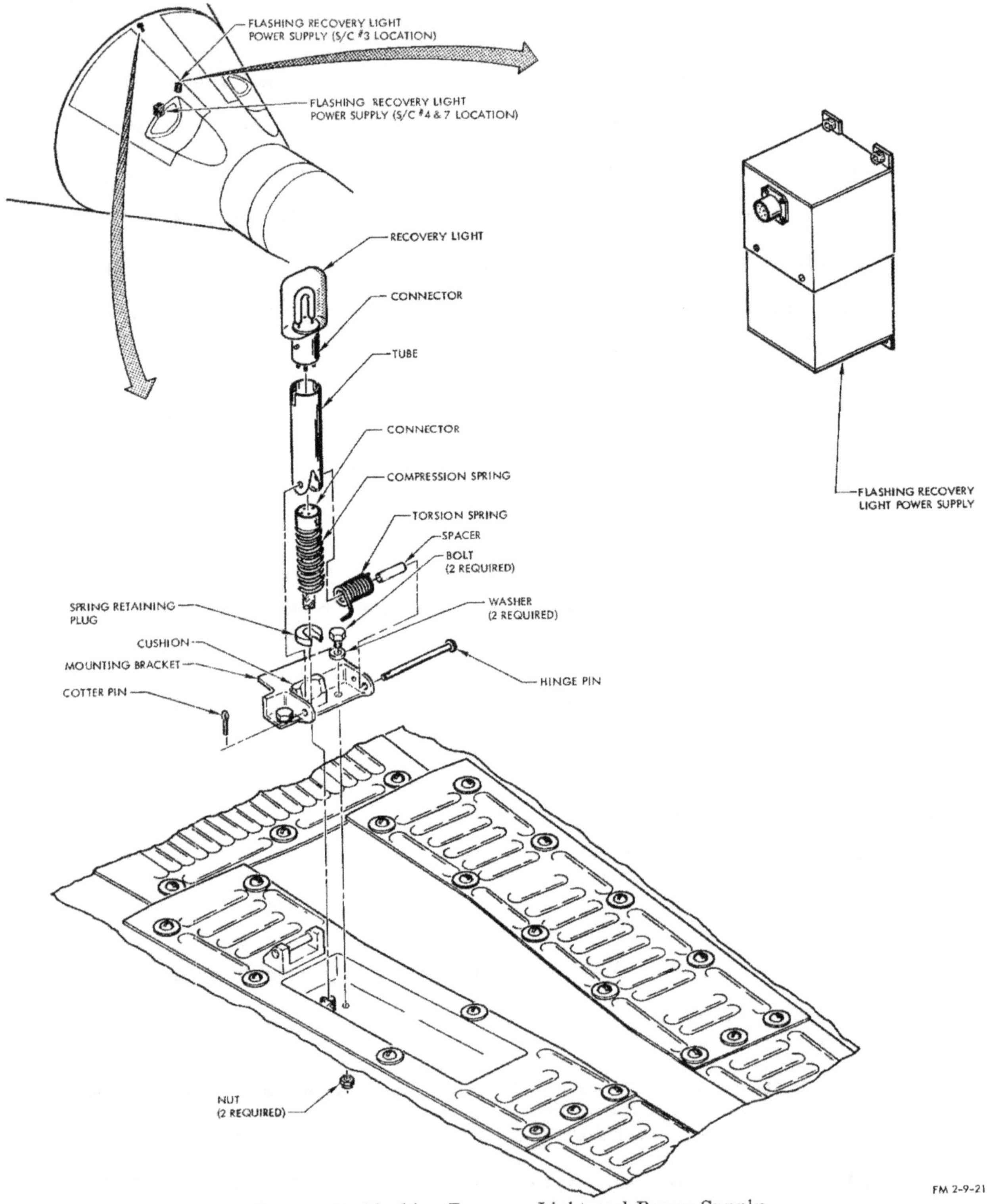

SEDR 300

PROJECT GEMINI

Figure 9-22 Flashing Recovery Light and Power Supply

SEDR 300

PROJECT GEMINI

the light is kept in place by a compression spring. The recovery light will be automatically extended at the time the main parachute is jettisoned.

The flashing recovery light power supply is mounted in the cabin aft of the ejection seats. The power supply is approximately 7 inches long, 4 inches wide, 3 inches deep and contains one connector. The flashing recovery light is approximately 1.25 inches wide, 0.75 inches thick, and 3.25 inches high excluding tube and erecting mechanism. The overall length of the light and erecting mechanism is approximately 6.5 inches.

Electrical Characteristics: On S/C 3 the recovery light, while being extended, is energized through contacts of a relay energized at main parachute jettison. On S/C 4 and 7 the extended recovery light is energized by positioning the RESC BEACON CONTROL switch to ON.

The power supply consists of a battery pack and transverter. The battery pack consists of several Mercury cells to comprise a power source of 6.75 VDC to a DC-DC converter whose output is fed to a voltage doubler and a capacitive network. The 450 VDC output of the voltage doubler is used to power the flashing light while the capacitive network in conjunction with a thyratron, provides trigger pulses to accomplish switching or flashing action of the light. The trigger pulses occur at a rate of 15 triggers per minute.

DIGITAL COMMAND SYSTEM

Purpose

The digital command system (DCS) provides a discrete command link and a digital

SEDR 300
PROJECT GEMINI

data updating capability for the spacecraft computer and time reference system (TRS).

The discrete command link enables the ground to control radar tracking beacons, selection of telemetry transmitters, instrumentation data acquisition, and abort indications.

The capability of digital data updating enables the mission control center to update the spacecraft computer and TRS to bring about a controlled re-entry at a pre-determined point, and allows timed shutdown of equipment controlled by DCS relays.

Physical Characteristics

The DCS consists of a receiver/decoder package and two relay boxes as illustrated in Figures 9-23 and 9-24, respectively. The three components are located in the electronic module of the adapter equipment section.

The receiver/decoder package is approximately 8 inches high, 8 inches wide, and 12 inches long. Both relay boxes are identical. Each relay box is approximately 2.25 inches wide, 5 inches high, and 3 inches deep. The combined weight of the received/decoder package and the two relay boxes is approximately 23 pounds. The receiver/decoder package contains two UHF receivers and a decoder while each of the two relay boxes contain eight relays.

General Description

The DCS receives phase shift keyed (PSK) frequency modulated (FM) signals composed of a reference and an information signal. The information signal is in

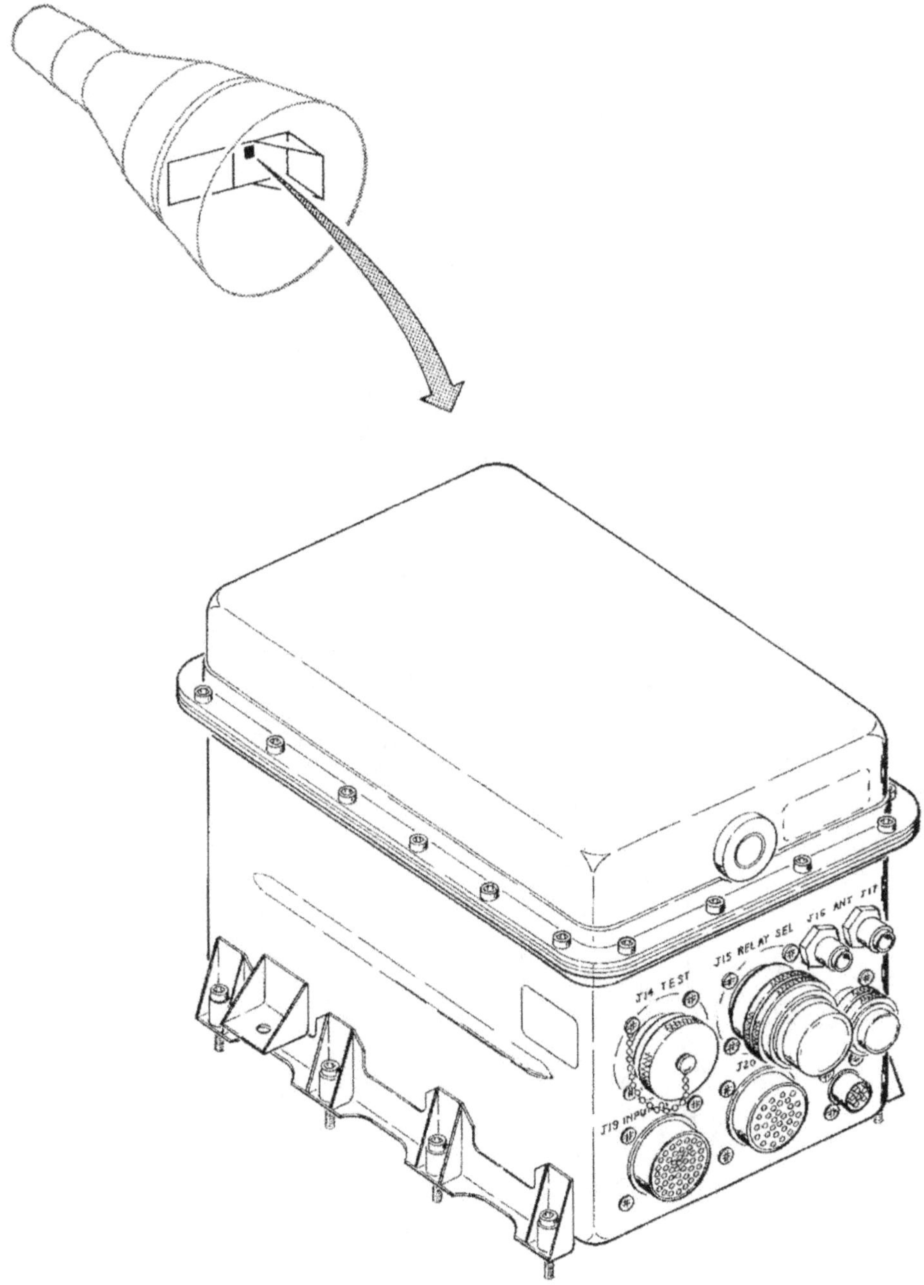

SEDR 300
PROJECT GEMINI

Figure 9-23 DCS Receiver/Decoder

 SEDR 300

PROJECT GEMINI

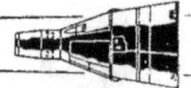

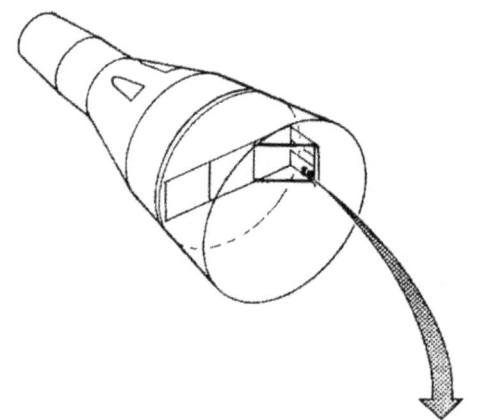

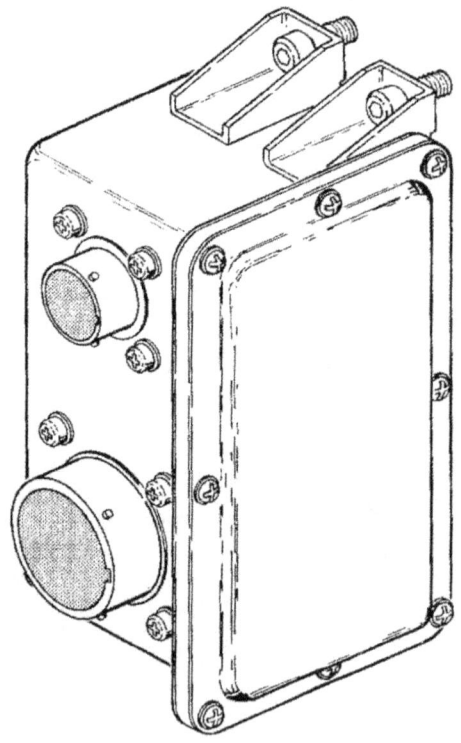

Figure 9-24 DCS Relay Box

phase with the reference for a logical "one" and 180° out of phase with the reference for a logical "zero"; thus establishing the necessary requirements for digital data.

Types of Commands

The DCS receives types of digital commands: real time commands (RTC) and stored program commands (SPC). RTC causes relays within the DCS to be actuated. Nine of the 16 relays available for RTC are utilized to perform the following functions:

(1) select the standby telemetry transmitter for real time transmission
(2) select the standby telemetry transmitter for delayed time transmission
(3) select real time telemetry and acquisition aid beacon transmission
(4) select real time and delayed time telemetry transmission
(5) actuate the re-entry C-band radar beacon
(6) actuate the adapter C-band radar beacon on S/C 4 and 7, or the S-band radar beacon on S/C 3.
(7) illuminate the abort indicators
(8) actuate the playback tape recorder
(9) initiate calibration voltage for the PCM programmer

The remaining seven relays are not utilized and perform no mission function. DCS Channel assignments for the nine functions listed above may be different on each spacecraft.

When the spacecraft goes out of range of the ground station, equipment controlled by DCS channels may be shut-down by a signal applied from the TRS to reset the

DCS relays. (The ABORT channel is not controlled by the shut-down signal.) This condition is known as salvo. The DCS relays in one relay box may be reset by the Flight Crew momentarily positioning the TAPE PLY BK switch to RESET.

Message Format and Modulation

The ground station transmits a 30 bit message for SPC's and a 12 bit message for RTC's. Each bit consists of five sub-bits. The five sub-bits are coded to represent a logical "one" or "zero". The first three bits of each message designate the vehicle address. If the vehicle address is not correct, the DCS will reset itself and will not accept the message. If the vehicle address is accepted the sub-bit code will be automatically changed for the remainder of the message to reduce the probability of accepting an improper message.

The second three bits of each message designate the system address and identifies the remainder of the message as being a RTC or one of the following SPC: computer update, TRS TTG to T_R, or TRS TTG to T_X. If the message is a SPC, the last 24 bits will be a data word. If the SPC is a TRS TTG to T_X command, the last eight bits are ignored by the TRS. In case of a computer message, six bits of the data word contains the internal computer address and the remaining 18 bits contains information. Since a RTC consists of 12 bits, the six bits following the system address contain a five-bit relay number and a one bit relay set/reset discrete.

The PSK modulation signals are a 1 KC reference and a 2 KC information signal. The receiver output is the composite audio of the 1 KC and the 2 KC signals.

The composite audio output is filtered to recover the 1 KC and the 2 KC signals. The phase comparator compares the 2 KC to the 1 KC signal. The output of the phase comparator is used to trigger a flip-flop to produce either a logical "one" or "zero" sub-bit. The 1 KC reference signal is used to synchronize the DCS.

Operational Description

A block diagram of the DCS receiver/decoder is shown in Figure 9-25. Basically, the block diagram consists of a receiver, a decoder, and a power supply common to both sections.

The audio outputs of the two receivers are linearly summed in an emitter follower of the sub-bit detector module. The sub-bit detector converts the audio to sub-bits. The 5 stage shift register provides buffer storage for the output of the sub-bit detector. The states of the five stages of the shift register represent the sub-bit code. When a proper sub-bit code exists in the shift register, the bit detector produces a corresponding "one" or "zero" bit. The output of the bit detector is applied to the 24 stage shift register. The operation for RTC and SPC is identical up to the input to the 24 stage shift register.

The sub-bit sync counter produces a bit sync output for every five sub-bits. The bit sync is used to gate the 24 stage shift register.

When a message is received, the vehicle address is inserted into the first three stage of the 24 stage shift register. If the vehicle address is correct, the vehicle address decoder circuit will produce an output to the bit detector

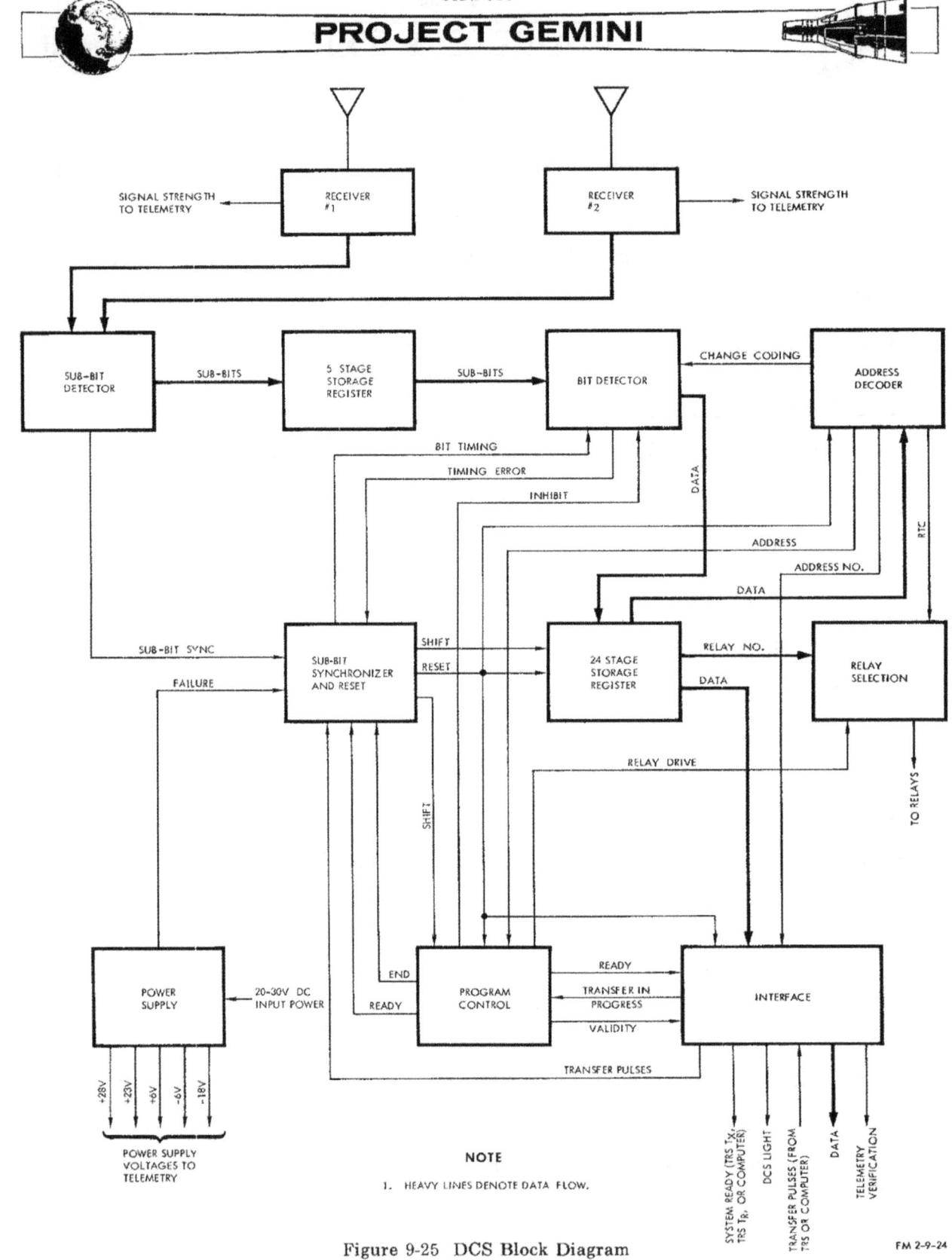

Figure 9-25 DCS Block Diagram

which changes the acceptable sub-bit code for the remainder of the message. The next three bits of the message, system address, are inserted into the first three stages of the 24 stage shift register, displacing the vehicle address to the next three stages. The system address decoder circuit identifies the specific address and sets up the DCS to handle the remainder of the message.

When the system address is recognized to be a RTC, the message is inserted into the first six stages of the 24 stage shift register and the system address and vehicle address are shifted into the next six stages. The real time command selection circuit recognizes the first stage of the 24 stage shift register to be either a relay set or reset function and will apply a positive voltage to all set or reset relay coils, as applicable. The real time command selection gates, select the proper relay from the relay number stored in the 24 stage shift register and provides an output which applies a power return to the coil of the selected relay.

When the system address is a SPC, the six address bits in the 24 stage shift registers are cleared and the remaining 24 bits of the message are placed into the register.

Assuming that the system address recognizes a RTS TTG to T_R message, the data flow would be as follows: the TRS T_R isolation amplifier, in the interface circuit, will apply a "ready" pulse to the TRS. The "ready" pulse sets up the TRS to transfer TRS TTG to T_R data from the DCS. When the TRS is ready to accept the data, it sends 24 shift pulses at the TRS data rate to the TRS input of the DCS. The data in the 24 stage shift register is then shifted out of the register through the DCS data isolation amplifier to the TRS. The DCS operations for computer updating and TRS TTG to T_X messages are similar to TRS TTG to T_R operations.

Salvo occurs when TRS TTG to T_X reaches zero. At $T_X = 0$, the TRS applies a signal to the TRS T_X input line of the DCS which causes the real time command selection circuits to reset the DCS relays.

After a SPC or RTC has been carried out by the DCS, a verification signal is supplied to the telemetry system for transmission to a ground station. The DCS indicator, on the pilots instrument panel, illuminates when a SPC is transferred to the appropriate system.

Upon completion of data transfer or if the system to which the data was transferred fails to respond within 100 milli-seconds, the DCS will reset in preparation for the next message. The DCS will also reset in the event of a timing error in transmission of data, or if the DCS power supply voltages become out of tolerance.

INSTRUMENTATION SYSTEM

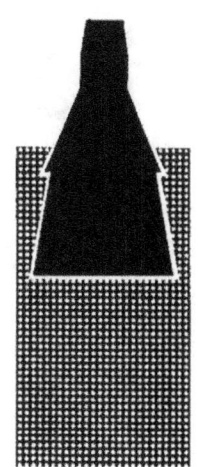

TABLE OF CONTENTS

Section X

TITLE	PAGE
SYSTEM DESCRIPTION	10-5
SYSTEM OPERATION	10-6
SEQUENTIAL SYSTEM PARAMETERS	10-10
ELECTRICAL POWER SYSTEM PARAMETERS	10-15
ECS PARAMETERS	10-18
INERTIAL GUIDANCE SYSTEM PARAMETERS	10-22
ACME PARAMETERS	10-25
OAMS PARAMETERS	10-27
RE-ENTRY CONTROL SYSTEM PARAMETERS	10-29
AERODYNAMIC AND CREW CONTROL PARAMETERS	10-32
COMMUNICATION SYSTEM PARAMETERS	10-32
INSTRUMENTATION SYSTEM PARAMETERS	10-35
PHYSIOLOGICAL PARAMETERS	10-38
SYSTEM UNITS	10-40
PRESSURE TRANSDUCERS	10-40
TEMPERATURE SENSORS	10-42
SYNCHRO REPEATERS	10-44
CO_2 PARTIAL PRESSURE DETECTOR	10-46
ACCELEROMETERS	10-46
INSTRUMENTATION PACKAGES	10-49
MULTIPLEXER/ENCODER SYSTEM	10-51
TRANSMITTERS	10-56
PCM TAPE RECORDER	10-56
DC-DC CONVERTERS	10-60
BIO-MED TAPE RECORDERS AND POWER SUPPLY	10-62

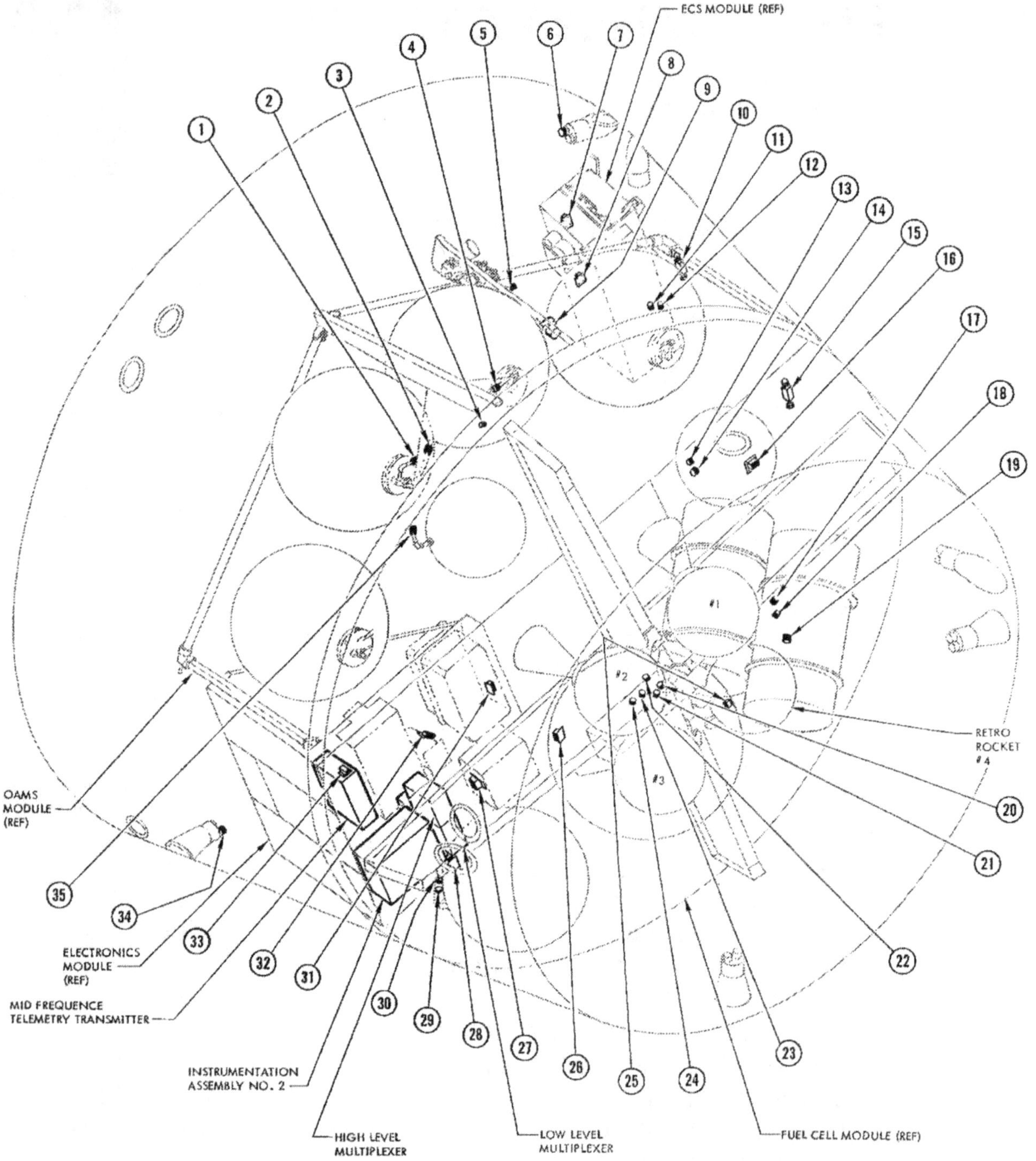

SEDR 300
PROJECT GEMINI

Figure 10-1 Instrumentation System Components (Sheet 1 of 3)

PROJECT GEMINI

SEDR 300

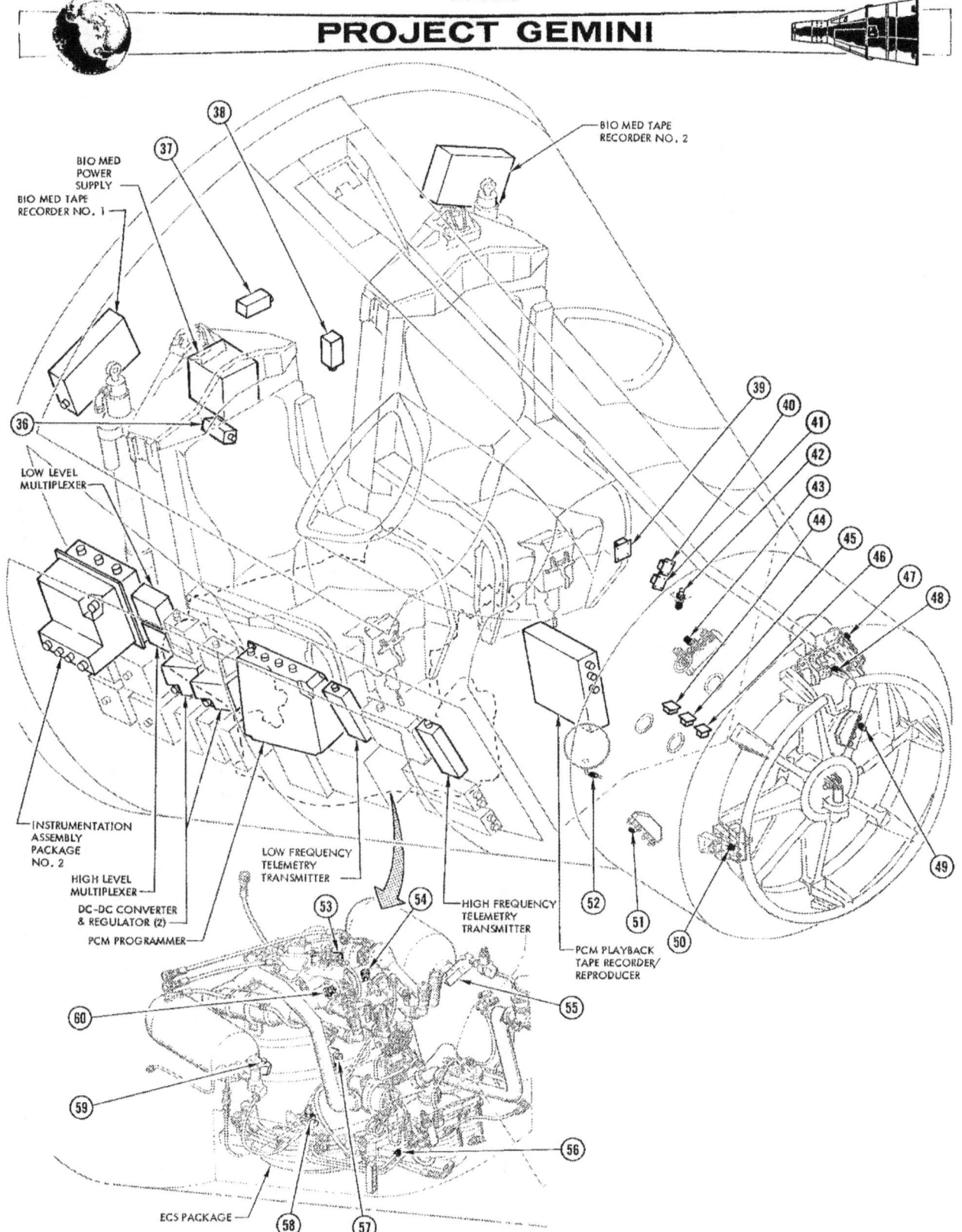

Figure 10-1 Instrumentation System Components (Sheet 2 of 3)

SEDR 300

PROJECT GEMINI

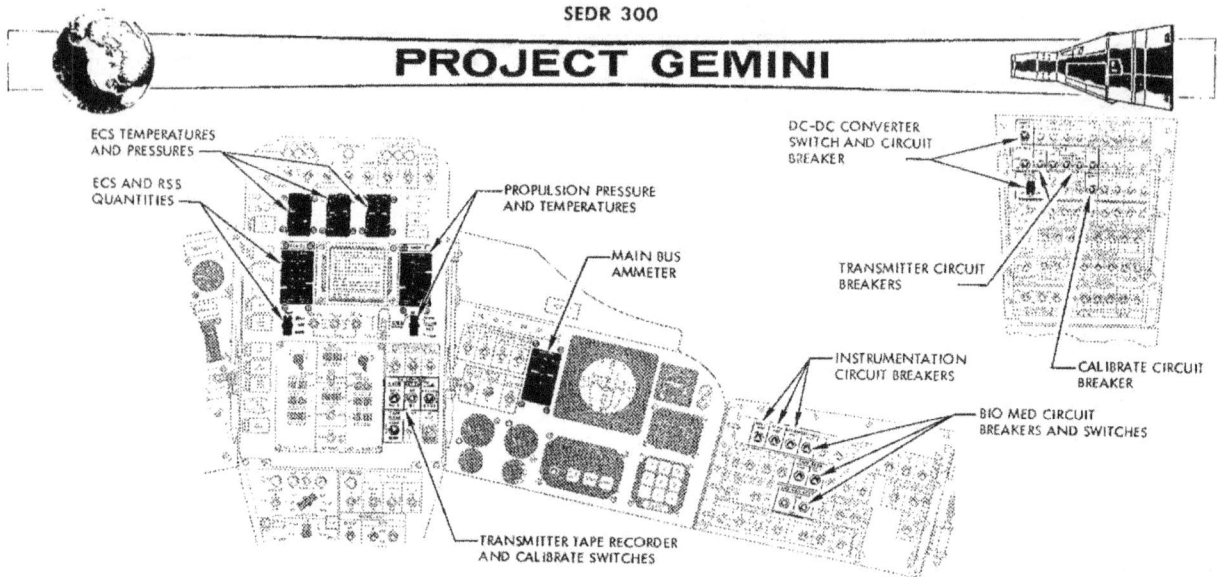

LEGEND

ITEM	PARAMETER	NOMENCLATURE
1	GC04	REG He AT OXID TANK TEMPERATURE
2	GB02	OXIDIZER FEED TEMPERATURE
3	GB01	FUEL FEED TEMPERATURE
4	GC03	REG He AT FUEL TANK TEMPERATURE
5	GC01	SOURCE He PRESSURE
6	GD08	TCA #7 HEAD TEMPERATURE
7	CJ01	PRIMARY COOLANT PUMP INLET PRESSURE
8	CJ02	SECONDARY COOLANT PUMP INLET PRESSURE
9	GC05	REGULATED He PRESSURE
10	CH03	SECONDARY COOLANT RADIATOR OUTLET TEMPERATURE
11	CD04	SECONDARY COOLANT TEMP AT OUTLET OF RADIATOR
12	CD03	PRIMARY COOLANT TEMP AT OUTLET OF RADIATOR
13	CA06	PRIMARY ECS O_2 SUPPLY BOTTLE TEMPERATURE
14	CA02	PRIMARY ECS O_2 TANK PRESSURE
15	CH02	PRIMARY COOLANT RADIATOR OUTLET TEMPERATURE
16	CA09	CRYO MASS QUANTITY (RSS-ECS)
17	CD02	SECONDARY COOLANT INLET TO F.C. SECT 2 TEMP
18	CD01	PRIMARY COOLANT INLET TO F.C. SECT 1 TEMP
19	CL01	WATER PRESSURE
20	BC03	F.C. H_2 TEMP AT HEAT EXCHANGER OUTLET
21	BA04	HYDROGEN TANK PRESSURE
22	BB05	F.C. O_2 TEMP AT HEAT EXCHANGER OUTLET
23	BA06	RSS H_2 SUPPLY BOTTLE TEMPERATURE
24	CA09	CRYO MASS QUANTITY (RSS-ECS)
25	HH01	RETRO ROCKET CASE TEMPERATURE
26	CA09	CRYO MASS QUANTITY (RSS-ECS)
27	LC09	ADAPTER C-BAND BCN PACKAGE TEMPERATURE
28	CA09	CRYO MASS QUANTITY (RSS-ECS)
29	BA05	RSS O_2 SUPPLY BOTTLE TEMPERATURE
30	BA02	OXYGEN TANK PRESSURE

LEGEND

ITEM	PARAMETER	NOMENCLATURE
31	LA05	DCS PACKAGE TEMPERATURE
32	LD01	ACQ AID BCN CASE TEMPERATURE
33	MC02	MID FREQ TM XMTR CASE TEMPERATURE
34	GD07	TCA #3 HEAD TEMPERATURE
35	GC02	SOURCE He TEMPERATURE
36	KA01	Z ACCELERATION
37	KA02	X ACCELERATION
38	KA03	Y ACCELERATION
39	CB07	FWD COMPARTMENT ABSOLUTE PRESS
40	KB02	STATIC PRESSURE
41	CB01	CABIN PRESS TO FWD COMP.
42	CB02	CABIN AIR TEMPERATURE
43	HC03	REG N_2 PRESSURE-SYST A
44	DQ07	PITCH ATTITUDE-SYNCHRO REPEATER
45	DQ08	ROLL ATTITUDE-SYNCHRO REPEATER
46	DQ09	YAW ATTITUDE-SYNCHRO REPEATER
47	HC04	REG N_2 PRESSURE-SYST B
48	HC06	SOURCE N_2 PRESS-SYST B
49	HC02	N_2 SOURCE PRESS-RCS SYST B
50	HA02	RCS OXIDIZER FEED TEMP-SYST A
51	HC01	N_2 SOURCE PRESS-RCS SYST A
52	HC05	SOURCE N_2 PRESS-SYST A
53	CC06	CO_2 PARTIAL PRESSURE SENSOR
54	CC03	LEFT SUIT INLET AIR TEMP
55	CA03	ECS O_2 SUPPLY PRESS NO. 1-SEC
56	CK06	SUIT HEAT EXCHANGER INLET TEMP-PRI
57	CC01	LEFT SUIT PRESSURE
58	CC02	RIGHT SUIT PRESSURE
59	CA04	ECS O_2 SUPPLY PRESS NO. 2-SEC
60	CC04	RIGHT SUIT INLET AIR TEMP

Figure 10-1 Instrumentation System Components (Sheet 3 of 3)

 SEDR 300

PROJECT GEMINI

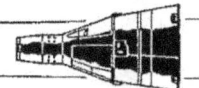

SECTION X INSTRUMENTATION SYSTEM

SYSTEM DESCRIPTION

The instrumentation system provides a means of data acquisition with respect to the performance and operation of the spacecraft throughout its entire mission. Data acquisition is defined as the sensing of specific conditions or events on board the spacecraft, displaying the derived data from these inputs to the spacecraft crew and ground operation personnel, and recording and later processing this data for use in post flight reports and analysis. In this respect the data acquisition function is shared by all spacecraft systems, the ground operational support system, and the data processing facility.

Basically, the instrumentation parameters are divided into two categories: operational and non operational. Operational parameters are those which are necessary for determining the progress of the mission, assessing spacecraft status, and making decisions concerning flight safety. Non operational parameters are those which are required for post mission analysis and evaluation.

The basic components comprising the instrumentation system are: sensors, signal conditioners, multiplexers and encoders, and transmitters. Because the system is used to sense parameters of every spacecraft system, its components are located throughout the entire spacecraft as shown in Figure 10-1.

 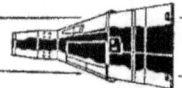

SEDR 300

PROJECT GEMINI

SYSTEM OPERATION

The purpose of the instrumentation system is data acquisition during the entire spacecraft mission necessitating its operation throughout the entire mission. The instrumentation system provides the capability of data acquisition and transmission to ground station while the data source is provided by all the spacecraft system. The basic functions by which the system fulfills its purpose are: to sense the various conditions and functions, convert them to proportional electrical signals (if applicable); condition the resulting signal (when necessary) to make it compatible with the encoding and multiplexing equipment, display pertinent data in the spacecraft cabin, record data for delayed time (data dump) transmission, and provide signals for real time transmission to the ground station. An overall block diagram of the instrumentation system is shown in Figure 10-2 and the power distribution is shown in Figure 10-3.

The system senses the prescribed parameters through the use of sensors which may be contained within the instrumentation system or which may be an integral part of the data source system. Typical sensors include pressure transducers, accelerometers, and temperature sensors. Signals may also be obtained from such functions as switch and relay actuations, and from electronic package monitor points. Sensors and signal sources are shown in block diagram form on the applicable data source system illustration.

The majority of the signals acquired are usable for the spacecraft cabin indicators and/or the encoding equipment without alteration. Some of them, however, are routed to signal conditioning packages (instrumentation assemblies) where

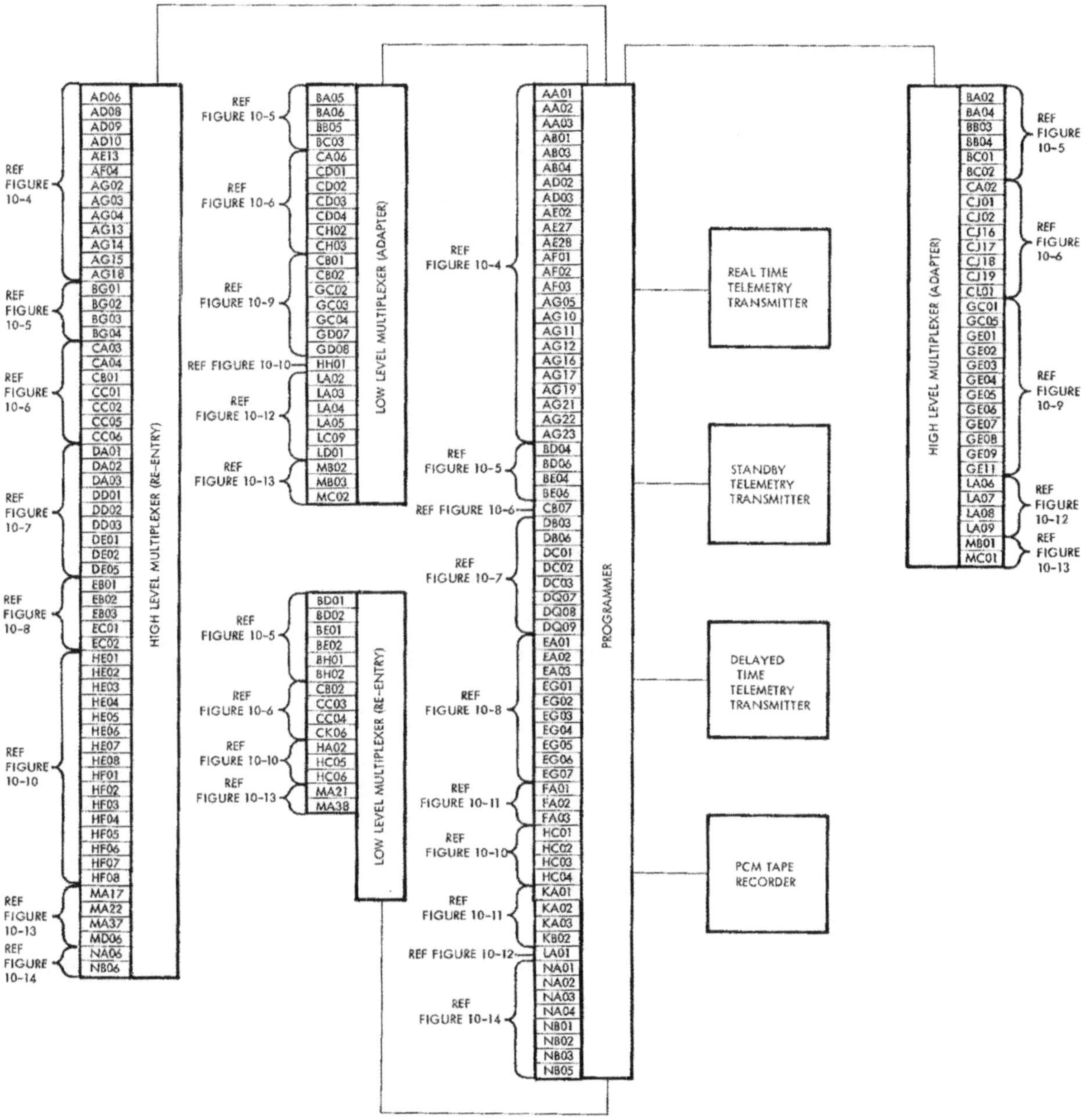

SEDR 300
PROJECT GEMINI

Figure 10-2 Instrumentation System Signal Flow Block Diagram

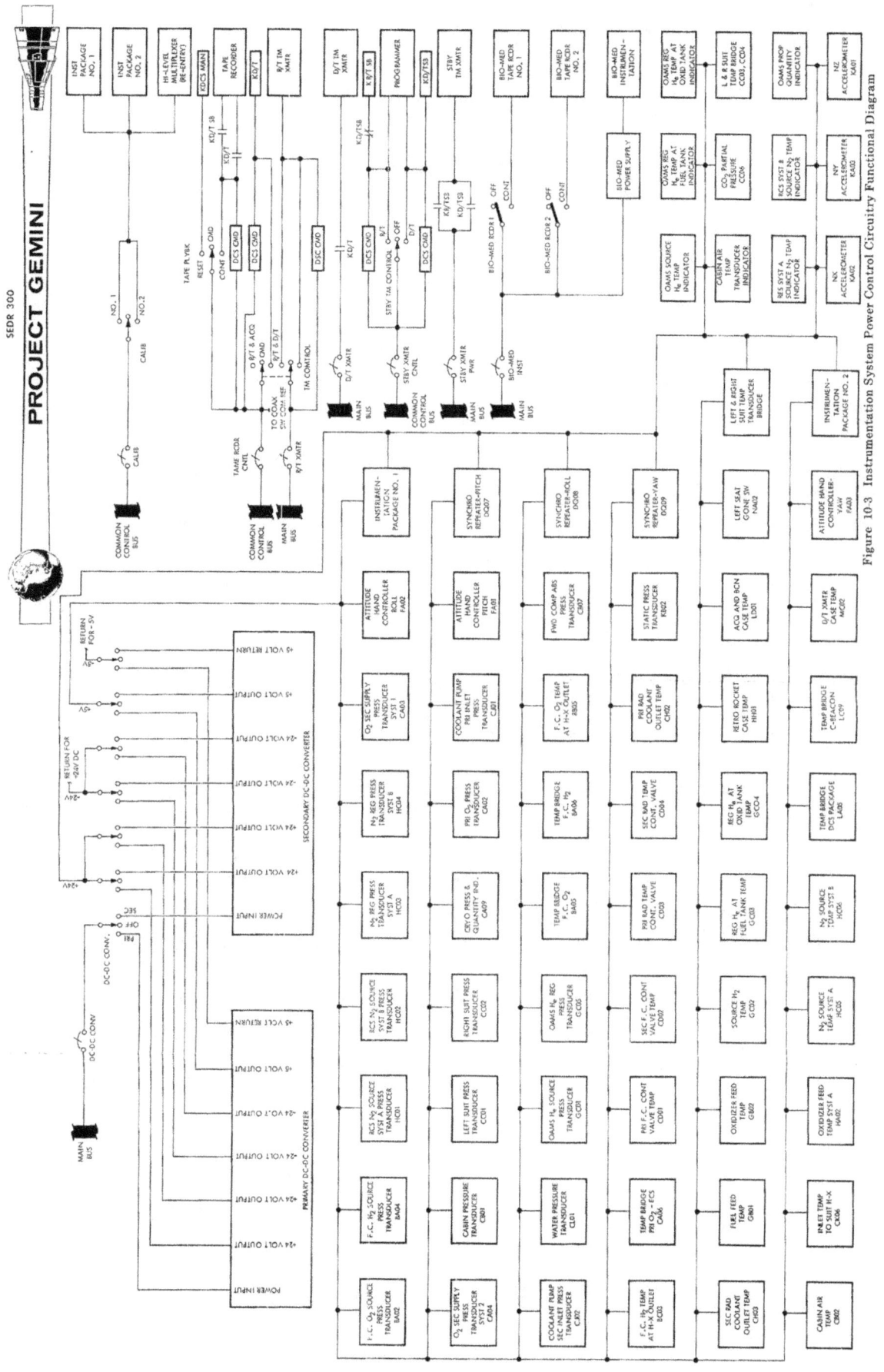

Figure 10-3 Instrumentation System Power Control Circuitry Functional Diagram

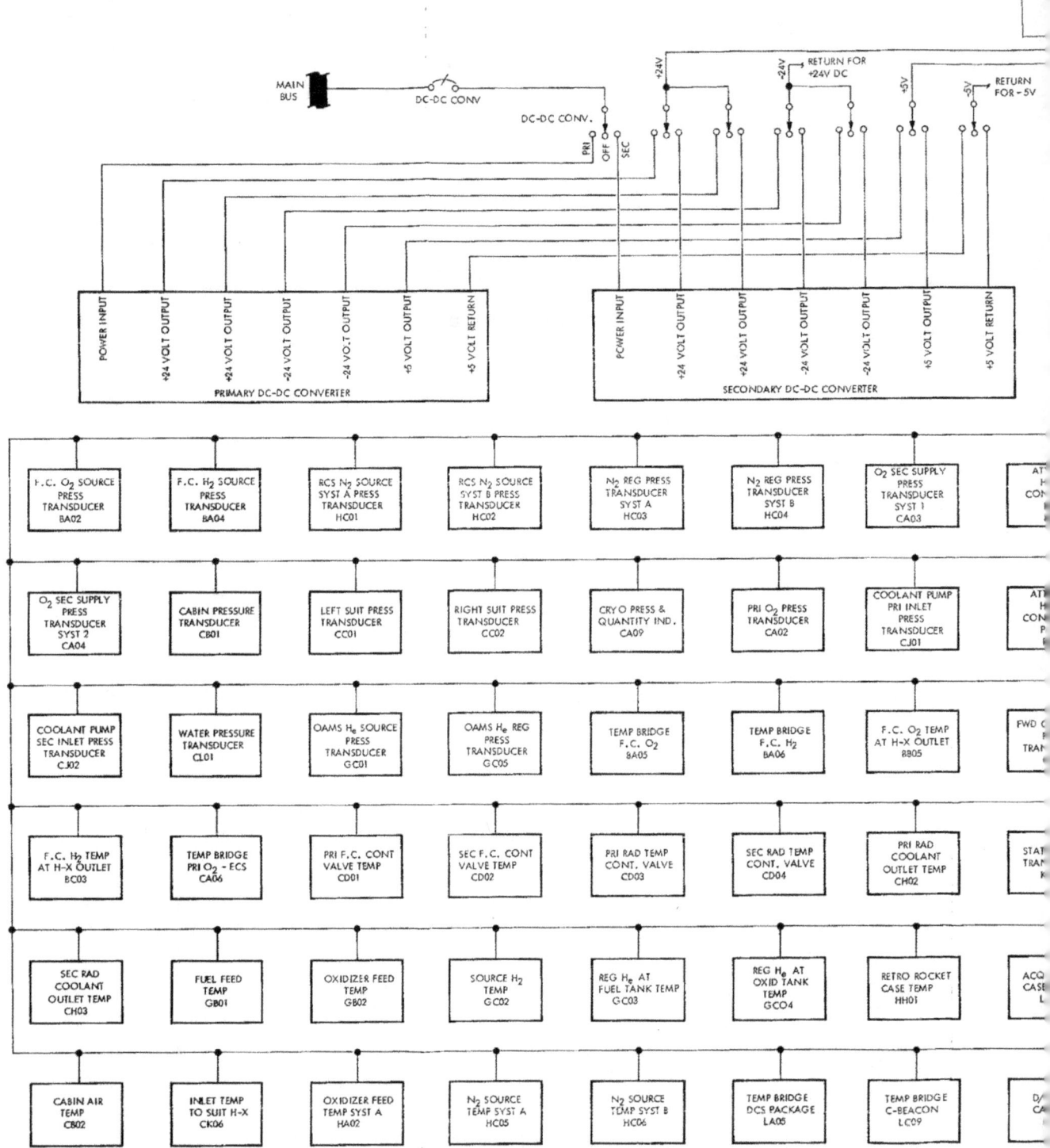

10-8

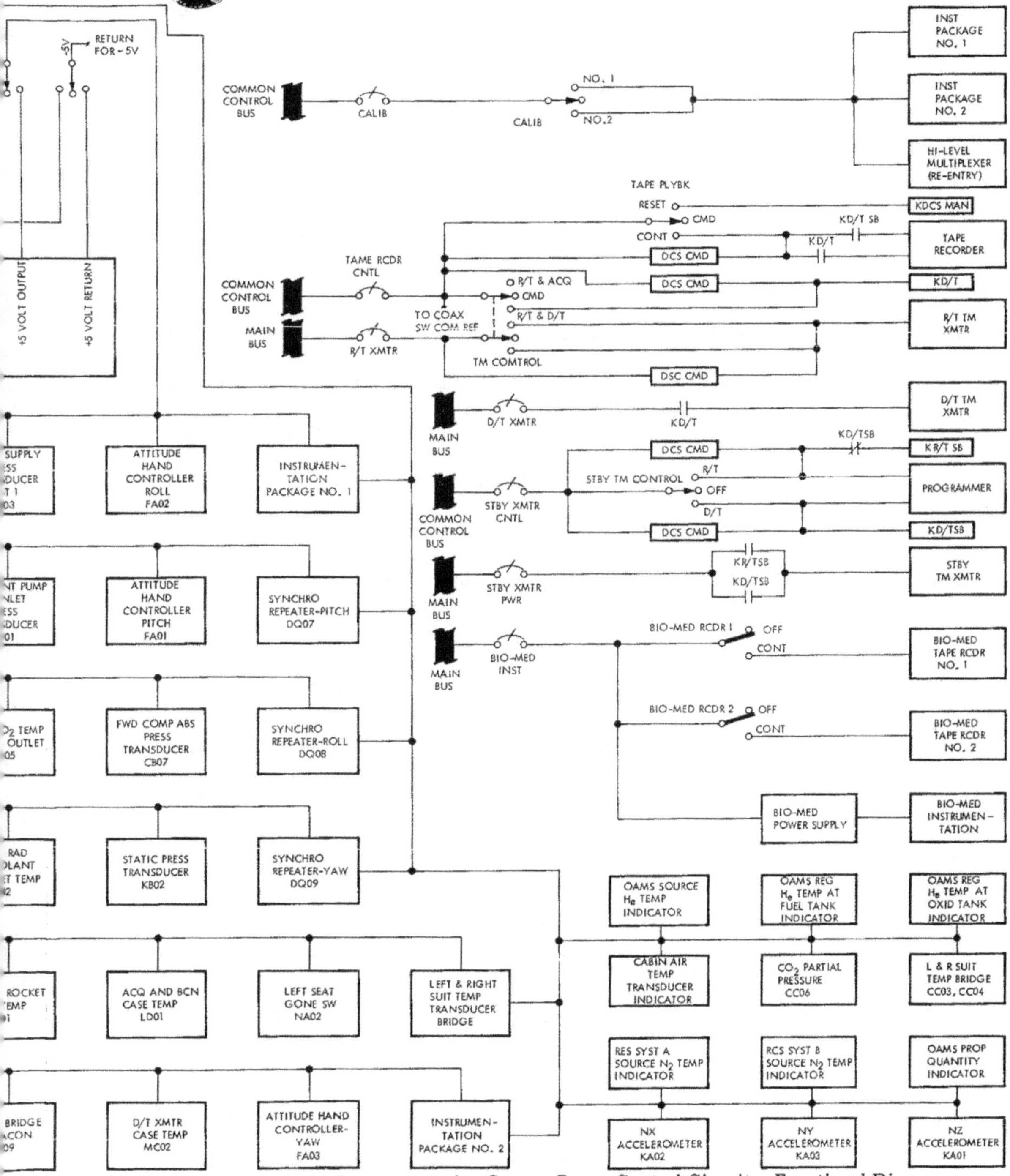

SEDR 300
PROJECT GEMINI

Figure 10-3 Instrumentation System Power Control Circuitry Functional Diagram

their characteristics and/or amplitudes are changed. The resulting signals, as well as those from the other sensors, are of four basic types: low-level (0-20 MV DC), high-level (0-5 VDC), bi-level (0 or 28 VDC), and bi-level pulse (0 or 28 VDC). Signals of selected parameters are supplied to the cabin indicators, while signals of all parameters are supplied to the mulitplexer/encoder system. The multiplexer/encoder system converts the various spacecraft analog and digital signals to a serial binary-coded digital signal for presentation to the data-dump tape recorder and the real-time telemetry transmitter. The tape recorder records a portion of the real-time data from the programmer at a tape speed of 1 7/8 inches per second and, upon command, will play back the data for transmission to a ground station, at a speed of 41.25 IPS (22 times the recording speed).

Five physiological functions are monitored for each pilot. All of the measurements are supplied as real-time data, while only one is supplied as delayed-time data. In addition, some of the measurements are recorded by two special (BIO-MED) tape recorders.

During pre-launch operations, data acquisition is accomplished by use of hardlines attached to the spacecraft umbilical and by telemetry. Between launch and orbital insertion, data acquisition is via the real time telemetry transmitter. While the spacecraft is in orbit, data is acquired via the real time telemetry transmitter for the period while the spacecraft is within range of a ground station. Data during the period while the spacecraft is out of range of a ground station is recorded on the PCM recorder and played back via the delayed time telemetry transmitter while the spacecraft is within range of a ground station. A more detailed description of the telemetry transmitters is given in Section IX.

The paragraphs to follow present a brief description of all instrumentation parameters. The parameters are described in groups identified by their applicable data source system. It should be noted that although most of the parameters are also applicable to S/C 3 and 4, the following list of parameters is for S/C 7 specifically.

SEQUENTIAL SYSTEM PARAMETERS

A functional diagram showing the sequential system parameters is presented in Figure 10-4. The instrumentation system monitors 39 sequential events and sequential system parameters. Each parameter is described below either individually, or as part of a group of related parameters.

The sequential system time reference system (TRS) provides three 24-bit digital words to the 24-bit shift register of the PCM programmer. These three signals are: time since lift-off (AA01, AA02) and time to retrograde (AA03). Time since lift-off is referenced to the launch vehicle lift-off signal and provides time correlation for the data tape recorders. Time to retrograde (AA03) indicates the time remaining before retrofire initiation by the TRS. This signal is used to verify that the correct retrofire time has been inserted into the TRS by ground command or by the Pilots.

Launch vehicle second stage cutoff (AB01) is monitored for ground station indication of this event. This parameter is provided by a signal from the spacecraft IGS computer to a bi-level channel of the programmer.

SEDR 300

PROJECT GEMINI

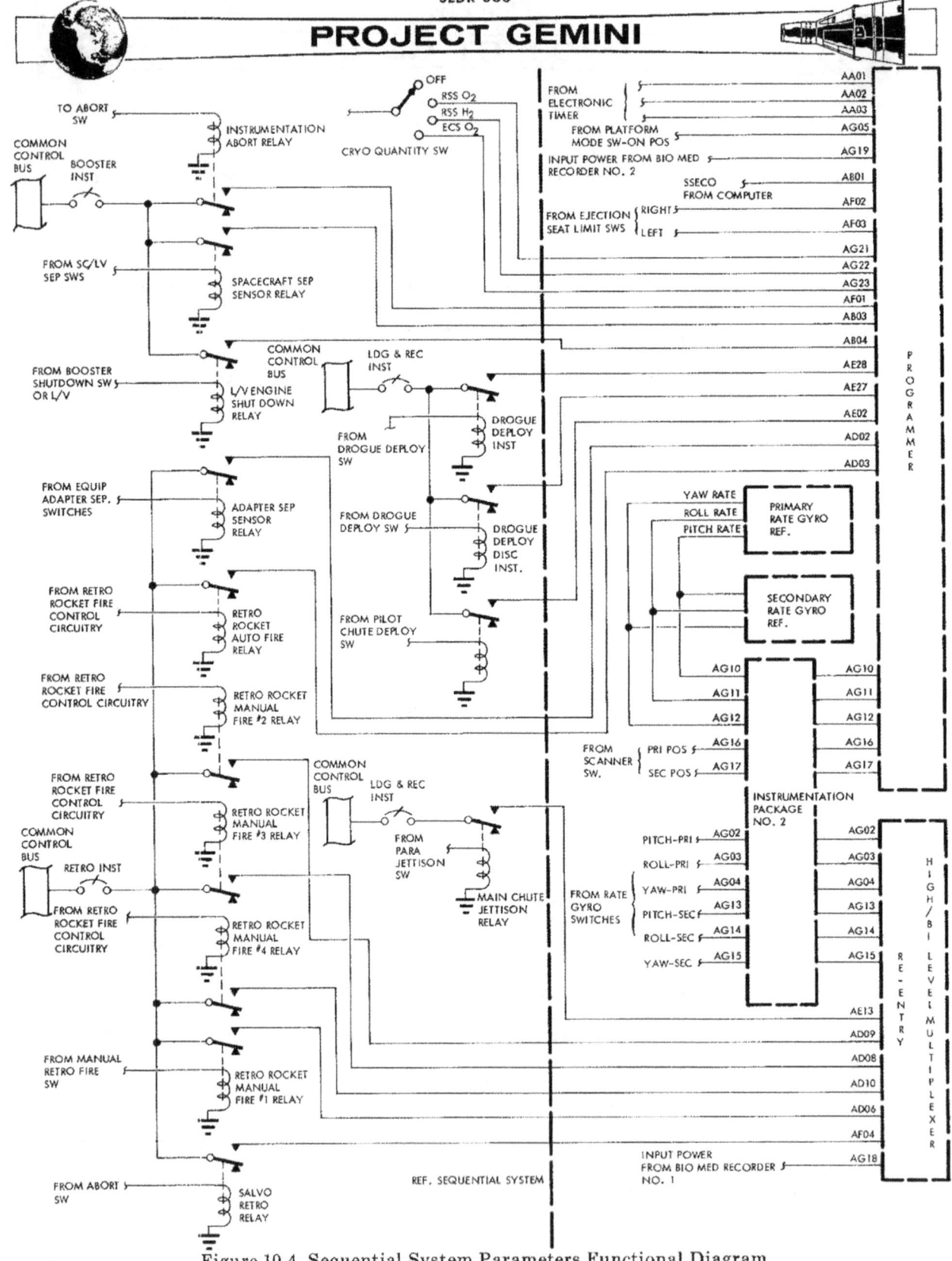

Figure 10-4 Sequential System Parameters Functional Diagram

 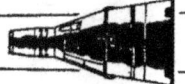

PROJECT GEMINI

Launch vehicle/spacecraft separation (AB03) is indicated to the ground station when any two of the three spacecraft/launch vehicle limit switches close energizing the spacecraft separation relays. Actuation of either of the two relays applies 28 VDC to a bi-level channel of the programmer.

Equipment section separation (AD02) is monitored to indicate a safe condition for retrograde prior to manual initiation or ground command of retro fire as a back up to the automatic system. This signal is originated when any two of the three separation sensors close, energizing the equipment section separation relays. Actuation of either of the two relays applies 28 VDC to a bi-level channel of the programmer.

The retro rocket ignition commands are monitored by ground stations to provide data for calculation of expected re-entry trajectory. Automatic (AD03) and manual (AD06) ignition commands are monitored. Parameters are obtained from the ignition command of the four retro rockets individually; AD09, rocket 2; AD08; rocket 3; AD10, rocket 4. The manual and automatic retro fire commands indicate retro rocket 1 fire. The signals, 28 VDC, are applied to the re-entry high/bi-level multiplexer.

Indication of a pilot actuated abort (AF01) is provided to the ground station. The signal is originated when the abort handle is moved to the ABORT position actuating a limit switch which energizes the instrumentation abort relays. Actuation of one of the relays applies a signal to a bi-level channel of the programmer.

In case of pilot ejection during an abort left (AF03) and right (AF02) ejection seat gone signals are provided for the ground station. The signals are originated at the time the ejection seats leave the spacecraft closing the corresponding limit switch and applying the signals to the bi-level channels of the programmer.

Confirmation of salvo retro fire is given to the ground station in case of an abort. A signal is applied to a bi-level channel of the re-entry high/bi-level multiplexer when the salvo retro relay is energized.

Indication of booster cutoff command (AB04) is given to the ground station when pilots move the abort handle to the SHUTDOWN position, actuating a limit switch. This energizes a relay applying 28 VDC to a bi-level channel of the programmer.

Ground indication of pilot chute deployment (AE02) is provided via a bi-level channel of the programmer. The signal is originated when a lanyard from the chute actuates a toggle switch, energizing the pilot chute deployed instrumentation relay.

The parachute jettisoned (AE13) signal is initiated when the pilot depresses the CHUTE JETT switch energizing redundant main chute jettison relays. The relays apply a 28 VDC signal to a bi-level channel of the re-entry high/bi-level multiplexer.

PROJECT GEMINI

SEDR 300

Platform mode selection (AG05) is indicated to the ground station. Any position other than OFF on the PLATFORM mode switch will apply a signal to a bi-level channel of the programmer.

Primary (AG16) or secondary (AG17) horizon scanner operation can be monitored by the ground station via bi-level channels of the programmer.

Primary pitch (AG02), roll (AG03), and yaw (AG04) and secondary pitch (AG13), roll (AG14), and yaw (AG15), gyro operation is monitored to indicate on or off condition. Each signal is applied to a signal conditioner whose output is applied to a bi-level channel of the high/bi-level multiplexer.

Pitch (AG10), roll (AG11), and yaw (AG12) attitude gyro (primary or secondary depending which is operational) outputs are applied to three signal conditioners. Each of the signal conditioners is a transistor switch providing no output for an input of 0-0.325 volts and a 16.5 volt output for an input greater than 0.325 volts. The conditioned signals are applied to bi-level input channels of the programmer.

Bio-medical tape recorder on-off signals (AG18, AG19) are used for time correlation of the recorded bio-medical data with the telemetry data. An on-off indication is provided to the playback recorder and to telemetry by a bi-level signal to the programmer (AG19) and re-entry high level multiplexer (AG18).

Drogue chute deployment (AE27) and drogue release (AE28) can be verified by the ground station via bi-level channels of the programmer. The signals are initiated when the HI-ALT DROGUE switch is depressed.

10-14

The selected cryogenic quantity switch position is indicated to the ground station by AG21 (RSS O_2), AG22 (RSS H_2), and AG23 (ECS O_2) to allow the ground station to identify the reading of CA09 described under environmental control system.

ELECTRICAL POWER SYSTEM PARAMETERS

Figure 10-5 shows a functional diagram of the electrical power system parameters. Approximately 24 electrical power system parameters are monitored by the instrumentation system. The parameters are listed and described in the following sub-paragraphs.

Fuel cell oxygen (BA02) and hydrogen (BA04) tank pressures are monitored by dual potentiometer pressure transducers installed as part of the fuel cell system. Each dual transducer provides one output to the adapter high level multiplexer and the other output drives an indicator on the instrument panel in the cabin.

To evaluate proper operation of the fuel cell, stack 1A (BD01), 1B (BD02), 2A (BE01), 2B (BE02) and section 1 (BH01) and 2 (BH02) currents are monitored and transmitted to the ground station. Stack C currents are obtained mathematically by subtracting section A and B currents from the corresponding section current. The signals being monitored originate from 50 millivolt shunts. The shunts are installed at the main buses for the section, and in the lines from stacks A and B to the main buses for stack A and B currents. Each of these signals is conditioned to a 0 to 20 millivolt signal which is directly proportional to the input current and then applied to the re-entry low level multiplexer.

10-15

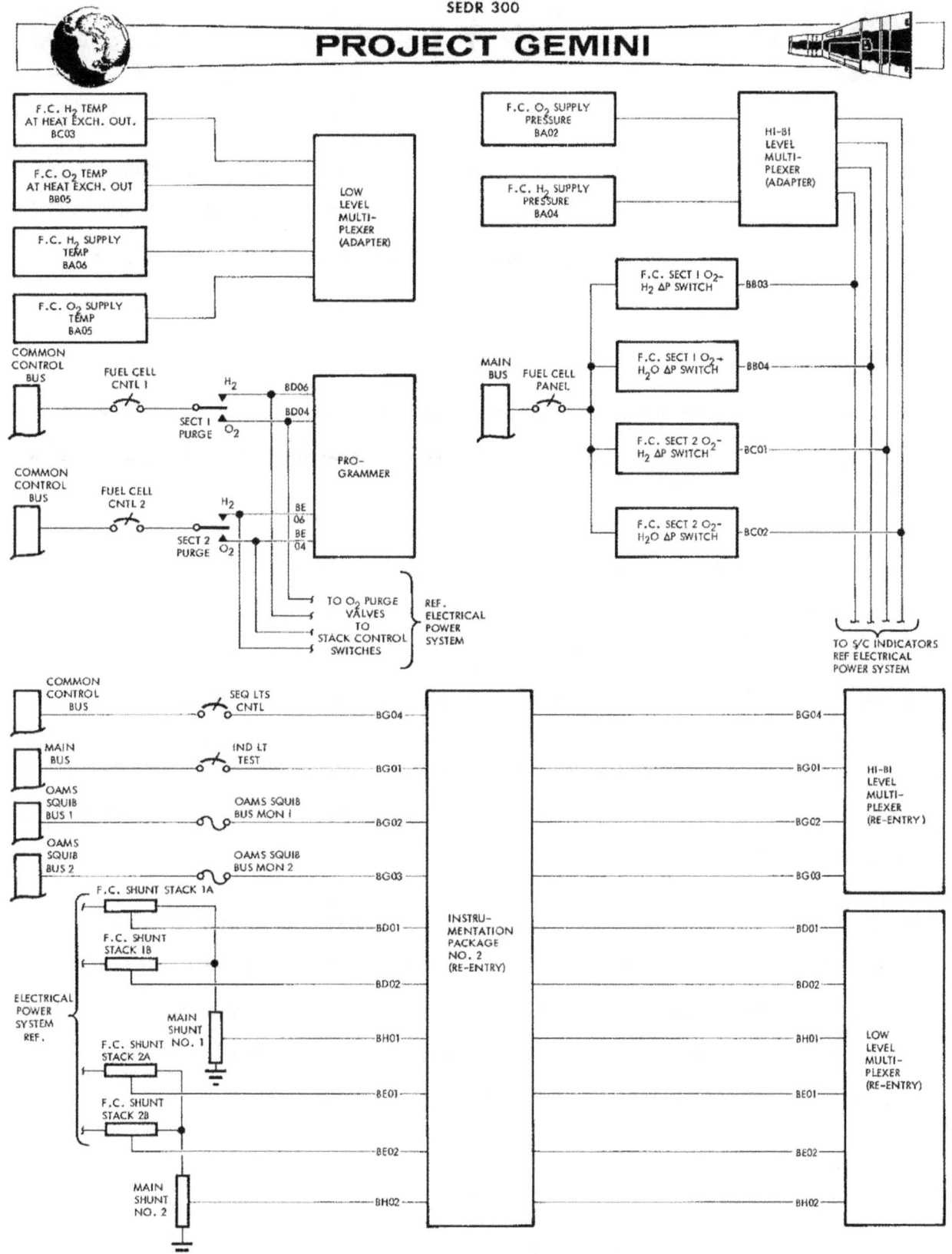

Figure 10-5 Electrical Power System Parameters Functional Diagram

PROJECT GEMINI

The following parameters relate to the ground station information regarding spacecraft main, squib and control bus voltages: BG01 (main), BG02 (squib 1), BG03 (squib 2), BG04 (control bus). Each of these parameters is conditioned and then applied to the re-entry high level multiplexer.

The reactant supply system (RSS) O_2 (BA05) and H_2 (BA06) supply bottle temperatures are monitored by means of two temperature sensors located on each supply bottle. The output of the sensors is applied to the adapter low level multiplexer.

Fuel cell section 1 O_2 to H_2O (BB04), section 1 H_2 to O_2 (BC01), and section 2 H_2 to O_2 (BC02) differential pressures are monitored by a pressure sensitive switch installed within the fuel cell to provide for safe operation monitoring capability of the fuel cell by the ground station. The outputs of the pressure switch is applied to bi-level channels of the adapter hi-level multiplexer.

Oxygen (BB05) and hydrogen (BC03) temperatures at the outlet of the heat exchanger outlets are monitored and relayed to the ground station via the adapter low level multiplexer.

To provide an aid in evaluating fuel cell operation by the ground station, section 1 O_2 (BD04), section 2 O_2 (BE04), section 1 H_2 (BD06), and section 2 H_2 (BE06) purging is monitored. The signals are actuated by the pilots by placing the corresponding section purge switch to the H_2 or O_2 position. The signals are applied to the bi-level channels of the programmer.

SEDR 300

 **PROJECT GEMINI**

ENVIRONMENTAL CONTROL SYSTEM PARAMETERS

A functional diagram showing the environmental control system (ECS) instrumentation parameters is presented in Figure 10-6. Twenty-seven parameters and RSS/ECS quantities associated with the ECS are monitored by the instrumentation system and relayed to the ground station for analysis.

The primary oxygen tank pressure (CA02) is telemetered to the ground station and displayed in the spacecraft cabin. The signals originate from a dual potentiometer pressure transducer installed as part of the ECS. The signal is relayed to the ground station via the adapter high level multiplexer.

A differential pressure transducer is used to sense cabin to forward compartment pressure differentials (CB01). The transducer has a dual output used for cabin indications and for transmission to ground via the re-entry high level multiplexer.

Left (CC01) and right (CC02) suit to cabin differential pressure is displayed in the spacecraft cabin and telemetered to the ground station. Dual potentiometer pressure transducers serve as the signal source. The output of each transducer is applied to the cabin indicator and to the re-entry high level multiplexer.

The ground station is informed of an O_2 high rate condition by CC05. This signal is originated when the spacecraft CABIN FAN switch is placed in the O_2 HI RATE position, when manual O_2 high rate is selected by the pilot, or when the suit

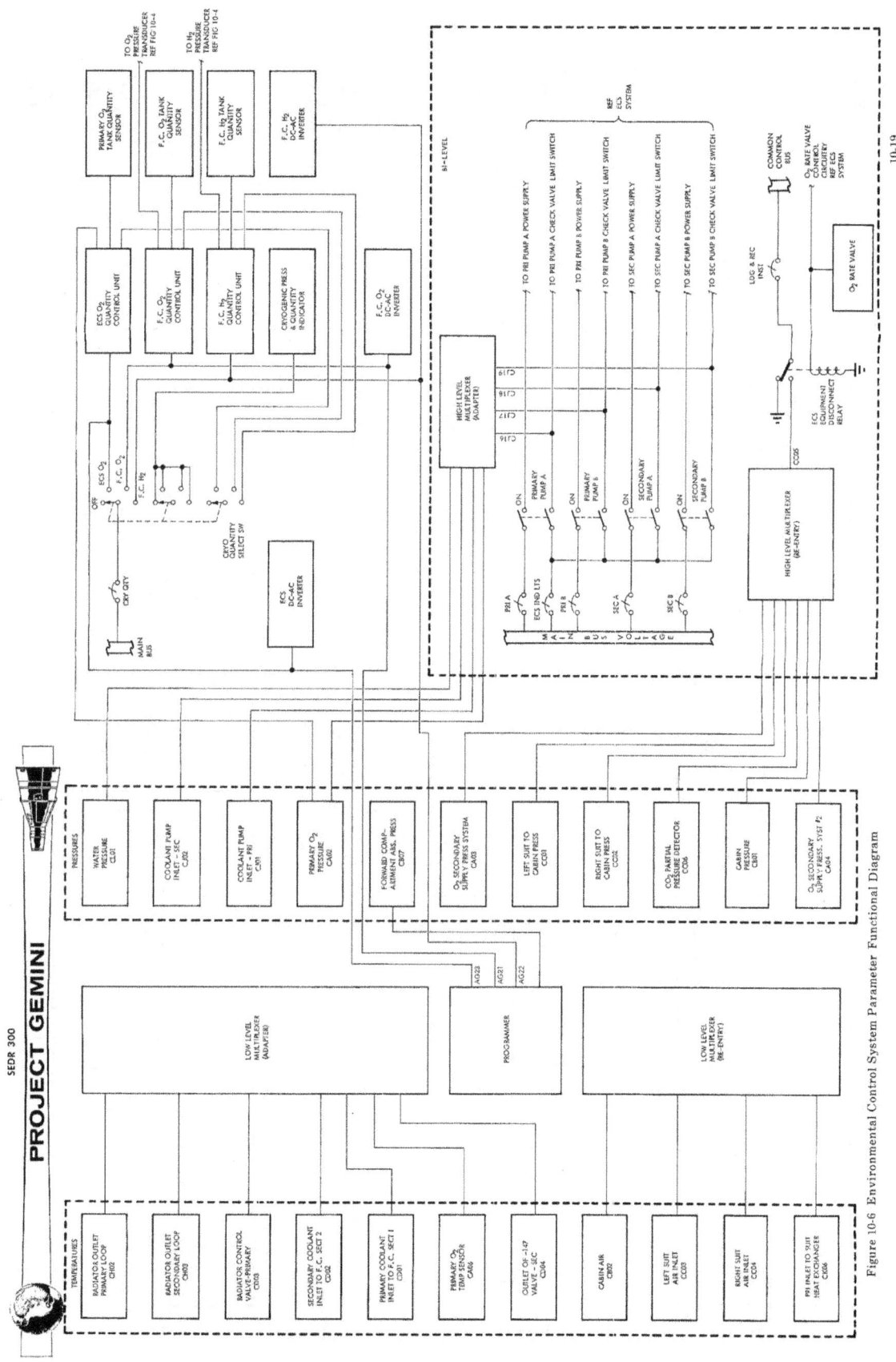

Figure 10-6 Environmental Control System Parameter Functional Diagram

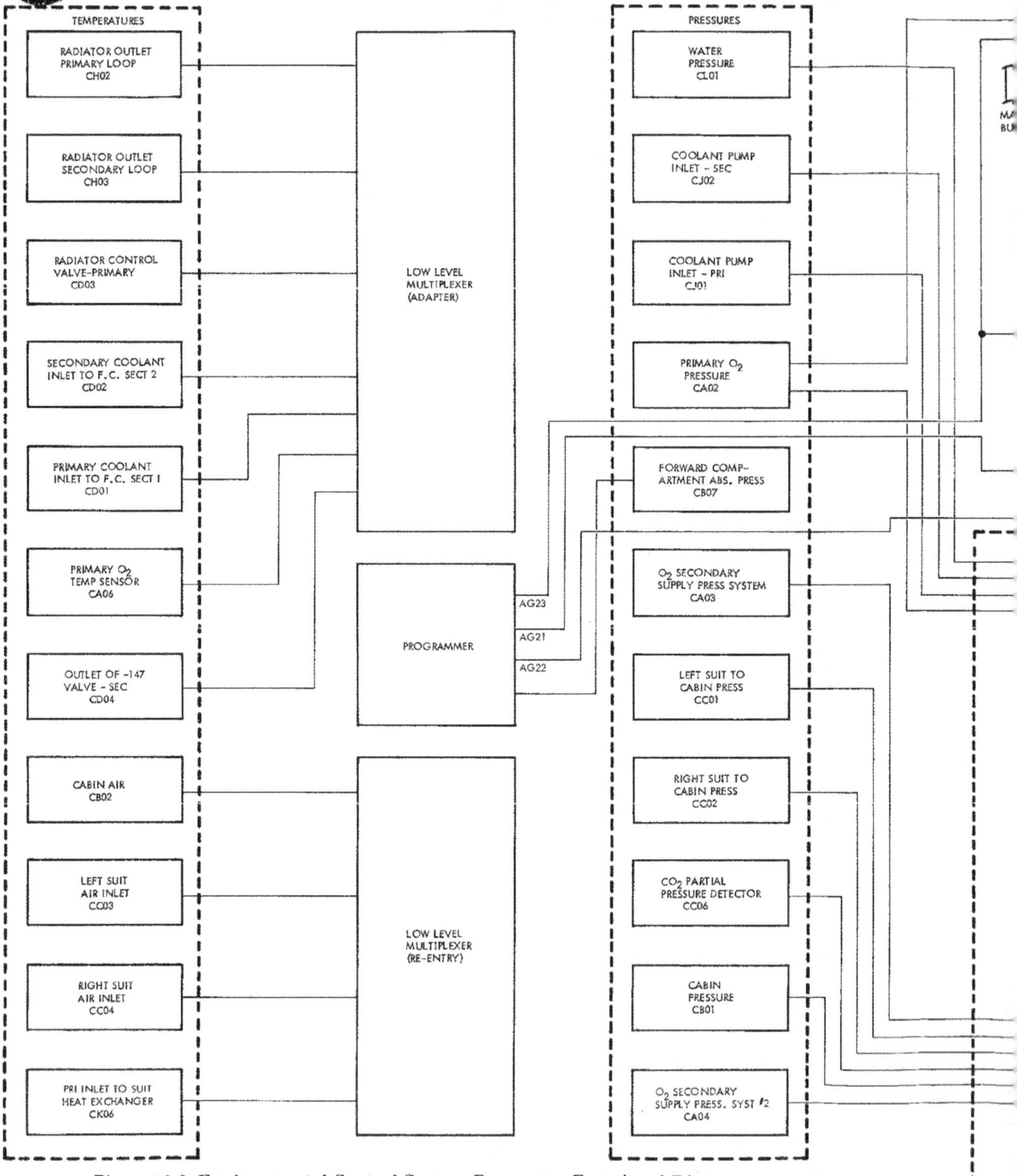

SEDR 300

PROJECT GEMINI

TEMPERATURES
- RADIATOR OUTLET PRIMARY LOOP — CH02
- RADIATOR OUTLET SECONDARY LOOP — CH03
- RADIATOR CONTROL VALVE-PRIMARY — CD03
- SECONDARY COOLANT INLET TO F.C. SECT 2 — CD02
- PRIMARY COOLANT INLET TO F.C. SECT 1 — CD01
- PRIMARY O$_2$ TEMP SENSOR — CA06
- OUTLET OF -147 VALVE - SEC — CD04
- CABIN AIR — CB02
- LEFT SUIT AIR INLET — CC03
- RIGHT SUIT AIR INLET — CC04
- PRI INLET TO SUIT HEAT EXCHANGER — CK06

PRESSURES
- WATER PRESSURE — CL01
- COOLANT PUMP INLET - SEC — CJ02
- COOLANT PUMP INLET - PRI — CJ03
- PRIMARY O$_2$ PRESSURE — CA02
- FORWARD COMPARTMENT ABS. PRESS — CB07
- O$_2$ SECONDARY SUPPLY PRESS SYSTEM — CA03
- LEFT SUIT TO CABIN PRESS — CC01
- RIGHT SUIT TO CABIN PRESS — CC02
- CO$_2$ PARTIAL PRESSURE DETECTOR — CC06
- CABIN PRESSURE — CB01
- O$_2$ SECONDARY SUPPLY PRESS. SYST #2 — CA04

LOW LEVEL MULTIPLEXER (ADAPTER)

PROGRAMMER — AG23, AG21, AG22

LOW LEVEL MULTIPLEXER (RE-ENTRY)

Figure 10-6 Environmental Control System Parameter Functional Diagram

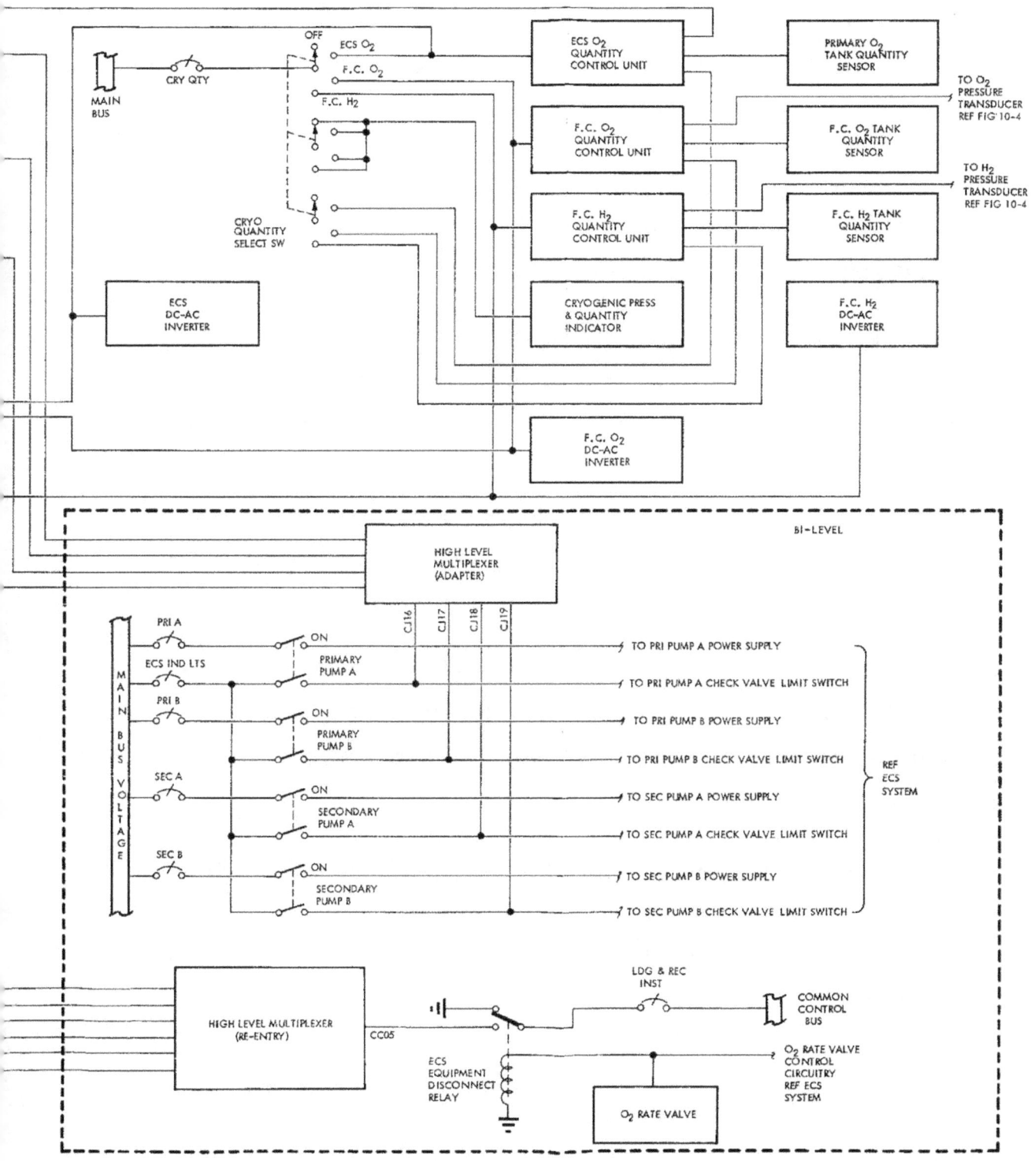

10-19

pressure drops below 3.3 psia and O_2 high rate is automatically selected. The signal is applied to a bi-level channel of the re-entry high level multiplexer.

To assure that sufficient oxygen is available to the pilots, CO_2 partial pressure (CC06) is monitored indicating the percentage of carbon dioxide with respect to the total pressure of gas in the suits. CO_2 partial pressure is displayed in the spacecraft cabin and is applied to the re-entry high level multiplexer.

Primary and secondary coolant temperatures are monitored at various locations within the coolant loop to evaluate system performance. Coolant temperatures are monitored at the primary coolant inlet to section 1 of the fuel cell (CD01), secondary coolant inlet to section 2 of the fuel cell (CD02), the radiator control valve in the primary loop (CD03), the secondary loop (CD04), radiator outlet in the primary loop (CH02), and radiator outlet in the secondary loop (CH03).

To relay information concerning proper operation of the coolant loop and pumps, to the ground station primary (CJ01) and secondary (CJ02) coolant pump inlet pressures are monitored. The outputs of the transducers are applied to the adapter high level multiplexer.

The condition of the primary and secondary coolant pumps is monitored by CJ16 (primary pump A), CJ17 (primary pump B), CJ18 (secondary pump A), and CJ19 (secondary pump B). The signal is originated when the corresponding coolant pump is actuated, and is applied to bi-level channels of the adapter high level multiplexer.

SEDR 300
PROJECT GEMINI

To insure safe operation of the fuel cell, water pressure (CL01) is monitored at the output of the fuel cell. The signal is applied to the adapter high level multiplexer.

The coolant inlet temperature to the suit heat exchanger (CK06) is monitored to relay to ground stations information concerning the environmental condition of the pilots. The output of the temperature sensor is applied to the re-entry low level multiplexer.

The position of the cryogenic quantity select switch is monitored to identify parameter CA09. The parameter CA09 indicates ECS O_2, RSS O_2, or RSS H_2 quantity depending upon the position of the cryogenic quantity select switch. The position of the selector switch is indicated to the ground station by AG21 (F.C. O_2), AG22 (F.C. H_2), and AG23 (ECS O_2). The signals are applied to bi-level channels of the programmer. The parameter CA09 is also applied to the programmer and is displayed in the spacecraft cabin.

Secondary O_2 supply pressures are monitored in the no. 1 (CA03) and no. 2 (CA04) systems. The transducers are installed as part of ECS secondary O_2 supply assemblies. The outputs of the pressure transducers are applied to the re-entry high level multiplexer.

As an aid in calculating ECS O_2 quantity, the primary O_2 supply bottle temperature (CA06) is monitored and applied to the adapter low level multiplexer.

10-21

 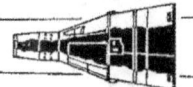

To provide the capability for the ground station to monitor the environmental condition of the cabin and to provide an aid for evaluating suit pressure, a cabin air temperature transducer (CB02), and a forward compartment absolute pressure transducer (CB07) is provided. Absolute pressure is applied to the programmer and cabin temperature is applied to the re-entry low level multiplexer. Cabin temperature is also displayed in the spacecraft cabin.

To further evaluate system performance and pilot environmental condition, the air entering the suit circuit is monitored with respect to temperature by 2 dual temperature sensors (1 for each suit circuit). The temperatures are displayed in the spacecraft cabin and are applied to the re-entry low level multiplexer as CC03 (left suit), and CC04 (right suit).

INERTIAL GUIDANCE SYSTEM PARAMETERS

Figure 10-7 shows a block diagram of the inertial guidance system (IGS) parameters except the digital computer functions. The instrumentation system monitors 18 IGS parameters and handles approximately 200 computer words.

The instrumentation system monitors the computer mode of operation; pre-launch, ascent, catch-up, rendezvous, re-entry, and touchdown. Important functions or parameters (approximately 200) are monitored during each mode of operation. This information is used during post mission analysis and is applied to the programmer.

In addition to the digital computer words, the instrumentation system monitors the following IGS parameters:

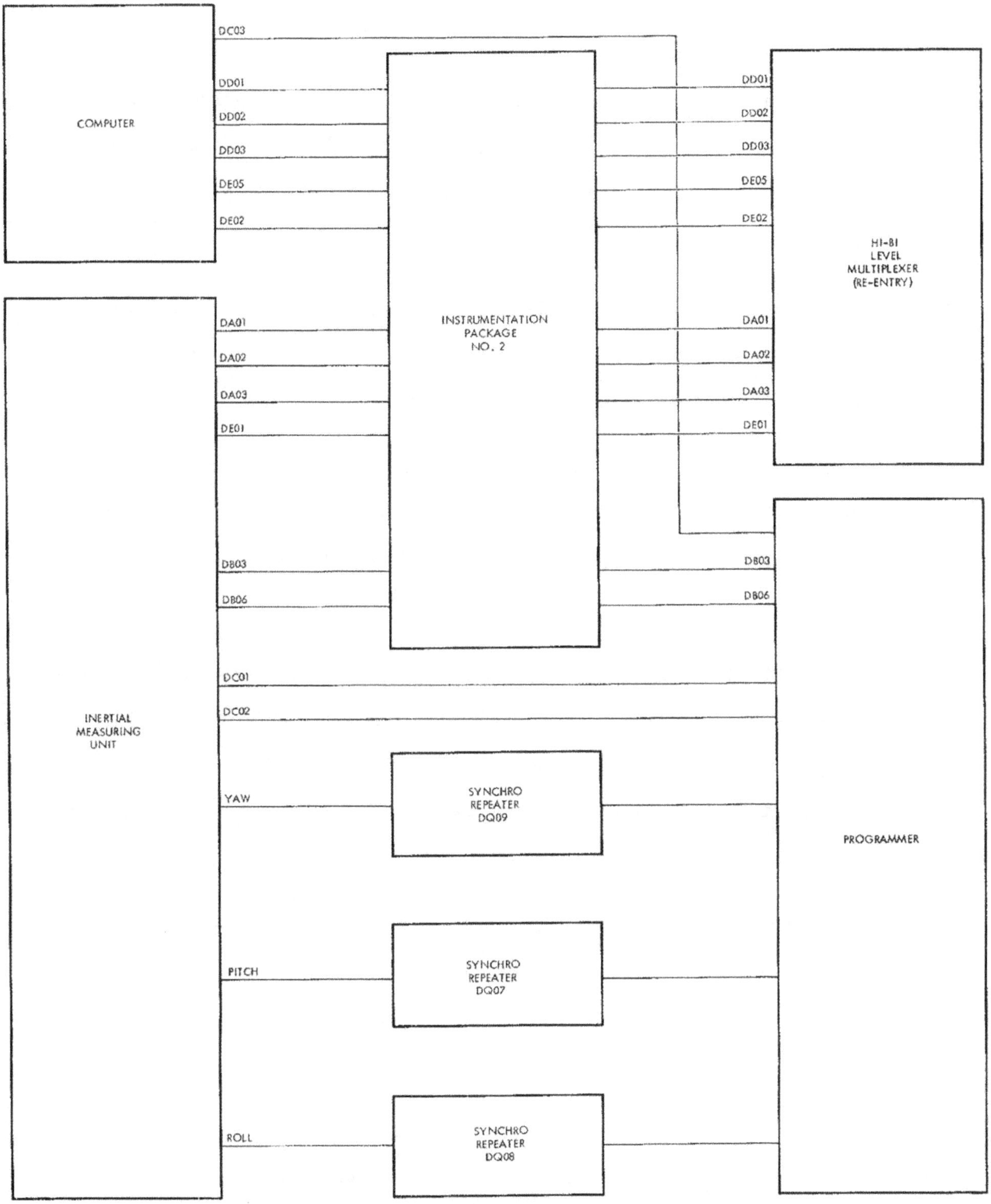

SEDR 300

PROJECT GEMINI

Figure 10-7 Inertial Guidance System Parameters Block Diagram

Inertial platform attitudes are monitored to provide ground stations with attitude data during flight. Roll (DQ08), pitch (DQ07), and yaw (DQ09) signals are taken from the inertial measuring unit (IMU), conditioned by synchro repeaters, and applied to the programmer.

Pitch (DA01), roll (DA02), and yaw (DA03) gyro torque currents are measured to verify platform alignment. The signals originate in the IMU, are conditioned and then applied to the re-entry high level multiplexer.

To verify the temperature environment of the temperature compensated components, X-axis gyro (DB06), and X-axis accelerometer (DB03) IMU temperature control amplifier outputs are monitored. These signals are conditioned and applied to the programmer.

Accelerometer (DC01), attitude (DC02), and computer (DC03), malfunction signals are monitored to detect malfunctions of the accelerometer and/or attitude reference system or the computer. These signals appear in conjunction with the malfunction lights displayed in the spacecraft cabin. The signals are applied to bi-level channels of the programmer.

Pitch (DD01), roll (DD02), and yaw (DD03) attitude errors are monitored to evaluate attitude control during critical flight periods. These signals originate at the computer, are conditioned, and applied to the re-entry hi-level multiplexer.

IGS regulated power is monitored at various points, 35 VDC (DE01), 28.9 VDC (DE02), 26 VAC (DE04), and 10.2 VDC (DE05). These voltages are conditioned and then applied to the re-entry bi-level multiplexer.

ATTITUDE CONTROL AND MANEUVERING ELECTRONICS PARAMETERS

A block diagram showing the attitude control and maneuvering electronics (ACME) instrumentation system parameters is shown in Figure 10-8. Fifteen ACME parameters are monitored by the instrumentation system.

Spacecraft rates in pitch (EA01), roll (EA02), and yaw (EA03) are monitored to allow evaluation of the rate control portion of the stabilization system. Each signal from the rate gyro package is conditioned by a phase sensitive demodulator and then applied to the high level channels of the programmer. Primary and secondary rate gyro signals are parallel summed and monitored on the same channels.

Horizon sensor operation is monitored with respect to pitch (EB01) and roll (EB02) outputs, and search mode of operation (EB03). Pitch and roll parameters are monitored to verify inertial platform alignment for the retrograde phase of the mission. These parameters (EB01, EB02) provide pitch and roll attitudes from the horizon scanner during orbital flights when the platform has been shut down to conserve electrical power. The signals originate when the SCANNER switch is in the PRI or SEC position. The pitch and roll outputs are conditioned and then applied to the re-entry high level multiplexer. The search mode of operation is monitored to determine whether the horizon scanner unit is in the search mode, or has sensed the horizon. This signal illuminates the SCANNER light in the cabin and is also applied to a bi-level channel of the re-entry high level multiplexer.

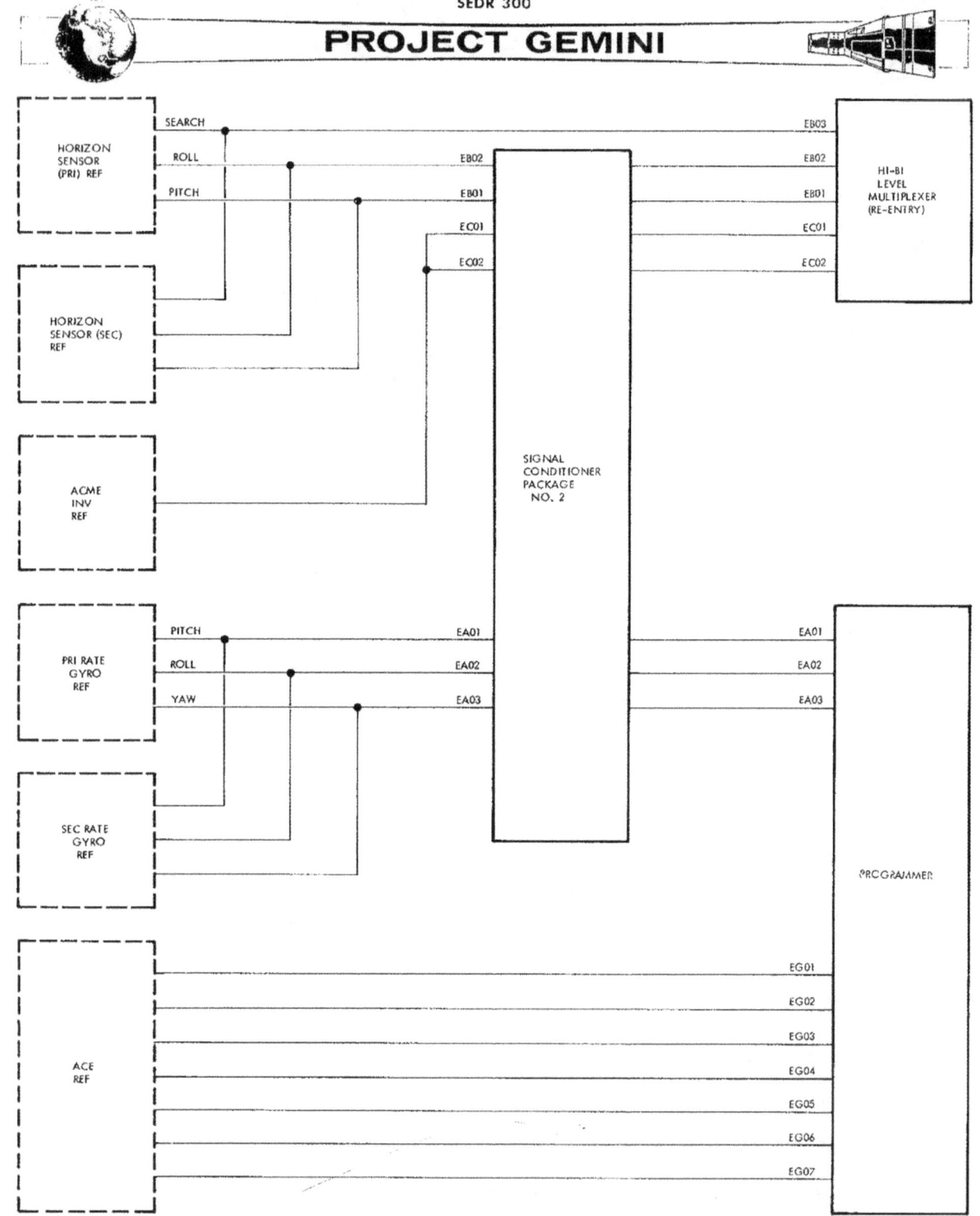

Figure 10-8 Attitude Control & Maneuvering Electronics Parameters Block Diagram

SEDR 300
PROJECT GEMINI

ACME inverter 26 VAC voltage (EC01) and frequency (EC02) is monitored for post mission analysis. The signals are conditioned and then applied to the re-entry high level multiplexer.

The following attitude control modes are monitored depending upon the position of the ATTITUDE CONTROL switch: HOR SCAN (EG01), RATE CMD ORBIT (EG02), DIRECT (EG03), PULSE (EG04), RATE CMD RNTY (EG05), RE-ENTRY (EG06), and PLATFORM (EG07). The signals are applied to bi-level channels of the programmer.

ORBIT, ATTITUDE AND MANEUVERING SYSTEM PARAMETERS

The orbit, attitude and maneuvering system (OAMS) instrumentation system parameters are shown in Figure 10-9 in block diagram form. A brief description of each of the 19 parameters is given in the paragraphs to follow.

To insure that adequate propellant pressure is available for OAMS, helium source pressure (GC01) is monitored. The signal originates from a dual potentiometer pressure transducer at the helium pressurant tanks. One output is applied to the adapter high level multiplexer and the other is used to drive an indicator in the spacecraft cabin.

The propellant feed temperature at the fuel (GB01) and oxidizer (GB02) feed lines is monitored to verify that propellant aboard is above freezing temperature and is available for use. The signals originate from two individual temperature sensors and are applied to the adapter low level multiplexer.

To allow monitoring capability of the helium source temperature (GC02), a temperature sensor is installed on the helium supply line at the supply tank.

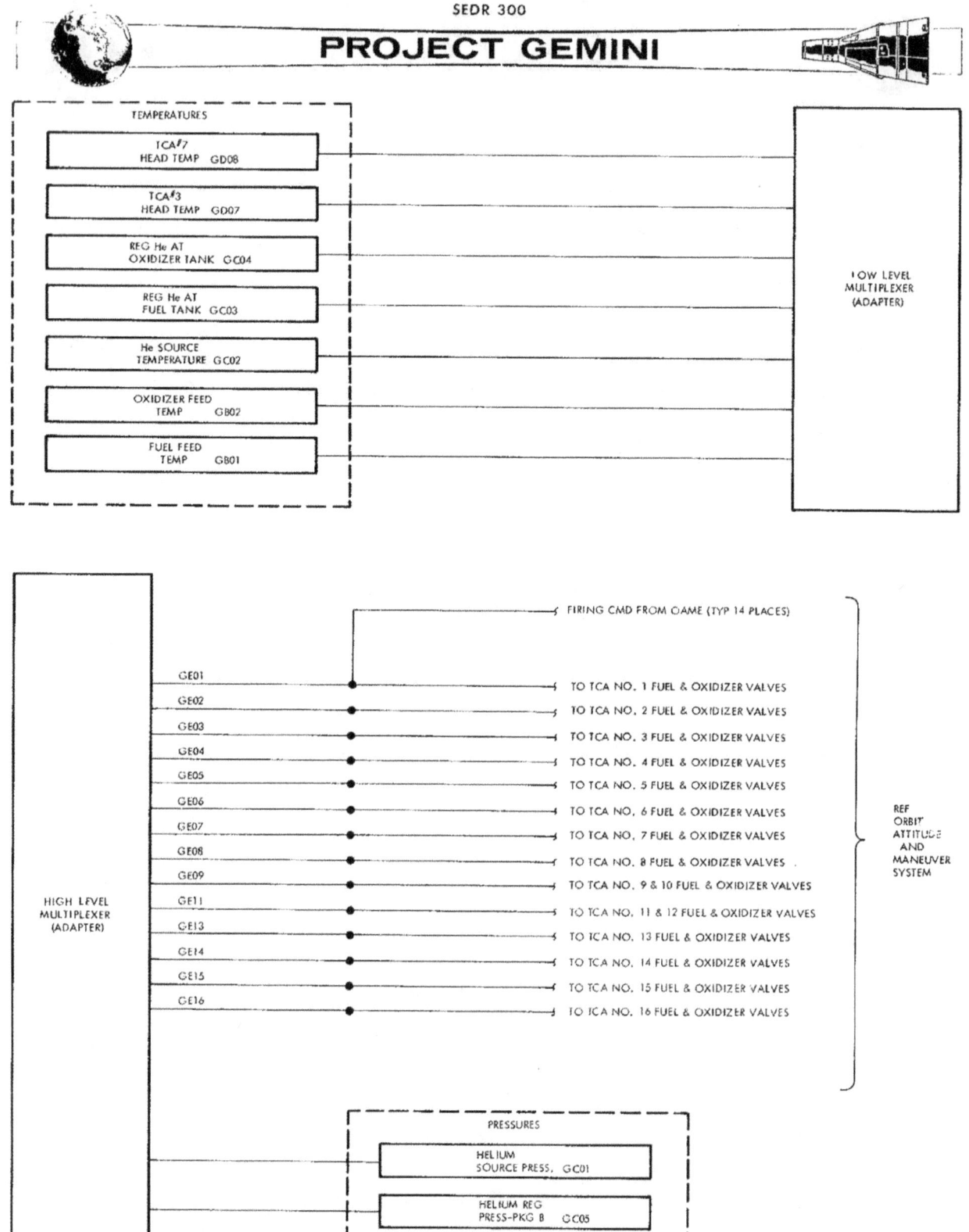

Figure 10-9 Orbit, Attitude & Maneuvering System Parameters Block Diagram

The output is applied to the adapter low level multiplexer. A separate sensor is installed to drive an indicator in the spacecraft cabin.

Temperature of the pressure regulated helium is monitored at the fuel (GC03) and oxidizer (GC04) tank inlet lines. The outputs of these temperature sensors is applied to the adapter low level multiplexer. Two additional sensors are installed to drive indicators in the spacecraft cabin.

Regulated helium pressure (GC05) is monitored by a dual potentiometer pressure transducer. One of the outputs is applied to the adapter high level multiplexer, and the other is used to drive a cabin indicator.

To provide an indication of maximum thrust chamber assembly (TCA) temperature, TCA 3 (GD07) and TCA 7 (GD08) injector head temperatures are monitored. These signals are applied to the adapter low level multiplexer.

To provide ground station monitoring capability of TCA firing, the following TCA solenoid command signals are applied to bi-level channels of the adapter high level multiplexer: GE01 (TCA 1), GE02 (TCA 2), GE03 (TCA 3), GE04 (TCA 4), GE05 (TCA 5), GE06 (TCA 6), GE07 (TCA 7), GE08 (TCA 8), GE09 (TCA 9, 10), GE11 (TCA 11, 12), GE13 (TCA 13), GE14 (TCA 14), GE15 (TCA 15), and GE16 (TCA 16).

RE-ENTRY CONTROL SYSTEM PARAMETERS

Figure 10-10 shows in block diagram form the re-entry control system parameters. Some 24 parameters are monitored by the instrumentation system to provide for ground station observation of proper system performance.

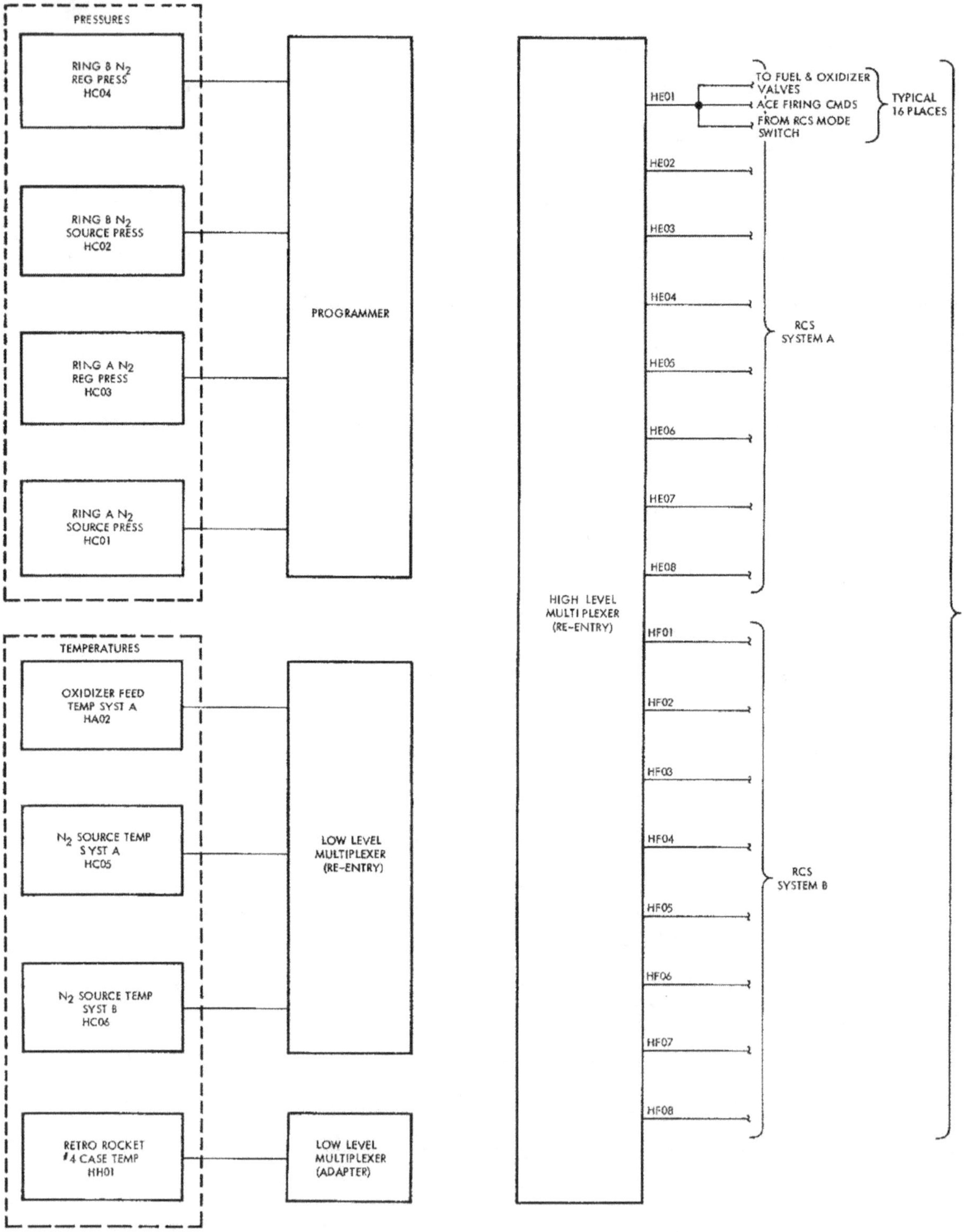

SEDR 300
PROJECT GEMINI

Figure 10-10 Re-Entry Control System Parameters Block Diagram

SEDR 300
PROJECT GEMINI

Nitrogen source pressure, HC01 (system A) and HC02 (system B), and nitrogen source temperature, HC05 (system A) and HC06 (system B) are monitored. Pressure is sensed by two dual pressure transducers. One of the outputs of each transducer is used to drive a cabin indicator, and the other is applied to the programmer. Outputs of the temperature sensors are applied to the re-entry low level multiplexer and are used to drive a spacecraft cabin indicator.

Because the oxidizer has a more critical temperature range than fuel, its temperature is measured to insure that both fuel and oxidizer are within the proper temperature range for use in the re-entry control system. The oxidizer feed temperature (HA02) is applied to the re-entry low level multiplexer.

Regulated nitrogen pressure is monitored for system A (HC03) and system B (HC04). The outputs of the pressure transducers is applied to the programmer.

To provide for ground station monitoring capability of proper RCS thrust chamber assembly (TCA) firing, firing commands are applied to bi-level channels of the re-entry high level multiplexer. RCS system A thrusters, 1A thru 8A have been assigned parameters HE01 thru HE08 respectively and system B thrusters 1B thru 8B are designated by HF01 thru HF08 respectively.

The retro rocket case temperature (HH01) is also monitored. The signal originates from a surface mounted temperature sensor located on retro rocket no. 4 and is applied to the adapter low level multiplexer.

AERODYNAMIC AND CREW CONTROL PARAMETERS

Aerodynamic and crew control parameters are monitored as shown in block diagram form in Figure 10-11.

Spacecraft longitudinal (KA01), lateral (KA02), and vertical (KA03) accelerations are monitored to provide ground station indications during the launch and re-entry phases of the mission. The accelerometer outputs are applied to the programmer.

Static pressure (KB02) is monitored by a potentiometer type absolute pressure transducer. Static pressure is obtained from four static pressure ports equally spaced around the forward part of the conical section and connected in parallel to the transducer. The transducer output is applied to the programmer.

Pitch (FA01), roll (FA02), and yaw (FA03) attitude control stick positions are monitored to indicate pilot manual control usage and to evaluate thruster operation. Signals originate from the attitude hand controller potentiometers and are applied to the programmer.

Two bi-level channels are reserved for events to be monitored as required by the experiments of each particular spacecraft mission. Electrical provisions for monitoring these parameters are provided at the right (FD01) and left (FE01) utility receptacles.

COMMUNICATION SYSTEM PARAMETERS

The instrumentation system monitors 11 communication system parameters. These parameters are shown in block diagram form in Figure 10-12. A brief description of each of the parameters is presented in the sub-paragraphs to follow.

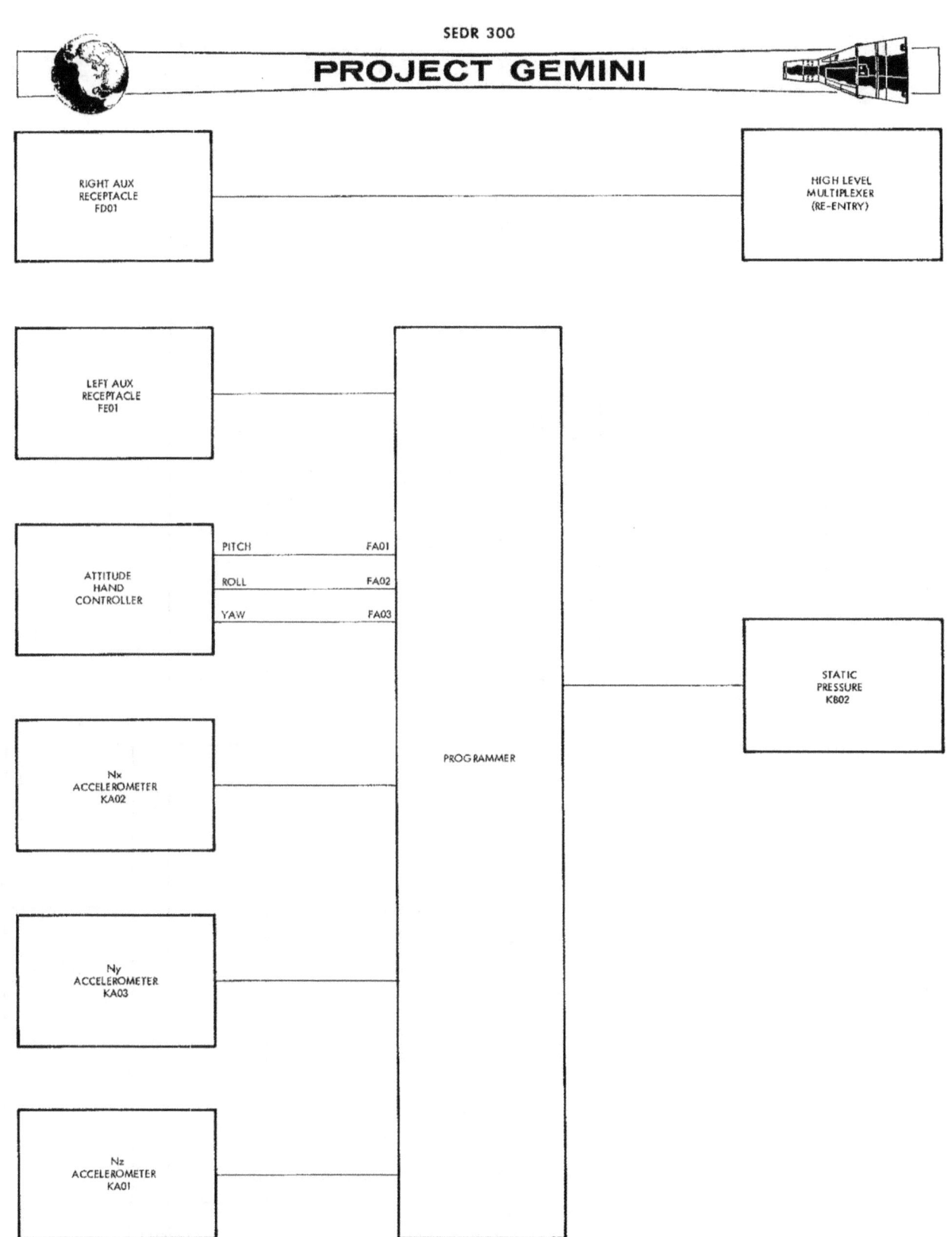

Figure 10-11 Aerodynamic and Crew Control Parameters Block Diagram

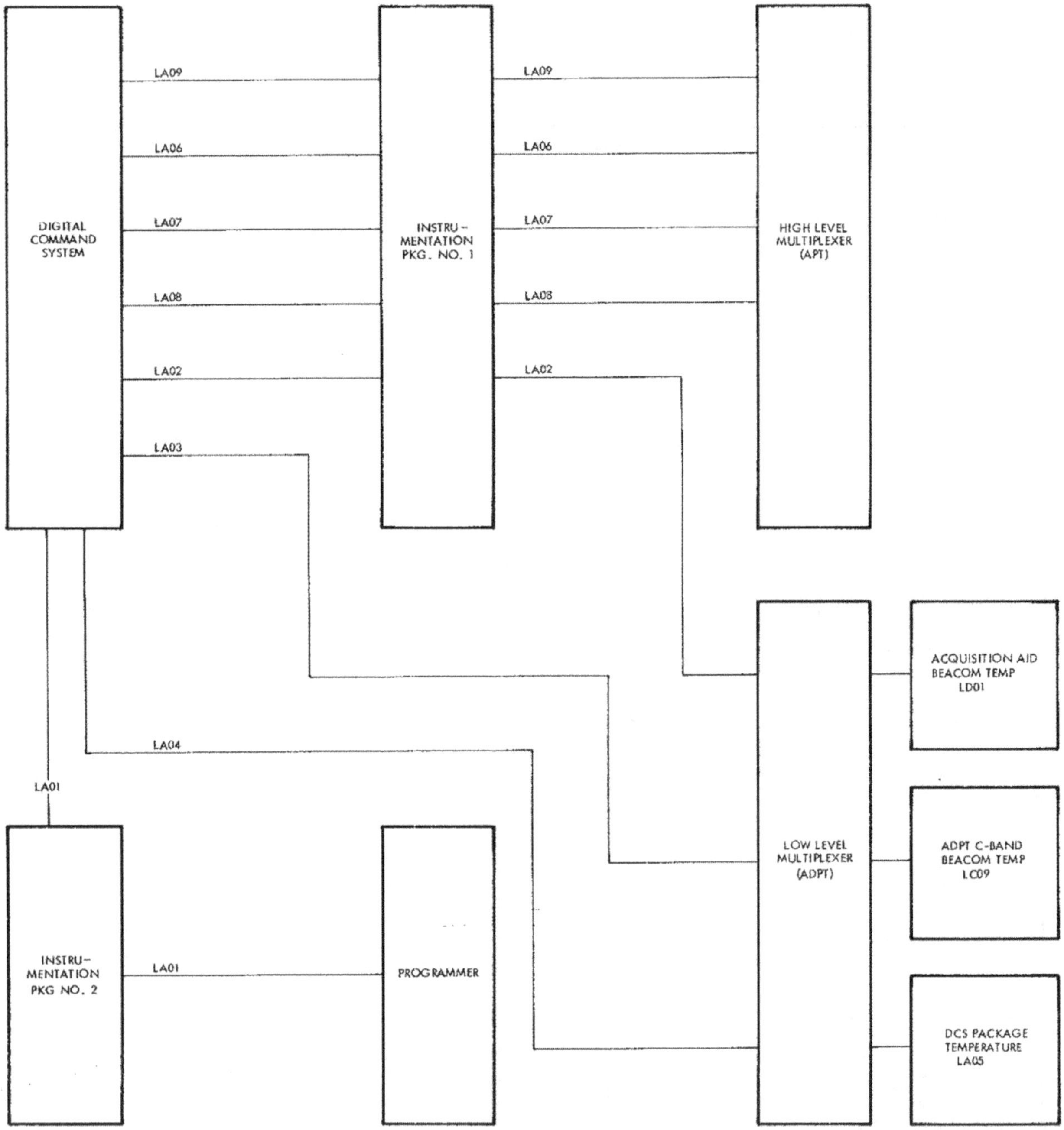

SEDR 300

PROJECT GEMINI

Figure 10-12 Communication System Parameters Block Diagram

10-34

The digital command system (DCS) verification signal (LA01) from the receiver decoder unit is monitored to provide automatic operation of the ground computers during insertion of information into the spacecraft computer. Verification is indicated by a signal originating at the receiver decoder, and is then conditioned, applied to the programmer, and transmitted to ground as an 8-bit digital word. Eight binary ones indicates no verification and eight binary zeros indicates verification.

To verify proper DCS performance and aid in malfunction isolation, the following DCS parameters are monitored: diplexer (LA04) and quadriplexer (LA03) receiver signal strength, package temperature (LA05), 6 VDC regulated power (LA02), 28 VDC regulated power (LA06), -18 VDC regulated power (LA07), 23 VDC regulated power (LA08), and -6 VDC regulated power (LA09). Parameters LA02, LA06, LA07, LA08, and LA09 are conditioned and then applied to the adapter high level multiplexer except LA02 which is applied to the adapter low level multiplexer. Parameters LA03, LA04, and LA05 are applied directly to the adapter low level multiplexer.

Acquisition aid beacon (LD01) and adapter C-band beacon (LC09) case temperatures are also monitored to assure proper equipment performance. These temperature signals are applied to the adapter low level multiplexer.

INSTRUMENTATION SYSTEM PARAMETERS

To insure proper operation of the instrumentation system, various reference voltages and other pertinent data is telemetered to ground stations for analysis. The instrumentation system parameters are shown in Figure 10-13 in block diagram

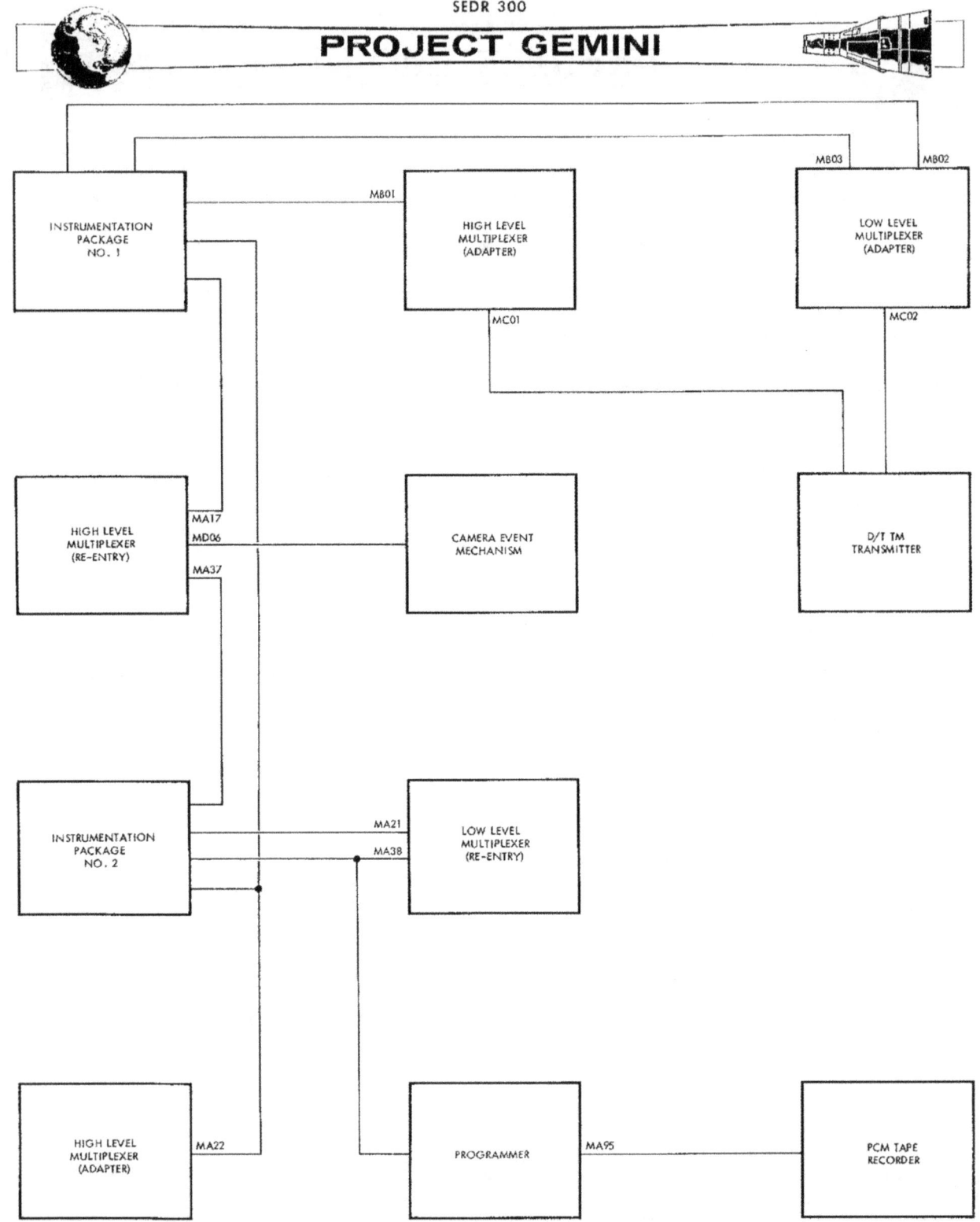

Figure 10-13 Instrumentation System Parameters Block Diagram

form. A brief description of each parameter follows.

High (MA17) and low (MA38) level zero reference voltages are monitored to insure that proper scaling is being employed by the multiplexing and encoding systems. The low level zero reference originates from the 5 VDC output of the DC-DC converter which is attenuated by a signal conditioner to 3 millivolts (the zero reference point) and is then applied to a channel in each of the two low level multiplexers. This signal is also applied to the programmer as MA38. For the high level reference, signal return is monitored on a high level channel of the high level commutator.

High (MA37) and low (MA21) level full scale reference voltages, as the zero reference voltages, are required to insure that proper scaling is being employed by the PCM multiplexing and encoding systems. The 5 VDC output of the DC-DC converter is attenuated to 4.5 VDC and to 15 millivolts prior to application to channels of the high and low level multiplexers, respectively. These parameters are required to provide a measurement of the reference voltage for potentiometer type transducers and resistive element temperature sensors.

To provide monitoring of the high level full scale reference voltage during re-entry, MB01 is provided. Parameters MB01 and MA37 provide the same information except that the signal conditioners for MA37 is located in the adapter, and for MB01 in the re-entry module.

Parameter MA22 (calibrate signal) is provided to indicate that a calibration voltage is being applied, thus eliminating the confusion between a data and a calibrate signal. Parameter MA22 will exist whenever the CALIB switch in the spacecraft cabin is actuated or a calibration is commanded by the DCS.

An indication of proper functioning of the PCM tape recorder is provided by monitoring tape motion (MA95). This is accomplished by providing a signal to a bi-level channel of the programmer when the recorder drive motor is in motion.

Two additional low level zero reference voltages, MB02 and MB03, are provided for instrumentation package 1. These parameters are similar to the ones described earlier.

The RF power output (MC01) and the case temperature (MC02) of the delayed time telemetry transmitter is monitored to provide an indication of transmitter operation. The transmitter physically located in the adapter is chosen for these measurements because it is subject to more extreme environmental temperature changes than the other two transmitters. Temperature signals are applied to the adapter low level multiplexer, and the RF power output is applied to the adapter high level multiplexer.

A camera event (MD06) is indicated to the ground station when the pilot initiates the camera event mechanism on the onboard camera. This signal is applied to a bi-level channel of the re-entry high level multiplexer.

PHYSIOLOGICAL PARAMETERS

The physiological functions of the crew are monitored by sensors which are attached at various points to their skin. A block diagram showing the physiological parameters is shown in Figure 10-14. Signal conditioners, located in

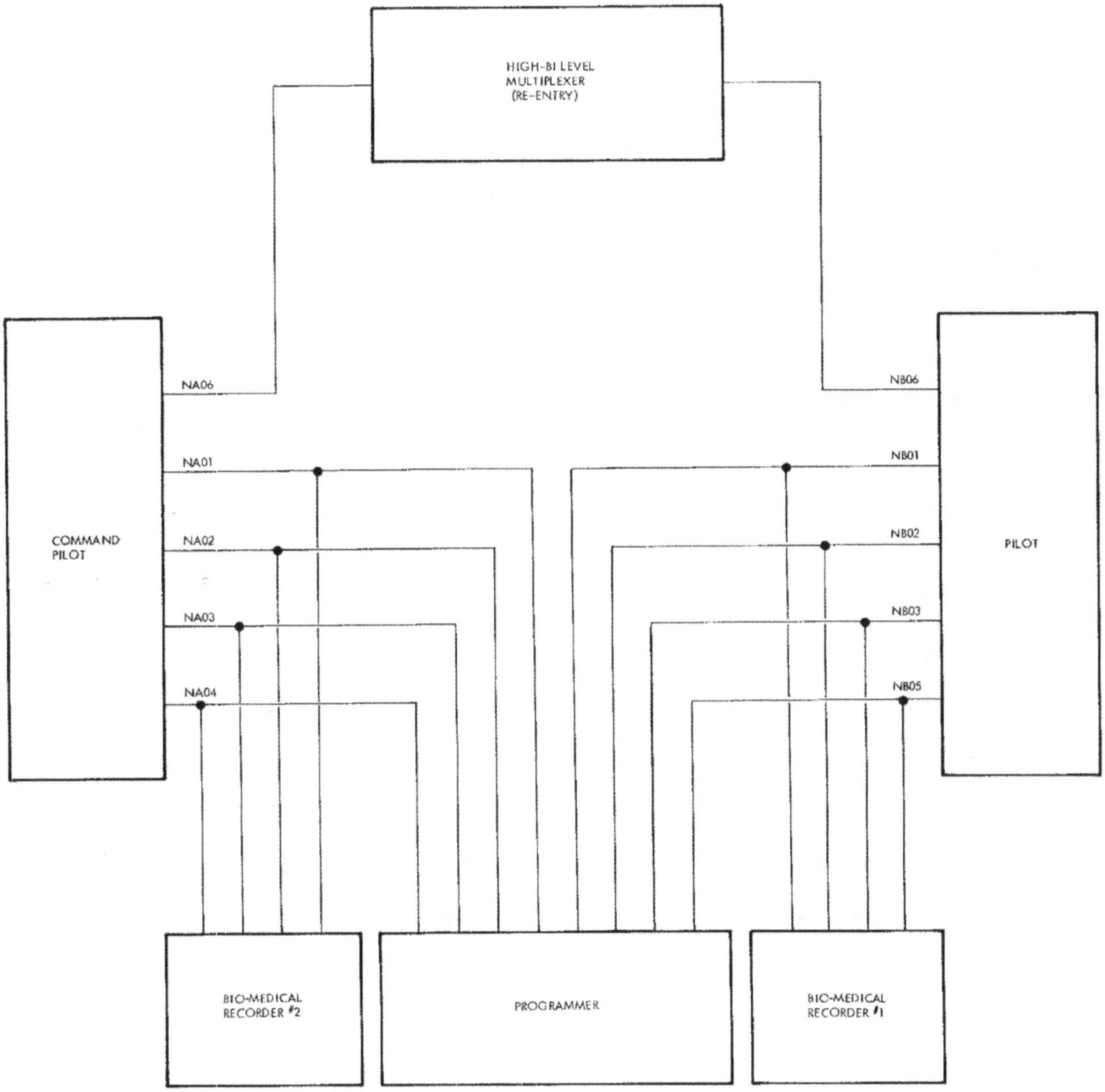

Figure 10-14 Physiological Parameters Block Diagram

pockets of the underwear, condition the signals from the sensors to make them compatible with the recording and multiplexing equipment. All parameters except the oral temperature are recorded on bio-medical recorders. Command pilot parameters are recorded on recorder no. 2 and pilot parameters are recorded on recorder no. 1. All signals, except oral temperature, in addition to being recorded, are applied to the programmer. Oral temperature is applied to the re-entry high level multiplexer. The following command pilot parameters are monitored: electrocardiogram no. 1 and no. 2 (NA01, NA02), respiration rate and depth (NA03), blood pressure (NA04), and oral temperature (NA06). The following pilot parameters are monitored: electrocardiogram no. 1 and no. 2 (NB01, NB02), respiration rate and depth (NB03), blood pressure (NB05), and oral temperature (NB06).

SYSTEM UNITS

PRESSURE TRANSDUCERS

The purpose of the pressure transducer is to sense pressure, and to convert this pressure into a proportional electrical signal. There are about six physically different configured pressure transducers as shown in Figure 10-15. Transducers have different physical appearances and different pressure range to accommodate the specific application or use. The numerical call outs below each transducer in Figure 10-15 identifies the location and application of the transducer as shown in Figure 10-1. The numbers correspond to those on Figure 10-1.

PROJECT GEMINI

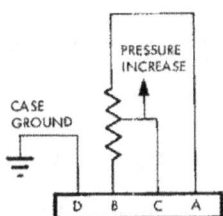

SINGLE POTENTIOMETER TRANSDUCER SCHEMATIC (TYPICAL)

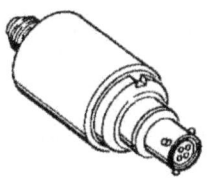

WATER PRESSURE TRANSDUCER
REF FIGURE 10-1 INDEX NO. 19 FOR LOCATION

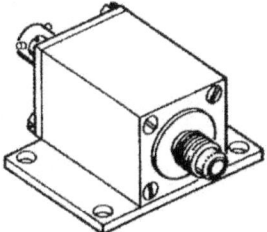

ABSOLUTE AND STATIC PRESSURE TRANSDUCER
REF FIGURE 10-1 INDEX NO. 39 & 40 FOR LOCATION

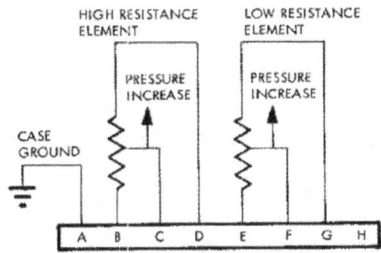

DUAL POTENTIOMETER TRANSDUCER SCHEMATIC (TYPICAL)

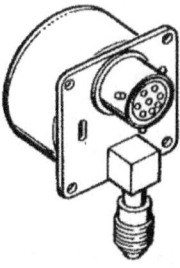

ECS SECONDARY SUPPLY PRESSURE TRANSDUCER
REF FIGURE 10-1 INDEX NO. 55 & 59 FOR LOCATION

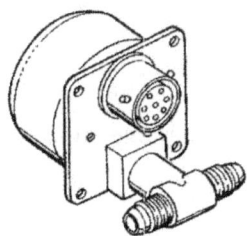

RSS AND PRIMARY ECS SUPPLY PRESSURE TRANSDUCER
REF FIGURE 10-1 INDEX NO. 14, 21 & 30 FOR LOCATION

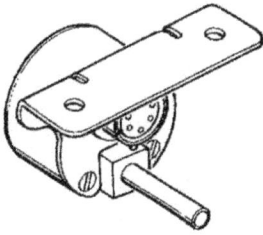

OAMS PROPELLANT QUANTITY

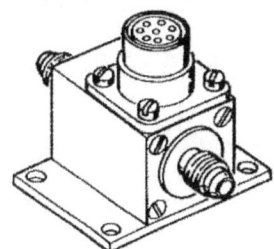

CABIN AND SUIT PRESSURE TRANSDUCER
REF FIGURE 10-1 INDEX NO. 41, 57 & 58 FOR LOCATION

Figure 10-15 Pressure Transducers

The sizes of the units vary from about 1 1/4" x 1 1/4" x 3" to approximately 2 1/2" x 2 1/2" x 4"; the weights vary from approximately .45 lb. to 2 lb. The unit construction utilizes a bellows or bourdon tube which varies the wiper position of a potentiometer, proportionally, with the input pressure. Two potentiometers are used in the dual-output units to separate the cabin indicator circuit from the multiplexer/encoder (telemetry) circuit, thus avoiding a possible loading error in the latter. With one exception, pressure transducer outputs range from 0 to 5 VDC. The OAMS quantity system pressure-temperature sensor, driving a cabin indicator, has an output of 0 to 24 VDC.

TEMPERATURE SENSORS

Temperature sensors are used to convert temperatures into directly proportional electrical signals. Basically there are two types of temperature sensors: a probe type and a surface mounted type. Variations exist within each type to accommodate specific mounting requirements. Nine physically different types of temperature sensors are shown in Figure 10-16. With respect to temperature range, approximately 20 different sensors are used. The numbers beneath each sensor in Figure 10-16 corresponds to the sensors locations and application as shown in Figure 10-1.

Spacecraft temperatures are monitored by platinum element temperature sensors. The sensors vary somewhat in size but are roughly .4" x .75" x 2.0". There are two types of resistive-element sensors: a probe type and a surface-mounted type. Probes are used to monitor fluid temperatures, and surface-mounted sensors are used to monitor surface temperatures. Both types utilize a fully-annealed pure-platinum wire, encased in ceramic insulation. The sensors form one leg

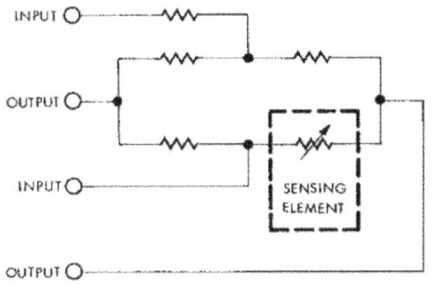

TYPICAL SCHEMATIC

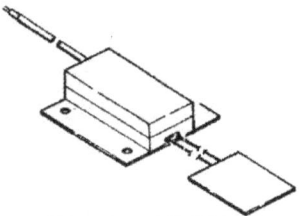

SURFACE MOUNTED SENSOR AND BRIDGE
REFER TO FIGURE 10-1,
INDEX NO. 6, 27, 31 & 34
FOR LOCATION

PROBE AIR TEMPERATURE SENSOR AND BRIDGE
REFER TO FIGURE 10-1
INDEX NO. 42, 54, & 60
FOR LOCATION

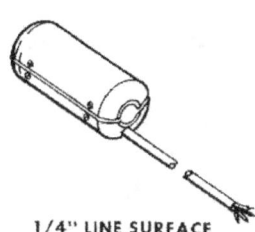

1/4" LINE SURFACE MOUNTED SENSOR AND BRIDGE
REFER TO FIGURE 10-1
INDEX NO. 2, 3, 11, 17, 20, 22, 35, 48, 50, 52 & 56
FOR LOCATION

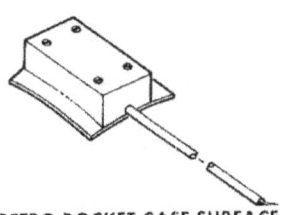

RETRO ROCKET CASE SURFACE MOUNTED SENSOR AND BRIDGE
REFER TO FIGURE 10-1
INDEX NO. 1, 4, 25, 32 & 33
FOR LOCATION

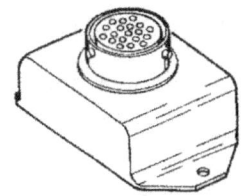

BRIDGE PACKAGE
REFER TO FIGURE 10-1
INDEX NO. 54 & 60
FOR LOCATION

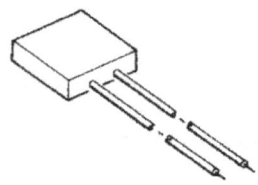

SURFACE MOUNTED SENSOR ELEMENT
OAMS QUANTITY

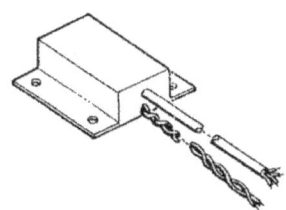

BRIDGE PACKAGE
REFER TO FIGURE 10-1
INDEX NO. 13, 23 & 29
FOR LOCATION

SUPPLY BOTTLE SENSING PROBE
REFER TO FIGURE 10-1
INDEX NO. 13, 23, 29
FOR LOCATION

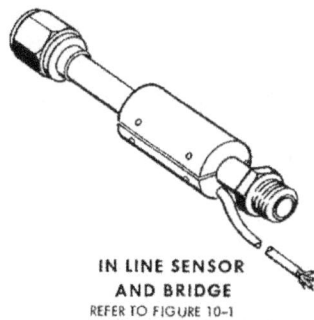

IN LINE SENSOR AND BRIDGE
REFER TO FIGURE 10-1
INDEX NO. 10, 12, 15 & 18
FOR LOCATION

Figure 10-16 Temperature Sensors

SEDR 300

PROJECT GEMINI

of a bridge network whose unbalance will produce an output of either 0 to 20 MV DC or 0 to 400 MV DC. The 0-20 MV DC outputs are used for data transmission purposes and the 0-400 MV DC outputs for cabin displays.

In some applications, mounting and space requirements necessitate that the bridge is remotely located from the sensing element. In most cases, however, the bridge and sensing element are housed in the same case.

Regardless of how the bridge and sensing element are housed, combined, they comprise a schematic as shown in Figure 10-16.

SYNCHRO REPEATERS

Three synchro repeater assemblies, mounted in the upper portion of the left landing gear well as shown in Figure 10-17, monitor the synchros on the inertial guidance system (IGS) platform gimbals. Each synchro repeater output is a DC signal proportional to the spacecraft roll, yaw and pitch attitude in terms of platform coordinates. Two outputs are available per repeater; a coarse output, which provides 0-5 VDC output for 0-350 degrees of synchro travel and a fine output, which gives 0-5 VDC output for every 35 degrees of synchro travel; only the coarse output is monitored as shown in Figure 10-17. A dead band of 10 degrees (max.) exists, centered around the 135-degree position in the synchro repeater potentiometers. Control of the synchro repeaters is achieved by pilot actuation of the platform mode select switch.

 SEDR 300
PROJECT GEMINI

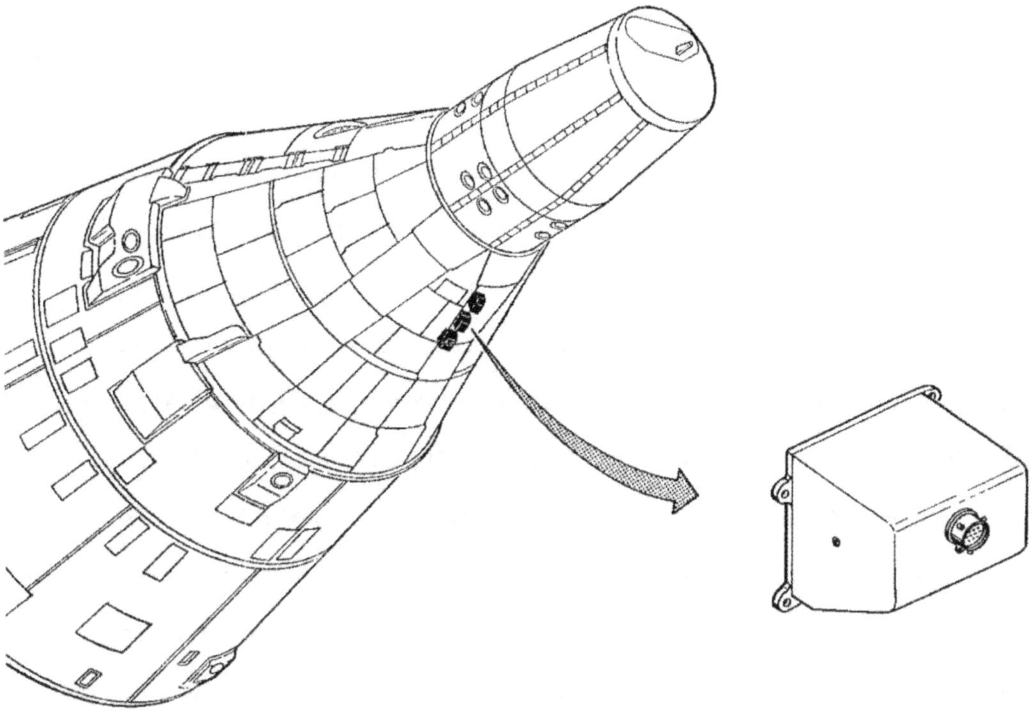

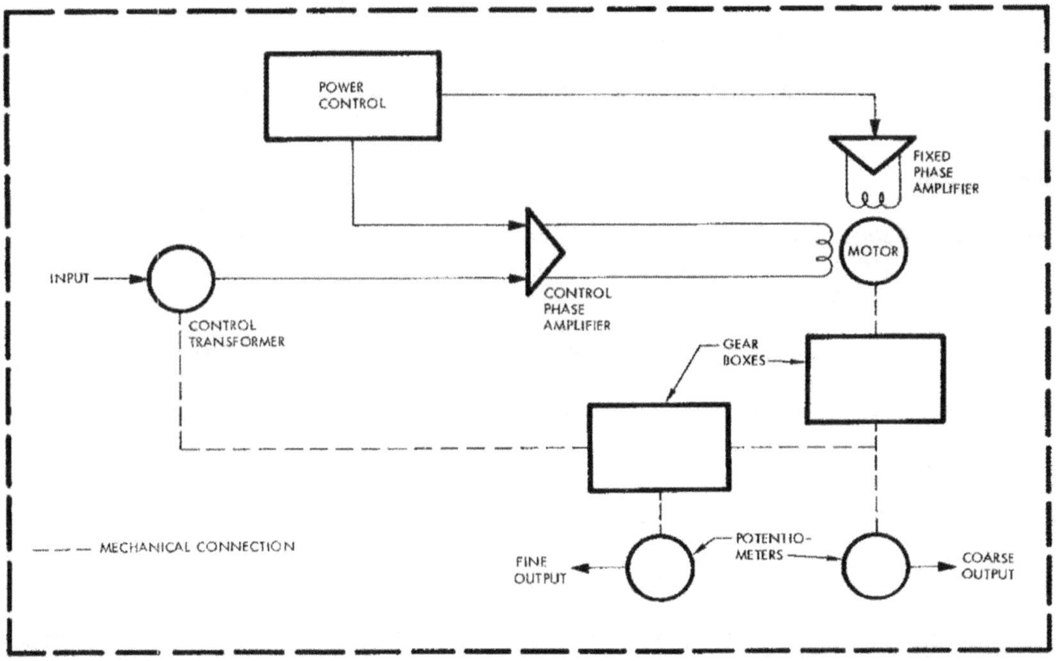

Figure 10-17 Synchro Repeaters and Schematic Diagram

PROJECT GEMINI

CO_2 PARTIAL PRESSURE DETECTOR

A carbon dioxide partial pressure detector, as shown in Figure 10-18, is utilized to insure that there is a safe level of CO_2 in the pilots suit circuits. The detector is located in the environmental control system (ECS) module. The gaseous mixture to be sampled is obtained as it exists from the ECS carbon dioxide and odor absorbers. The sample stream is divided through two separate passages, both filtering water vapor, but only one filtering carbon dioxide. The streams then pass into identical ion chambers which are polarized with $\pm$ 50 VDC obtained from a DC-DC converter contained in the detector assembly; there a radio active source ionizes the gases. The difference of the electrical outputs is amplified and provides a voltage which is proportional to the partial pressure of the mixture. The gas is then returned to the inlets of the suit compressors. The system provides two outputs: 0-5 VDC into a nominal 2.5 megohm load for telemetry use and 0-100 micro amps into a 4000 ohm cabin indicator.

ACCELEROMETERS

Three linear accelerometers are provided to measure the accelerations along each of the spacecraft axes. The units are approximately 1.2" x 1.2" x 3" and are pictured in Figure 10-19. The accelerometers are electrically-damped, force-balance, servo-type units with outputs of 0-5 VDC. The unit which is used for longitudinal measurements has a range of -3 to +19 G's and the other two have ranges of $\pm$3 G's. The accelerometer is a torque balanced, closed loop system with a pendulous mass supported by an extremely low friction jewel bearing. The schematic of the accelerometer is shown in Figure 10-19. An electromagnetic position detector notes the slightest movement of the mass and supplies a directly proportional electrical signal to a servo amplifier. The output of

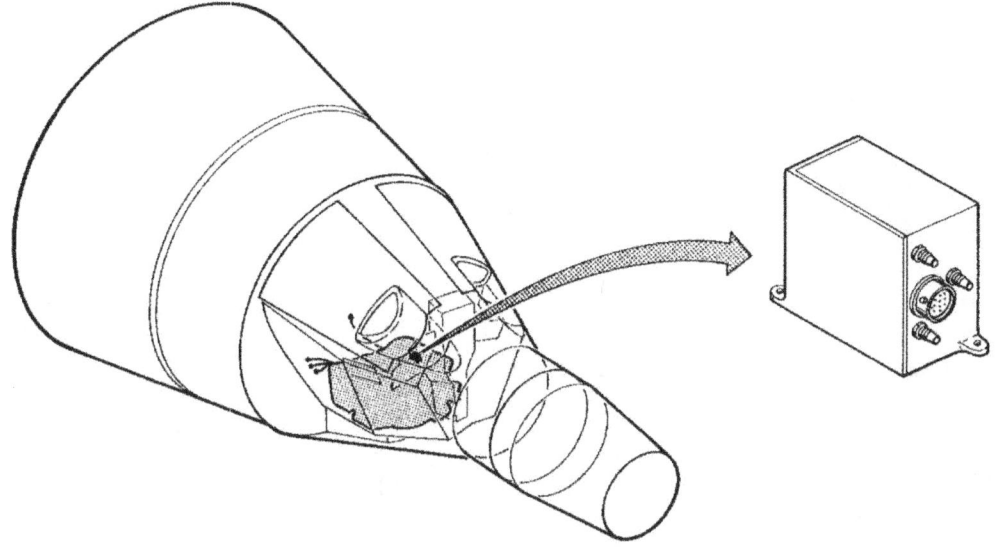

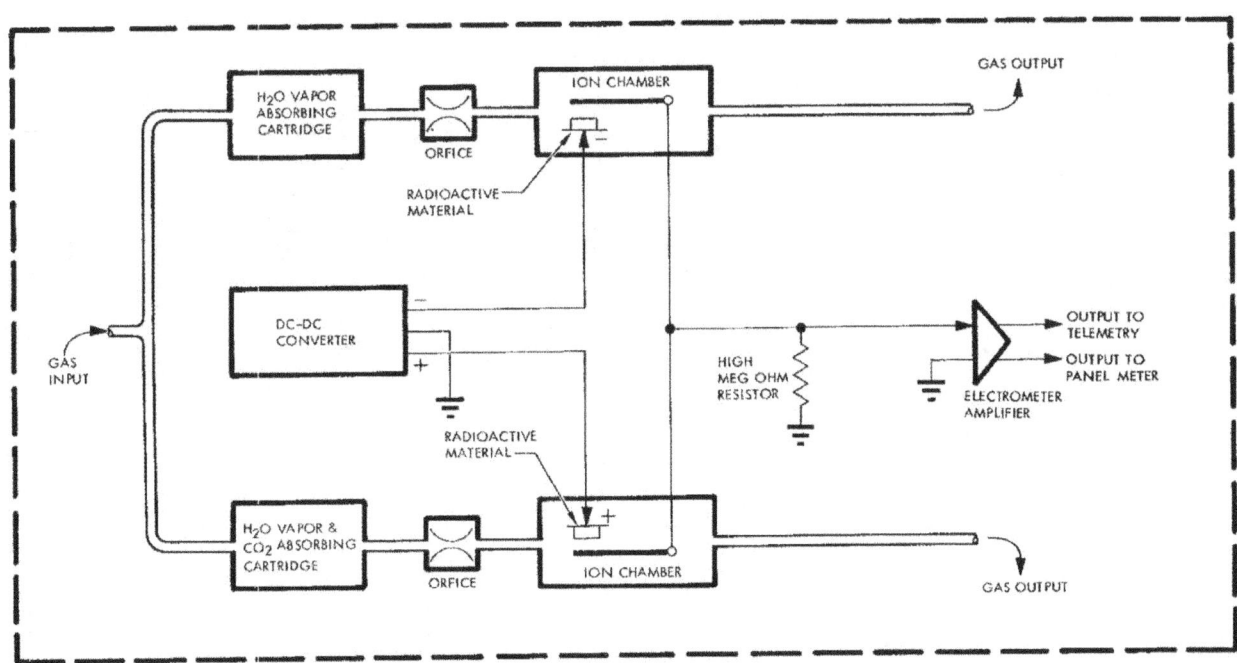

Figure 10-18 CO$_2$ Partial Pressure Detector and Schematic Diagram

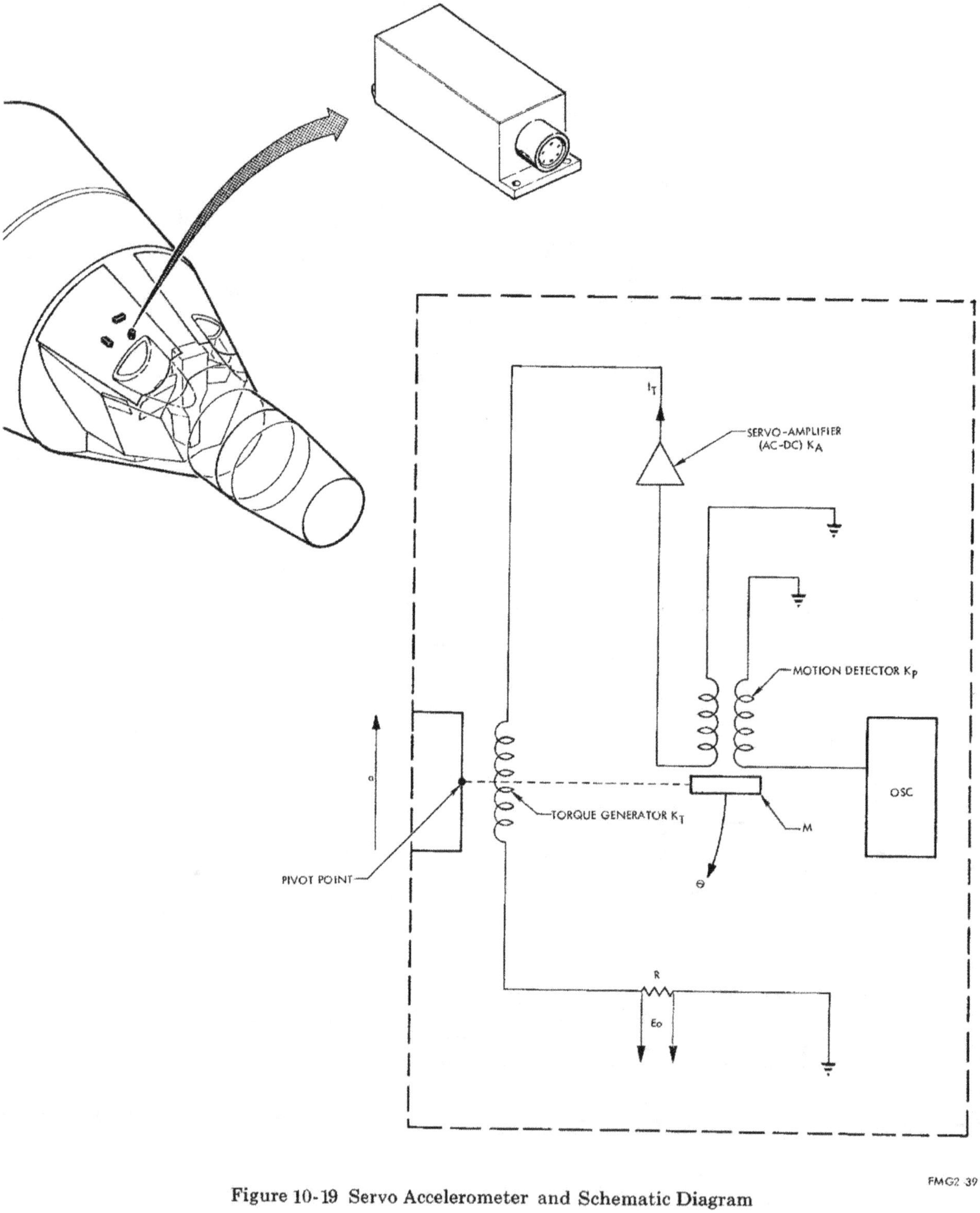

Figure 10-19 Servo Accelerometer and Schematic Diagram

the servo amplifier is applied to a torque generator which tends to restore the mass to its equilibrium position. The output of the accelerometer is obtained by sensing the voltage drop across the resistor in the system loop.

INSTRUMENTATION PACKAGES

A number of the signals in the various spacecraft systems are not compatible with the instrumentation circuitry, and therefore, must be conditioned for their use. Two signal-conditioning packages (instrumentation assemblies) are provided for this purpose. Instrumentation assembly No. 1 is approximately 8" x 10" x 3" and is located in the adapter section. Instrumentation assembly No. 2 is approximately 10" x 10" x 8" and is located in the upper right hand equipment bay of the re-entry section. Both units utilize sealed containers with an operating pressure of 4.5 PSID and are shown in Figure 10-20. The assemblies employ a modular construction with plug-in modules that may be replaced, individually. A module consists of one or two standard printed circuit boards with the necessary component parts and a connector for attachment to a mother board within the package. There are 18 modules spaced in assembly No. 1 and 51 in assembly No. 2. Some modules provide for one data channel and others for two. There are six basic types of modules, and several of these have additional variations for different signal handling capabilities.

There are six variations of the phase sensitive demodulators (PSD). Basically, the PSD accepts two input voltages: one signal voltage and one reference. It provides a DC output of five volts for a full scale input signal that is in phase with the reference and an output of 0 volts for a full scale signal that

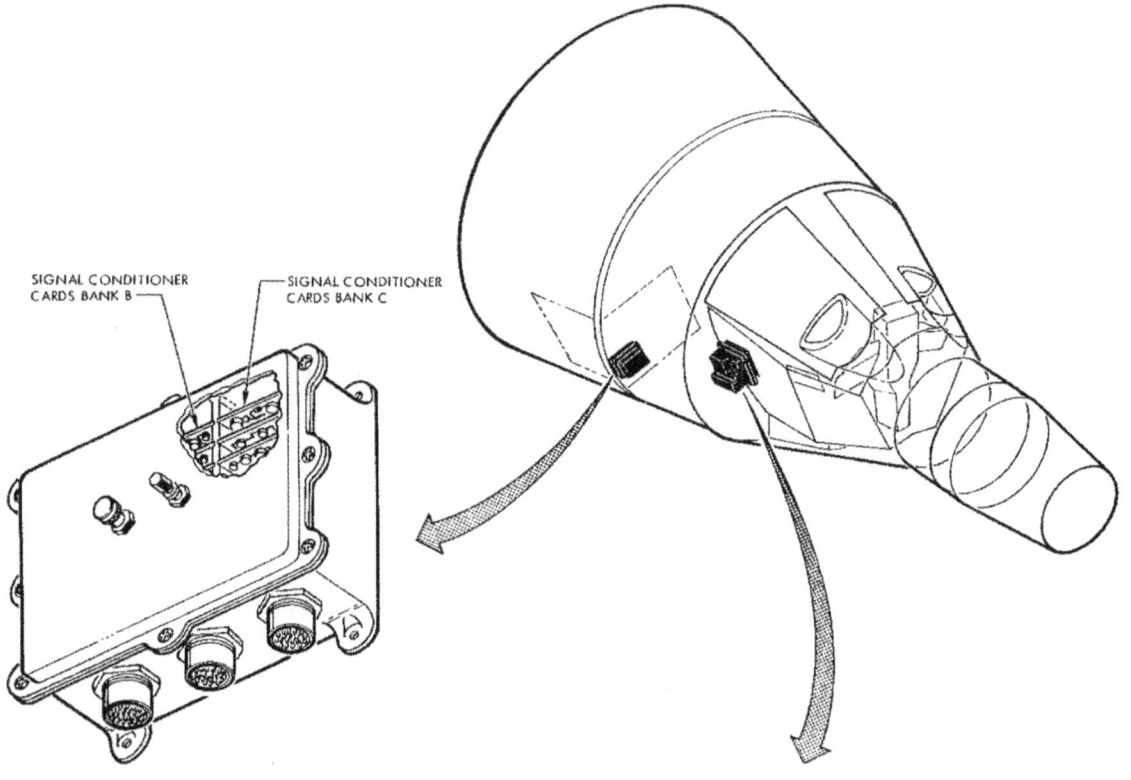
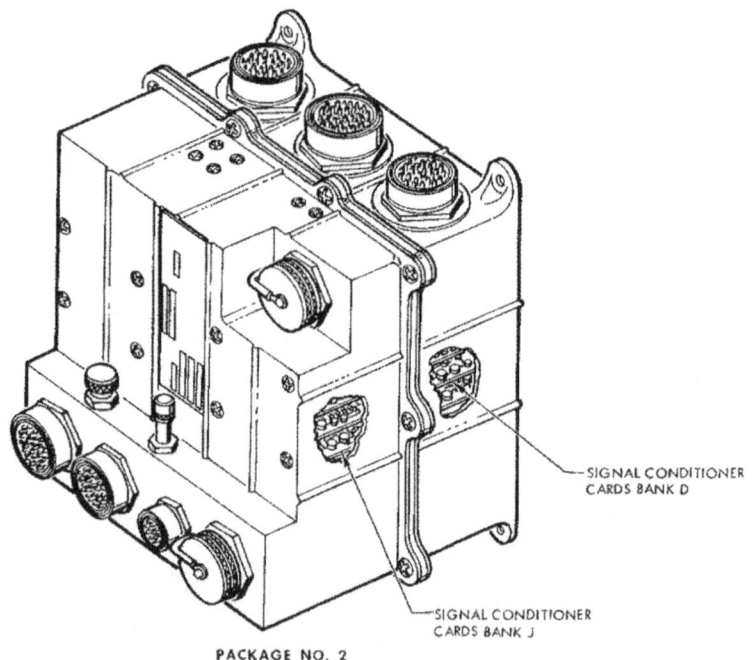

Figure 10-20 Instrumentation Package Assemblies

is out of phase with the reference. The various configurations of this unit provide different full-scale sensitivities including special calibration curves for rate gyros.

The twelve types of DC voltage monitors are designed to accept various positive and negative DC voltage inputs and provide outputs of 0 to 5 VDC.

The AC voltage monitor accepts a signal ranging from 23 to 29 volts rms over a frequency range of 380 to 420 cycles. The output is from 0-5 VDC, varying only with the input voltage.

There are nine types of attenuator modules. These modules have various DC inputs which are changed to signals in either the 0-20 MV DC range or the 0-5 VDC range. Some attenuator modules contain two data channels.

The DC millivolt monitor accepts an input of 0 to 50 MV DC and provides a proportional output of 0 to 20 MV DC.

The AC frequency sensor provides a 0 to 5 VDC output proportional to an input frequency varying from 380 cps to 420 cps. The voltage level of the input is 26 volts rms and does not affect the output of the module.

MULTIPLEXER/ENCODER SYSTEM

The multiplexer/encoder is divided into five packages to allow the data signals to be sampled near their sources. The units are shown in their respective locations in Figures 10-21 and 10-22. Two identical low-level multiplexers, one in the re-entry section and the other in the adapter, each sample 32 low-level signals. (0-20 MV DC) Two high-level multiplexers in simular locations

 SEDR 300
PROJECT GEMINI

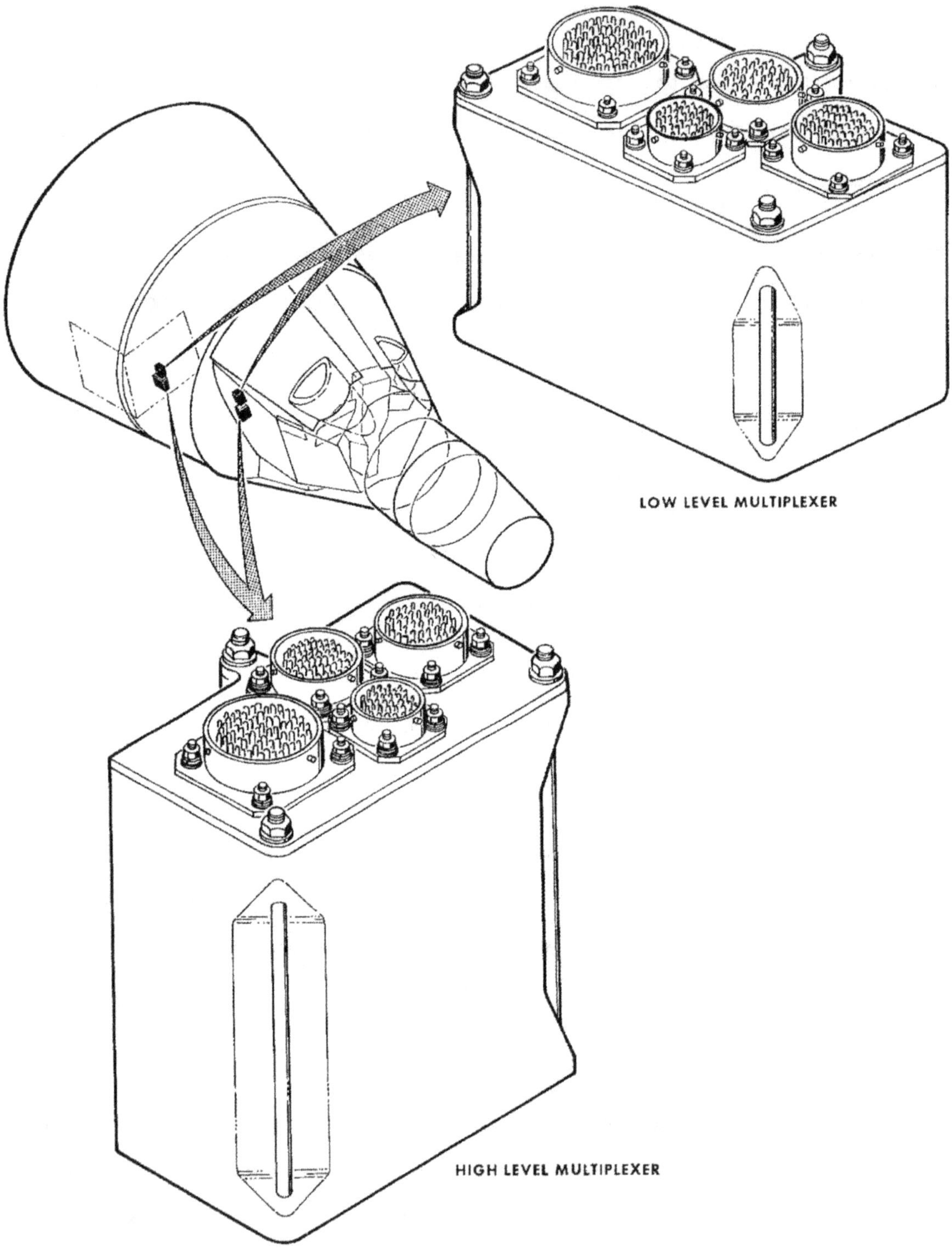

LOW LEVEL MULTIPLEXER

HIGH LEVEL MULTIPLEXER

Figure 10-21 Instrumentation System Multiplexers

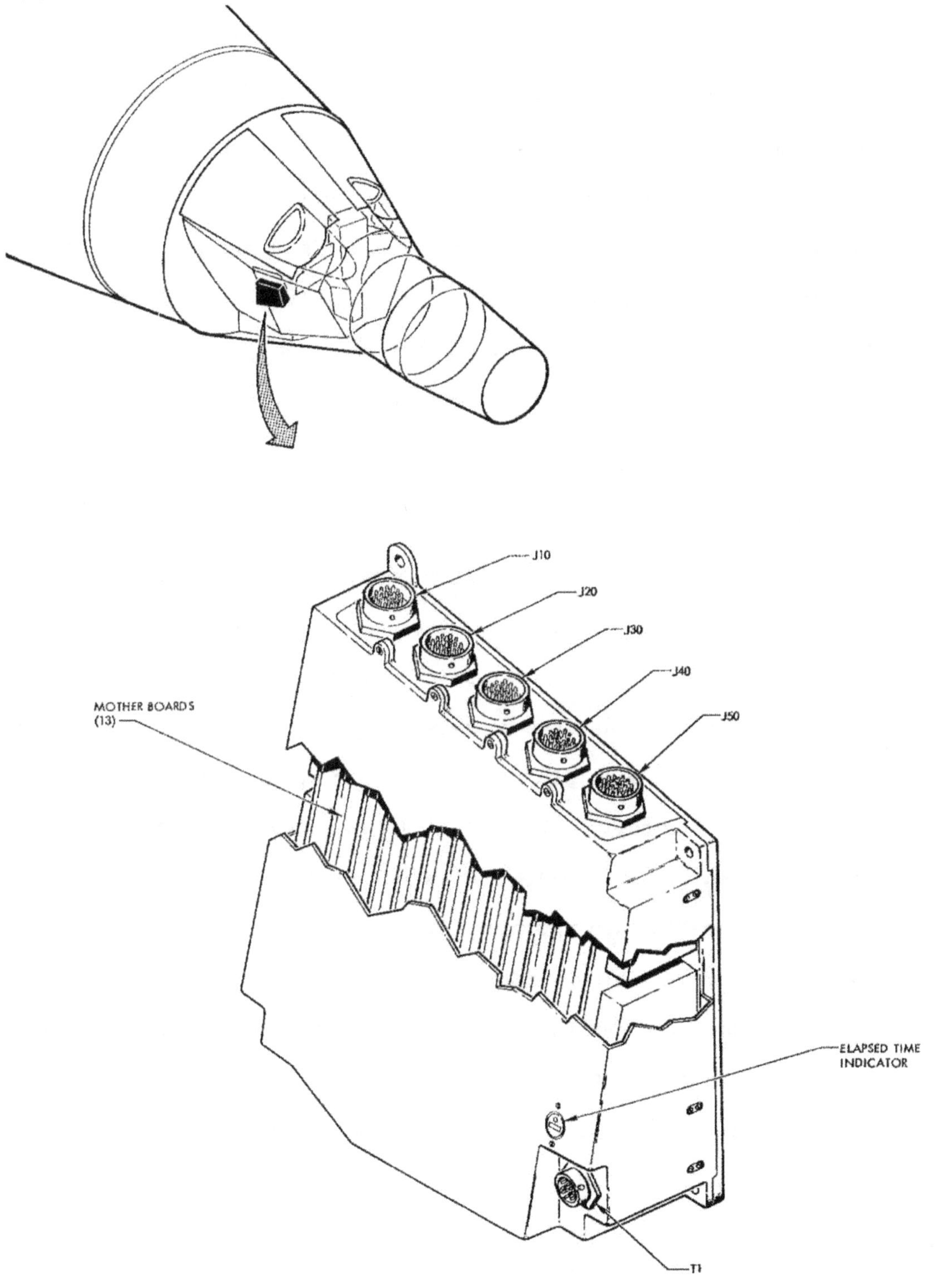

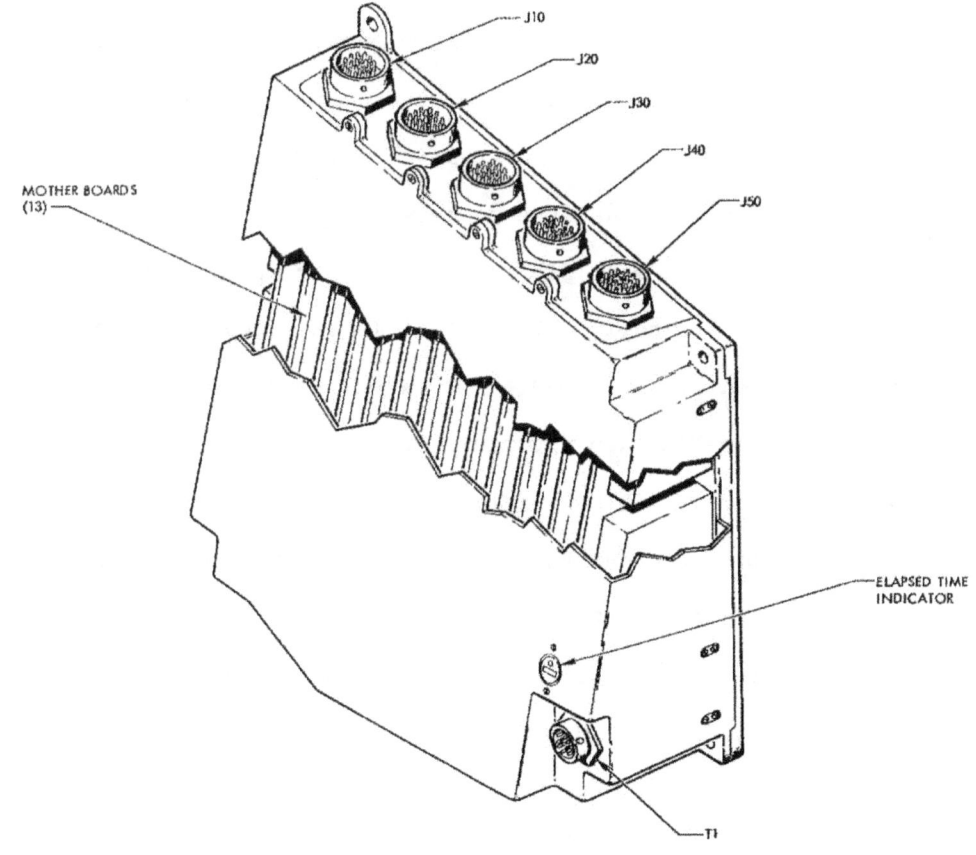

Figure 10-22 Instrumentation System Programmer

each accept 32 high-level (0-5 VDC) and 40 bi-level (0 or 28 VDC) signals. Each high-level multiplexer has 16 bi-level pulse gates which provide an output whenever an inverted plus (+28 VDC to 0 VDC) of at least ten milliseconds duration is applied. The gate is reset after sampling. The programmer, located in the re-entry module, contains the balance of the multiplexing circuitry, the analog-to-digital converter, program generators, sync generator, address generator, output shift registers, clock rate generators, digital shift regulator and a tape recorder converter unit.

The sampling rates of multiplexer inputs are established by the timing chain from the programmer. In the low-level multiplexer, a group of eight input gates operate at 1.25 samples per second and one group of 24 input gates operate at .416 samples per second. In the high-level multiplexer, all analog channels have 1.25 samples per second, and the bi-level and bi-level-pulsed signals are sampled in sets of eight at a sample rate of ten per second. For the bi-level signals, a binary one (Nominally 28V., but at least 15 VDC) may indicate that an event or function either has or has not taken place. For example, the indication that the Bio-Med tape recorders are on is a one but the indication that the computer is on is a binary zero (Nominally zero V., but less than 5 VDC). For bi-level-pulsed signals, 15 VDC or more represents a binary zero, while 5 VDC or less for at least ten milliseconds is a binary one. The pulse conditioning circuitry in the multiplexer senses these pulses and holds the voltage level until it is sampled by the programmer.

The serial outputs, provided from the programmer, all have positive voltages for ones and zero or negative voltages for zeros. The output for the tape recorder is a 5.12 K bit per second serial return-to-zero (RZ) signal with a +5 volt transition for data ones and a -5 volt transition for data zeros. A clock signal at 5.12 K bits per second is also provided for the tape recorder. This output is a pulse train of 50% duty cycle at a peak amplitude of 5 volts. The timing of the positive excursion is coincident with data one pulses. The programmer output for the real time transmitter is a 51.2 K bit non-return-to-zero (NRZ) signal with a voltage which is adjustable between .1 volt and 1.0 volt peak. Separate hardline outputs are provided to enable various test equipment to be used without degradation of either the transmitter or tape recorder outputs. The hardline outputs are real time PCM signal, basic PCM clock rate signal, and master reset pulse signal. The signals are two volts peak to peak and are fed over twinex coaxial or video cables.

The programmer message format includes a master frame for the real-time transmitter output and a prime subframe for the tape recorder output. The master frame consists of 160 words, each word consisting of eight data bits, sampled 40 times per second. Ninety-six master frames are required to completely cycle through all data inputs. Every tenth word in the master frame contains prime subframe data. The prime subframe consists of 64 words sampled ten times per second. Twenty-four prime subframes are required to cycle through all data inputs of this part of the system. Information bits are obtained from analog data, arranged with the most significant bit first, digital data, broken into groups of eight bits with the most significant bit first, or bi-level data grouped as eight consecutive data bits (Referred to as a bi-level set).

TRANSMITTERS

Three telemetry transmitters are used to transmit the instrumentation system data to the ground stations. Although the transmitters serve the instrumentation system, its antennas, and associated switching is part of the communication system; therefore, the transmitters are described in detail in Section IX, Communication System.

PCM TAPE RECORDER

The tape recorder is designed for monitoring and for producing a recording of the signals received from the PCM programmer. The tape recorder records PCM data at a tape speed of 1 7/8 inches per second and playback, on command, of this recorded data at twenty-two times the recorded speed. The recorder, if recording data, will, on command, stop, reverse tape direction and playback the recorded data at a tape speed of 41.25 inches per second. Erasure of data will occur only during record mode. The power control circuitry is described in detail in Section IX, Communication System.

Telemetered signals recorded are return to zero (RZ). The PCM tape recorder reproducer shown in Figure 10-23 consists of one completely enclosed tape recorder which is approximately 4.3 inches high, 10.0 inches wide and 10.0 inches deep. The tape recorder consists of the cover assembly, case assembly, capstan drive assembly and tape transport assembly. Connectors on the side of the case assembly provide signal connections, power connections and test connections.

SEDR 300
PROJECT GEMINI

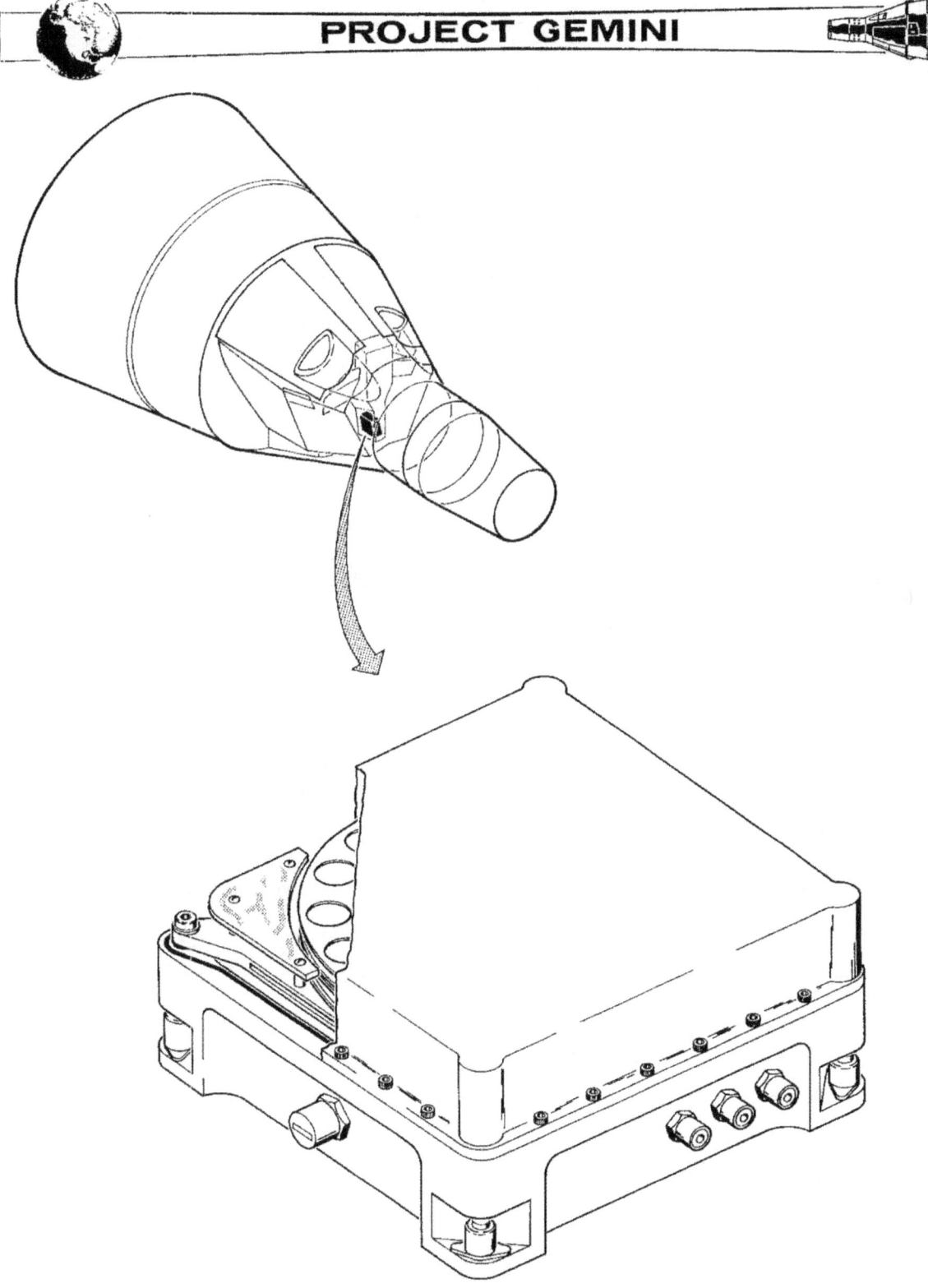

Figure 10-23 PCM Tape Recorder

10-57

Record

The magnetic tape recorder is capable of providing a minimum of four hours of recording time at a tape speed of 1 7/8 inches per second. Two tracks of simultaneous PCM data can be recorded at 1 7/8 inches per second. Four hours of return to zero (RZ) data at 5120 bits per second can be recorded at 1 7/8 inches per second.

Playback

On command, the Recorder is to re-wind the tape onto the supply reel at 22 times the record speed (41.25 inches per second) while reading and playing back the information recorded on the tape. Final output of recorded data is in non-return to zero (NRZ) form.

Diphase System

The diphase signal processing technique permits the maximum tape utilization efficiency, while avoiding certain serious problems encountered with use of conventional NRZ recording at high packing density. It involves the encoding of the digital information prior to recording and decoding of the playback and conversion of the reproduced signal into standard NRZ form.

The diphase technique is essentially a phase-modulated carrier process. The digital data format to be recorded in RZ (return to zero) with an accompanying clock, and the desired output in the reproduce mode is of the standard NRZ (non-return to zero) form.

The diphase signal to be recorded is created in the following manner. Inverted RZ data and clock signal are Or gated into a binary flip-flop such that a transition of the flip-flop occurs on every negative going edge. A logical zero is represented in the diphase code by a square wave at 1/2 the data rate. Each time a logical one is received, a phase transition occurs in the center of the bit cell so that a logic one is represented by a square wave at the data rate. The output of the flip-flop is the diphase signal. This signal is then fed to the record amplifier which drives the diphase signal into the record head.

Record Mode

During the record mode, the input signal is sent to a preamplifier, encoder and amplifier. A clock signal is applied to the input of the triggerable flip-flop. The diphase code produced is recorded on magnetic tape.

The magnetic tape is DC erased prior to recording. The magnetic head utilized is a high-quality instrumentation recording head has a gap width approximately 1/3 the recorded wave length. The gap width is not critical, but if it is much wider than 1/3 of the recorded wave length, the high frequency playback components are attenuated and if it is much narrower, the high frequency components are accentuated, causing a difficult equalization problem.

Reproduce Mode

During the reproduce mode, the signal is picked up by the magnetic head and applied to the playback amplifier, where it is amplified approximately 60 db, filtered, and equalized to compensate for the effects of the head to tape system. The equalized signal is then fed to an input coupler where approximately 40 db of hard limiting is provided, thus providing extremely high immunity from amplitude variation in the reproduced signal.

The ability of this system to operate satisfactorily through such a large variation in playback signal amplitude assures a high degree of reliability and extremely low data drop out. The outputs of the input coupler, the recorded diphase signal and its complement, are fed to the one shot timing extractor circuitry and simultaneously to the decoder circuitry.

The function of the timing extractor and decoder is to produce timing pulses from the amplified and limited diphase playback signal. This circuitry detects the data, using the timing pulses and diphase signal, and produces the final NRZ output.

The output filter is fed the decoder output and filters out some of the higher harmonics of the non-return-to-zero (NRZ) output signal. The hardline output amplifiers produce a hard-line output with good square wave characteristics at high frequencies.

DC-DC CONVERTERS

The two DC to DC converters (one of which is a standby unit) supply the instrumentation system with regulated DC power. The units are approximately 5.5" x 5.5" x 7", weigh approximately seven pounds each, and are located in the right-hand equipment bay of the re-entry section as shown in Figure 10-24. The converters are essentially voltage regulators which operate on 18 to 30.5 VDC and supply output voltages of +5 VDC, +24 VDC and -24 VDC.

 SEDR 300
PROJECT GEMINI

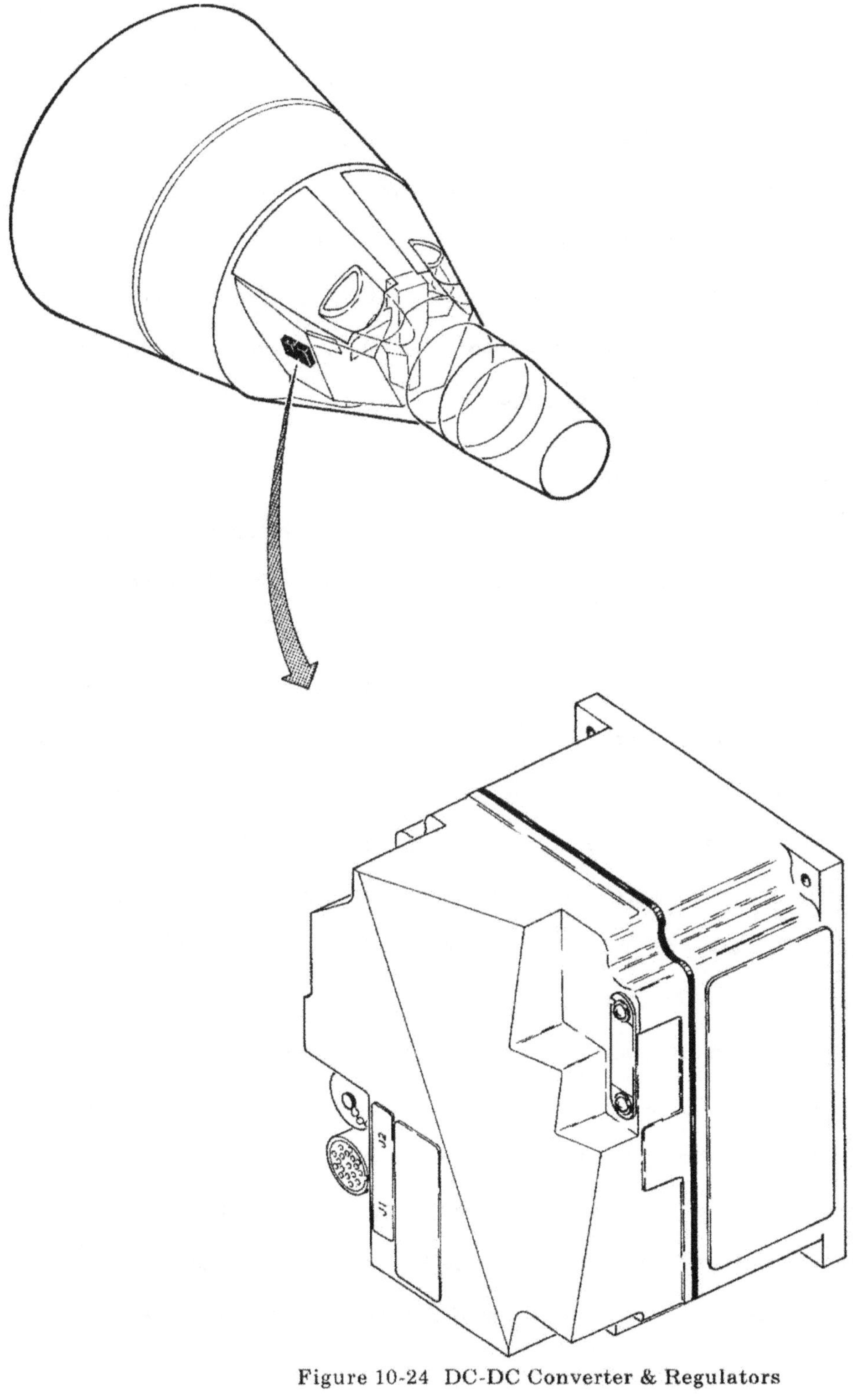

Figure 10-24 DC-DC Converter & Regulators

PROJECT GEMINI

The power control circuitry for the DC-DC converters is shown in Figure 10-3. Essentially, input power to the DC-DC converters is supplied through the on position of the DC-DC CONV circuit breaker on the overhead switch/circuit breaker panel. This arms the DC-DC CONV switch. Placing the DC-DC CONV switch on the overhead switch/circuit breaker panel, to the SEC or PRI position, will apply power to the corresponding converter. Usage of the DC-DC converter regulated output voltages is illustrated in Figure 10-3.

BIO-MED TAPE RECORDERS AND POWER SUPPLY

The two tape recorders used in the physiological instrumentation system are identical. Each one is approximately 9" x 6 1/2" x 1 3/4" (excluding connector and mounting projections) and weighs about three pounds. One external connector provides termination points for all inputs and outputs. The circuitry is made up of 19 printed circuit boards with solid-state components. The recorder uses recording tape with a width of 0.497 $\pm$ 0.001 inches. The reel capacity is 650 feet. All physiological functions, except oral temperature, of each pilot are recorded on separate tape recorders. Each recorder has six data channels and one timing channel. The timing input is a pulse coded wave train derived from the time reference system (TRS) through the time correlation buffer (TCB). This signal is used for time correlation during post mission analysis. The recorders will operate for a total of 75 hours at a normal tape speed of 0.0293 inches per second. Recorder operation is controlled by the crew during the mission without playing back the data. Upon completion of the mission, the recorders are removed from the spacecraft so that the tape can be removed and the data extracted. The total power requirement of each recorder is 1.2 watts at 24 VDC.

The electrical control circuitry for the bio-med instrumentation is shown in Figure 10-2 and the location of the components is shown in Figure 10-1. The recorders are government furnished equipment and are actuated from the spacecraft main bus through the BIO-MED INST circuit breaker and the CONT position of the BIO-MED RCDR switch (1 and 2).

The bio-med power supply, similar in construction to the DC-DC converters, supplies DC regulated voltage to the bio-med instrumentation. Input power for the converter is obtained from the main bus through the BIO-MED INST circuit breaker.

PYROTECHNICS and RETRO ROCKET SYSTEM

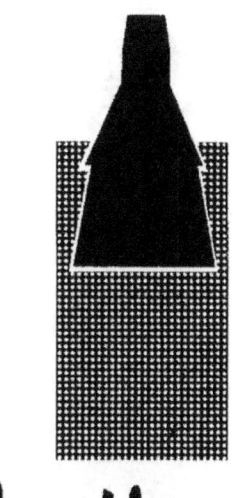

Section XI

TABLE OF CONTENTS

TITLE	PAGE
GENERAL INFORMATION	11-3
RECURRENT COMPONENTS	11-3
SEPARATION ASSEMBLIES AND DEVICES	11-8
EGRESS SYSTEMS AND DEVICES	11-35
PARACHUTE LANDING SYSTEM PYROTECHNICS	11-56
PYROTECHNIC VALVES	11-67
RETROGRADE ROCKET SYSTEM	11-69

 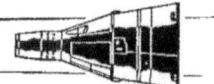

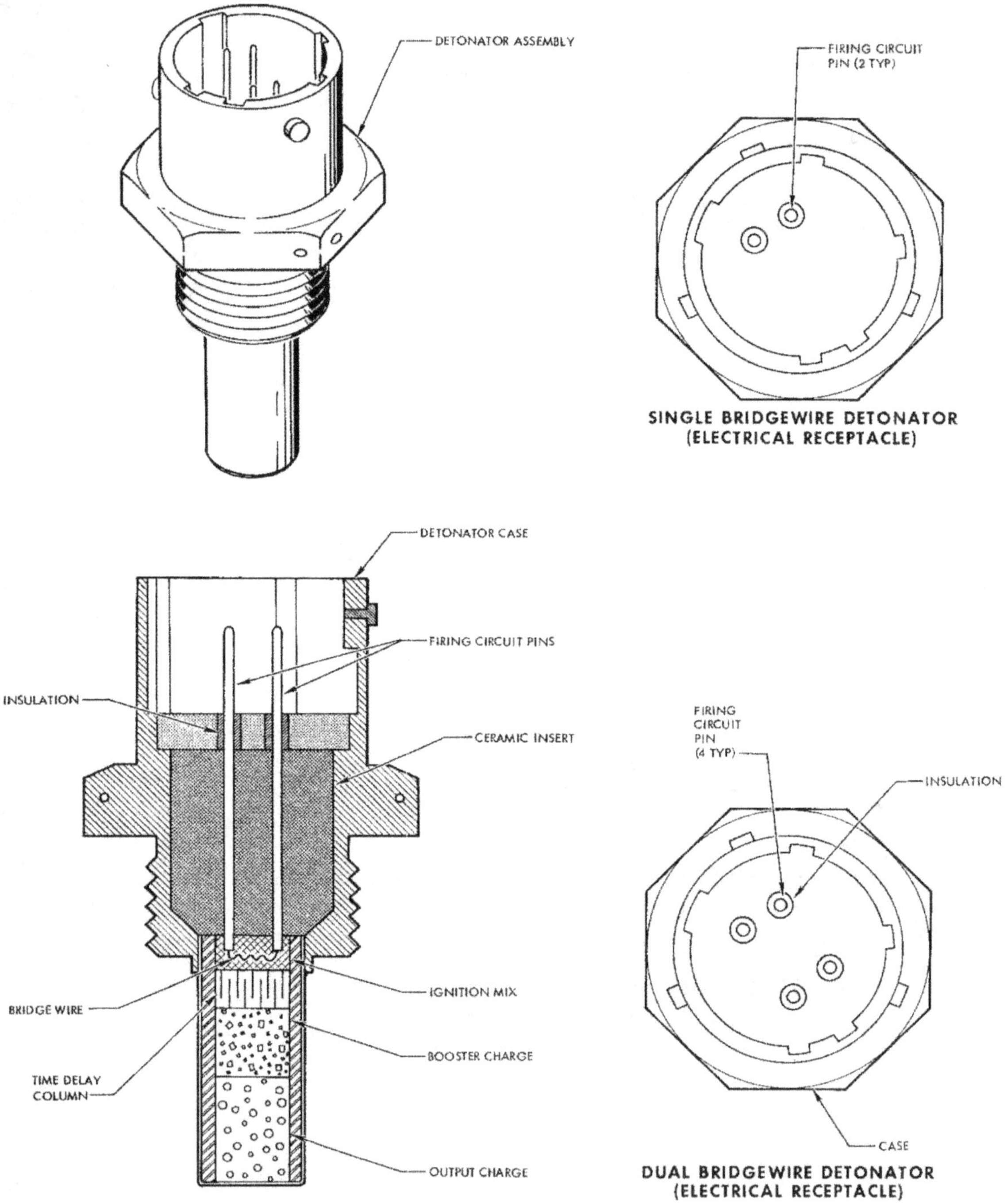

Figure 11-1 Detonator (Typical)

SEDR 300
PROJECT GEMINI

SECTION XI PYROTECHNICS AND RETROGRADE ROCKETS

GENERAL INFORMATION

The Pyrotechnic Devices and Retrograde Rockets, installed in the Gemini Spacecraft, perform a large part of the operational burden. They provide the escape system modes, enable and disable systems, and separate various sections and assemblies. Pyrotechnics are installed in each of the major sections, and in numerous locations throughout the spacecraft. The retrograde rockets retard the spacecraft's orbital velocity to initiate re-entry into the earth's atmosphere. The retrograde rockets are located in the retrograde section of the adapter.

RECURRENT COMPONENTS

Some pyrotechnic items are used extensively throughout the spacecraft. To avoid repetition in subsequent paragraphs, their description and operation will be presented at this time. When describing the various systems, these components shall be mentioned by name only.

DETONATOR

Description

The typical detonator (Figure 11-1) is a machined steel or aluminum cylinder containing an ignition mix, booster charge and an output charge. In some instances a pyrotechnic time delay column is used to provide a time delay between ignition and detonation. The cylinder is threaded at one end for installation purposes. An electrical receptacle is provided at the other end. Electrically, the detonators are provided in two different configurations. One incorporates two independent, identical firing circuits. The other incorporates only one firing circuit. The circuits of both detonators are insulated from and independent of the detonator body. Each firing circuit consists of two electrical connector pins, across which a bridge wire is incorporated. The detonator is used to initiate high explosive components.

Operation

Upon receipt of the proper electrical signal, the firing circuit or circuits will cause the detonator to function. Either circuit (detonators with dual circuits) will initiate the charge with the same performance characteristics as exist when both circuits are operative. The bridge wire ignites the ignition mix which in turn ignites the booster charge. The booster charge then propagates detonation to the output charge. If a delay column is installed, the ignition mix will ignite the delay column which ignites the booster charge. The output charge detonates and transmits the detonation wave to the assembly to which it is attached.

CARTRIDGE

Description

The typical cartridge (Figure 11-2) is a machined steel cylinder containing an ignition mix and an output charge. In some instances a pyrotechnic time delay column is used to provide a specific time delay between ignition and output. The cartridge is threaded at one end for installation purposes. An electrical receptacle is provided at the opposite end. Electrically, the cartridges are provided in two different configurations. One incorporates two independent, identical firing circuits. The other incorporates only one firing circuit. The circuits of both cartridges are insulated from and independent of the cartridge body. Each firing circuit consists of two electrical connector pins with a bridge wire

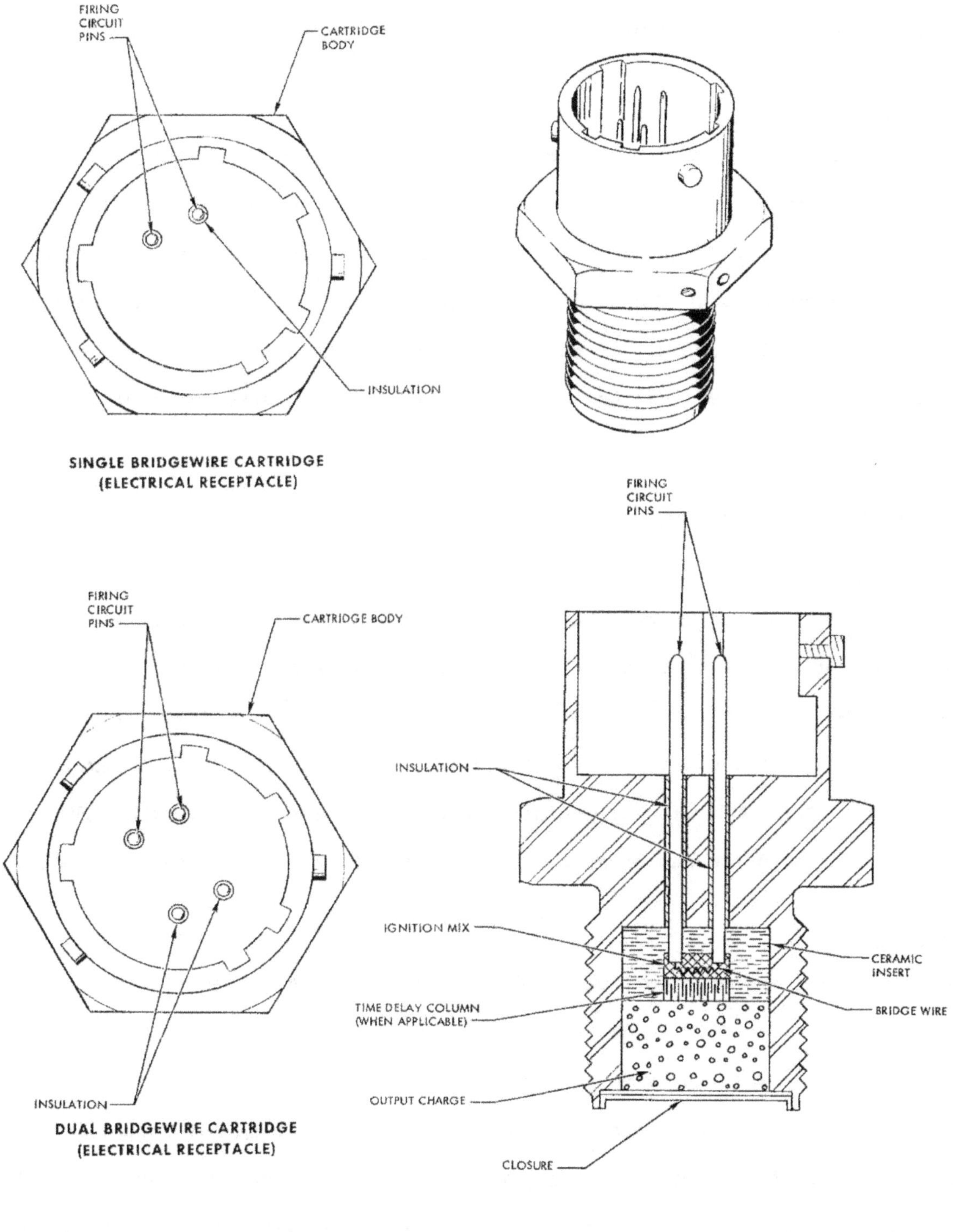

Figure 11-2 Cartridge (Gas Pressure)

attached between. The cartridge is used to produce hot gas pressure.

Operation

When initiated by the proper electrical signal, the firing circuits will cause the cartridge to function. Either circuit (cartridges with dual circuits) will fire the charge with the same performance characteristics as exist when both circuits are operative. The bridge wire ignites the ignition mix which propagates burning to the delay column, if applicable, and to the output charge. The output charge produces gas pressure that is used to operate the specific device in which the cartridge is installed.

FLEXIBLE LINEAR SHAPED CHARGE

Description

Flexible Linear Shaped Charge (FLSC) is a V-shaped, flexible lead sheathing containing a high explosive core. FLSC is used in separation assemblies to sever various types, thicknesses, and shapes of materials. The specific type, shape and thickness of the material to be separated, dictates the amount of explosive contained in the FLSC. In the Gemini Spacecraft & Agena Adapter, the FLSC is provided in four different core loadings: 7, 10, 20, and 25 grains per foot.

Operation

When installed, the open portion of the V-shaped FLSC is placed towards the item to be severed. The FLSC is detonated by a booster charge that has been initiated by a detonator. The explosive core of the FLSC detonates, resulting in collapse of the sheathing in the "V" groove, which produces a cutting jet composed of explosive products and minute metal particles. This jet produces extremely high localized pressures resulting in stress far above the yield strength of the target

SEDR 300

PROJECT GEMINI

material.

MILD DETONATING FUSE

Description

Mild Detonating Fuse (MDF) is a strand of high explosive encased in a lead sheathing with a circular cross section. MDF is used as a separation device and as an explosive interconnect. As a separation device, the strand contains 5 grains of explosive per foot. As an explosive interconnect, the strand contains 2 or 3.3 grains of explosive per foot. The interconnect type MDF is installed in either flexible woven steel mesh or nylon hose and rigid stainless steel tubing. Both rigid and flexible MDF have a small booster charge incorporated at each end. The booster charges are referred to as accepter and donor. The accepter being on the end that receives a detonation wave from an initiator. The donor being on the end that transmits a detonation wave to a component or another accepter. The interconnects are attached to various devices by AN type fittings or Bendix type quick disconnects.

Operation

The MDF used as a separation device is placed in a groove milled in a magnesium ring. The ring is formed to the shape of the items to be separated and is placed between the mating surfaces. The assembly to be jettisoned is attached to the main structure by frangible bolts. The bolts have been axially drilled to reduce tensile strength to a specified breaking point. When detonated, the MDF exerts a force against the mating surfaces greater that the tensile strength of the frangible bolts. The MDF, used as an explosive interconnect, is initiated when a detonator or booster charge propagates a detonation wave to the MDF booster. The

SEDR 300
PROJECT GEMINI

booster strengthens the wave and transmits it linearly through the length of the MDF strand. The booster, at the opposite end, propagates the detonation wave to the device to which it is attached.

SEPARATION ASSEMBLIES AND DEVICES

There are several different types of separation assemblies and devices used in the Gemini Spacecraft (Figure 11-3). These assemblies and devices are presented individually in the following paragraphs.

SPACECRAFT/LAUNCH VEHICLE SEPARATION ASSEMBLY

Description

The Spacecraft/Launch Vehicle Separation Assembly (Figure 11-4) separates the spacecraft from the launch vehicle by severing the mating ring. The separation assembly primarily consists of two flexible linear shaped charges (FLSC) installed around the periphery of the mating ring, three detonators, three detonator blocks, three dual boosters, a molded backup retainer and a back blast shield. The dual boosters are inserted in the detonator blocks. The dual booster protrude into the molded backup retainer, indexed directly above the FLSC, when the detonator blocks are installed. The detonators are inserted in the detonator blocks with the output charge adjacent to the dual boosters. The back blast shield attaches the molded backup retainer and FLSC to the mating ring.

Operation

Upon receipt of the proper electrical signal, the detonators transmit a detonation wave that is propagated to the dual boosters. The dual boosters strengthen the detonation wave to achieve proper detonation of the FLSC. The FLSC detonates and severs the mating ring redundantly. The backup retainer absorbs the shock in the

11-8

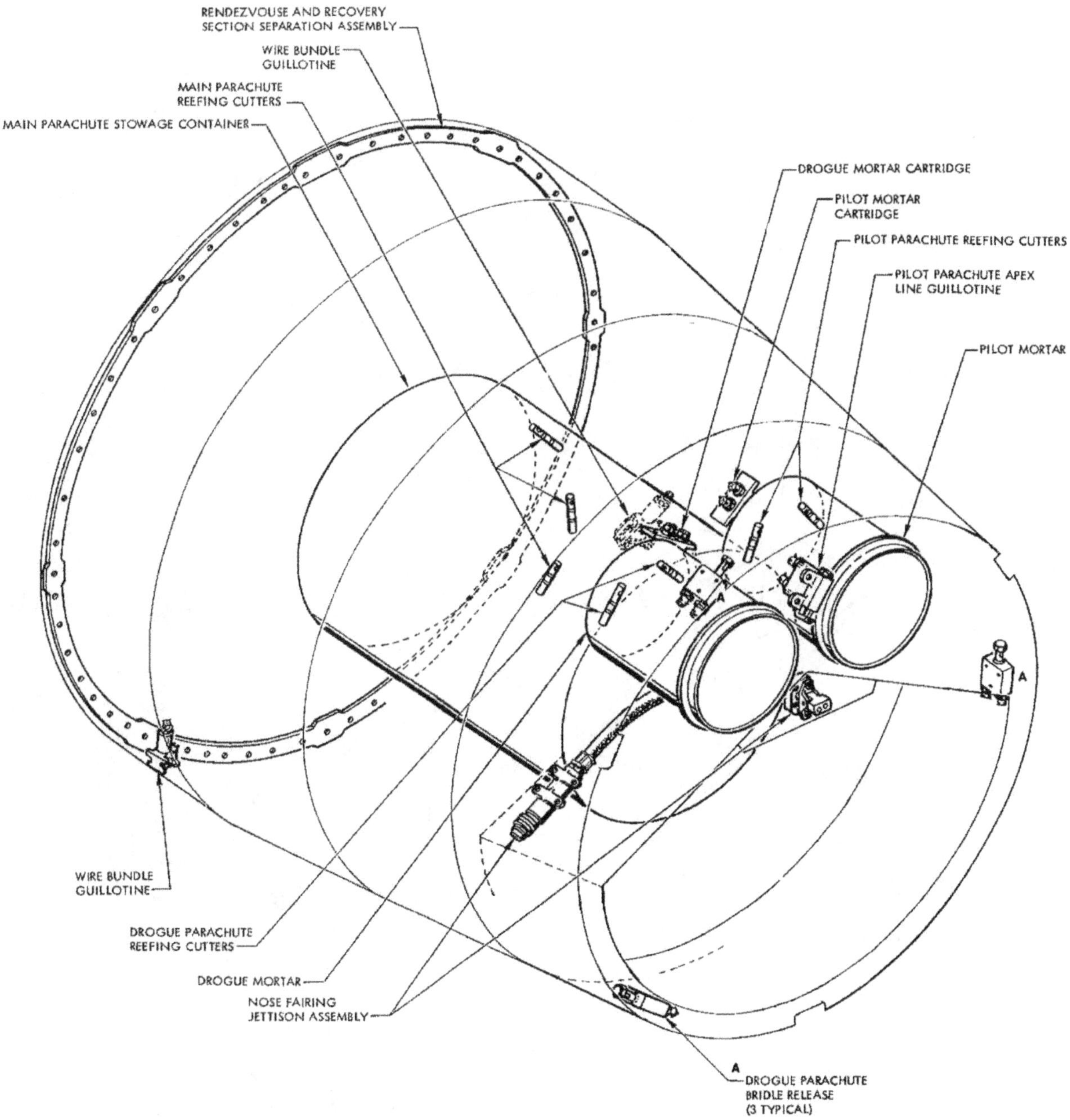

SEDR 300
PROJECT GEMINI

Figure 11-3 Spacecraft Pyrotechnic Devices (R & R Section) (Sheet 1 of 3)

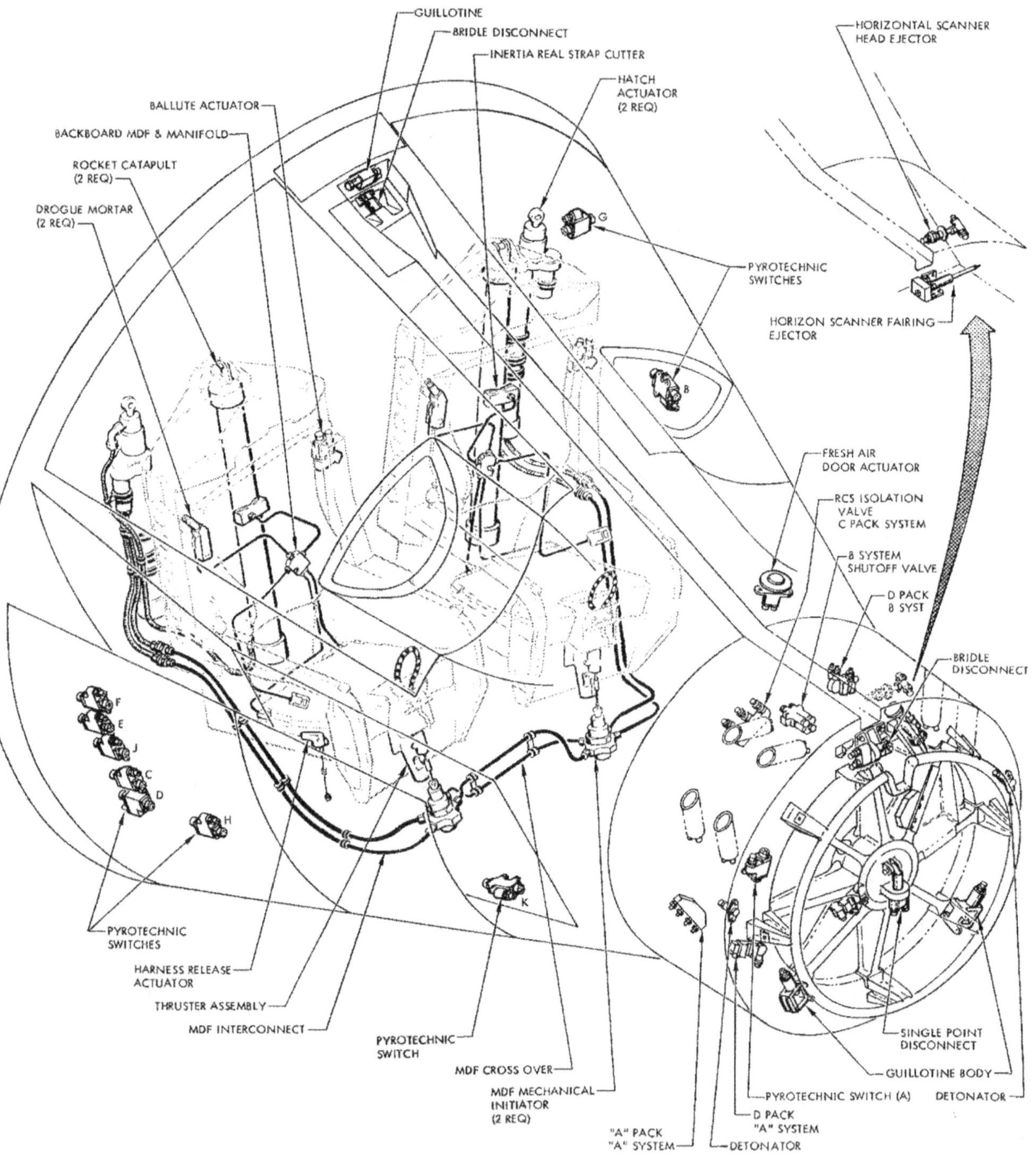

PROJECT GEMINI

SEDR 300

Figure 11-3 Spacecraft Pyrotechnic Devices (Landing Module) (Sheet 2 of 3)

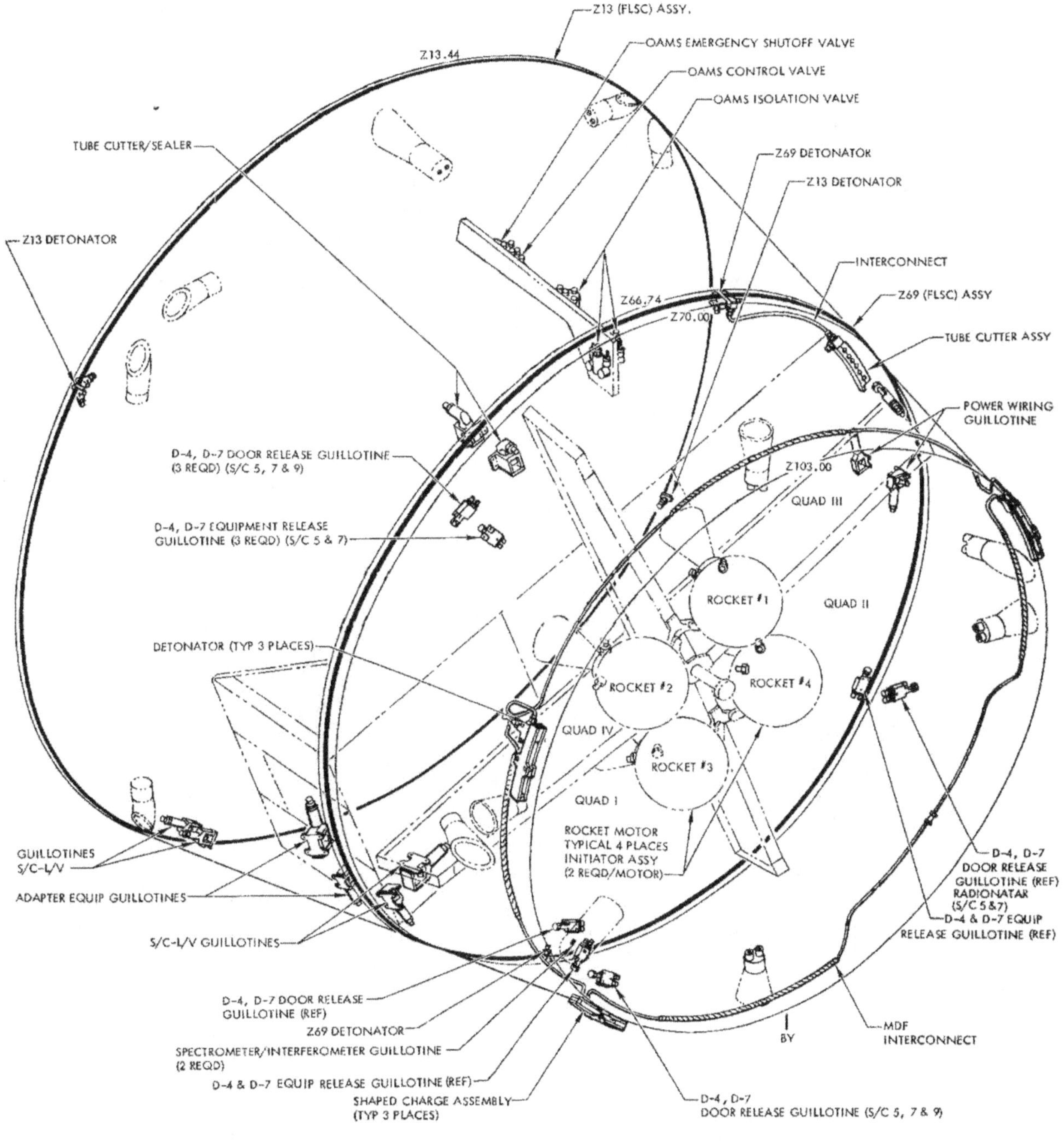

SEDR 300

PROJECT GEMINI

Figure 11-3 Spacecraft Pyrotechnic Devices (Adapter) (Sheet 3 of 3)

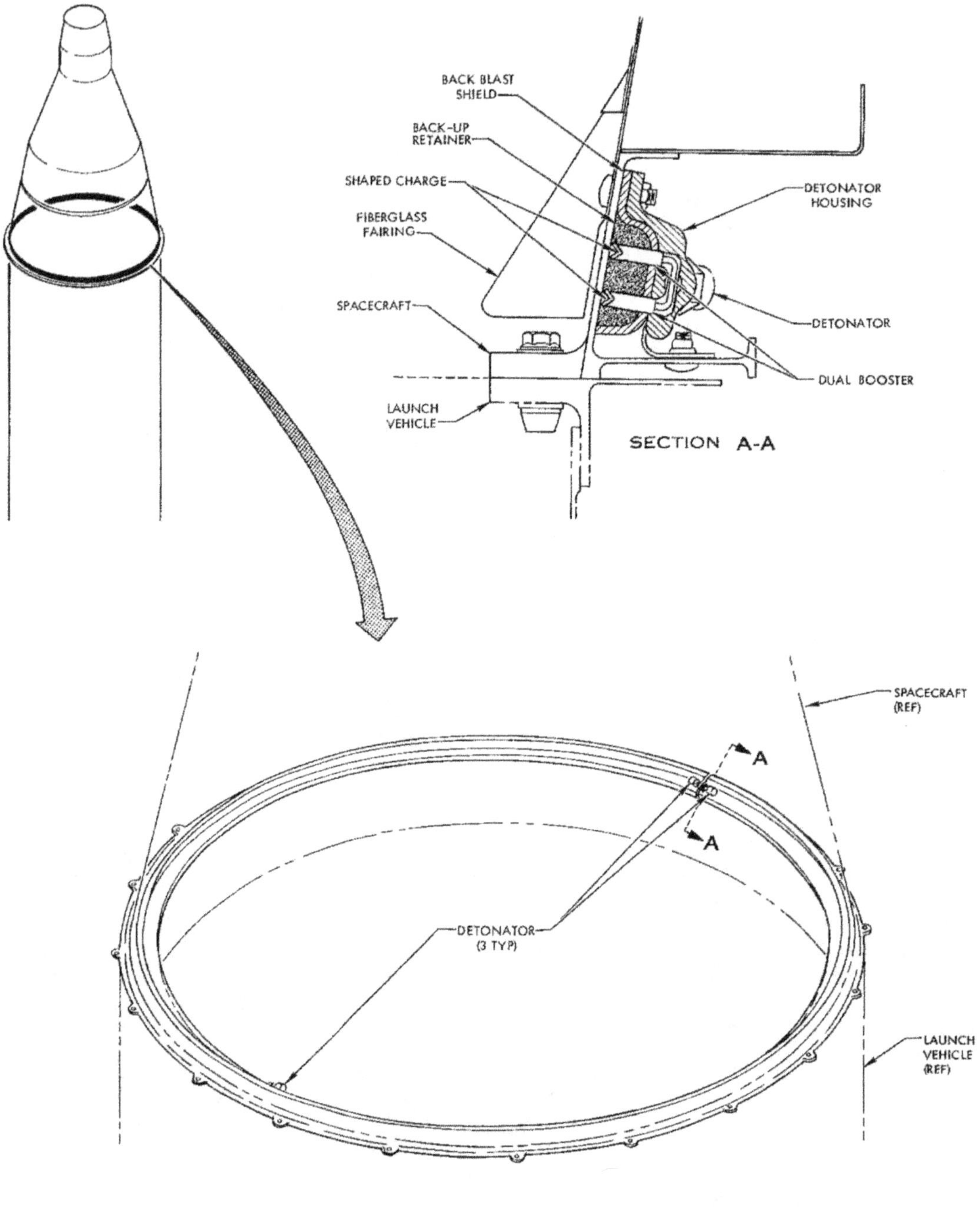

SEDR 300

PROJECT GEMINI

Figure 11-4 Spacecraft/Launch Vehicle Separation Assembly

back blast. The back blast shield protects the structure and equipment from shrapnel. Proper detonation of only one strand of FLSC is sufficient to sever the mating ring.

EQUIPMENT SECTION/RETROGRADE SECTION SEPARATION ASSEMBLY

Description

The Equipment Section/Retrograde Section Separation Assembly (Figure 11-5) separates the equipment section of the adapter from the retrograde section of the adapter. The assembly basically consists of two main units: the shaped charge assembly and the tubing cutter assembly. The shaped charge assembly primarily consists of two flexible linear shaped charges (FLSC), three detonator blocks, containing three crossovers and six boosters, three detonators, ten segmented backup strips and a molded backup retainer. The detonator blocks provide for installation of the detonators. One detonator block provides for the installation of the tubing cutter explosive interconnect. The tubing cutter assembly primarily consists of an explosive interconnect (MDF), two formed aluminum parallel housings, molded backup retainer, two flexible linear shaped charges with boosters attached, a detonator block and a detonator. The explosive interconnect (MDF) is a flexible nylon hose containing a strand of high explosive and end mounted booster charges. The interconnect has Bendix type connectors incorporated at each end for attaching the interconnect to the cutter and shaped charge detonator blocks. The interconnect is attached to the cutter detonator block with its booster charge adjacent to one of the boosters on the FLSC. The detonator is installed in the cutter detonator block with its output end adjacent to the other booster on the FLSC. The cutter assembly is bracket mounted to the inside of the retrograde section of the adapter forward of the parting line. The shaped charge assembly is

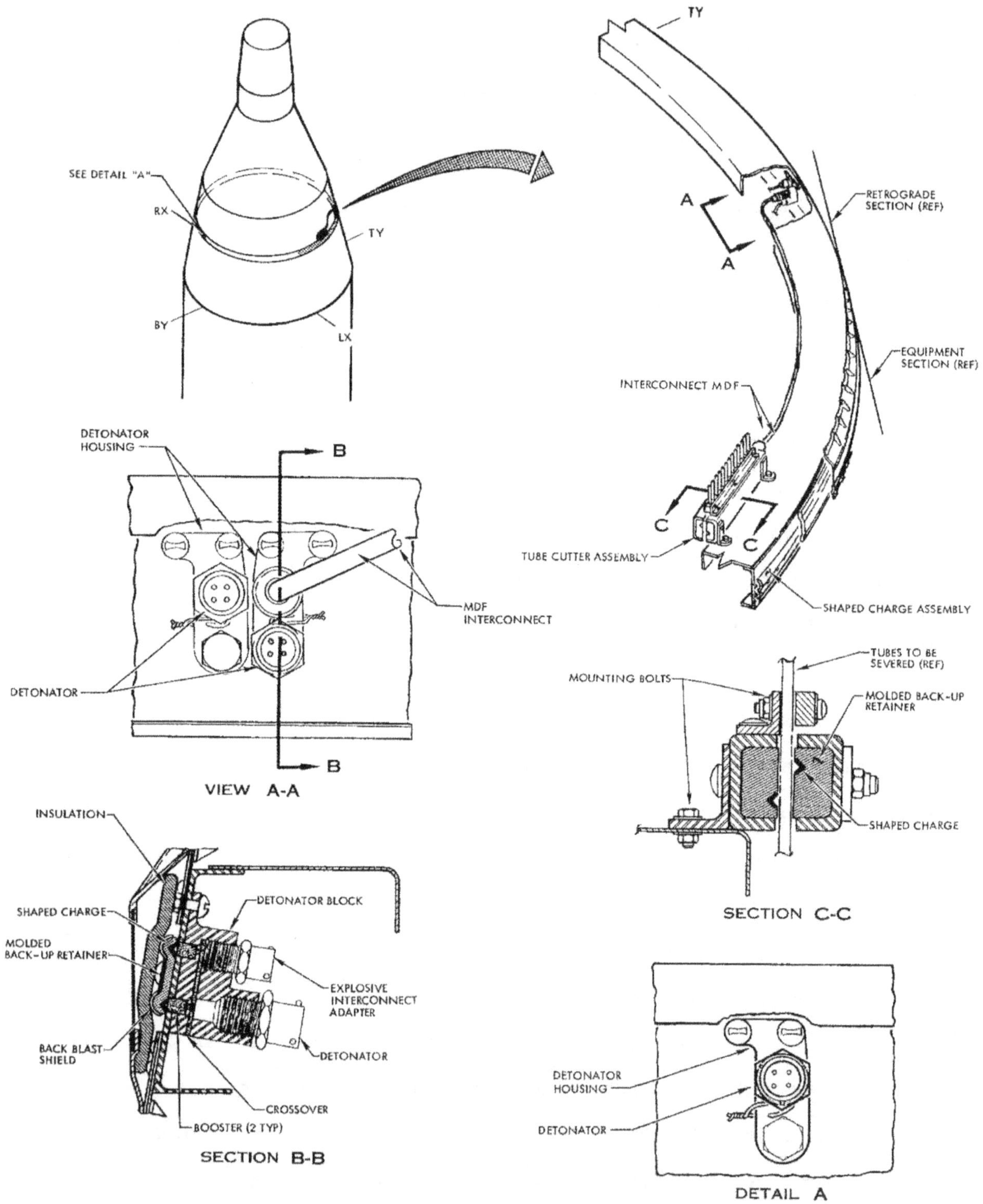

Figure 11-5 Separation Assembly-Equipment Section/Retrograde Section

installed around the outer periphery of the adapter at the equipment section/retrograde section parting line.

Operation

When initiated by the proper electrical signal, the detonators are caused to function. The detonators of the shaped charge assembly transmit a shock or detonation wave to the crossovers which in turn initiates the boosters. The boosters propagate the wave to the FLSC. The FLSC detonates and functions to sever the adapter at the parting line redundantly. The detonator of the tubing cutter assembly propagates detonation to the booster on one strand of FLSC in the cutter assembly. The explosive interconnect transmits detonation from the shaped charge assembly to the booster on the other strand of FLSC in the cutter assembly. The two boosters propagate the shock wave to the FLSC. The two strands of FLSC in the cutter assembly detonate and sever the twelve aluminum tubes and one nylon tube. Proper detonation of only one strand of FLSC, in both the shaped charge assembly and tubing cutter assembly, is sufficient to achieve separation.

RETROGRADE SECTION/RE-ENTRY MODULE SEPARATION ASSEMBLY

Description

The Retrograde Section/Re-entry Module Separation Assembly (Figure 11-6) functions to separate the retrograde section of the adapter from the re-entry module. Separation is accomplished by severing the three titanium straps and various tubes and wire bundles. The separation assembly primarily consists of three cutter assemblies, three detonator housings, three detonators, three parallel booster columns, six explosive interconnects, and three unions. The detonator housings contain a booster column and a parallel booster column. The cutter assemblies consists of

 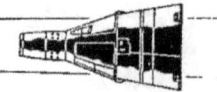

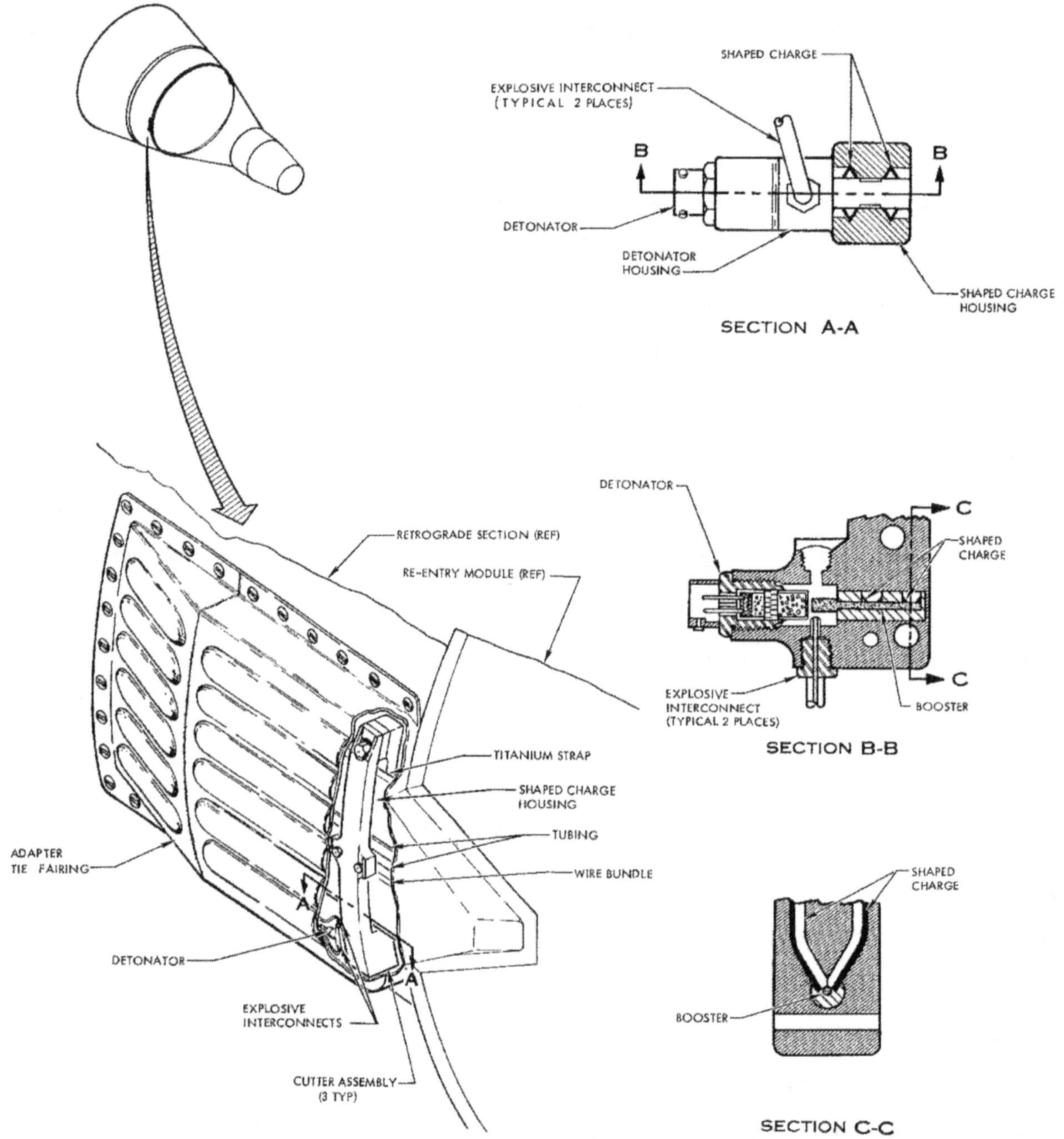

SEDR 300

PROJECT GEMINI

Figure 11-6 Retrograde Section/Re-Entry Module Separation Assembly

SEDR 300
PROJECT GEMINI

two parallel machined aluminum bars that contain four strips of FLSC. The bars are joined by the detonator housings with the parallel boosters. A detonator is installed in each of the three detonator housings. The cutter assemblies are located in three places around the parting line and are linked by the explosive interconnects.

Operation

When initiated by the proper electrical signal, the detonators propagate a detonation or shock wave to the boosters which relay propagation to cutter FLSC and simultaneously the shock wave is propagated to the explosive interconnects. The interconnects transmit the wave to all three cutter assemblies. This is to ensure detonation of all three cutters FLSC, in the event one or even two detonators do not function. Detonation of the cutter FLSC completely severs the titanium straps, wire bundles and tubing redundantly. Proper detonation of only two opposing strips of FLSC in each cutter is sufficient to achieve separation.

RENDEZVOUS AND RECOVERY SECTION SEPARATION ASSEMBLY

Description

The Rendezvous and Recovery Section Separation Assembly (Figure 11-7) separates the rendezvous and recovery (R & R) section from the re-entry control system (RCS) section. The assembly primarily consists of mild detonating fuse (MDF), MDF housing ring, two detonators, two detonator housings and two booster charges. Two strands of MDF are installed in parallel grooves milled in the housing ring face. The grooves intersect at the booster charges which are installed approximately 180° apart. The R & R section is attached to the RCS section by frangible bolts, with the MDF ring fastened to the R & R section at the mating surface. The detonator

11-17

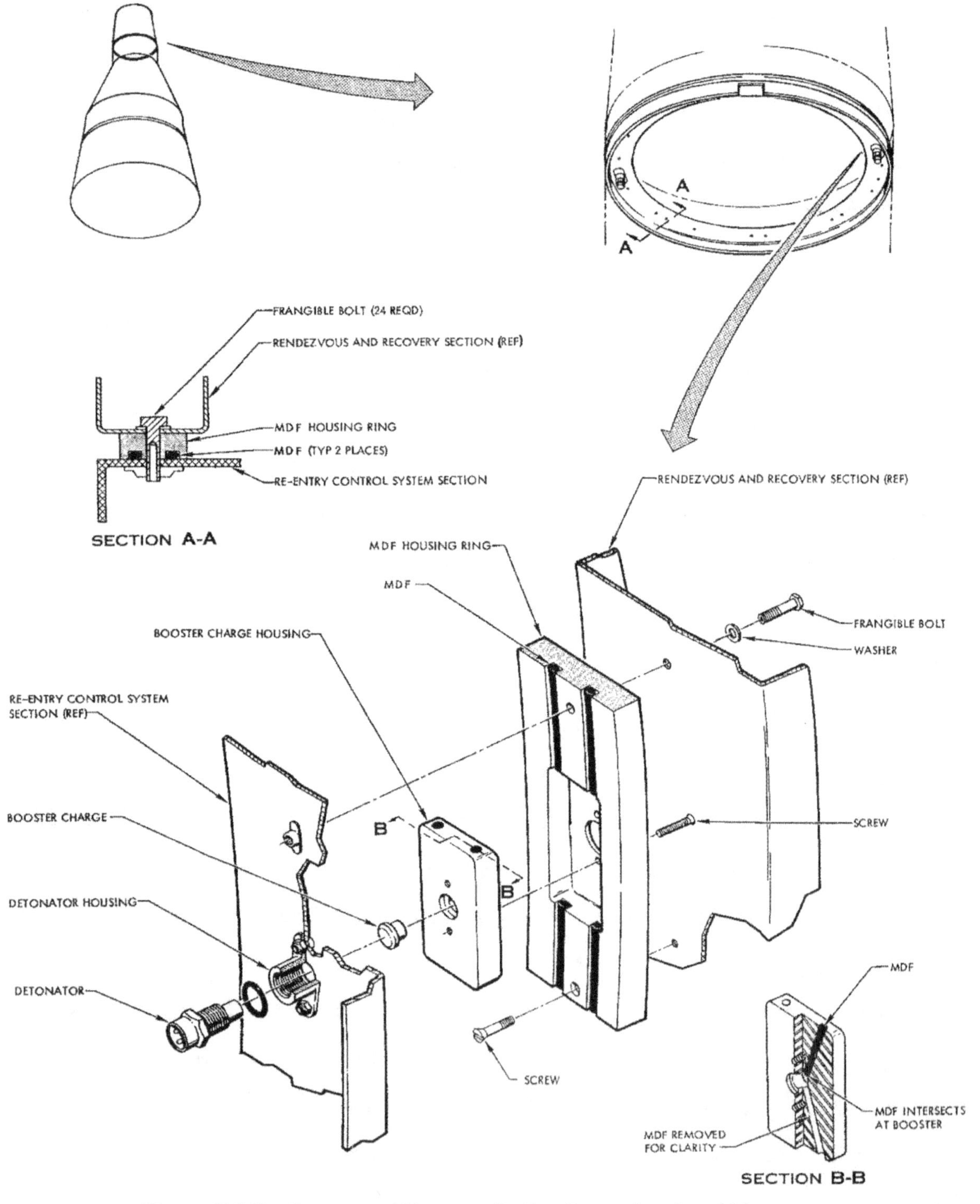

Figure 11-7 Rendezvous and Recovery Section Separation Assembly

FM2-11-5

SEDR 300
PROJECT GEMINI

housings are installed in the RCS section, with the detonators indexed directly above the booster charges, when the sections are mated.

Operation

When initiated by the proper electrical signal, the detonators propagate a detonation wave to the two booster charges. The booster charges strengthen the detonation wave and transmit it to the dual strands of MDF. The MDF detonates, exerting a force against the RCS and R & R section mating surfaces. The force breaks all the frangible bolts and allows the pilot chute to pull the R & R section free of the spacecraft. Satisfactory propagation of either strand of MDF will successfully separate the R & R section.

WIRE BUNDLE GUILLOTINE

Description

The Wire Bundle Guillotine (Figure 11-8) is used throughout the spacecraft to sever various sized bundles of electrical wires. The guillotines are used in two sizes. One size can sever a wire bundle up to one and one quarter inches in diameter and the other can sever a wire bundle up to two and one half inches in diameter. Both sizes are similar in design, appearance and operation. The guillotines primarily consist of a body, end cap or anvil, piston/cutter blade, shear pin(s) and an electrically fired gas pressure cartridge. The body houses the piston/cutter blade, provides for installation of the cartridge, and attachment of the anvil. The anvil is removable to facilitate removal and installation of either the guillotine or wire bundle. Two guillotines are used on a wire bundle, one on each side of the separation plane. Lugs, for attaching the guillotine to the spacecraft structure, are an integral part of the guillotine body.

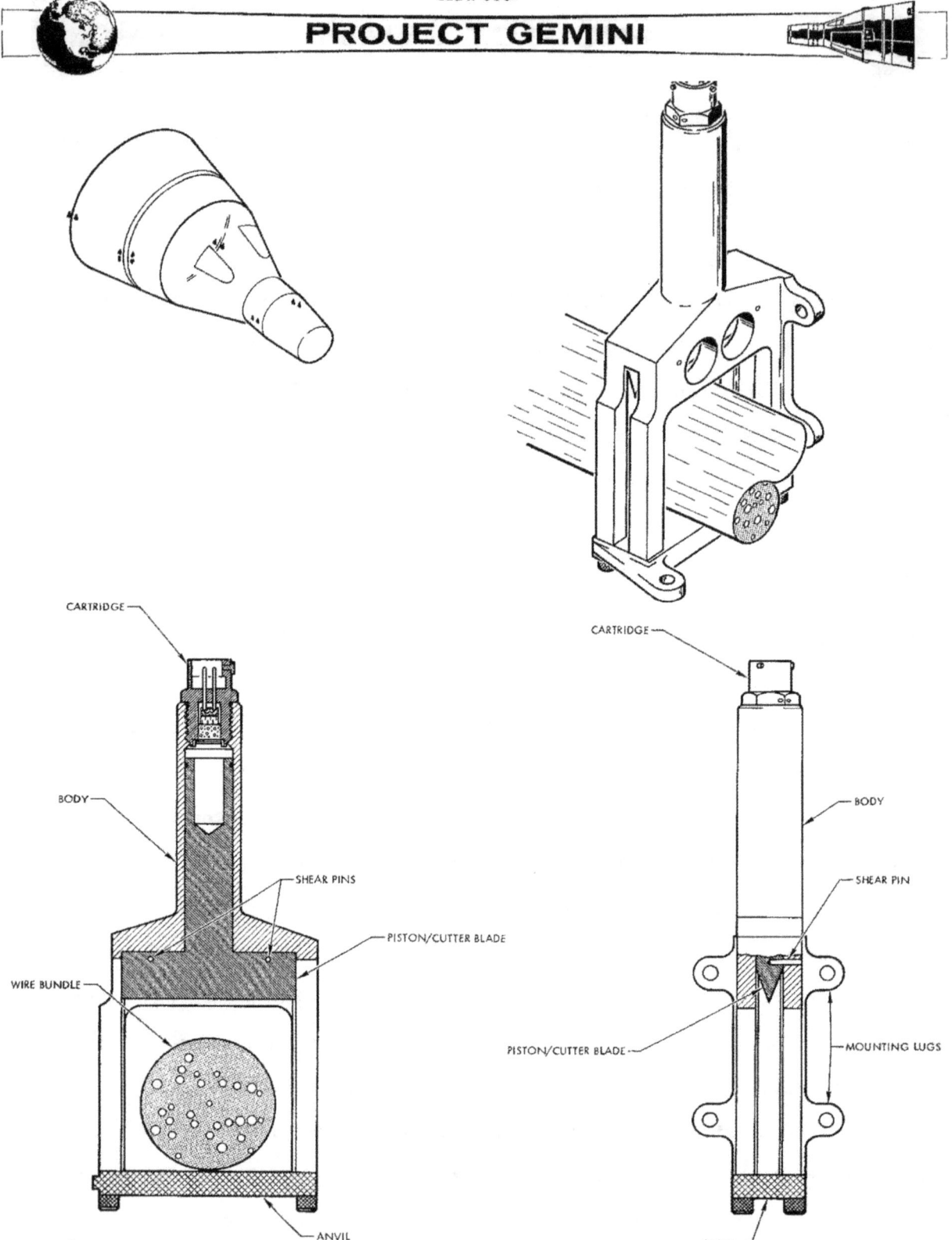

Figure 11-8 Wire Bundle Guillotine

SEDR 300
PROJECT GEMINI

Operation

When initiated by the proper electrical signal, the cartridge produces gas pressure. This gas pressure, exerts force on the piston/cutter blade. When sufficient force is applied, the piston/cutter blade will sever the shear pin(s). As the pin(s) shear, the piston/cutter blade strokes, and positively severs the wire bundle. The wire bundle is then free to pull out of the guillotine body.

WIRE BUNDLE GUILLOTINE (CABLE CUTTING)

Description

The Wire Bundle Guillotine (Cable Cutting) (Figure 11-9) is used to sever twisted stainless steel cables. The guillotine primarily consists of the body, piston/cutter blade, shear pin, anvil and end cap, and two electrically fired gas pressure cartridges. The body provides a piston actuation area and provides for cartridge installation. The anvil is retained in the barrel section of the body by the end cap. The anvil and end cap is removable to permit guillotine and cable installation and removal. Lugs, for attaching the guillotine to the spacecraft structure, are an integral part of the body. The shear pin is provided to retain the piston/cutter blade in a retracted position.

Operation

When initiated by the proper electrical signal, the two cartridges are caused to function and produce gas pressure. The gas pressure exerts force on the piston/cutter blade. When sufficient force is applied, the piston/cutter blade severs the shear pin. The piston/cutter blade travels the length of the barrel section and severs the cable installed in the guillotine. The cable is then free to pull out of the guillotine.

SEDR 300
PROJECT GEMINI

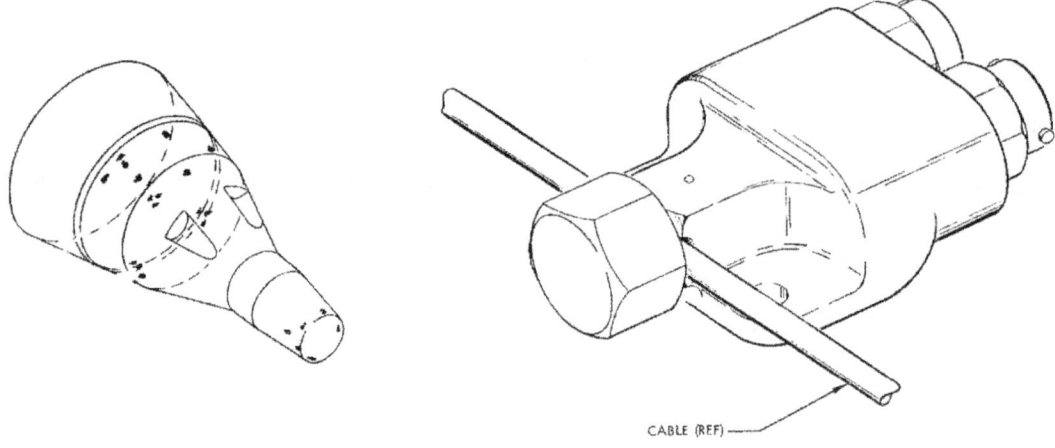

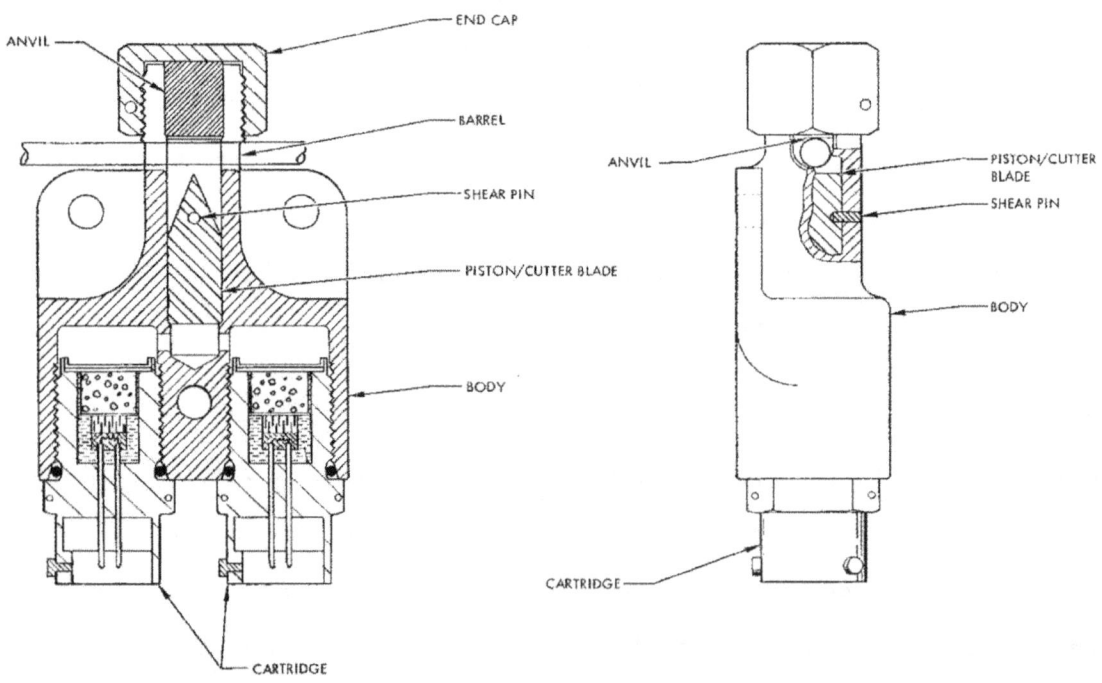

Figure 11-9 Wire Bundle Guillotine (Cable Cutting)

 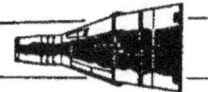

SEDR 300

PROJECT GEMINI

TUBING CUTTER/SEALER

Description

The Tubing Cutter/Sealer (Figure 11-10) is used to cut and seal two stainless steel, teflon lined tubes. The tubes contain hypergolic propellants used in the Orbit Attitude and Maneuvering System (OAMS). Two tubing cutter/sealer assemblies are located in the adapter, one on each side of the retrograde/equipment section separation line. The tubing cutter/sealer assembly primarily consists of the body, anvil, one electrically fired gas pressure cartridge, four shear pins and cutter assembly. The cutter assembly consists of the piston, crimper and blade. The crimper and blade are attached to the piston by two of the shear pins, (sequencing pins). The piston is secured in the body by the other two shear pins, (initial lock pins). The body provides for the installation of the cartridge, attachment of the anvil, and housing for the cutter assembly. Lugs, for attaching the tubing cutter/sealer to the spacecraft structure, are an integral part of the body.

Operation

When initiated by the proper electrical signal, the cartridge generates gas pressure. The gas pressure exerts a force on the piston of the cutter assembly. When sufficient force is applied to the piston, the initial lock pins are severed and the cutter assembly strokes to seal and cut the two tubes. The blade and crimper, extending past the end of the piston, contact the tubing first. As the cutter assembly moves down, the crimper flattens the tubing against the raised portion of the anvil. As the cutter assembly continues its travel, the sequencing pins are severed between the crimper and blade, stopping the travel of the crimper. The base of the piston and blade further crimp and seal the tubing with the blade severing the tubing. The sealed portion of the tubing remains in the tubing

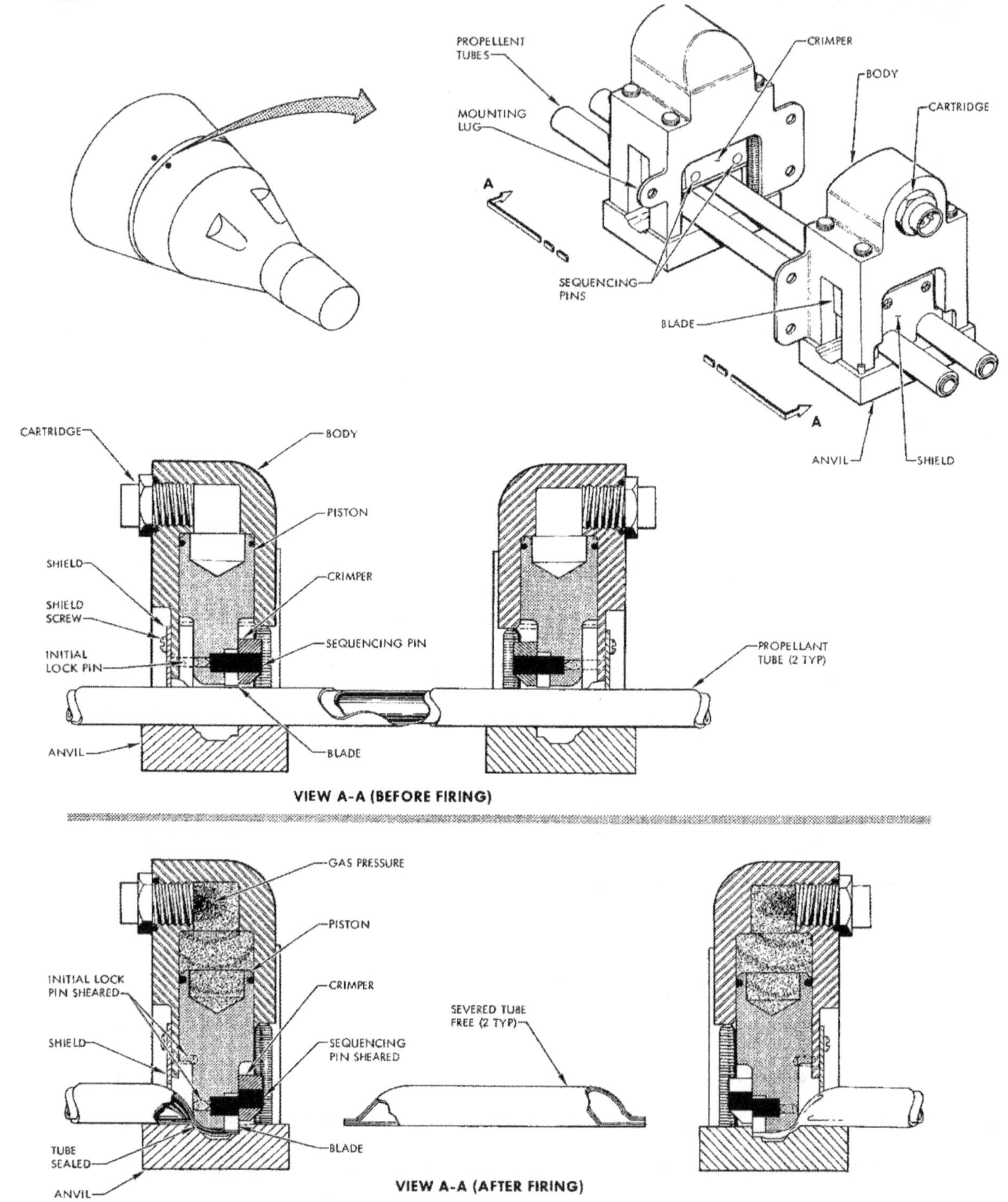

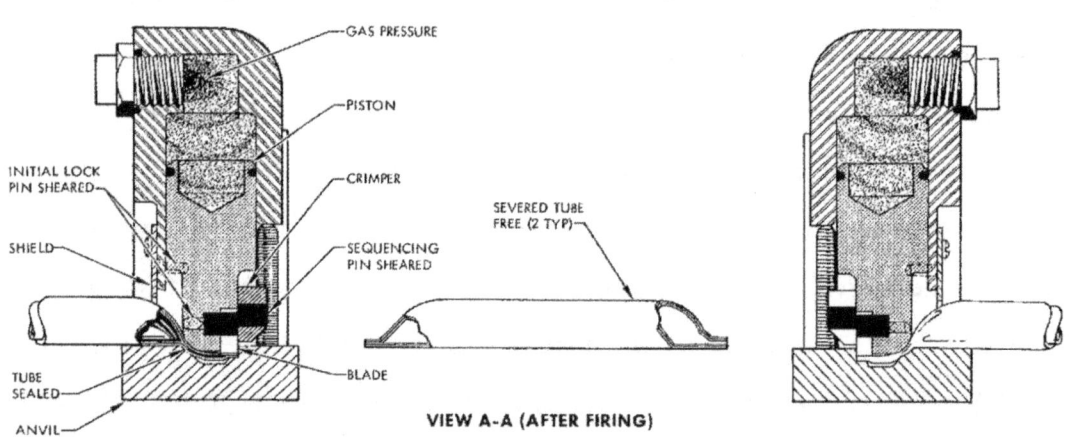

Figure 11-10 Tubing Cutter/Sealer

SEDR 300

PROJECT GEMINI

cutter/sealers at adapter separation. The severed portion of the tubing between the tubing cutter/sealers is free to pull out at adapter separation.

PYROTECHNIC SWITCH

Description

The Pyrotechnic Switch (Figure 11-11) functions to positively open electrical circuits and prevent current flow in various wire bundles prior to their being severed. The switches are located in various places throughout the re-entry module. The switches primarily consist of the body, actuator (piston), shear pin, spring lock, and electrically fired gas pressure cartridge. The shear pin secures the actuator in the switch closed position prior to switch actuation. Incorporated in opposite ends of the switch body are two electrical receptacles. The end mounted receptacles contain hollow spring leaf contacts. The contacts are axially connected by pins mounted in the actuator. All switches are identical in design and operation with the exception of the number of contacts in the receptacles. One model contains 41 contacts, and the other model contains 55 contacts. Lugs, for attaching the switch to the spacecraft structure, are an intergral part of the body.

Operation

When initiated by the proper electrical signal, the cartridge generates gas pressure that is ported through the switch body to the actuator. The pressure exerts a force against a flange of the actuator. The force causes the actuator to sever the shear pin and move axially in the body. As the actuator moves, the connecting pins mounted in the actuator are disengaged from the hollow contacts at one end and are driven further into the hollow contacts at the other end. The spring lock drops into place behind the actuator and prevents it from returning to its original

 SEDR 300

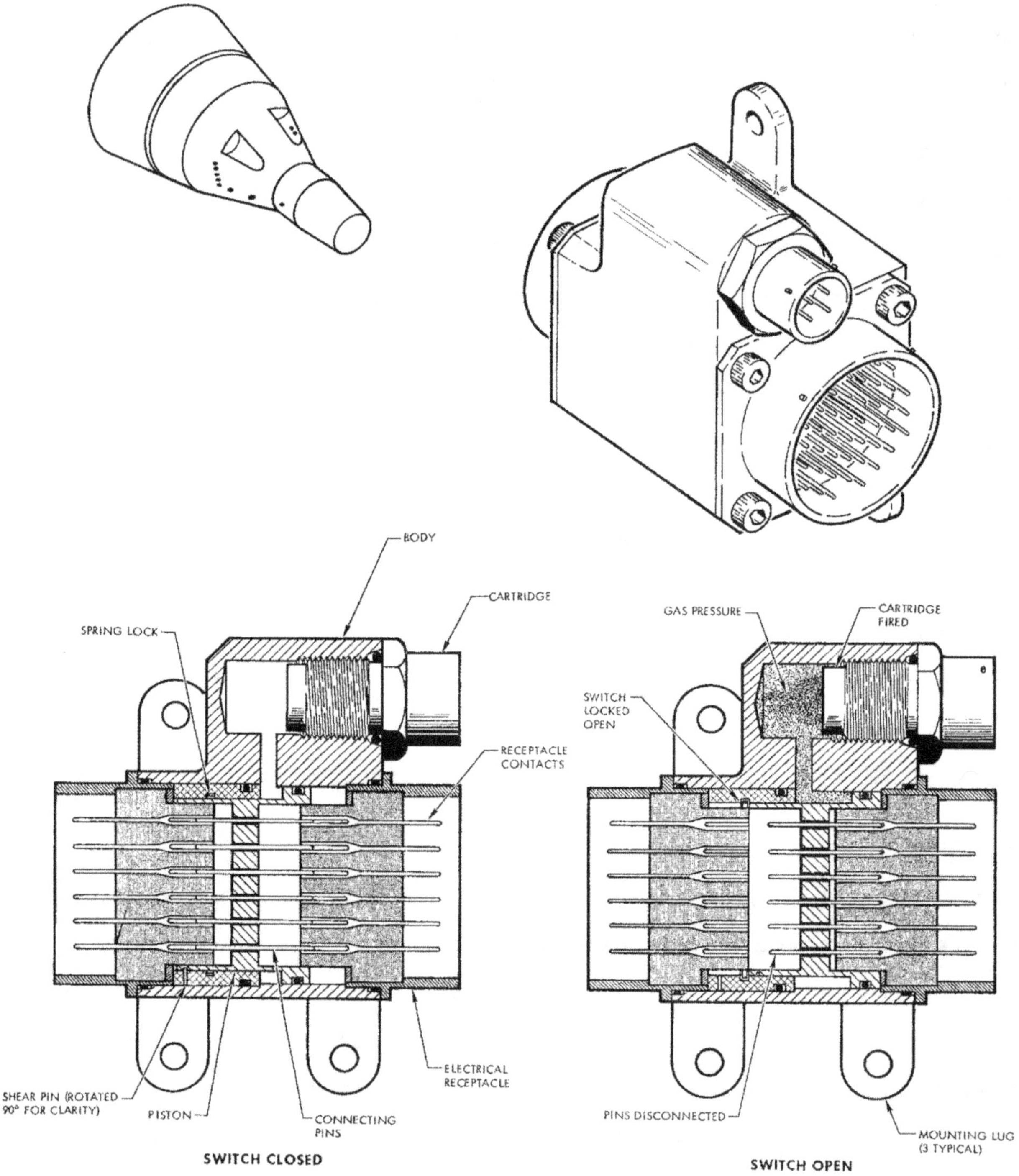

PROJECT GEMINI

Figure 11-11 Pyrotechnic Switch

SEDR 300

PROJECT GEMINI

position. The actuator is thus held in the "switch open" position.

HORIZON SCANNER FAIRING RELEASE ASSEMBLY

Description

The Horizon Scanner Fairing Release Assembly (Figure 11-12) secures the horizon scanner fairing to the spacecraft, and when initiated, jettisons the fairing. The assembly primarily consists of the actuator housing, actuator, actuator extension, main piston, release piston, eight locking pins and two electrically fired gas pressure cartridges. The actuator extension forms a positive tie between the actuator and the scanner fairing. The actuator is locked to the main piston by four locking pins. The main piston is locked in the base of the actuator housing by four locking pins, that are held in place by the release piston. The release piston is spring energized in the locked position. The actuator housing provides for installation of the cartridges and mounting for the assembly.

Operation

When initiated by the proper electrical signal, the cartridges produce gas pressure. The pressure is ported through a milled passage in the actuator housing, to the base of the piston. The gas pressure moves the release piston forward, which enables the four locking pins to cam inblard, releasing the main piston. The gas pressure causes the main piston, with attached actuator, to move through the length of the actuator housing. As the piston reaches the end of the housing, a shoulder stops the piston's travel. The four locking pins, securing the actuator extension to the piston, cam outboard into a recess and release the actuator extension. The actuator extension being thus freed is jettisoned with the scanner fairing attached.

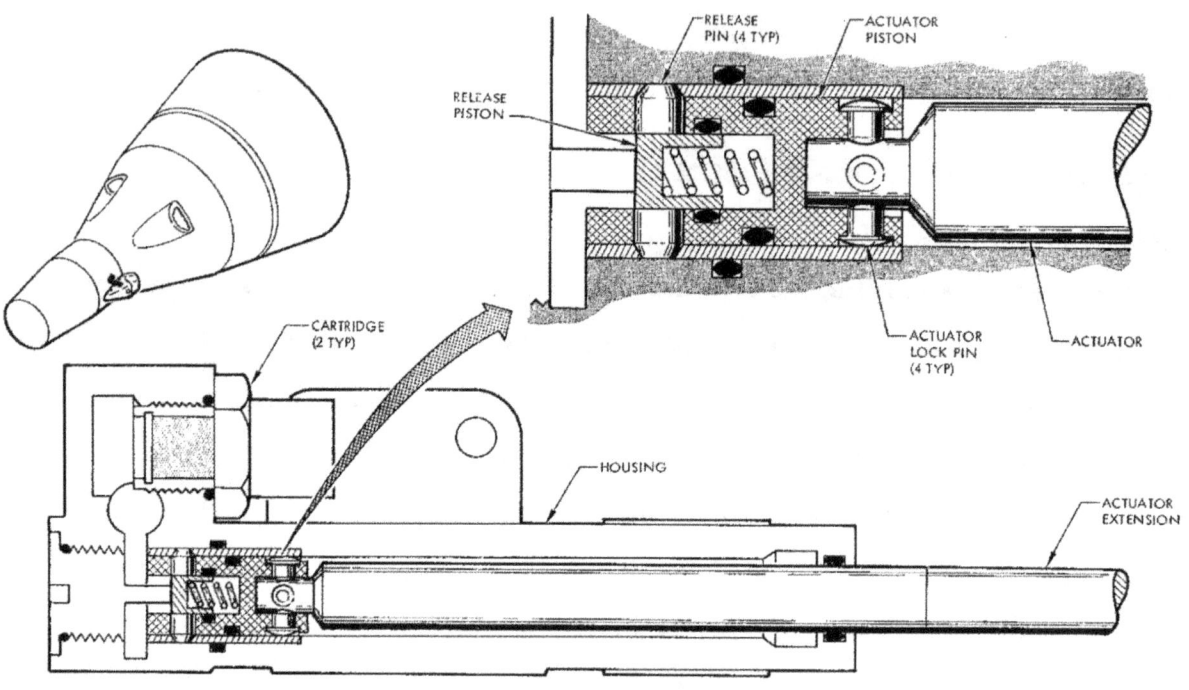
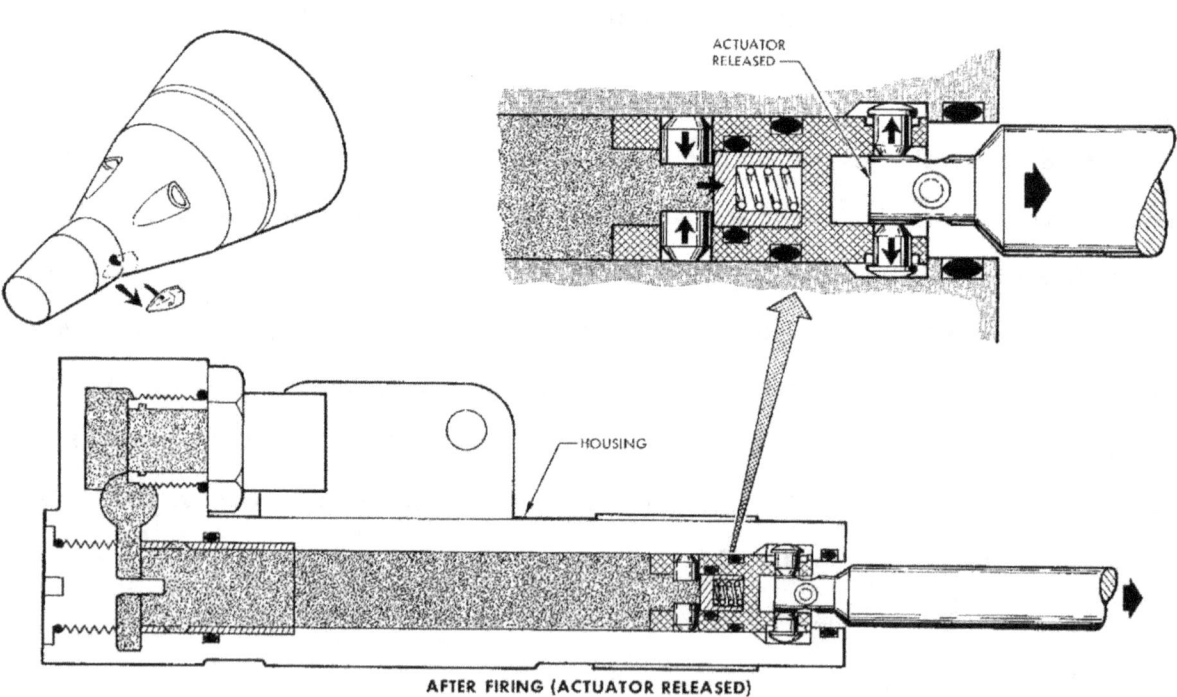

Figure 11-12 Horizon Scanner Fairing Release Assembly

HORIZON SCANNER RELEASE ASSEMBLY

Description

The Horizon Scanner Release Assembly (Figure 11-13) secures the horizon scanners to the spacecraft and jettisons the scanners when initiated. The horizon scanner release assembly primarily consists of the actuator housing, actuator, locking mechanism, cartridge housing, and two electrically fired gas pressure cartridges. The actuator is secured in the actuator housing by the locking mechanism. The locking mechanism consists of a tang lock, tang lock retainer and a shear pin. The tank lock is secured to and is located in the base of the actuator housing. The actuator housing is attached to and becomes a part of the spacecraft structure. The scanner base support and mounting platform are attached to the actuator prior to installing the cartridge housing on the actuator. The two cartridges are installed in the cartridge housing.

Operation

When initiated by the proper electrical signal, the two cartridges produce gas pressure which is ported through the hollow actuator to the base of the actuator housing. Slots in the tang lock allow the gas pressure to flow to the base of the tang lock retainer. The gas pressure exerts a force against the base of the retainer. The retainer moves axially in the actuator housing, severing the shear pin and exposing the tines of the tang lock. The tines cam open, releasing the actuator and allowing the gas pressure to jettison the actuator and horizon scanners.

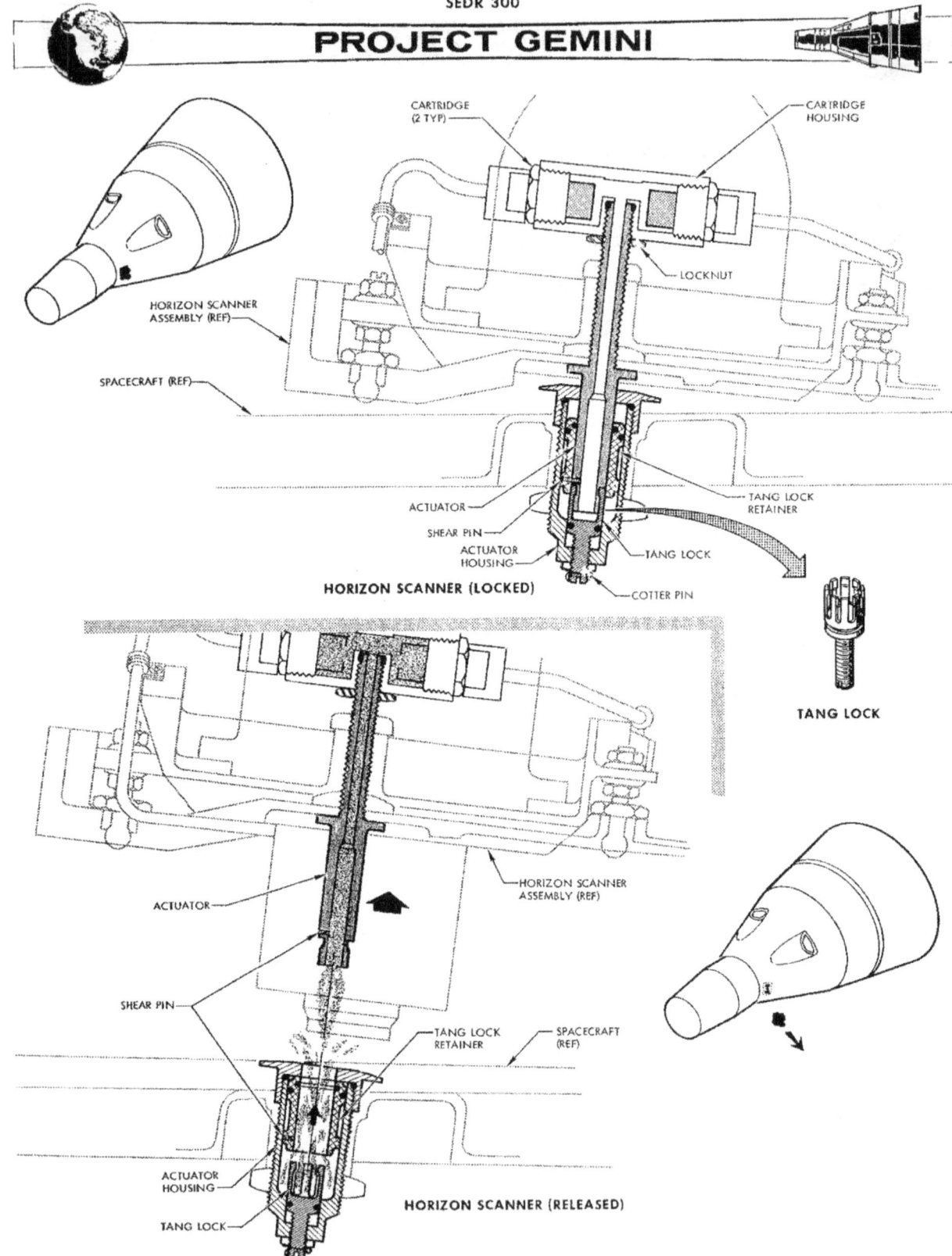

Figure 11-13 Horizon Scanner Release Assembly

SEDR 300

PROJECT GEMINI

FRESH AIR DOOR ACTUATOR

Description

The Fresh Air Door Actuator (Figure 11-14) is provided to retain the fresh air door to the spacecraft and to eject the door when initiated. The fresh air door actuator is located forward of the egress hatches, to the left of the spacecraft centerline and below the outer mold line. The actuator primarily consists of the breech, plunger, screw and two electrically fired gas pressure cartridges. The plunger forms a positive tie between the fresh air door and the breech. The plunger is retained in the breech by the screw which acts as a shear pin. The breech provides for installation of the two cartridges. Lugs, for attaching the actuator to the spacecraft structure, are an integral part of the breech.

Operation

When initiated by the proper electrical signal, the cartridges are caused to function. The cartridges generate gas pressure that exerts a force on the plunger. Where sufficient force is applied, the plunger severs the screw and is ejected out of the breech. The plunger and fresh air door are then jettisoned free of the spacecraft.

NOSE FAIRING EJECTOR

Description

The Nose Fairing Ejector (Figure 11-15) is used to secure the Rendezvous and Recovery nose fairing to the spacecraft until the proper electrical signal initiates a pyrotechnic response. When initiated by a proper electrical signal, the ejector shall positively jettison the nose fairing. The nose fairing ejector

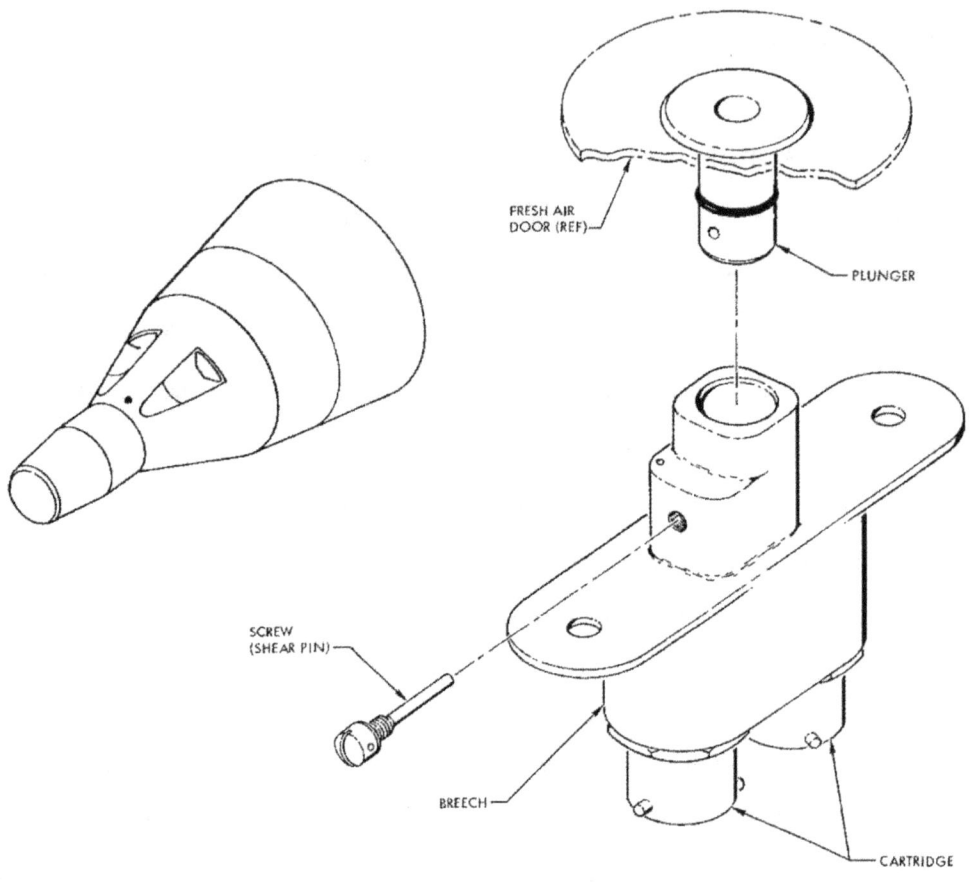

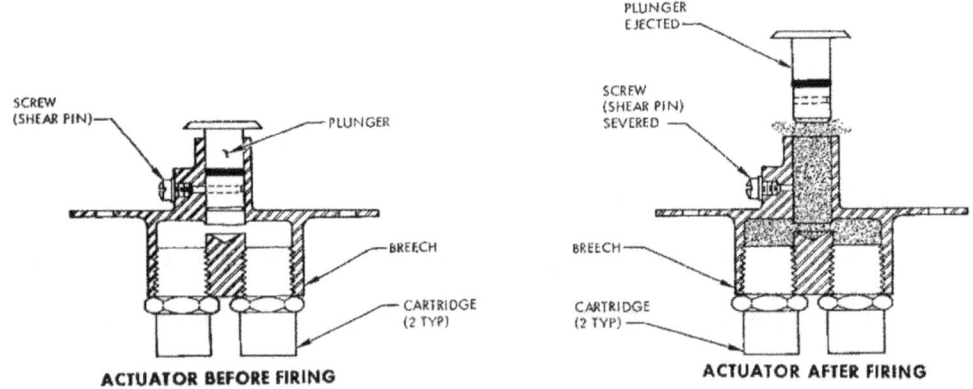

Figure 11-14 Fresh Air Door Actuator

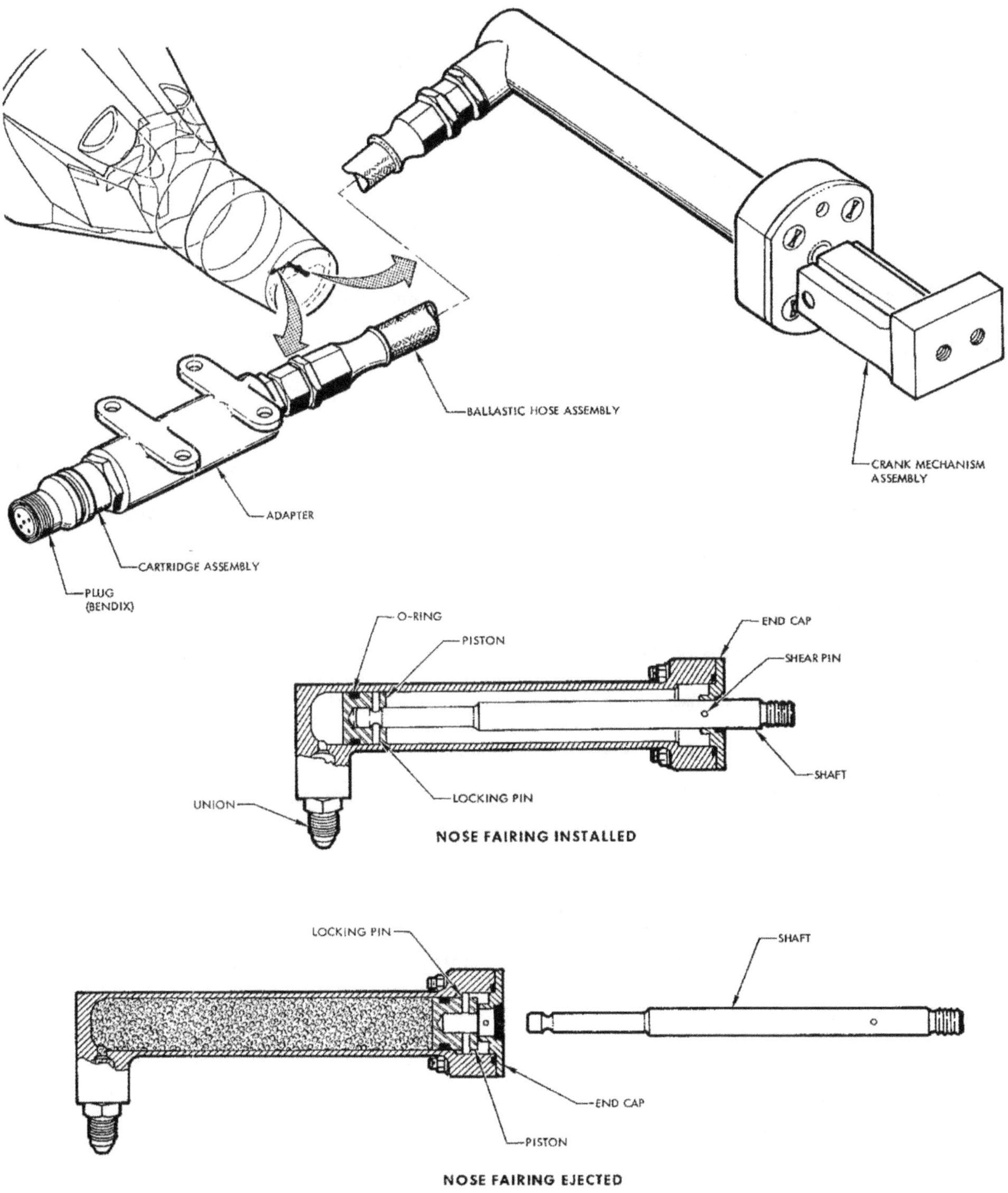

Figure 11-15 Nose Fairing Ejector Assembly

assembly consists of a cartridge, actuator assembly, breech assembly, hose assembly and a crank assembly. The cartridge is installed in the breech assembly and is positioned approximately nine inches from the actuator. The actuator is installed on the antenna support and fairing ejector fitting of the R & R section, and is located on the "X" axis, five inches up from "Y" zero. The crank assembly is installed on the nose fairing and secured to the actuator assembly.

Operation

When initiated by a proper electrical signal, the cartridge generates gas pressure that is transferred through a ballistic hose to the actuator. The pressure exerted on the piston causes the piston and retractable pin to move axially in the actuator body. As the piston approaches the opposite end of the actuator cylinder, the piston forces the retaining pins out of a detent on the shaft. The enlarged cylinder diameter allows the retaining pins to move away from the shaft and the piston continues to travel to the end of the actuator cylinder. Separation of the piston from the retractable pin allows the shaft to jettison with the fairing.

The rapid accelerating force of the shaft is transferred through the crank assembly and to the nose fairing. Inertia causes the nose fairing to continue its movement away from the R & R section. A hinge on the nose fairing, located on the outer mold line, releases and directs the path of the fairing away from the spacecraft.

EGRESS SYSTEMS AND DEVICES

The Egress Systems and Devices (Figure 11-16) provide the pilots with a rapid and positive method of escaping the spacecraft, should an emergency arise. The system is manually initiated and is used below an altitude of 70,000 feet only. Each system and device is presented in the sequence of their operation.

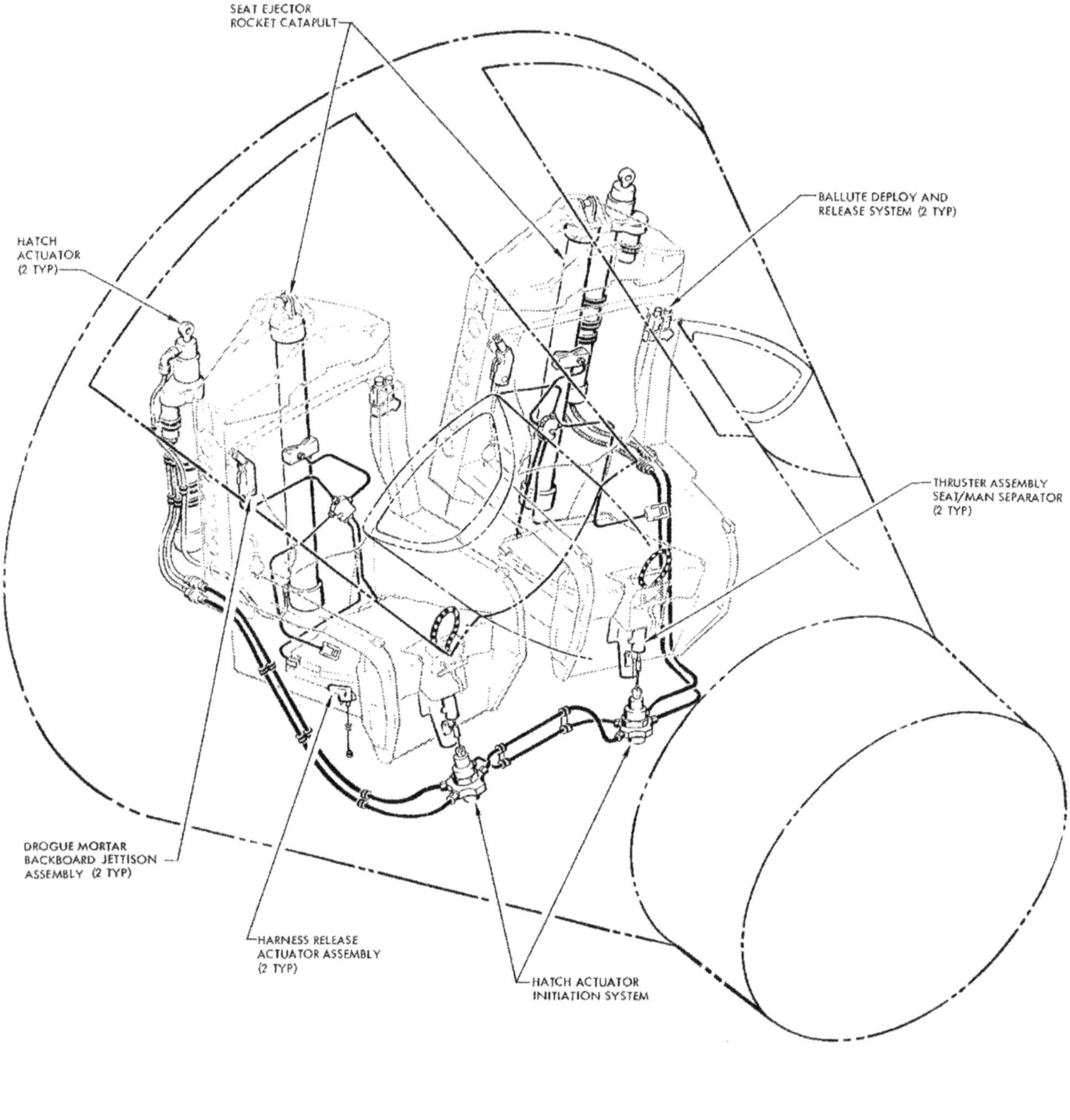

PROJECT GEMINI

SEDR 300

Figure 11-16 Egress System and Devices

PROJECT GEMINI

HATCH ACTUATOR INITIATION SYSTEM (MDF)

Description

The Hatch Actuator Initiation System (Figure 11-12) is used to initiate the firing mechanisms of both hatch actuators. The system is manually activated by either pilot. The system primarily consists of 8 MDF interconnects, two MDF crossovers and two manual firing mechanisms. The interconnects consists of four rigid and four flexible MDF assemblies that connect the firing mechanisms to the hatch actuators. The two crossovers are rigid MDF assemblies that cross connect the two initiation system firing mechanisms. The firing mechanisms each contain dual firing pins, dual percussion primers, and a booster charge. The firing mechanism is drilled and tapped for installing two MDF interconnects and two crossovers. The MDF interconnects and crossovers are installed so that the small booster on the end of each MDF is adjacent to the booster charge of the firing mechanism. The firing mechanism is attached to the spacecraft structure, located below the pilot's feet.

Operation

The hatch actuator initiation system is activated when either pilot pulls the ejection control handle ("D" ring) located between the pilot's knees and connected to the firing mechanism. Approximately one-half inch travel of the lanyard connecting the ejection control to the firing mechanism will cock and release the dual firing pins. The firing pins strike the dual percussion primers, causing the booster charge to detonate. The firing mechanism booster charge propagates detonation to the four MDF ends. The interconnecting MDF propagates the detonation wave to the firing pins of the hatch actuator breech assembly. The crossover MDF propagates the detonation wave to the other pilot's firing mechanism. This

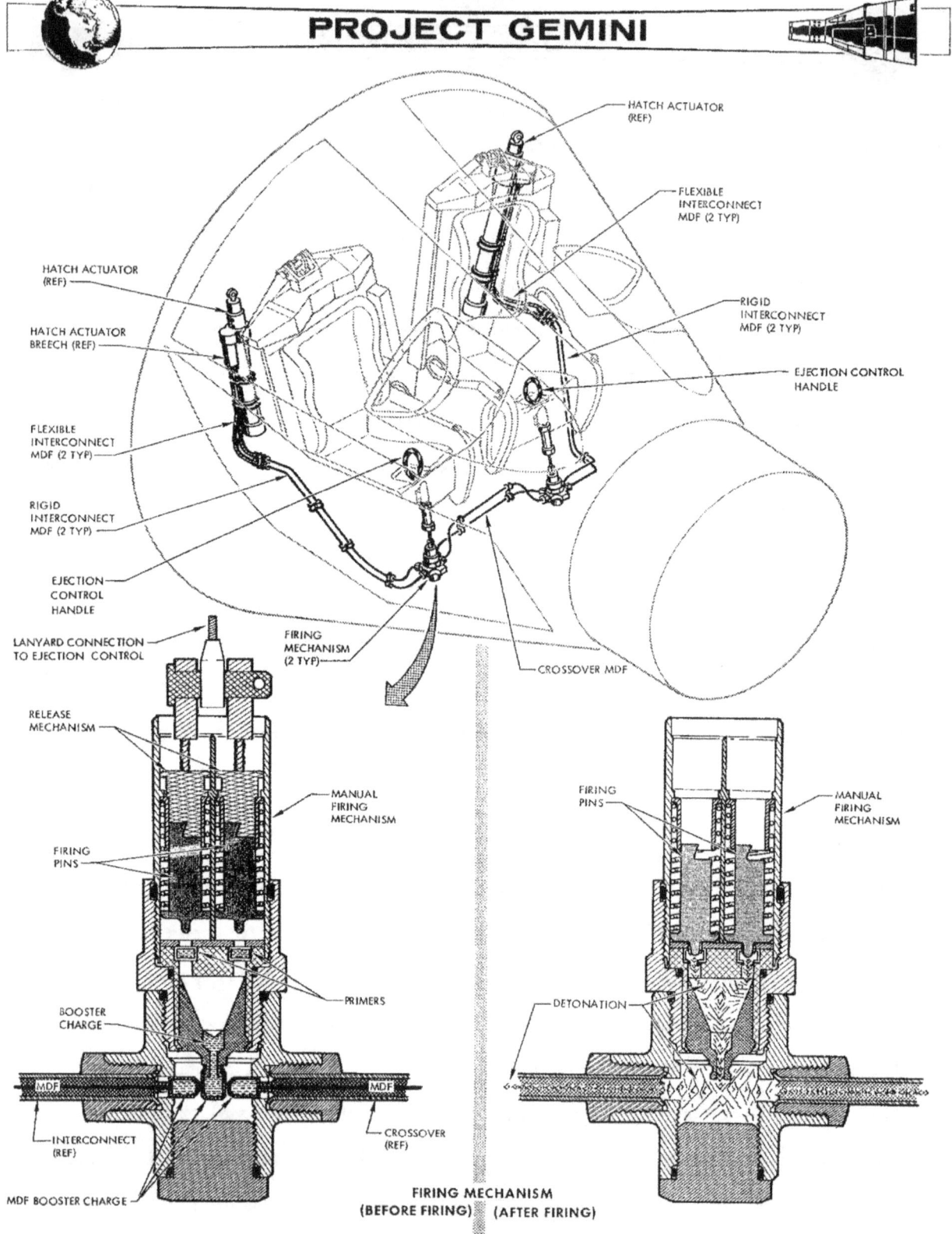

Figure 11-17 Hatch Actuator Initiation System

SEDR 300
PROJECT GEMINI

insures initiation of both hatch actuators.

HATCH ACTUATOR ASSEMBLY

Description

The Hatch Actuator Assembly (Figure 11-18) unlocks, opens and mechanically restrains the egress hatch in the open position. The assembly also furnishes sufficient pressure to initiate the firing mechanism of the seat ejector rocket catapult. The assembly primarily consists of the Breech End Cap, Breech, Cylinder, Stretcher Assembly, End Cap (Base) and Rod End Assembly. The breech end cap assembly contains the locking mechanism for mechanically restraining the egress hatch in the open position; provides for installation of the seat ejector rocket catapult ballistic hose; provides for installation of the breech assembly, and is thread mounted to the top of the cylinder. The breech contains two firing pins, two percussion fired cartridges, and a gas producing propellant charge. Two interconnects, from the hatch actuator initiation system, are attached to the breech adjacent to the firing pins. The stretcher assembly primarily consists of the piston and stretch link, and is located inside the cylinder. One end of the stretch link is attached to a web inside the piston. The other end is attached to the rod end assembly. The rod end assembly connects the stretcher assembly to the egress hatch. The end cap is attached to the lower end of the cylinder, and provides for attaching the hatch actuator assembly to the spacecraft structure. The end cap contains a latch piston that actuates the egress hatch unock mechanism.

Operation

The hatch actuator functions when initiated by the initiation system MDF interconnects. The shock wave, propagated by the MDF interconnects, causes the two

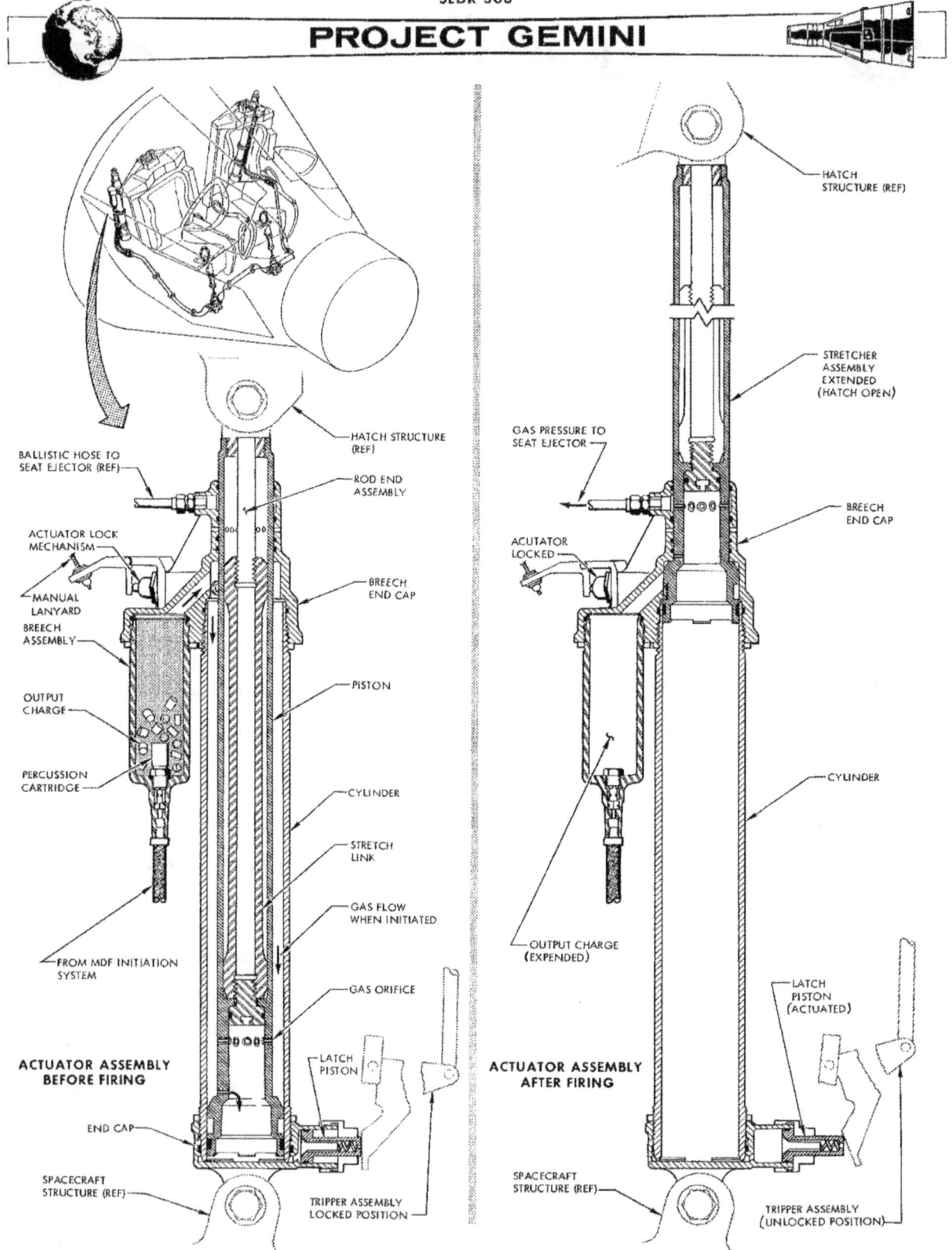

Figure 11-18 Hatch Actuator Assembly

firing pins of the breech assembly to sever shear pins and strike the primers of the two percussion fired cartridges. The cartridges ignite and generate hot gas which ignites the main propellant charge of the breech. The propellant charge produces a large volume of high pressure gas. The gas pressure is exhausted into the area between the piston of the stretcher assembly and the cylinder. Orifices in the lower end of the piston wall admit the gas pressure to the base of the stretcher assembly. The gas pressure is ported through a drilled passage to the latch piston. The gas pressure extends the latch piston, which unlocks the egress hatch through a bellcrank/pushrod mechanism. The gas pressure then acts on the base of the stretcher assembly, moving it through the length of the cylinder. Immediately prior to the stretcher assembly reaching full extension, gas pressure is exhausted through a port to the ballistic hose. The ballistic hose delivers the pressure to the firing mechanism of the seat ejector - rocket catapult. As the stretcher assembly reaches the fully extended position, the lock pin of the locking mechanism engages the piston of the stretcher assembly and holds the hatch open. The locking mechanism is also operative when the hatch is fully opened by hand. A lanyard, attached to the locking mechanism, permits the hatch to be unlocked, when manually actuated.

SEAT EJECTOR-ROCKET CATAPULT

Description

The Seat Ejector-Rocket Catapult (Figure 11-19) is used to eject the man-seat mass from the spacecraft. The seat ejector-rocket catapult basically consists of the catapult assembly and the rocket motor assembly. The catapult assembly primarily consists of the catapult housing, firing mechanism, main charge (gas producer), and locking assembly. The catapult housing contains all of the listed components

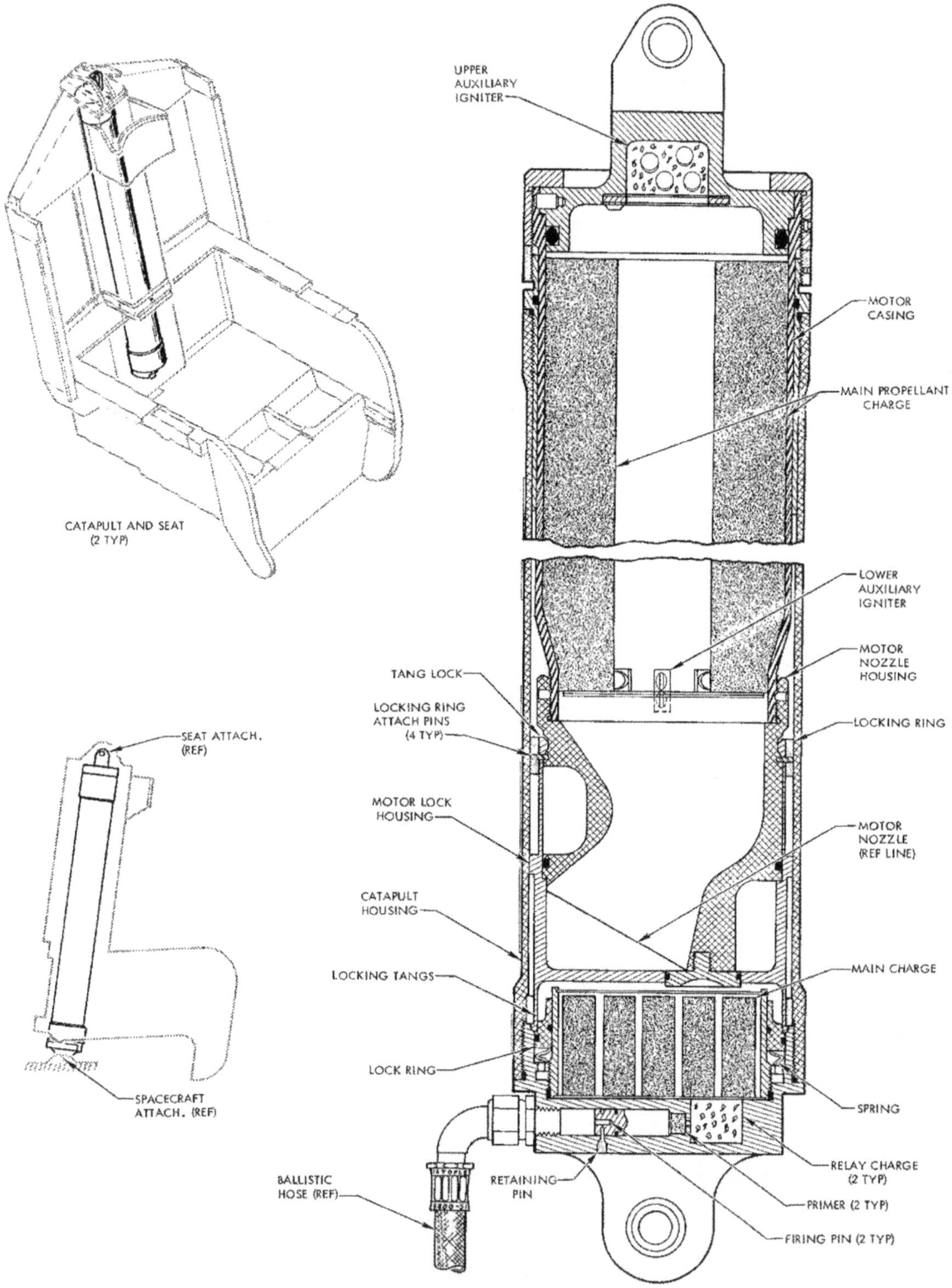

SEDR 300

PROJECT GEMINI

Figure 11-19 Seat Ejector - Rocket Catapult

SEDR 300
PROJECT GEMINI

in its base. The firing mechanism consists of dual firing pins, dual percussion fired primers, and a relay charge. The firing pins are secured in place by retaining pins. The locking assembly consists of the lock ring and a spring to hold the ring in place. The base of the catapult assembly is attached to the spacecraft structure. The rocket motor assembly primarily consists of the motor case, nozzle, motor lock housing, lock ring, shear pins, upper and lower auxiliary igniters, and the main propellant charge. The nozzle is threaded to the motor case and is secured by four set screws. The nozzle is secured to the motor lock housing by locking tangs. The locking tangs are held in place by a lock ring that is retained by four shear pins. The motor lock housing is secured in the base of the catapult by tang locks. The tangs are held in place by the lock ring of the catapult. The main propellant charge is located in the motor case with an auxiliary igniter at each end of the charge. The top end of the rocket motor assembly is attached to the seat.

Operation

The seat ejection cycle is initiated when gas pressure is received via the ballistic hose from the hatch actuator. Sufficient gas pressure will cause the dual firing pins to shear their retaining pins and strike the dual percussion primers. The primers ignite the relay and main charges. Hot gas pressure, produced by the main charge, releases the motor lock housing by displacing the lock ring against the spring through piston action. With the motor lock housing released, the gas pressure propels the rocket motor through the length of the catapult housing. Prior to complete ejection from the catapult housing, the lock ring of the motor lock housing makes contact with a stop which severs its four shear pins. The tang locks of the motor lock housing cam open and release the rocket motor. Separation of the rocket motor from the motor lock housing allows the hot gas from the catapult main charge to enter the rocket motor nozzle opening. The hot gas ignites the auxiliary igniters and the main propellant charge. The rocket motor is thus free to perform its basic function of obtaining optimum trajectory of the man-seat for spacecraft clearance.

HARNESS RELEASE ACTUATOR ASSEMBLY

Description

The Harness Release Actuator Assembly (Figure 11-20) which is installed on the seat structure, is provided to actuate the restraint harness release mechanism and to initiate the firing mechanism of the thruster assembly. The actuator assembly primarily consists of the actuator housing, firing mechanism, percussion fired gas pressure cartridge and the unlatch rod. The unlatch rod is installed in the actuator housing with an external spring preloading the rod in the latched position. The mated cartridge and firing mechanism is installed in the base of the actuator assembly. A lanyard attaches the firing mechanism to the spacecraft structure. A fitting is provided in the actuator housing for attaching the thruster ballistic hose.

Operation

The firing mechanism is initiated by lanyard pull when the seat rises on the ejection rails. The firing pin is cocked and released to strike the cartridge. The cartridge incorporates a time delay, to allow the seat adequate time to clear the spacecraft before actuating the harness release. The gas pressure, generated by the cartridge, is routed through passages in the actuator housing to the unlatch rod. The unlatch rod actuates the mechanical linkage to release the restraint harness. As the unlatch rod approaches the end of its travel, a port is exposed that allows gas pressure to enter the ballistic hose. The gas pressure travels through the

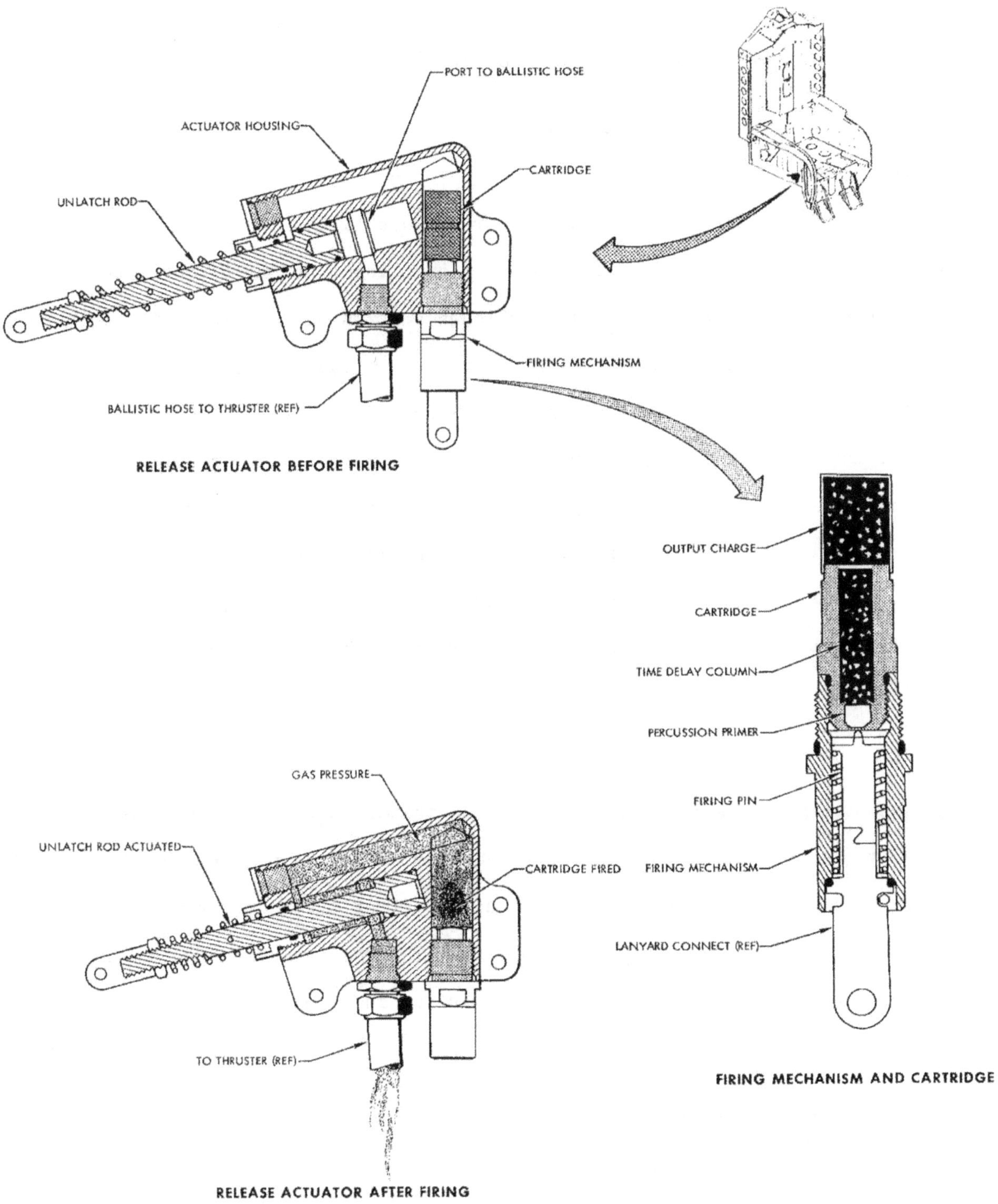

Figure 11-20 Harness Release Actuator Assembly

ballistic hose to initiate the thruster assembly.

THRUSTER ASSEMBLY - SEAT/MAN SEPARATOR

Description

The Thruster Assembly - Seat/Man Separator (Figure 11-21) is only a part of the seat/man separation assembly. The thruster supplies a stroke of adequate length and power to a webbed strap that accomplishes seat/man separation. The thruster assembly primarily consists of the Thruster Body, Thruster Piston, Firing Mechanism and Percussion Fired Gas Pressure Cartridge. The cartridge and firing mechanism is installed in the upper end of the thruster body. The firing mechanism contains a firing pin, retained by a shear pin. The ballistic hose from the harness release actuator is attached to the firing mechanism. The thruster piston is located in the thruster body and is retained in the retracted position by a shear pin. The thruster body is mounted on the front of the seat structure, between the pilot's feet.

Operation

High pressure gas from the harness release actuator is transmitted through the ballistic hose to the thruster firing mechanism. The gas pressure causes the firing pin to sever its shear pin and strike the primer of the cartridge. The cartridge is ignited and generates gas pressure. The gas pressure exerts force on the thruster piston, causing the piston to sever its shear pin. As the piston extends out of the thruster body, the strap is pulled taut effecting seat/man separation.

BALLUTE DEPLOY AND RELEASE SYSTEM

Description

The Ballute Deploy and Release System (Figure 11-22) primarily consists of the

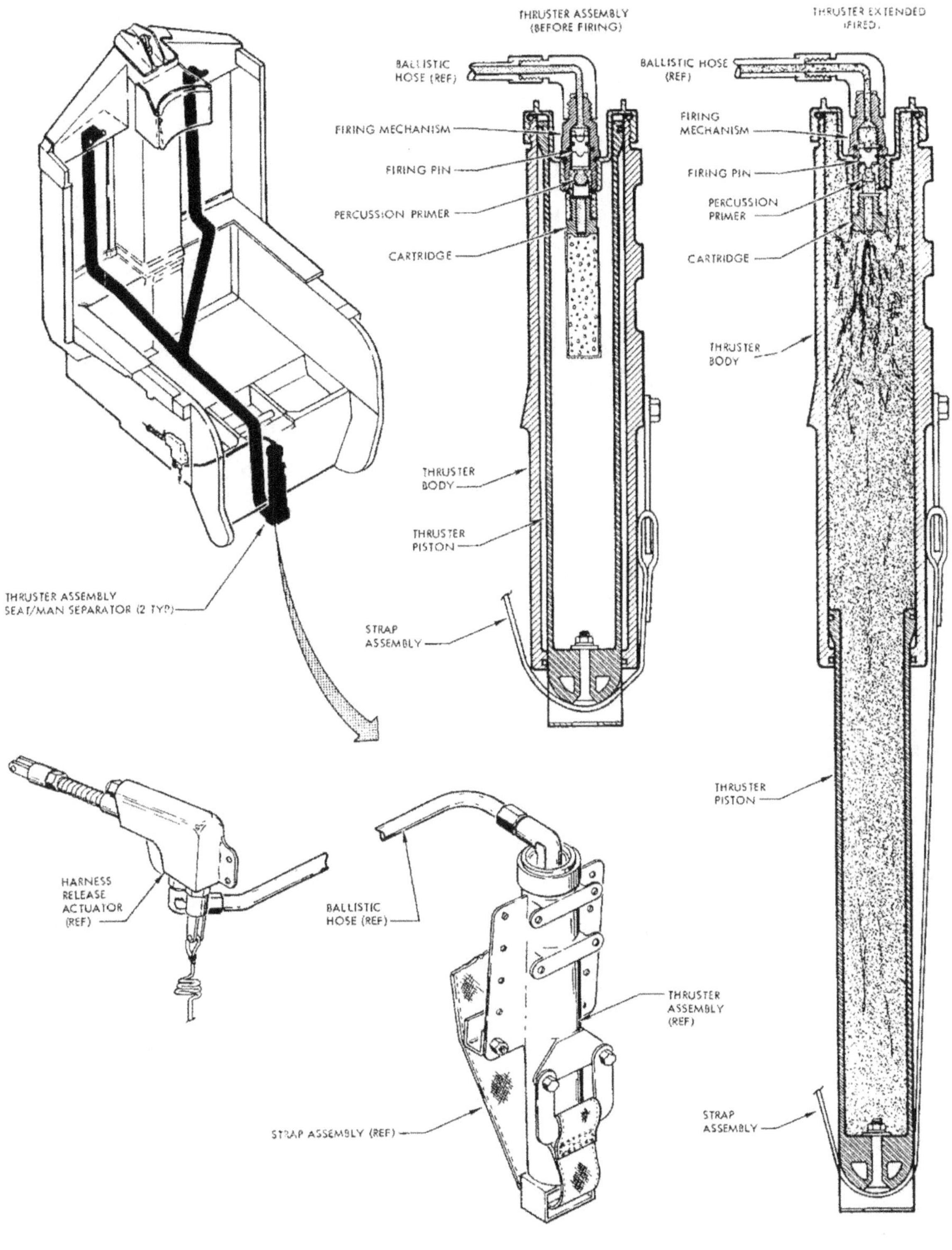

Figure 11-21 Thruster Assembly-Seat/Man Separator

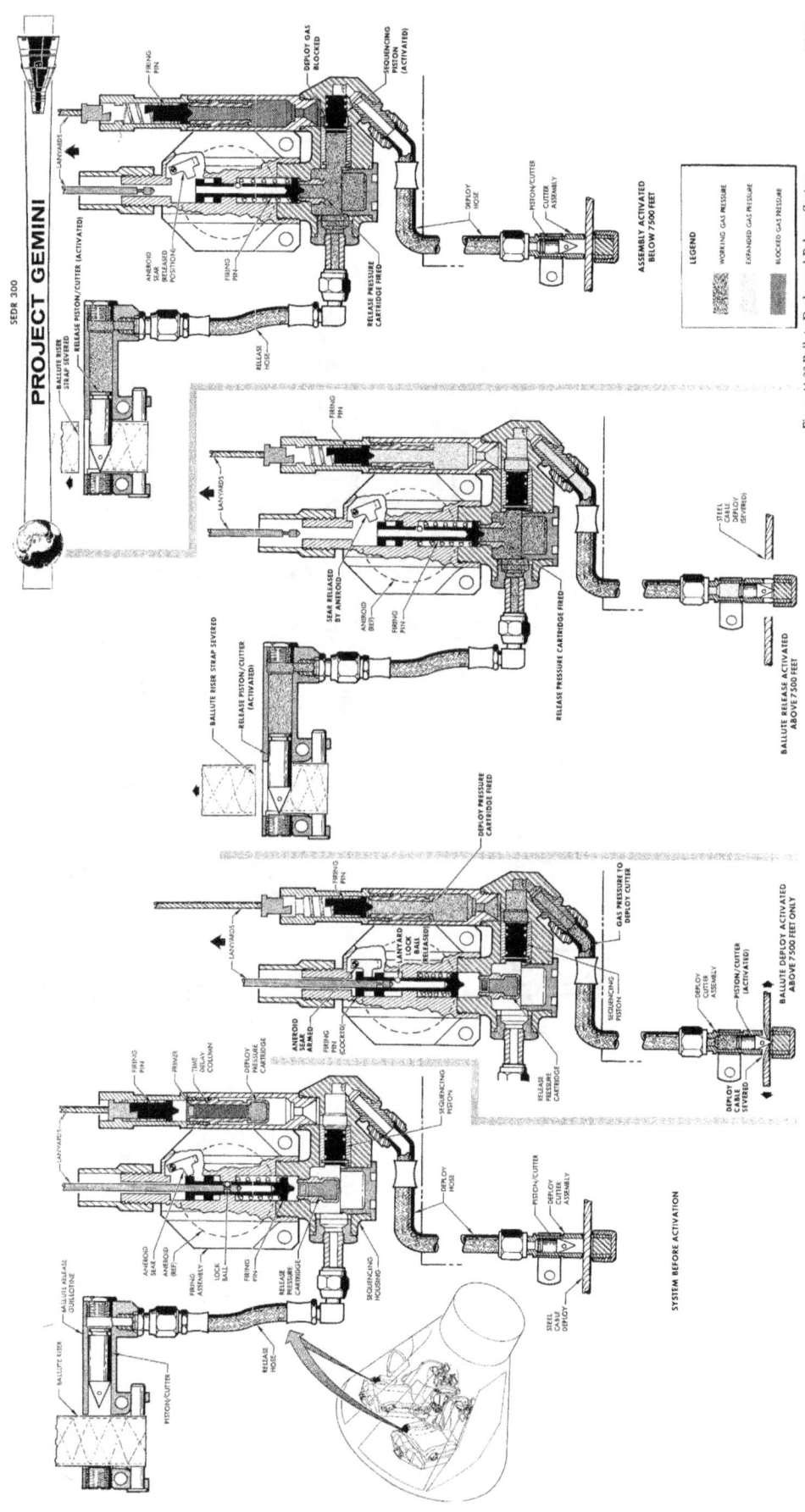

Figure 11-22 Ballute Deploy and Release System

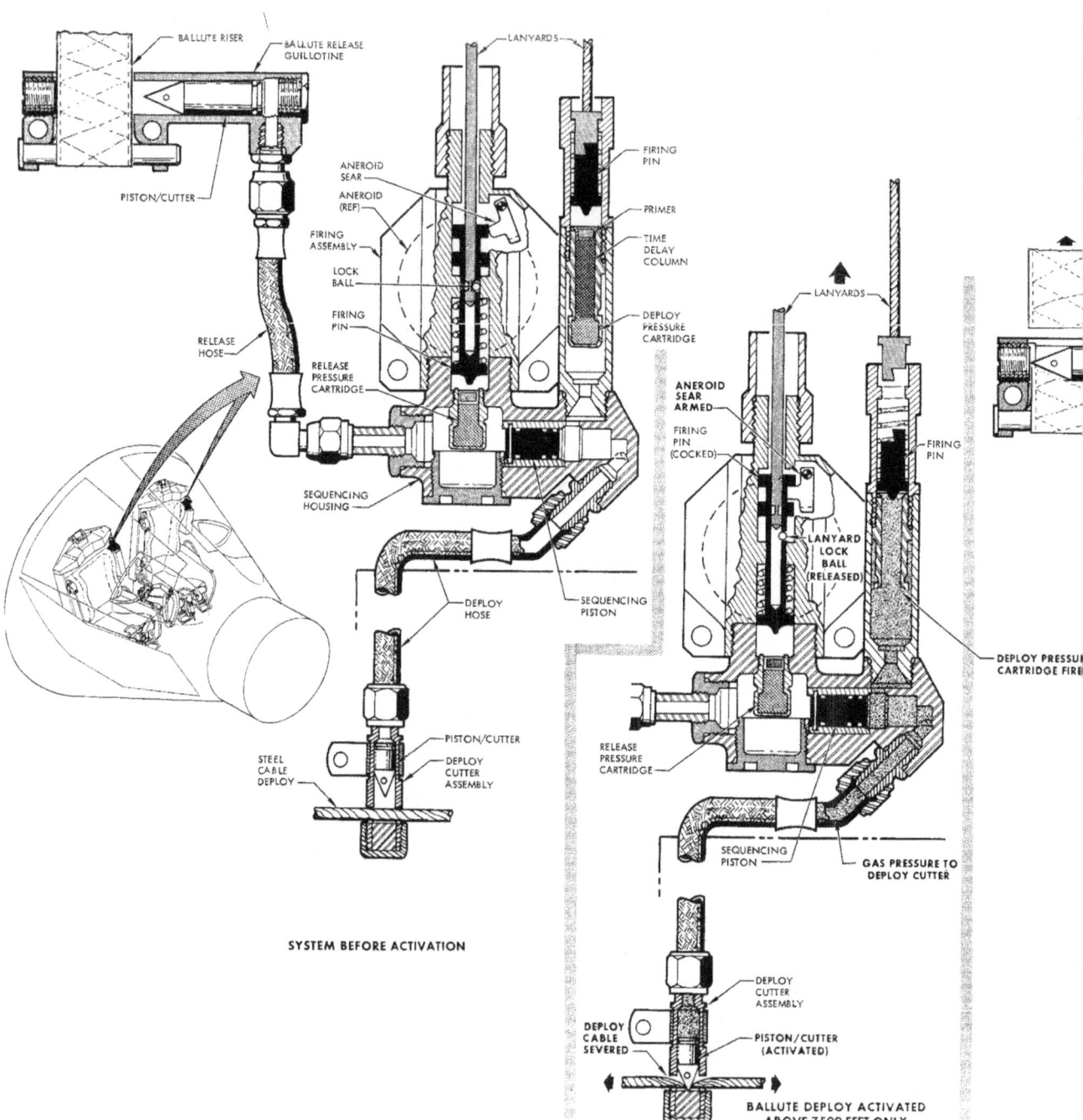

SYSTEM BEFORE ACTIVATION

BALLUTE DEPLOY ACTIVATED ABOVE 7500 FEET ONLY

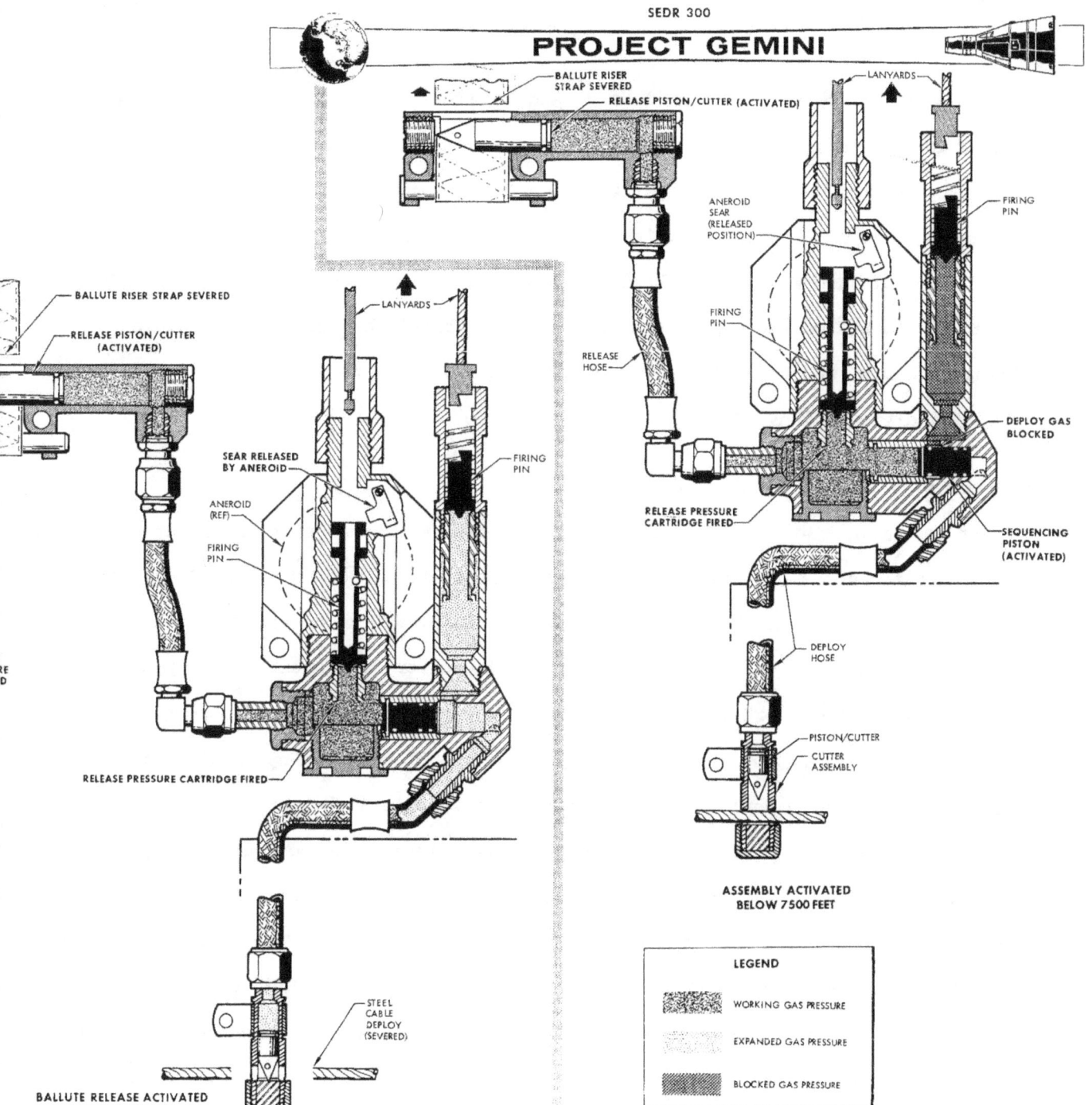

Figure 11-22 Ballute Deploy and Release System

Firing Assembly, Deploy Cutter and Hose, and Release Guillotine and Hose. Contained within the firing assembly, is the Release Aneroid Firing Mechanism and Cartridge, the Deploy Firing Mechanism and Cartridge, and the Sequencing Housing and Piston. The basic function of the system is to deploy and release the ballute between specified altitudes and prevent ballute deployment below specified altitudes. The system is located on the upper left side of each pilot's backboard. The deploy firing mechanism and the release aneroid firing mechanism is linked to the pilot's seat by individual lanyards.

Operation

The system is initiated by the lanyard pull as seat/man separation is effected. When initiated above 7500 feet, the release aneroid is armed and the deploy firing mechanism is activated. The firing pin of the deploy firing mechanism strikes the primer of the cartridge and causes ignition. The cartridge generates gas pressure after burning through the time delay column. The pressure is ported through the deploy hose to the deploy cutter assembly. The cutter severs a nylon strap that allows the ballute to deploy. The armed aneroid functions when an altitude pressure level of 7500 feet is reached. The aneroid sear releases the cocked firing pin of the ballute release firing mechanism. The firing pin strikes the primer, which ignites the cartridge and causes it to generate gas pressure. The pressure is ported through the release hose to the release guillotine. The guillotine severs the ballute riser strap and allows the ballute to be carried away. When the system is activated by the lanyard pull below 7,500 feet, both cartridges are immediately initiated. The time delay incorporated in the deploy cartridge permits the release cartridge to generate gas pressure first. The pressure is ported through the release hose to the release guillotine, which severs the ballute riser strap. Simultaneously gas pressure is ported to the sequencing housing and sequencing piston. The piston is actuated, causing it to block the gas exit of the deploy cartridge. The gas pressure, generated by the deploy cartridge, does not reach the deploy cutter, preventing deployment of the ballute.

DROGUE MORTAR - BACKBOARD JETTISON ASSEMBLY

The Drogue Mortar - Backboard Jettison Assembly is provided to deploy the personnel drogue parachute and to separate the backboard and egress kit from the pilot.

Description

Drogue Mortar

The Drogue Mortar (Figure 11-23) functions to eject a weighted slug with sufficient velocity to forcible deploy the personnel parachute and to initiate the backboard jettison assembly firing mechanism. The drogue mortar primarily consists of the mortar body, mortar barrel, drogue slug, main cartridge (gas pressure), initiator cartridge (detonator), aneroid assembly, main lanyard, manual lanyard, and the main and manual firing mechanisms. The mortar barrel is threaded into the mortar body and contains the drogue slug. The drogue slug is retained in the barrel by a shear pin. The aneroid assembly is attached to the mortar body and contains the main firing mechanism. The main lanyard in enclosed in a rigid housing to prevent inadvertant pulling of the lanyard. The housing is attached to the main firing mechanism housing at one end and to a take-up reel at the other. The main lanyard, a fixed length of cable, is attached to the main firing mechanism at one end and to the take-up reel at the other. The take-up reel incorporates an extendable cable that is attached to the ejection seat. The main cartridge is threaded into the mortar body, with the primer end, adjacent to the main firing mechanism, and the output end

11-50

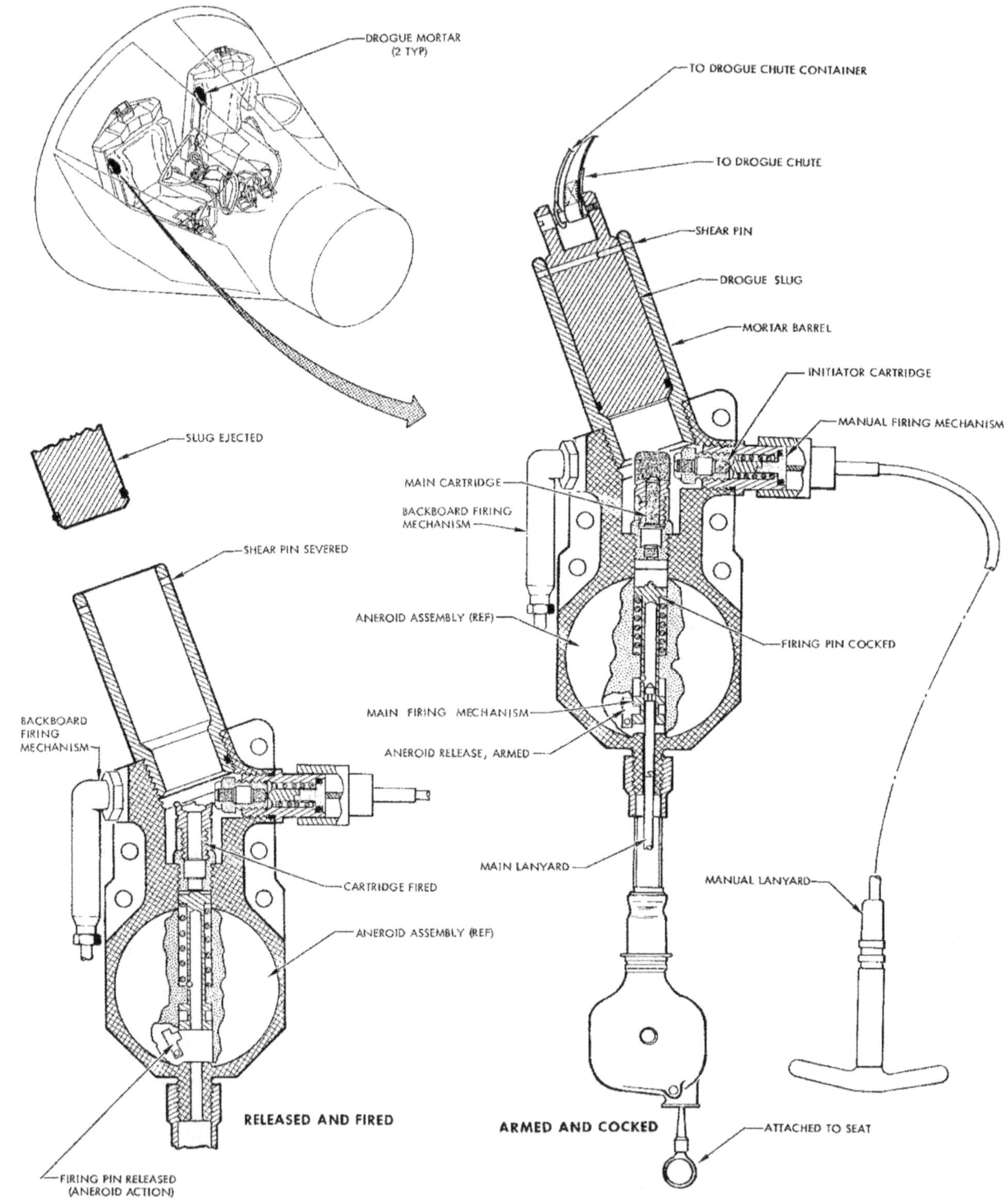

Figure 11-23 Drogue Mortar

in the mortar body pressure cavity. The manual lanyard is enclosed in a flexible conduit to prevent inadvertant pulling of the lanyard. The lanyard is attached to the manual firing mechanism at one end and to a manual pull handle at the other. The manual firing mechanism is threaded into the nortar body. The primer end of the (detonator) is threaded into the manual firing mechanism, with its output end 90° and adjacent to the main cartridge output area. The drogue mortar is attached to the upper right side of each pilot's backboard.

Backboard Jettison Assembly

The Backboard Jettison Assembly (Figure 11-24), functions to separate the backboard and egress kit from the pilot, when initiated by the pressure from the drogue mortar. The backboard jettison assembly primarily consists of the MDF firing mechanism, MDF time delay cartridge (DETONATOR), interconnect (time delay MDF), MDF manifold assembly, jetelox release pin, interconnect (jetelox pin MDF), lap belt disconnect, interconnect (belt disconnect MDF), restraint strap cutter (FLSC), and interconnect (strap cutter MDF). The MDF firing mechanism is attached to the drogue mortar body and contains a shear pin retained firing pin. The MDF time delay cartridge is a percussion fired cartridge and is installed in the MDF firing mechanism. The interconnect (time delay MDF) is connected to the MDF firing mechanism and the MDF manifold. The interconnect (jetelox pin MDF) is connected to the MDF manifold and the jetelox release pin. The interconnect (belt disconnect MDF) is connected to the MDF manifold and the lap belt disconnect. The interconnect (strap cutter MDF) is connected to the MDF manifold and the restraint strap cutter (FLSC). The three component interconnects terminate in the MDF manifold with their accepter end adjacent to the interconnect (time delay MDF) donor end. The jetelox release pin retains the jetelox joint to the egress kit until initiated.

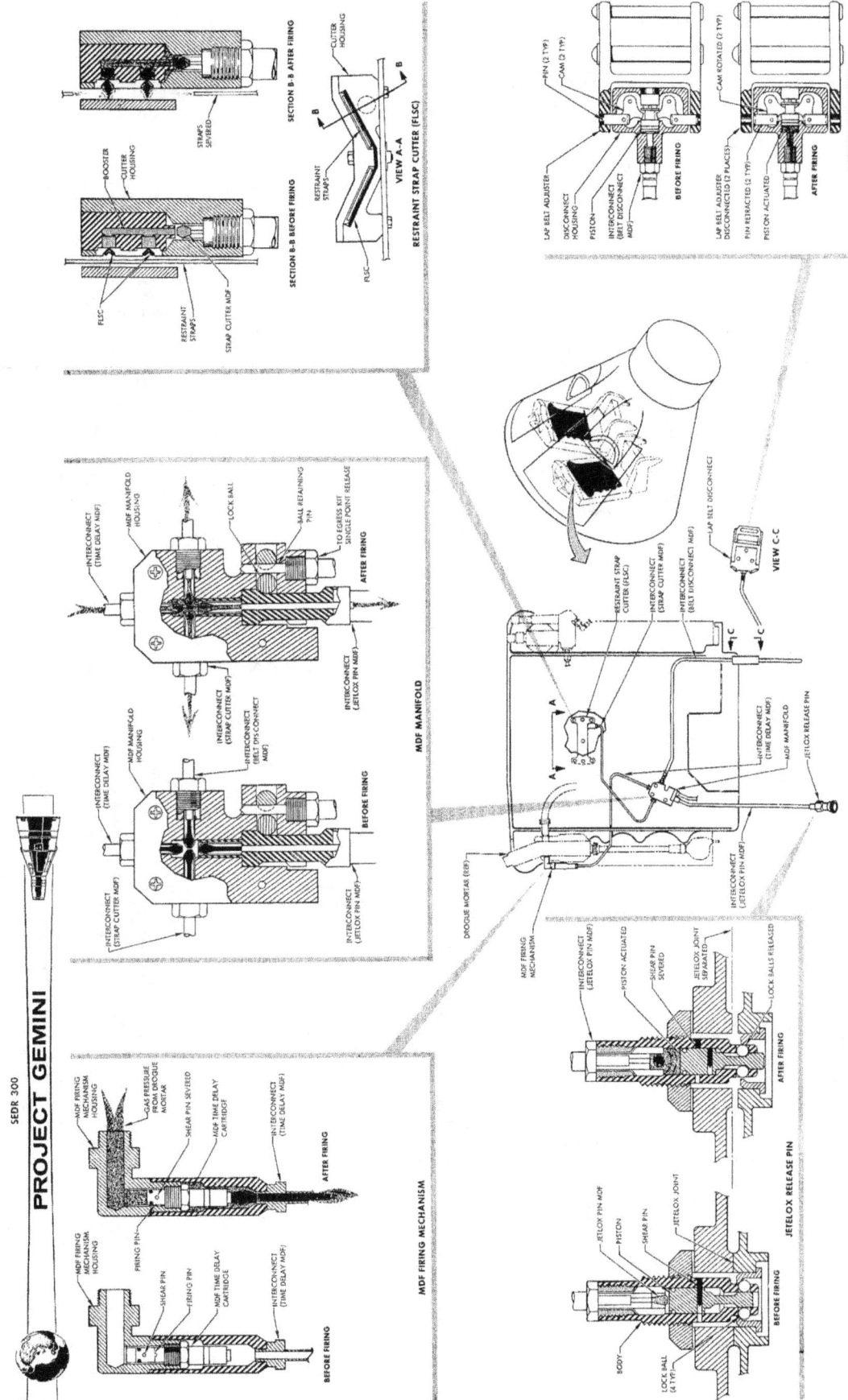

Figure 11-24 Backboard Jettison Assembly

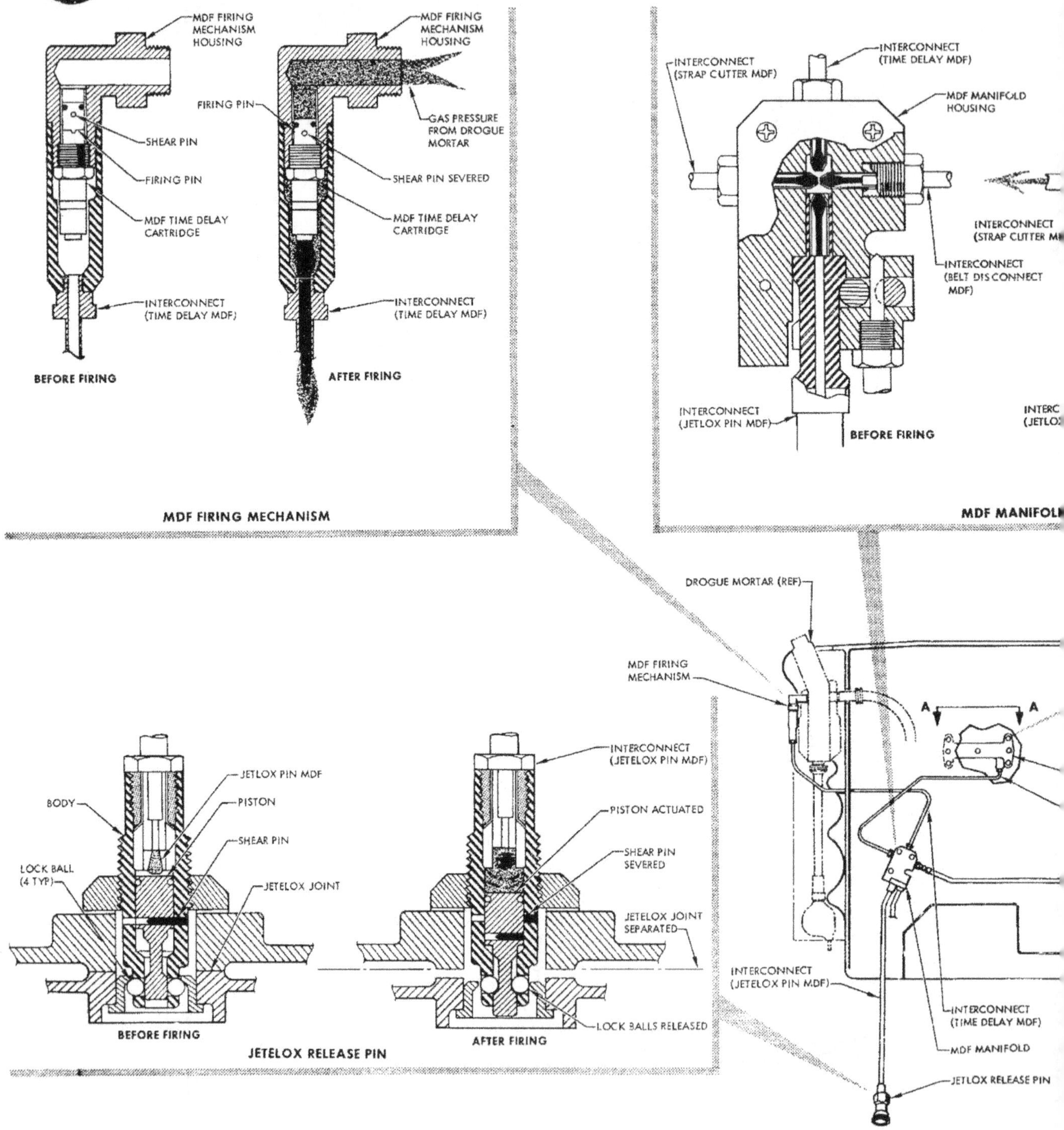

Figure 11-24 Backboard Jettison Assembly

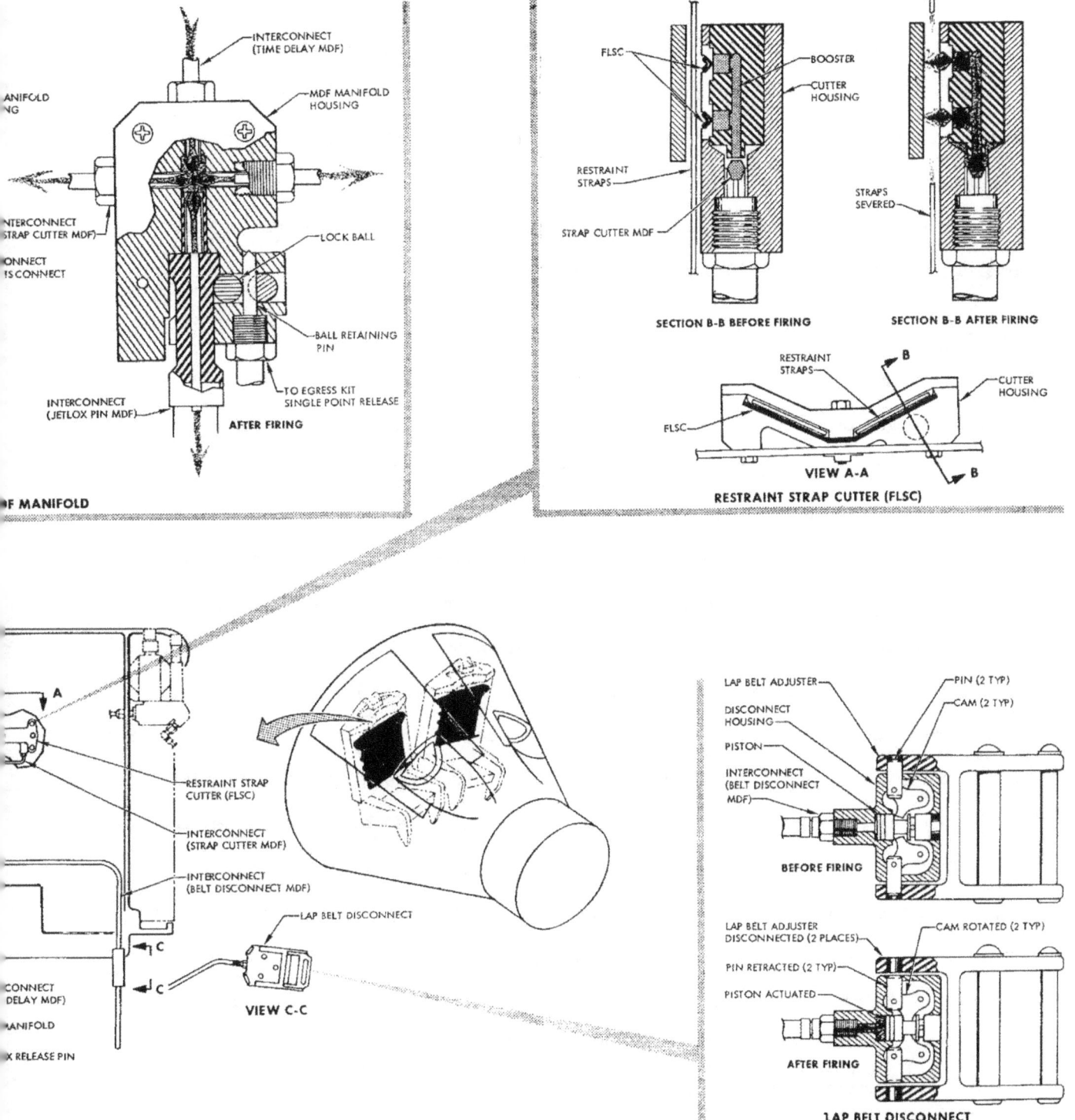

11-53

SEDR 300

PROJECT GEMINI

The jetelox release pin primarily consists of the body, piston, four lock balls, and a shear pin. The lap belt disconnect is provided to unfasten the lap belt when properly initiated. The lap belt disconnect primarily consists of the housing, two lock pins, two cams, piston and a shear pin. The restraint strap cutter is provided to sever the pilot's shoulder harness. The cutter primarily consists of the housing, two strips of FLSC and a booster.

Operation

Drogue Mortar

The drogue mortar is initiated by the pull of the main lanyard, at seat/man separation. The extendable cable, attached to the seat, uncoils from the take-up reel. Upon reaching the end of its travel, the cable pulls the take-up reel free of the rigid housing. The fixed length main lanyard attached to the reel is pulled, and if in excess of 5700 feet, cocks the main firing mechanism and arms the aneroid. At an altitude pressure level of 5700 feet, the aneroid releases the cocked main firing pin. The firing pin strikes the primer and ignites the main cartridge, which produces gas pressure. The gas pressure causes the dorgue slug to sever its shear pin and travel out of the mortar barrel. Simultaneoulsy, the gas pressure initiates the backboard firing mechanism. When initiated by the main lanyard below 5600 feet, the main firing mechanism is cocked and immediately released to fire the main cartridge. The aneroid is in the release position because of the altitude pressure level, therefore is not armed and does not delay the cartridge firing. The drogue mortar may be initiated manually by pulling the manual lanyard handle at any altitude. The lanyard cocks and releases the manual firing pin, which strikes the primer of the initiator cartridge (detonator). The initiator cartridge detonates and ignites the output charge of the main cartridge, which produces the gas pressure for drogue slug ejection and backboard firing mechanism initiation.

Backboard Jettison Assembly

The Backboard Jettison Assembly is caused to function when the main cartridge of the drogue mortar is fired. Gas pressure from the drogue mortar main cartridge, causes the firing pin of the backboard firing mechanism, to sever its shear pin and strike the primer of the time delay cartridge. After the proper time delay, the cartridge propagates a detonation wave to the MDF interconnect, which transmits the wave simultaneously, to the three MDF interconnects attached to the MDF manifold assembly. Simultaneously, the detonation wave is propagated by the three MDF interconnects to the restraint strap cutter (FLSC), lap belt disconnect, and the jetelox release pin. The detonation wave propagated by the interconnect (jetelox pin MDF) acts upon the piston of the jetelox pin, causing it to sever the shear pin. As the piston moves, a recess in the piston is aligned with the lock balls. The pressure exerted by the jetelox joint, forces the lock balls into the piston recess, and releases the jetelox joint and egress kit. The detonation wave, propagated by the interconnect (belt disconnect MDF), is directed against the piston of the lap belt disconnect. The detonation wave moves the piston causing it to sever the shear pin. As the piston moves, the cams rotate and retract the pins from the lap belt adjuster. The lap belt separates and permits the pilot to be partially free of the backboard. The detonation wave, propagated by the interconnect (strap cutter MDF), is transmitted to the booster of the restraint strap cutter (FLSC). The booster strengthens and increases the reliability of the detonation wave for proper detonation of the two strips of FLSC. The FLSC detonates and severs the two restraint straps allowing the pilot to be completely free of the backboard. The seat may be released manually by the pilot actuating the seat single point release handle. Effective spacecraft 5 and 6 a cable from the single point release, pulls a ball retaining pin from the MDF manifold. The pressure of the interconnect (jetelox pin MDF), moves the ball aside and pulls out of the MDF manifold.

PARACHUTE LANDING SYSTEM PYROTECHNICS

The Parachute Landing System (Figure 11-25) is provided to safely recover and land the re-entry module, after its entry into the earths atmosphere. The pyrotechnic portion of the system consists of the drogue, pilot, and main parachute reefing cutters; the drogue and pilot parachute mortars; the drogue parachute bridle release guillotines; the pilot parachute apex line guillotine; and the main parachute disconnects. Each of these pyrotechnic devices are presented in the following paragraphs.

DROGUE PARACHUTE MORTAR ASSEMBLY

The drogue parachute mortar assembly is provided to positively deploy the drogue chute. The assembly is similiar to the pilot parachute mortar assembly (Figure 11-25) in design and operation.

DROGUE PARACHUTE REEFING CUTTERS

The drogue parachute reefing cutters are provided to disreef the drogue chute. The cutters are similiar to the pilot parachute reefing cutters (Figure 11-26) in design, operation, and number.

DROGUE PARACHUTE BRIDLE RELEASE GUILLOTINES

The drogue parachute bridle release guillotines are provided to sever the bridle

 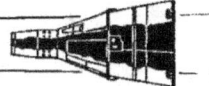

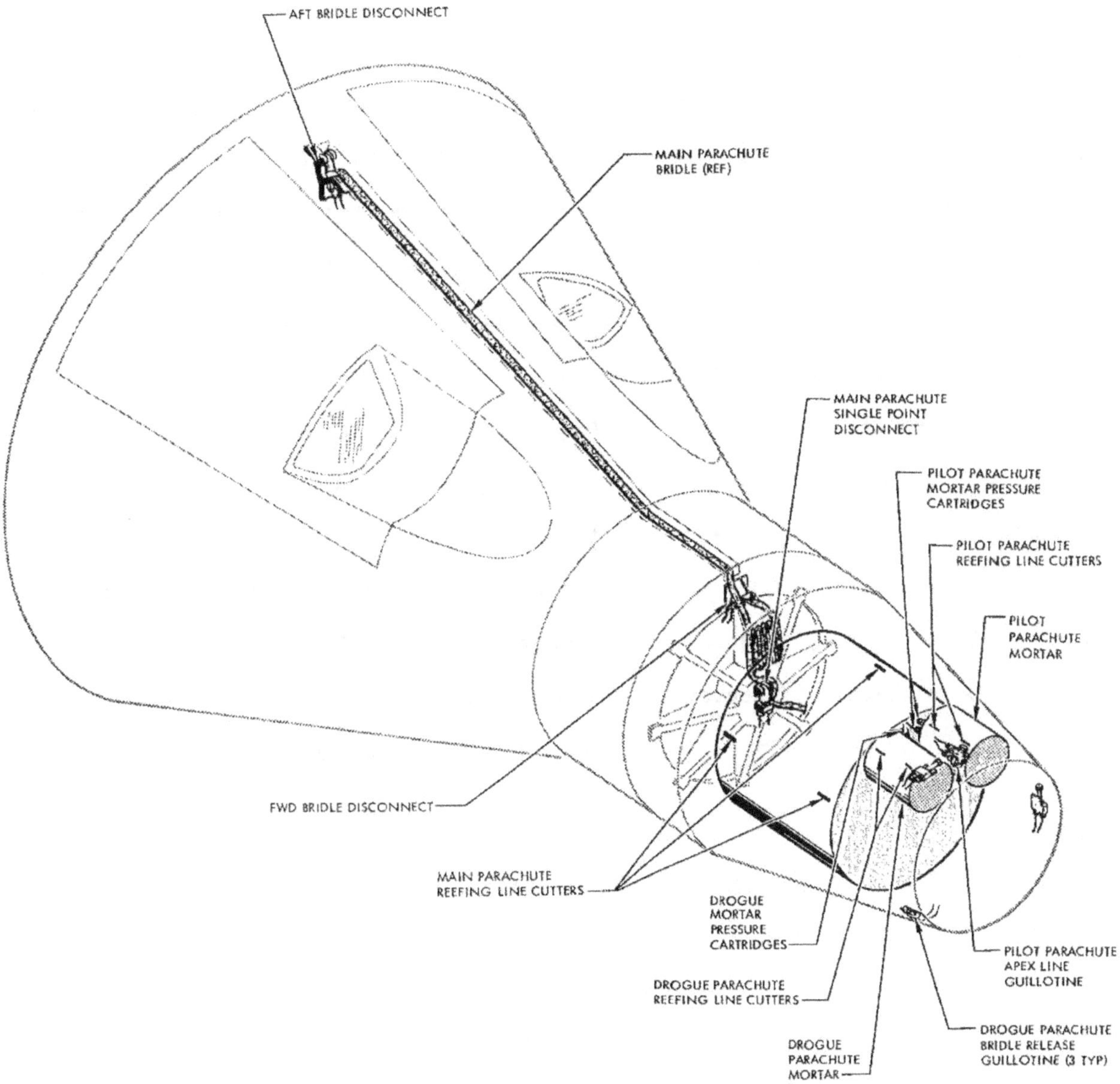

SEDR 300

PROJECT GEMINI

Figure 11-25 Parachute Landing System Pyrotechnics

SEDR 300
PROJECT GEMINI

at its three attach points. The release guillotines are similar in design and operation to the cable and wire bundle guillotine (Figure 11-9).

PILOT PARACHUTE MORTAR ASSEMBLY

Description

The pilot parachute mortar assembly (Figure 11-26) functions to deploy the pilot parachute in the event of a malfunctioning drogue chute. The mortar assembly is located in the forward end of the rendezvous and recovery section. The mortar assembly primarily consists of the mortar tube, sabot, breech, orifice, frangible bolt, washer and two electrically fired gas pressure cartridges. The base of the mortar tube is attached to the breech and the breech is attached to the rendezvous and recovery section structure. The flanged orifice passes through the base of the mortar tube and breech. The orifice is secured beneath the breech by a locknut. One end of the breech is drilled and tapped to provide for installation of the two cartridges. The sabot is located in the lower section of the mortar tube and is secured by a washer and a frangible bolt. The frangible bolt passes through the washers, the center of the sabot mortar tube base, and is threaded into the base of the orifice. The pilot chute is installed in the sabot.

Operation

The mortar functions when the cartridges are initiated by a 28 vdc electrical signal. The cartridges generate gas pressure that is ported through the breech and orifice to the base of the sabot. When sufficient pressure is exerted on the sabot, the frangible bolt will part and release the sabot. The gas pressure propels the sabot and pilot chute out of the mortar tube, thus effecting positive chute deployment.

SEDR 300
PROJECT GEMINI

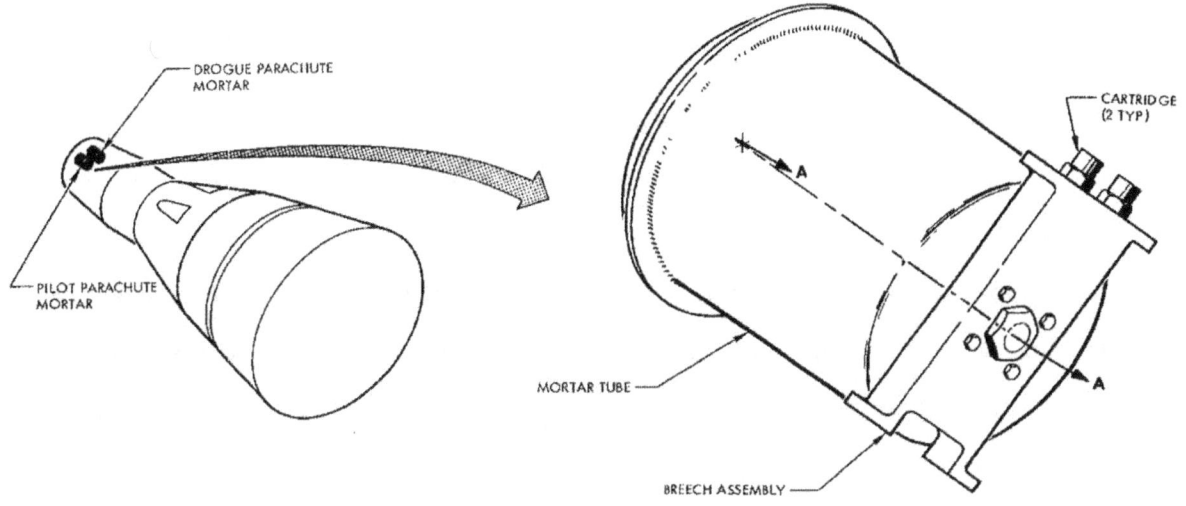

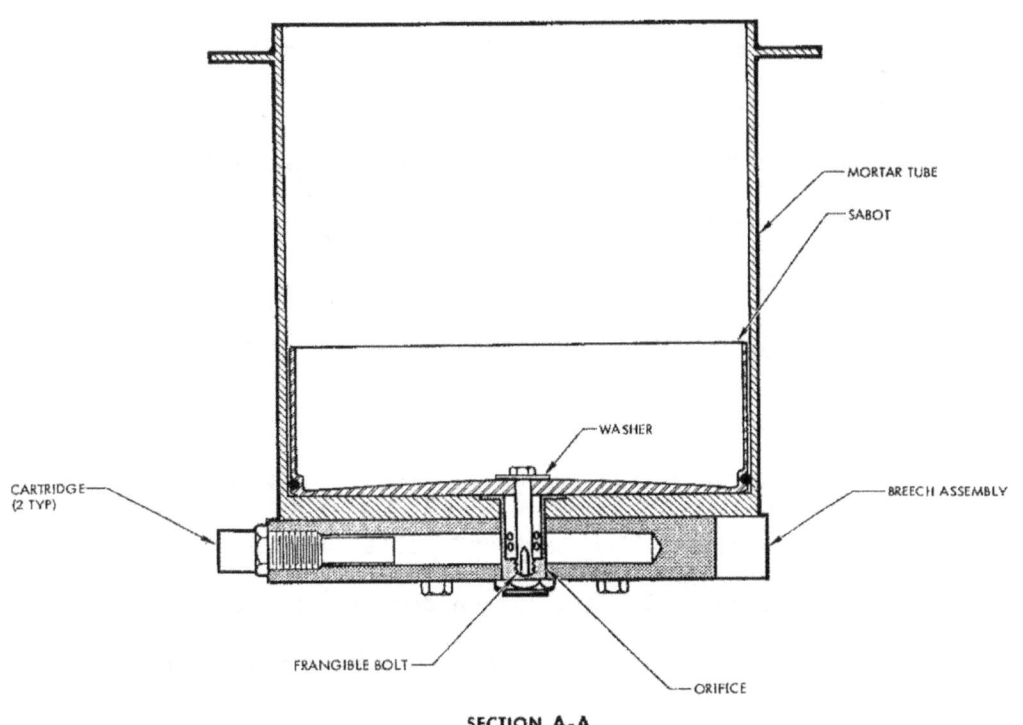

Figure 11-26 Pilot Parachute Mortar Assembly

11-59

 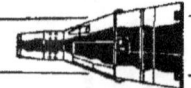

PILOT PARACHUTE REEFING CUTTERS

Description

The pilot parachute reefing cutters (Figure 11-27) are provided to disreef the pilot chute by severing the reefing line. The reefing cutters primarily consist of the cutter body, cutter blade, firing mechanism and percussion fired time delay cartridge. Two cutter assemblies are sewn to the inside of the parachute skirt band 180 degrees apart. The reefing cutter is a tubular device, with all its components contained within the cutter body. The firing mechanism is contained in one end of the cutter body and consists of a firing pin, lock ball, spring, and sear pin. The firing pin is retained in the cocked position by the lock ball. The lock ball is held in place by the sear pin. A lanyard is attached to the sear pin and to the parachute canopy. The spring is precocked and energizes the firing pin when the sear pin is pulled. The cartridge is installed in the center portion of the cutter body and is roll crimped in place. The cartridge consists of a percussion primer, time delay column and output charge. The cutter blade is stake locked in the cutter body, below and adjacent to the output end of the cartridge. A washer is crimp locked in the end of the cutter body, and serves as the anvil and stop for the cutter blade. A hole in each side of the cutter body, between the cutter blade and washer, permits installation of the reefing cable.

Operation

Deployment of the pilot chute causes the reefing cutters to be initiated. As the canopy extends, the lanyard is pulled taut and pulls the sear pin from the firing mechanism. The lock ball moves inboard and unlocks the firing pin. The spring energized firing pin is driven into the primer of the cartridge and ignites the

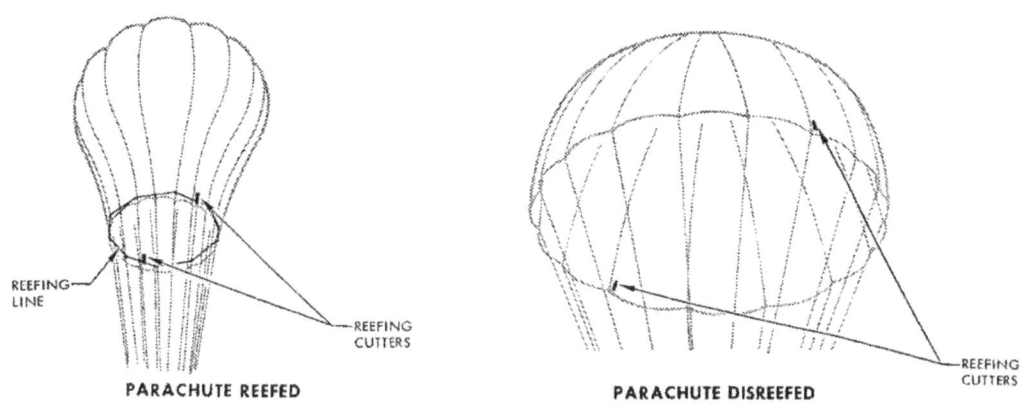
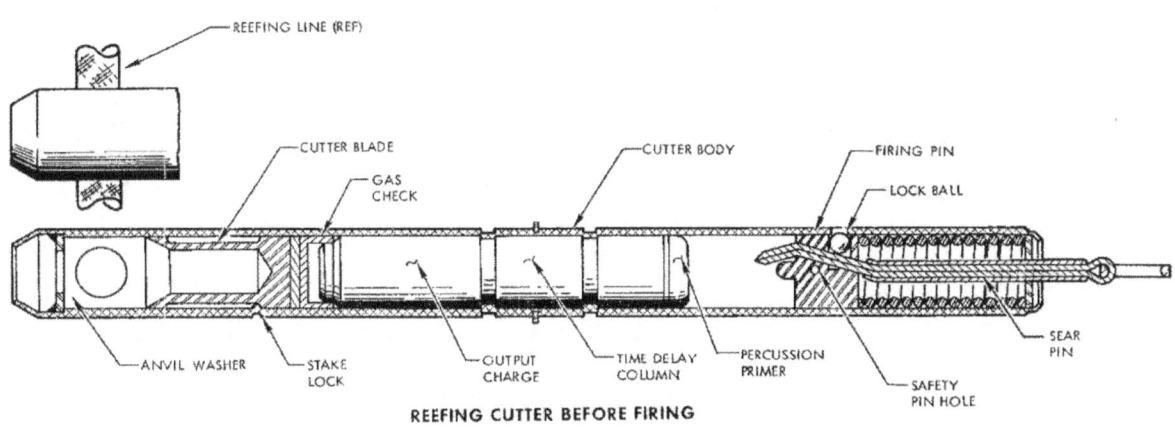
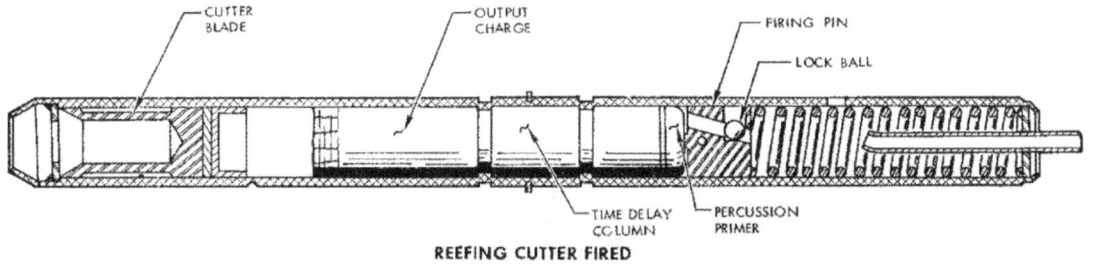

Figure 11-27 Pilot Parachute Reefing Cutters

time delay column. After the specified time delay, the cartridge produces gas pressure that exerts force on the cutter blade. When sufficient force is exerted, the cutter blade shears the stake lock and strokes to sever the reefing cable. Proper functioning of only one of the two cutters is sufficient to perform pilot chute disreefing.

PILOT PARACHUTE APEX LINE GUILLOTINE

Description

The pilot parachute apex line guillotine (Figure 11-28) is provided to sever the pilot chute apex line, in the event of a drogue chute malfunction. The guillotine primarily consists of the body, cutter blade, and two electrically fired gas pressure cartridges. The cutter blade is retained by a shear pin. The body provides for the installation of the two cartridges, and incorporates drilled passages from the cartridges to the cutter blade. Design of the guillotine allows the apex line to pull free, when the pilot chute is deployed by the drogue chute. The guillotine is located in the forward section of the rendezvous and recovery section.

Operation

When initiated by a 28 vdc electrical signal, the cartridges produce gas pressure. The pressure is ported through a drilled passage to the head of the cutter blade. Sufficient pressure causes the cutter blade to sever the shear pin and stroke to cut the apex line. The apex line is thus free of the malfunctioned drogue chute, permitting mortar deployment of the pilot chute.

MAIN PARACHUTE REEFING CUTTER

The main parachute reefing cutter (Figure 11-29) is provided to disreef the main

PROJECT GEMINI

SEDR 300

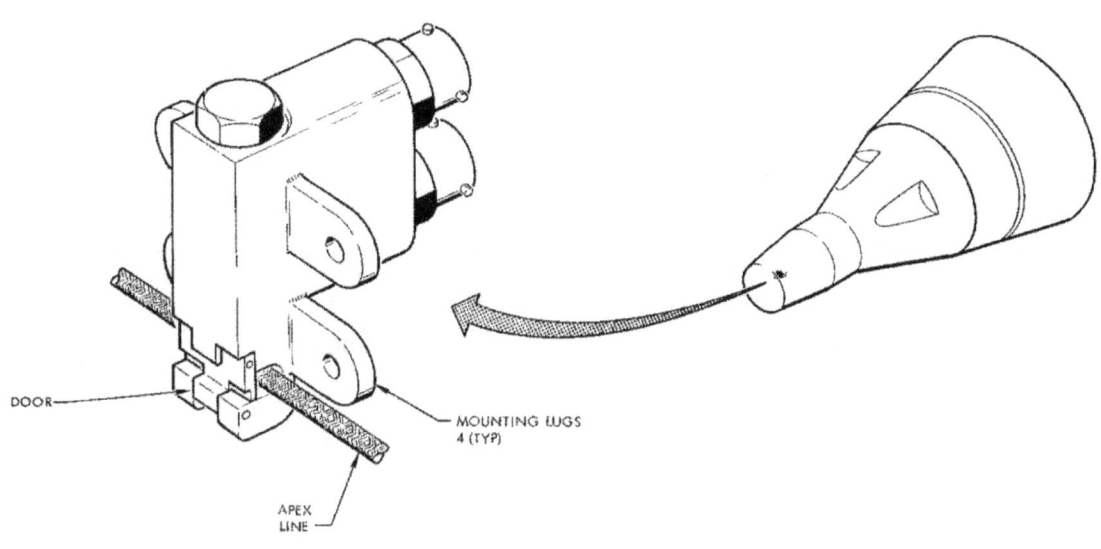

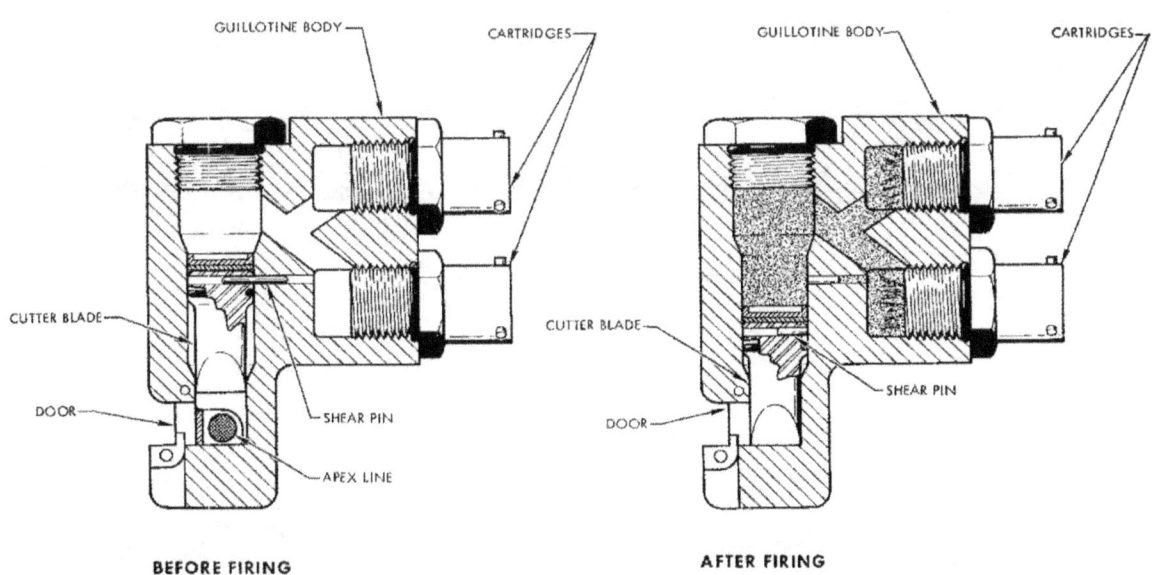

Figure 11-28 Pilot Parachute Apex Line Guillotine

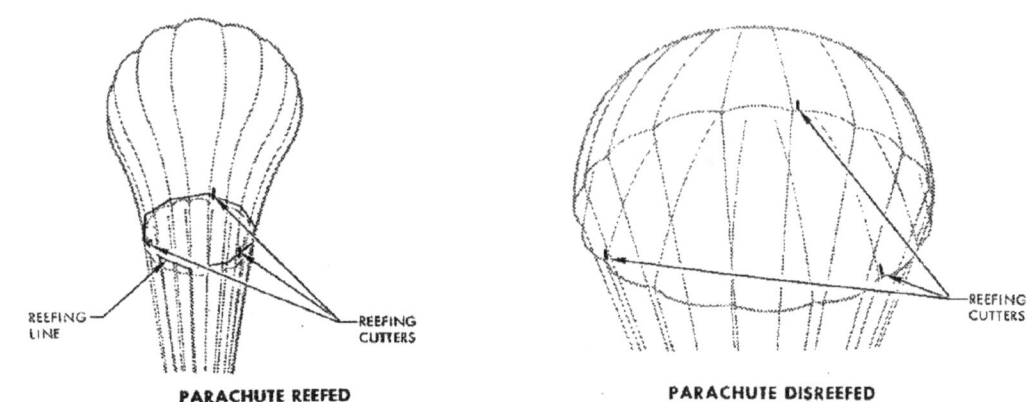
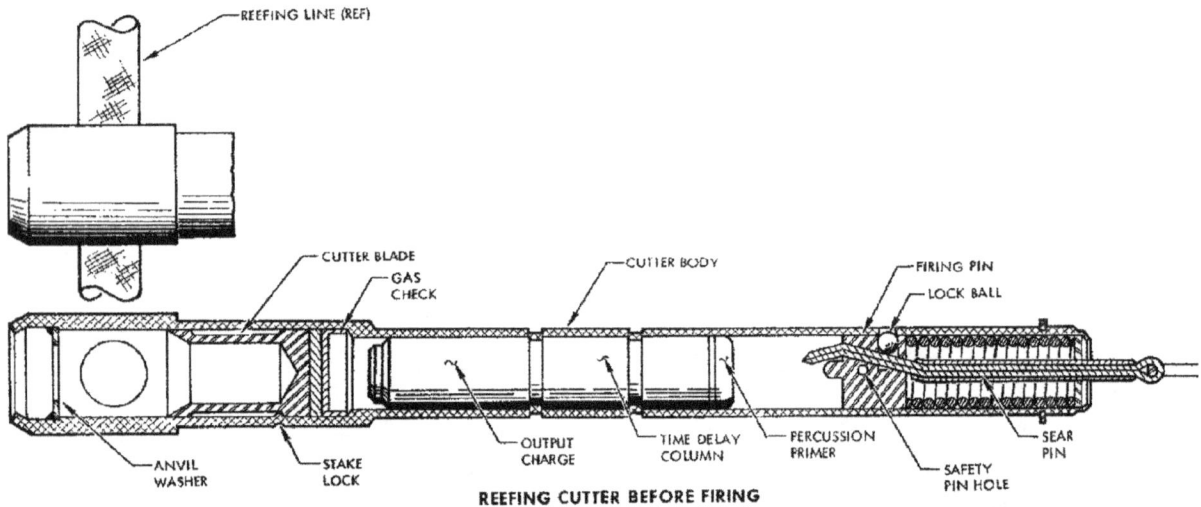

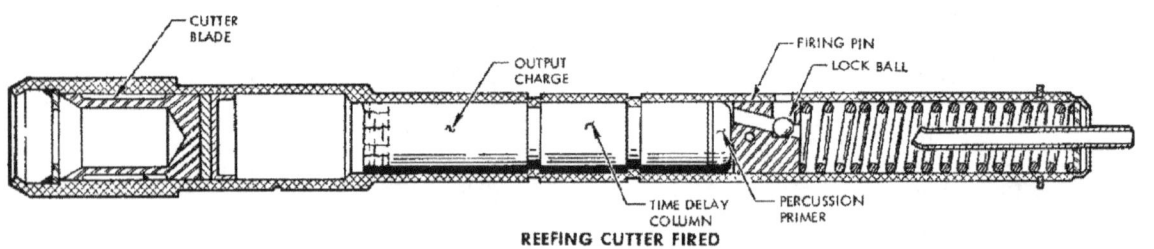

Figure 11-29 Main Parachute Reefing Cutter

parachute. Three reefing cutters are located on the inside of the canopy skirt band 120° apart. The reefing cutters are similar in design and identical in operation to the pilot chute reefing cutters. Proper operation of only one of the three cutters is sufficient to disreef the parachute.

MAIN PARACHUTE DISCONNECT

Description

The Main Parachute Disconnects (Figure 11-30) include the single point disconnect assembly and the forward and aft bridle disconnect assemblies. The disconnect assemblies are identical, in design and function. The disconnect assemblies primarily consist of the breech assembly, arm, and two electrically fired gas pressure cartridges. The breech assembly consists of the adapter, piston, lead slug, snubber disc, and plunger. The piston is retained in the adapter by a shear pin. The lead slug is located on the end of the piston. The snubber disc is located under the head of the plunger. The adapter is threaded into the spacecraft structure with the piston extending into the arm. The breech is threaded onto the adapter and the cartridges are installed in the breech. The single point disconnect is mounted on the hub of the main parachute adapter assembly. The adapter assembly is located on the forward ring of the re-entry control system section. The forward bridle disconnect is mounted at the top of the forward ring of the re-entry control system section. The aft bridle disconnect is located forward of the heat shield between the egress hatches.

Operation

When initiated by the proper electrical signal, the disconnect cartridges are ignited. The cartridges produce gas pressure that is ported through drilled passages

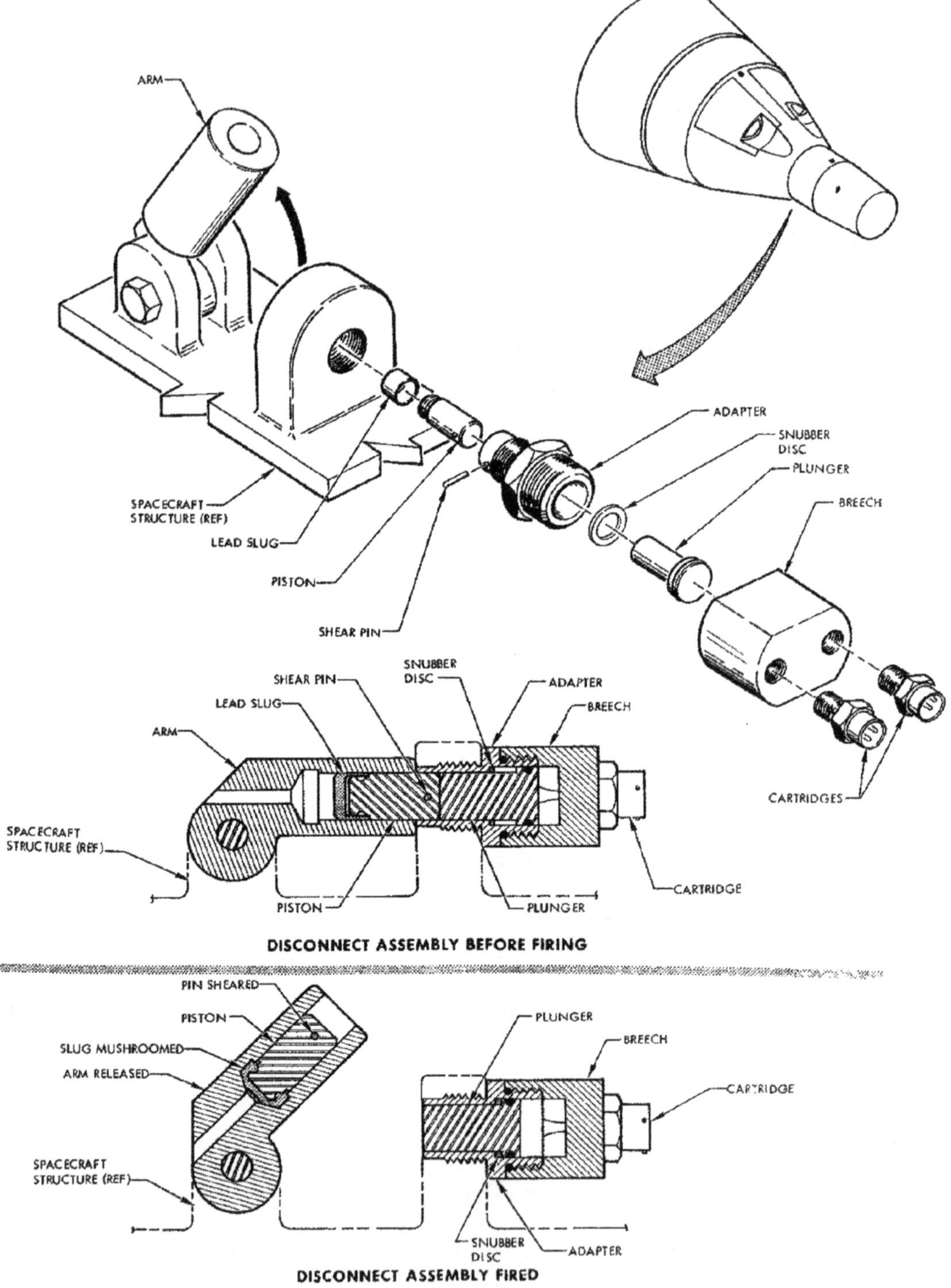

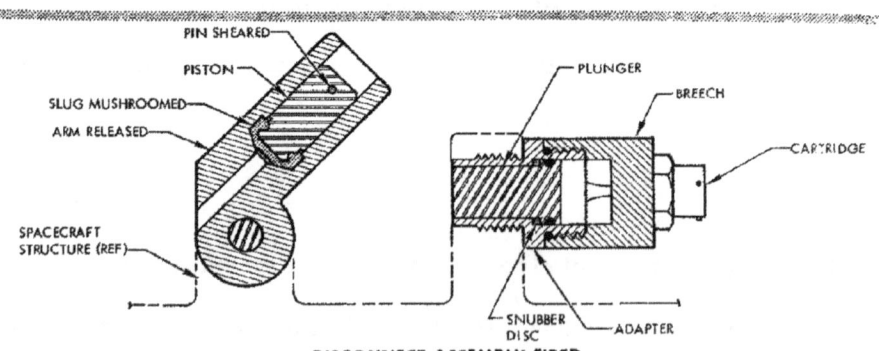

Figure 11-30 Main Parachute Disconnects

in the breech, to a common chamber at the head of the plunger. The gas pressure exerts a force on the head of the plunger, which in turn propels the piston by physical contact. The piston severs the shear pin and is driven into the arm. The plunger is prevented from following the piston, since the head of the plunger strikes a shoulder in the adapter. The sunbber disc provides a cushioning effect, to prevent shearing the plunger head. As the piston strikes the back of the arm, the lead slug at the end of the piston mushrooms. Mushrooming of the slug, retains the piston in the arm, preventing the piston from hindering arm operation. The pull of the parachute causes the arm to cam open, thus releasing the riser or bridle attached thereto.

PYROTECHNIC VALVES

DESCRIPTION

Pyrotechnic Valves (Figure 11-31) are installed in the Orbit Attitude and Maneuvering System and in the Re-entry Control System. The valves are one time actuating devices, used to control the flow of fluids. The spacecraft contains pyrotechnic valves that consist of the electrically fired high explosive cartridge, valve body, mipple, ram, seal, and screw. The nipple, either open or closed depending on the particular valve, is installed in the valve body and welded into place. The ram, incorporating the seal and screw at its head is located in the valve body, indexed directly above the center of the nipple. The cartridge is installed in the valve body at the top of the ram head. Two type of valves are used; normally open and normally closed. The "A" packages of the RCS and OAMS, contain a normally closed non-replaceable valve. The "E" package of the OAMS contain a noramlly open, and a normally closed, non-replaceable valve. If the valves in the "A" and "E"

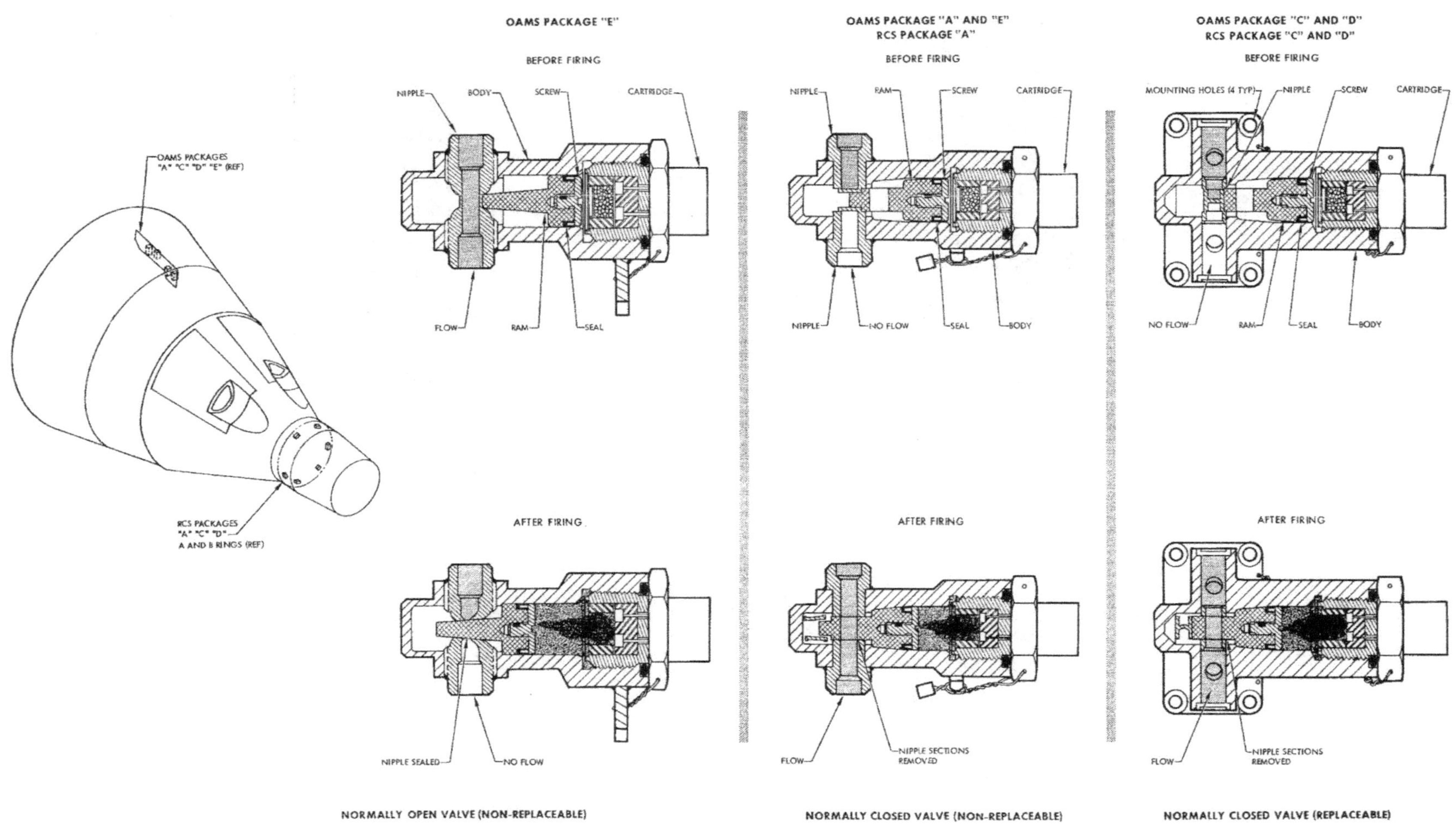

Figure 11-31 Pyrotechnic Valves

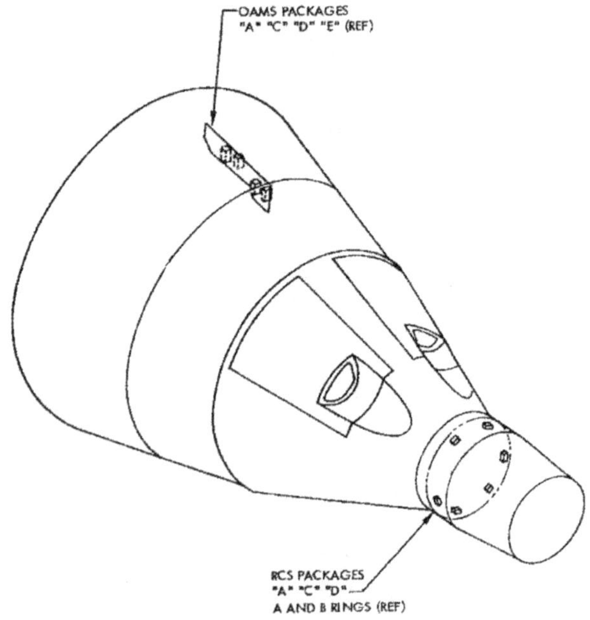

OAMS PACKAGE "E"

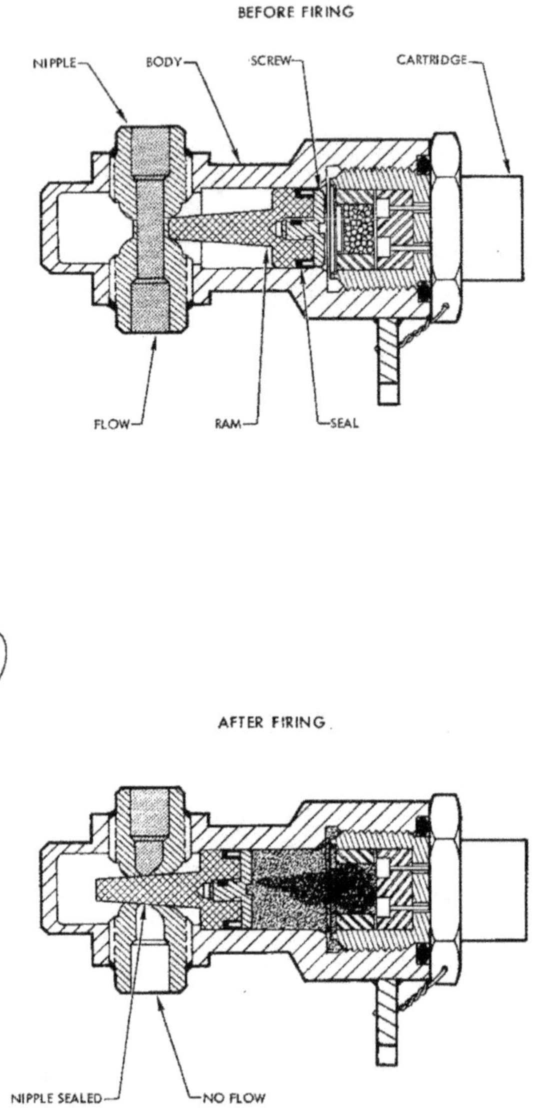

BEFORE FIRING

NIPPLE — BODY — SCREW — CARTRIDGE

FLOW — RAM — SEAL

OAMS PACKAGES "A" "C" "D" "E" (REF)

RCS PACKAGES "A" "C" "D" A AND B RINGS (REF)

AFTER FIRING

NIPPLE SEALED — NO FLOW

NORMALLY OPEN VALVE (NON-REPLACEABLE)

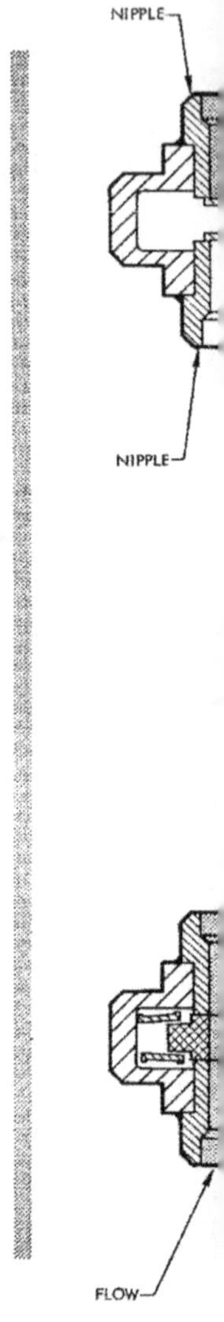

NIPPLE

NIPPLE

FLOW

NORM

11-68

 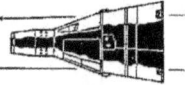

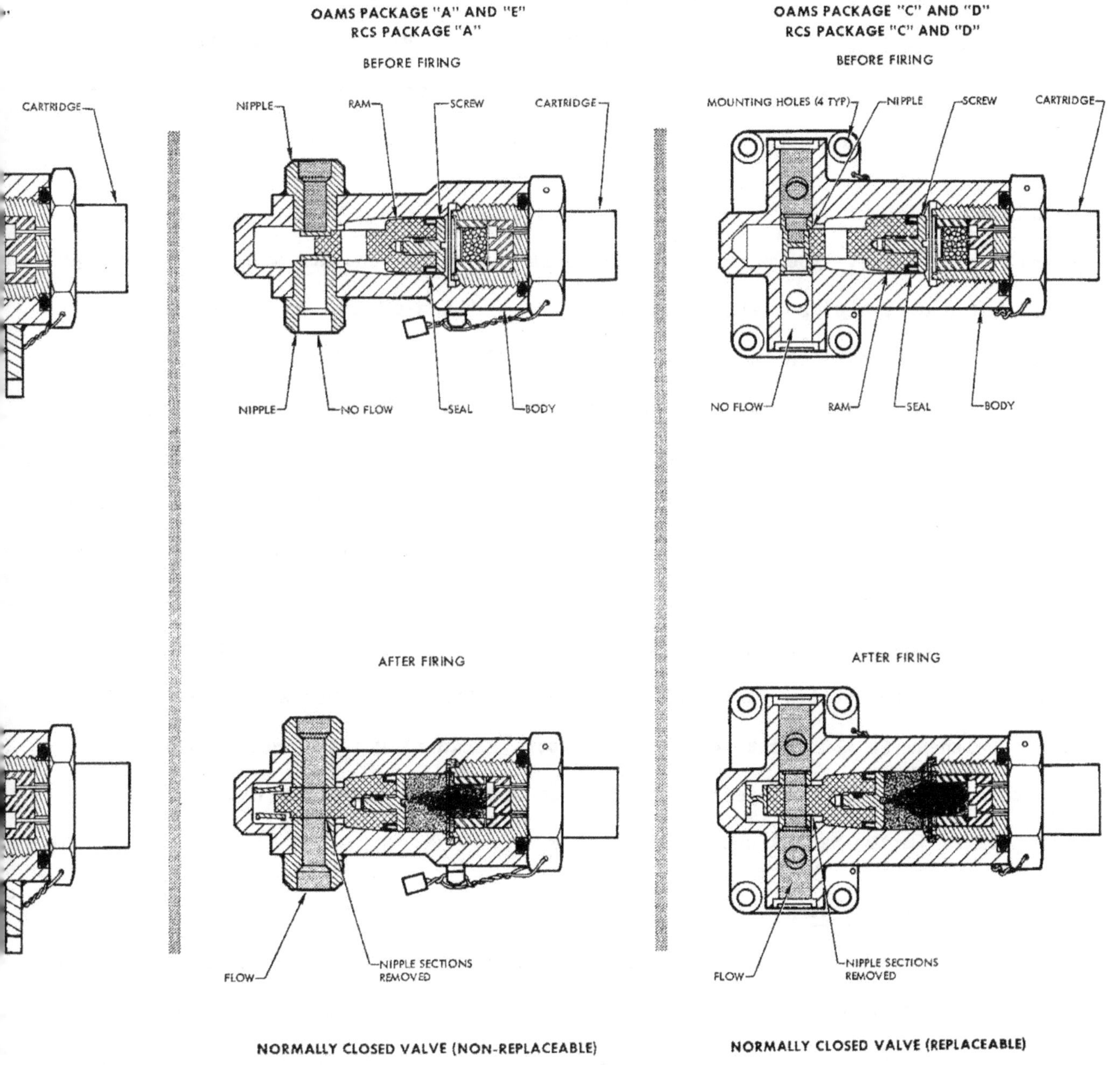

Figure 11-31 Pyrotechnic Valves

SEDR 300
PROJECT GEMINI

packages are defective, or the cartridge has been fired, the packages must be changed. The "C" and "D" packages of the RCS and the OAMS contain normally closed replaceable valves. These valves are attached to the exterior of the package, and if defective or the cartridge fired, may be changed individually.

OPERATION

Normally open valve; The pyrotechnic valve is caused to function when the cartridge is initiated by the proper electrical signal. Ignition of the cartridge produces gas pressure that acts on the head of the ram. The ram is driven down on the nipple, severing and removing a section of the nipple. The ram, having a tapered cross section, is wedged in the nippled opening, completely sealing the nipple, thus stopping the flow of fluid. Normally closed pyrotechnic valves are all basically identical except for nipple design. The non-replaceable valve has two closed end nipples butted together. The ram severs and removes the end of each nipple and wedges itself between the ends. A hole is incorporated in the ram, allowing fluid flow after ram actuation. The replaceable pyrotechnic valve has a nipple installed with a bulkhead in the cross section that stops fluid flow. The ram removes the section of the nipple containing the bulkhead and wedges itself in place. A hole incorporated in the ram allows fluid to flow.

RETROGRADE ROCKET SYSTEM

The Retrograde Rocket System (Figure 11-32) primarily consists of four solid propellant rocket motors and eight igniter assemblies. The retrograde rockets are provided to retard spacecraft orbital velocity for re-entry and to provide distance and velocity to clear the launch vehicle in the event of an abort during ascent. The rocket motors are symmetrically located about the longitudinal axis of the

 SEDR 300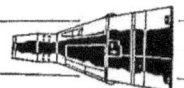

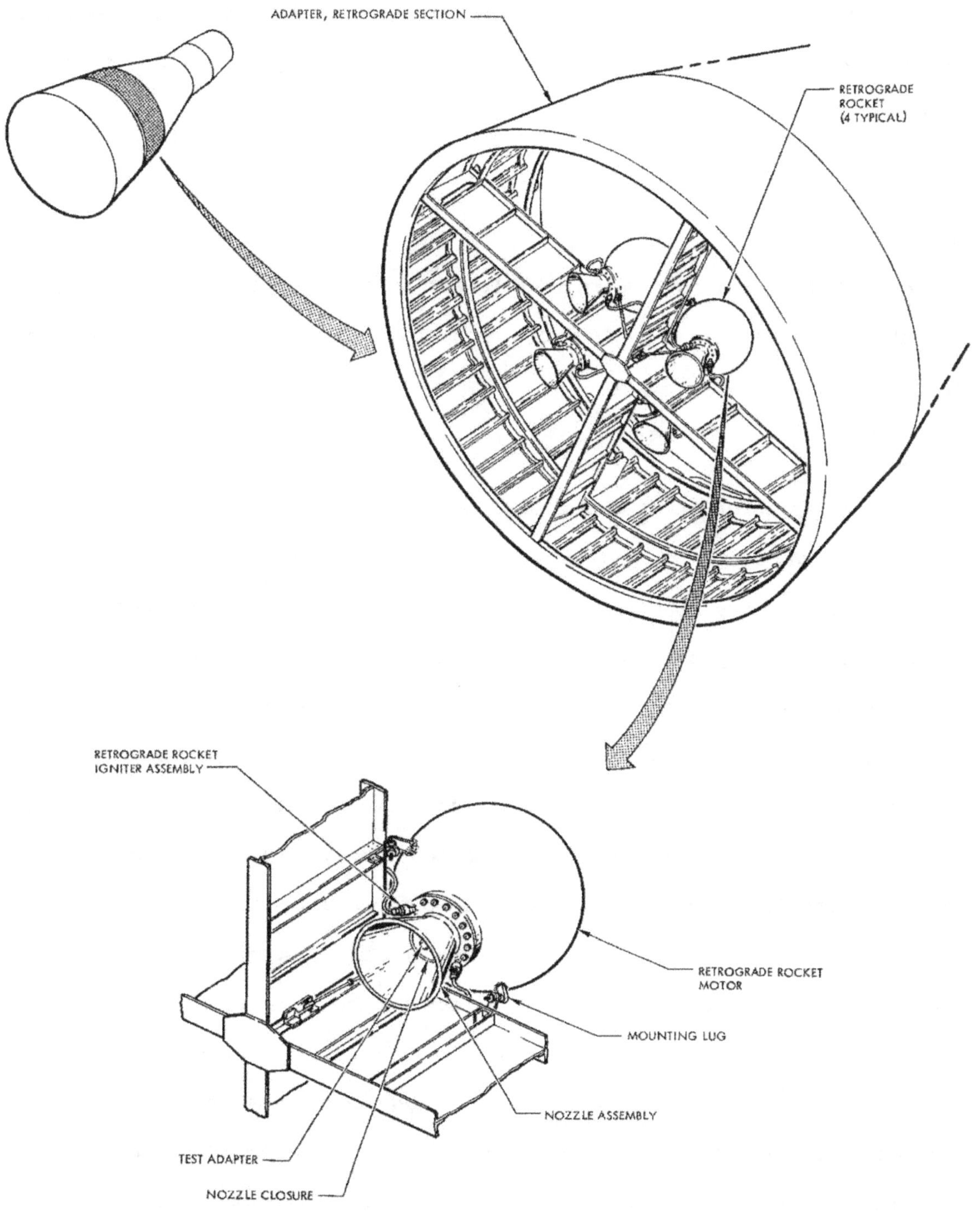

PROJECT GEMINI

Figure 11-32 Retrograde Rocket System

SEDR 300
PROJECT GEMINI

spacecraft and are mounted in the retrograde section of the adapter. The rocket motors are individually, optically aligned prior to mating the adapter to the reentry module.

RETROGRADE ROCKET MOTOR ASSEMBLY

Description

The spacecraft contains four Retrograde Rocket Motor Assemblies (Figure 11-33) that are identical in design and performance, spherical in shape, and are approximately 13 inches in diameter.

Rocket Motor Case

The motor case is formed from titanium alloy in two hemispherical halves. The halves are forged, machined, and welded together at the equator. Each hemisphere is insulated to reduce heat transfer during motor operation. The aft hemisphere is drilled and tapped to provide a mating flange for the nozzle assembly. The nozzle assembly, a partially submerged type, consists of the expansion cone, throat insert and the nozzle bulkhead. The nozzle bulkhead is a machined titanium alloy, bolted to the flange at the aft end of the motor case. The bulkhead is threaded to provide for expansion cone installation. The expansion cone is compression molded of vitreous silica phenolic resin and is threaded into the nozzle bulkhead. The throat insert is machined from high density graphite and is pressed into the nozzle bulkhead. The throat insert is insulated from the bulkhead by a plastic material to reduce heat transfer during motor operation. The throat insert is recessed into the motor case to reduce nozzle assembly length. A rubber nozzle closure is sandwiched between the throat insert and the nozzle bulkhead. The closure incorporates a shear groove that permits ejection at a predetermined internal

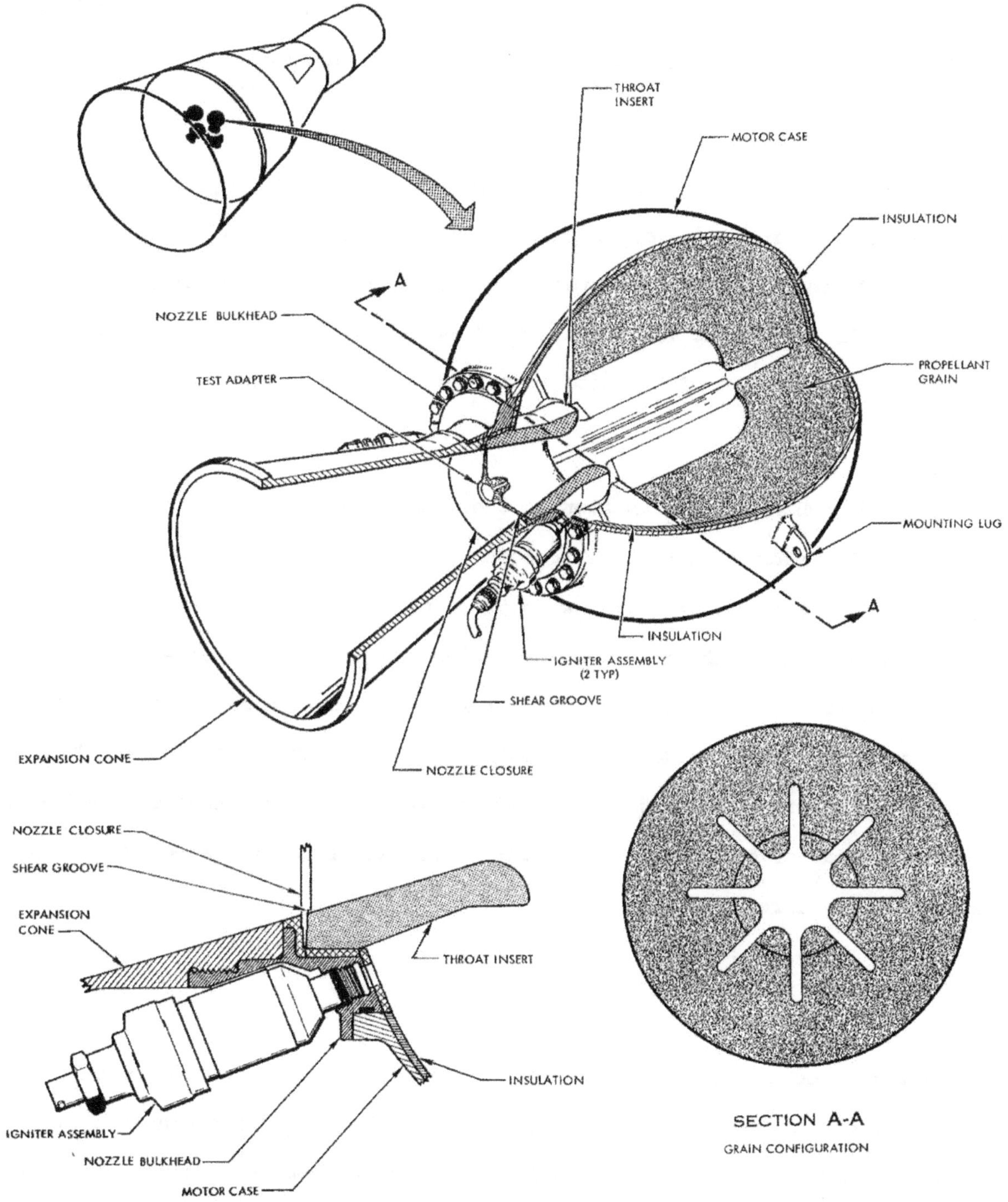

Figure 11-33 Retrograde Rocket Motor Assembly

pressure level, or basically at motor ignition. A test adapter fitting is incorporated in the closure to permit pressure checking of the rocket motor.

Rocket Motor Propellant

The motor case is lined with a rubber material that provides propellant grain to case adhesion. The rocket motor propellant is cast and cured in the motor case. The propellant grain is cast in an internal burning eight pointed star configuration. The propellant grain is ignited by the two igniter assemblies, mounted 180° apart on the aft end of the motor case, adjacent to the nozzle assembly.

Operation

The retrograde rocket motors function in two modes: Normal and abort. In the normal mode of operation, the rocket motors are used to initiate spacecraft reentry. The rocket motors are fired at 5.5 second intervals in 1-2-3-4 order. The propellant grain of the rocket motor is ignited by the hot gases from the igniter assemblies. The propellant grain burns over the entire surface of its eight pointed star configuration until exhausted. The thrust produced by the motors in transmitted to the spacecraft structure and retards spacecraft velocity. In the abort mode of operation, the rocket motors are fired in salvo or as mission requirements may direct.

RETROGRADE ROCKET IGNITER ASSEMBLY

Description

The Retrograde Rocket Igniter Assemblies (Figure 11-34) are used to ignite the propellant grain of the retrograde rocket motor. The spacecraft contains eight igniter assemblies that primarily consist of the case, head cap, grain, booster

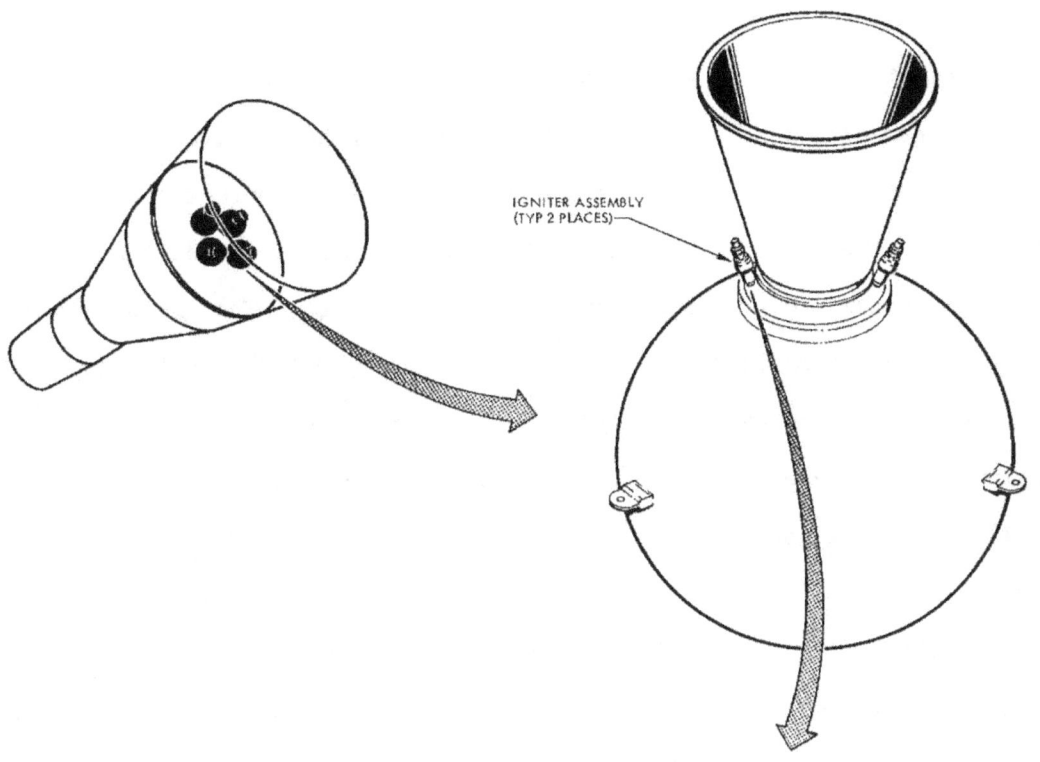

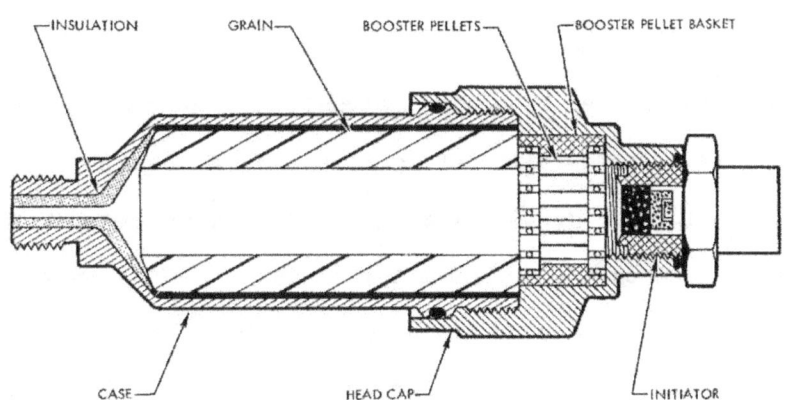

Figure 11-34 Retrograde Rocket Igniter Assembly

pellets, pellet basket, and initiator. The igniter assembly is essentially a small solid propellant rocket motor. The case and head cap are individually machined from a stainless steel alloy and have a threaded interface. On the internal surface of the case, at the gas exit, a silia-phenolic insulating material is bonded to reduce heat transfer during ingiter firing. The grain is cast and cured in a phenolic paper tube. The grain is inserted into the igniter case prior to case and head cap assembly. The booster pellets, consisting of boron potassium nitrate, are contained in the pellet basket, located in the head cap. The pellet basket is held in place by the head cap and is installed prior to case and head cap mating. The initiator cartridge consists of the body, one firing circuit (bridge wire), ignition mix, and output charge. The basic function of the initiator is to fire the igniter. The initiator is threaded into the head cap of the igniter at the time of igniter assembly build up at the vendors.

Operation

When initiated by the proper electrical signal, the initiator of the igniter assembly is activated. The initiator ignites the boron pellets, which boosts the burning to the igniter grain. The igniter grain generates hot gas which is exhausted into the retrograde rocket motor cavity. The hot gases provide the temperature and pressure for retrograde rocket motor propellant grain ignition. Either igniter is sufficient to initiate burning of the rocket motor.

LANDING SYSTEM

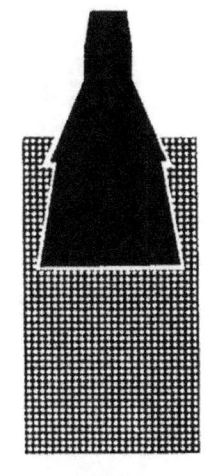

Section XII

TABLE OF CONTENTS

TITLE	PAGE
SYSTEM DESCRIPTION	12-3
SYSTEM OPERATION	12-3
EMERGENCY OPERATION	12-7
SYSTEM UNITS	12-7
DROGUE PARACHUTE ASSEMBLY	12-7
PILOT PARACHUTE ASSEMBLY	12-12
MAIN PARACHUTE AND RISER ASSEMBLY	12-14

 SEDR 300

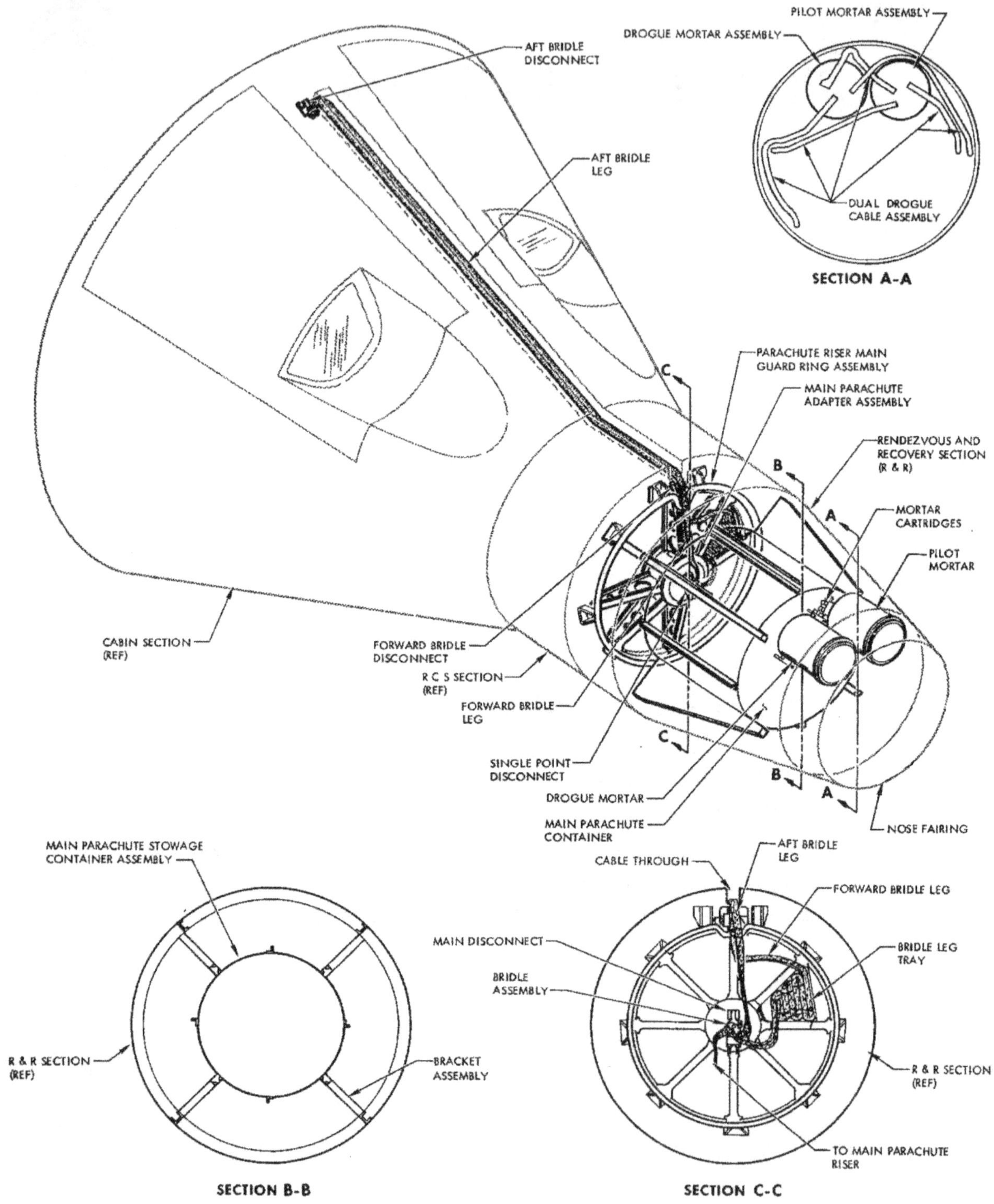

PROJECT GEMINI

Figure 12-1 Parachute Landing System

SEDR 300
PROJECT GEMINI

SECTION XII LANDING SYSTEM

SYSTEM DESCRIPTION

The Parachute Landing System (Figure 12-1) provides a safe rate of descent to return the re-entry module safely to the earth's surface and furnishes the proper attitude for a water impact. A system of three parachutes in series is utilized for stabilizing and retarding the velocity of the re-entry module. During the final stage of descent, the main parachute suspension is inverted from a single point to a two point suspension system in order to achieve a more favorable attitude for a water landing. The landing system consists of three parachute assemblies (a drogue parachute assembly, a pilot parachute assembly, and a main parachute assembly), two mortar assemblies, reefing cutters, disconnect assemblies, riser assemblies, and attaching hardware. The entire landing system, with the exception of the aft bridle leg and disconnect assembly, is located in the rendezvous and recovery section of the spacecraft. Figure 12-2 illustrates the sequence of events from re-entry to impact in block diagram form. Figure 12-3 illustrates the electrical sequence of the landing system.

SYSTEM OPERATION

Prior to re-entry, the landing and postlanding common control electrical buses are armed by positioning the LANDING switch to ARM. This also applies power to the two barometric pressure switches for illumination of the 10.6 K and 40 K warning indicators at the appropriate altitudes.

In order to stabilize the re-entry module, the drogue parachute is deployed at an altitude of 50,000 feet. The manually actuated HI-ALT DROGUE switch

SEDR 300

PROJECT GEMINI

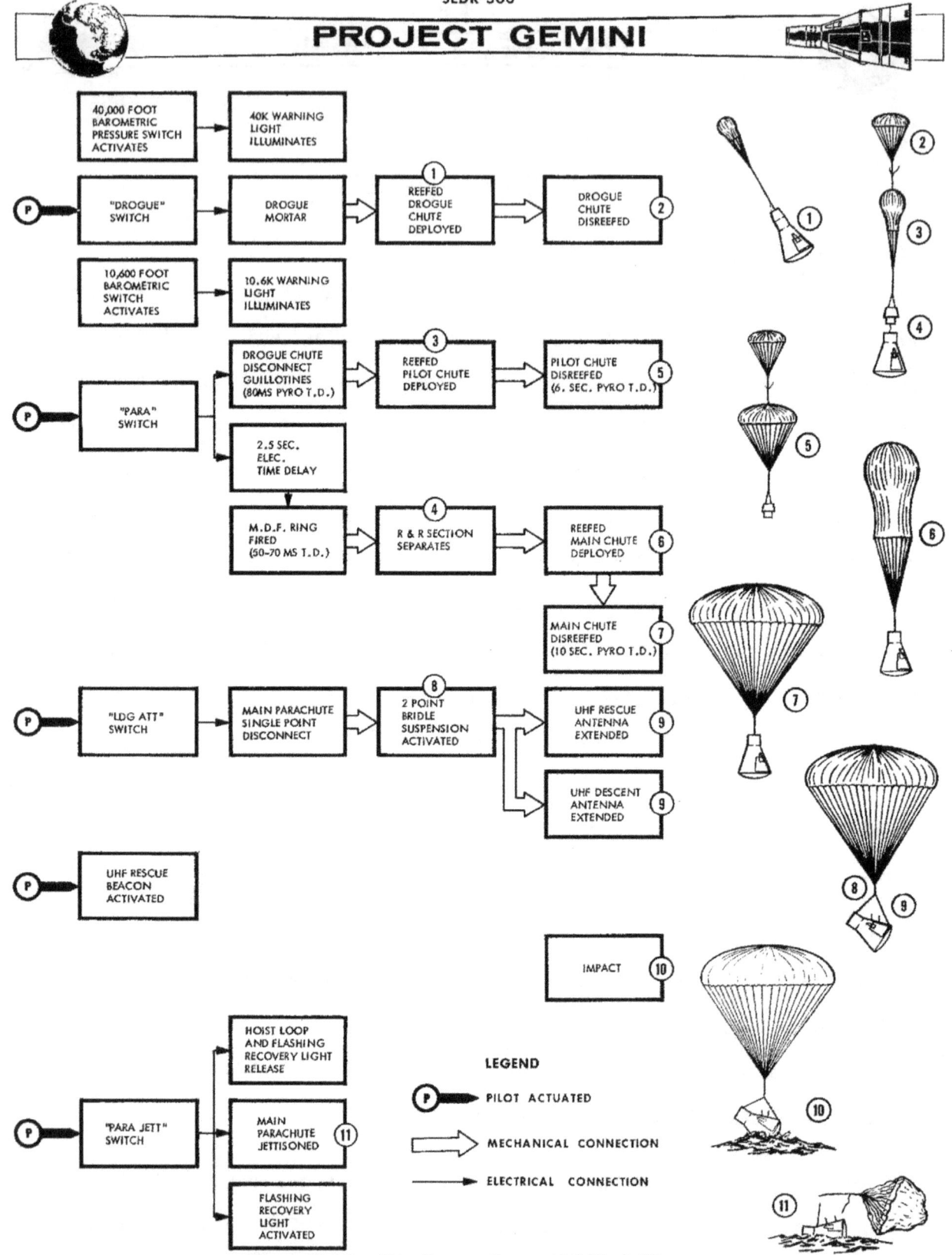

Figure 12-2 Landing System Sequential Block Diagram

12-4

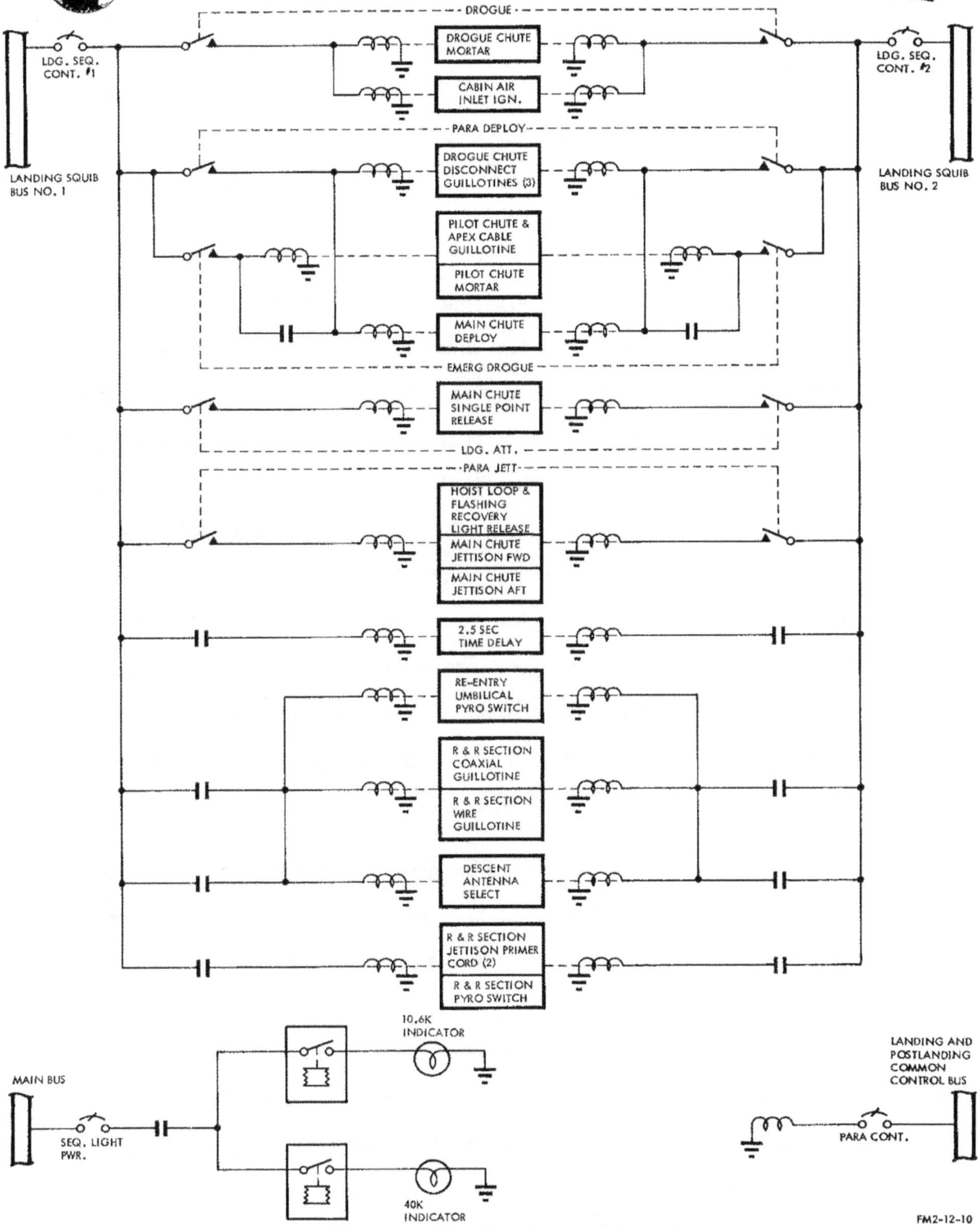

SEDR 300

PROJECT GEMINI

Figure 12-3 Landing System Schematic

12-5

PROJECT GEMINI

energizes two single pyrotechnic cartridges in the drogue mortar. To limit the opening shock load, the drogue parachute is deployed in a reefed condition. Sixteen seconds after deployment, two pyrotechnic reefing cutters disreef the drogue parachute.

As the re-entry module approaches an altitude of 11,000 feet, the PARA switch is activated. The PARA switch fires the three drogue cable guillotines and sets a 2.5 second time delay to the MDF ring detonators. After the drogue riser legs have been cut, the drogue parachute pulls away from the re-entry module extracting the pilot parachute from the pilot mortar tube with the apex line. When deployed, the pilot parachute is reefed to limit the initial shock load. Two pyrotechnic reefing cutters disreef the pilot chute six seconds after deployment. 2.5 seconds after the pilot chute has been deployed, the MDF ring fires separating the rendezvous and recovery section from the landing module. The pilot parachute functions to decelerate the re-entry module, remove the rendezvous and recovery section from the landing module, and deploy the main parachute.

As the landing module falls away from the rendezvous and recovery section, the reefed main parachute is pulled from the main parachute container located in the rendezvous and recovery section. Three pyrotechnic reefing cutters disreef the main parachute ten seconds after deployment. The two decelerations provided by the main parachute divide the retarding shock load. After the main parachute has been disreefed, the manually operated LDG ATT switch is actuated to change the single point suspension system to a two point suspension system.

SEDR 300
PROJECT GEMINI

The two point suspension system provides a more favorable attitude for impact than the one point suspension system. As soon as the landing module contacts the ocean surface, the PARA JETT switch is activated. The PARA JETT switch energizes the forward and aft bridle disconnects releasing the main parachute from the landing module. Upon completion of the landing, the landing module is prepared for transmitting data and recovery information.

EMERGENCY OPERATION

In the event the drogue parachute does not deploy or deploys improperly, the DROGUE EMERG 10.6 K switch is actuated. The closure of this switch fires the three drogue cable guillotines, the apex line guillotine, and the pilot parachute mortar and also starts the 2.5 second time delay to the MDF rings. The pilot mortar deploys the pilot parachute in a reefed condition. From this point to impact, the emergency sequence of events is exactly the same as used during a normal landing. Figure 12-4 illustrates the emergency sequence of events in block diagram form, and Figure 12-5 illustrates the emergency deployment.

SYSTEM UNITS

DROGUE PARACHUTE ASSEMBLY

The drogue parachute assembly (Figure 12-6) stabilizes the re-entry module and deploys the pilot parachute. This assembly consists of an 8.3 ft. diameter conical ribbon parachute with twelve 750-pound tensile strength suspension lines. A three legged riser assembly attaches the parachute assembly to the rendezvous and recovery section.

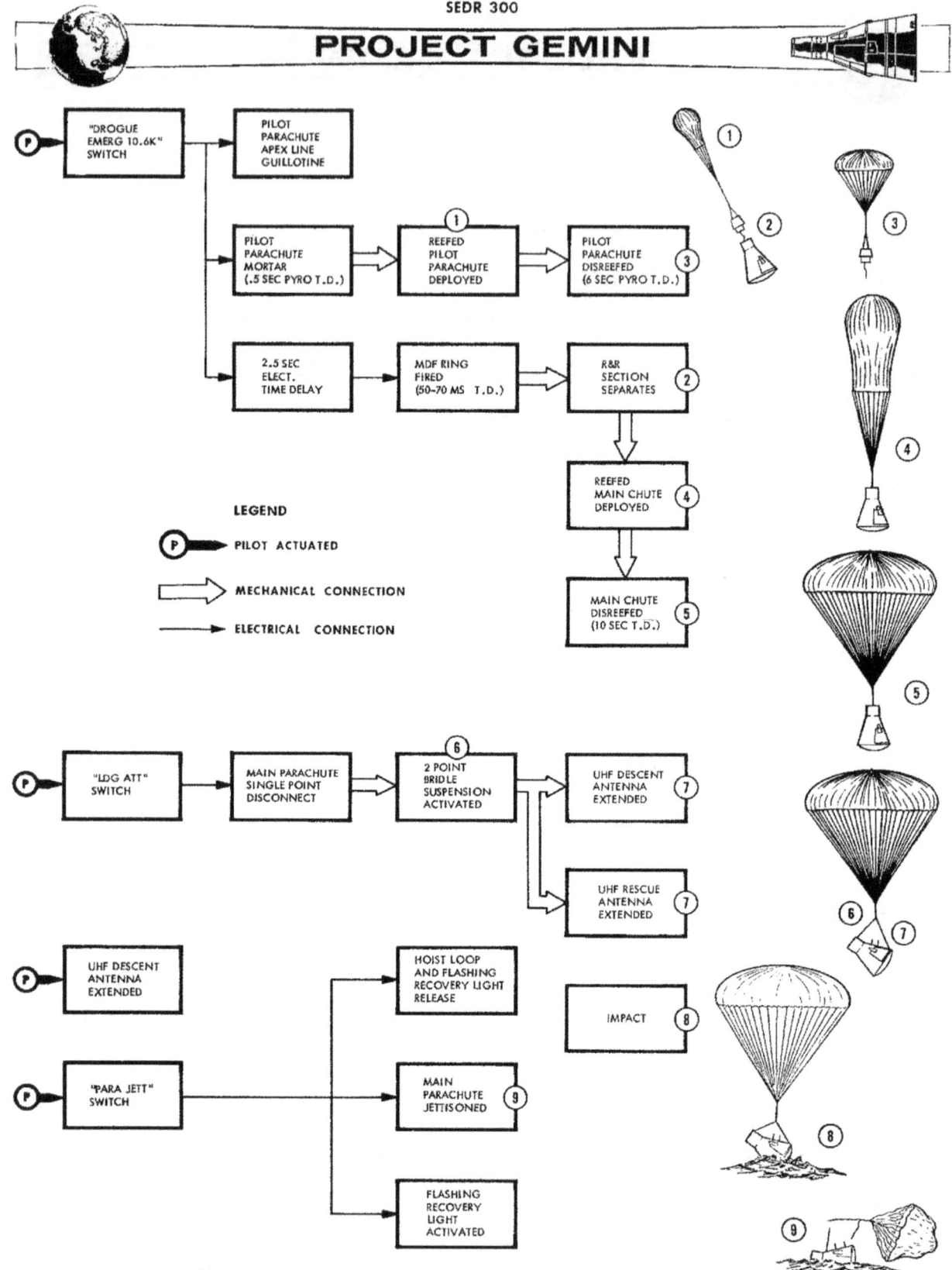

Figure 12-4 Emergency Landing Sequential Block Diagram

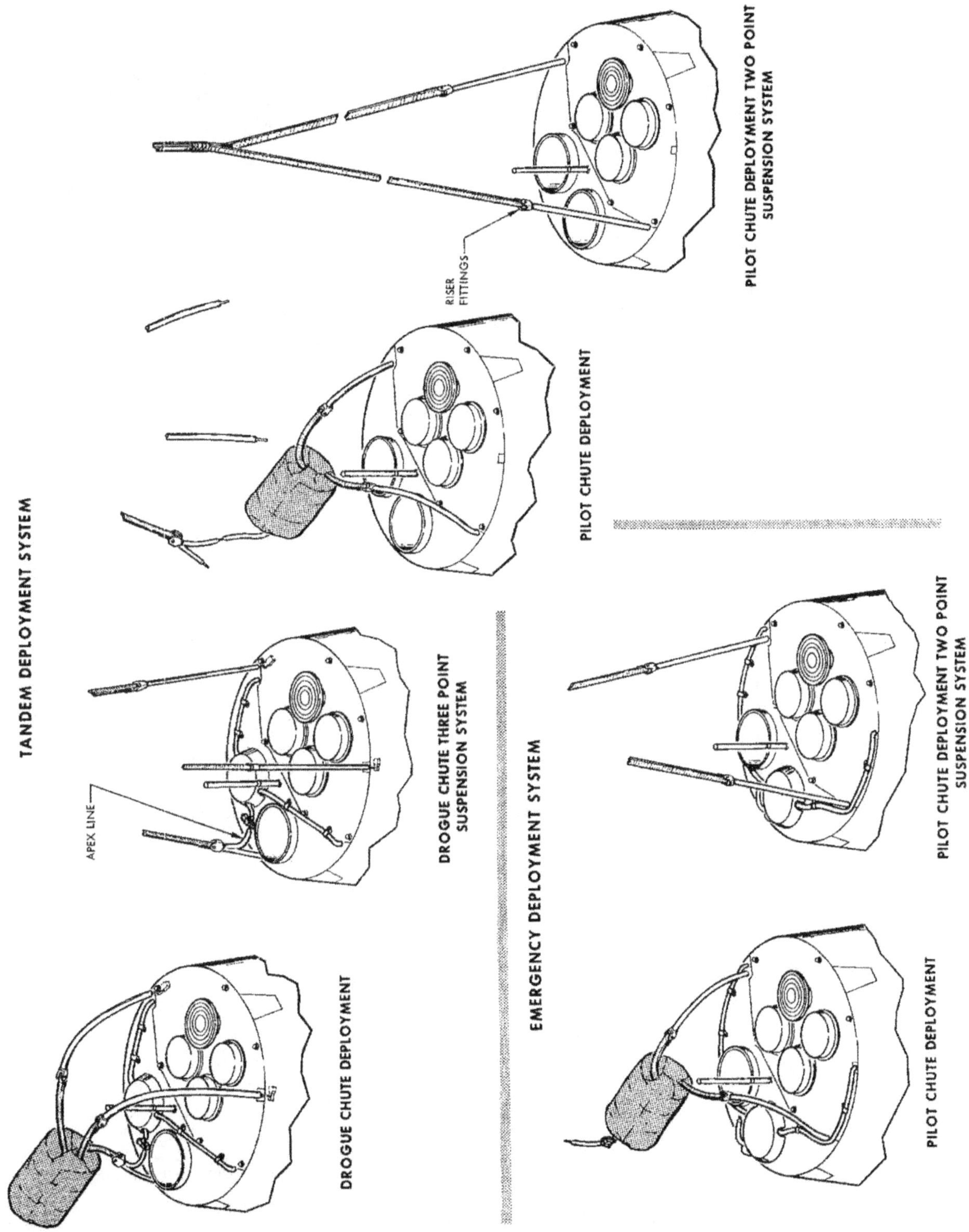

Figure 12-5 Tandem and Emergency Deployment System Operation

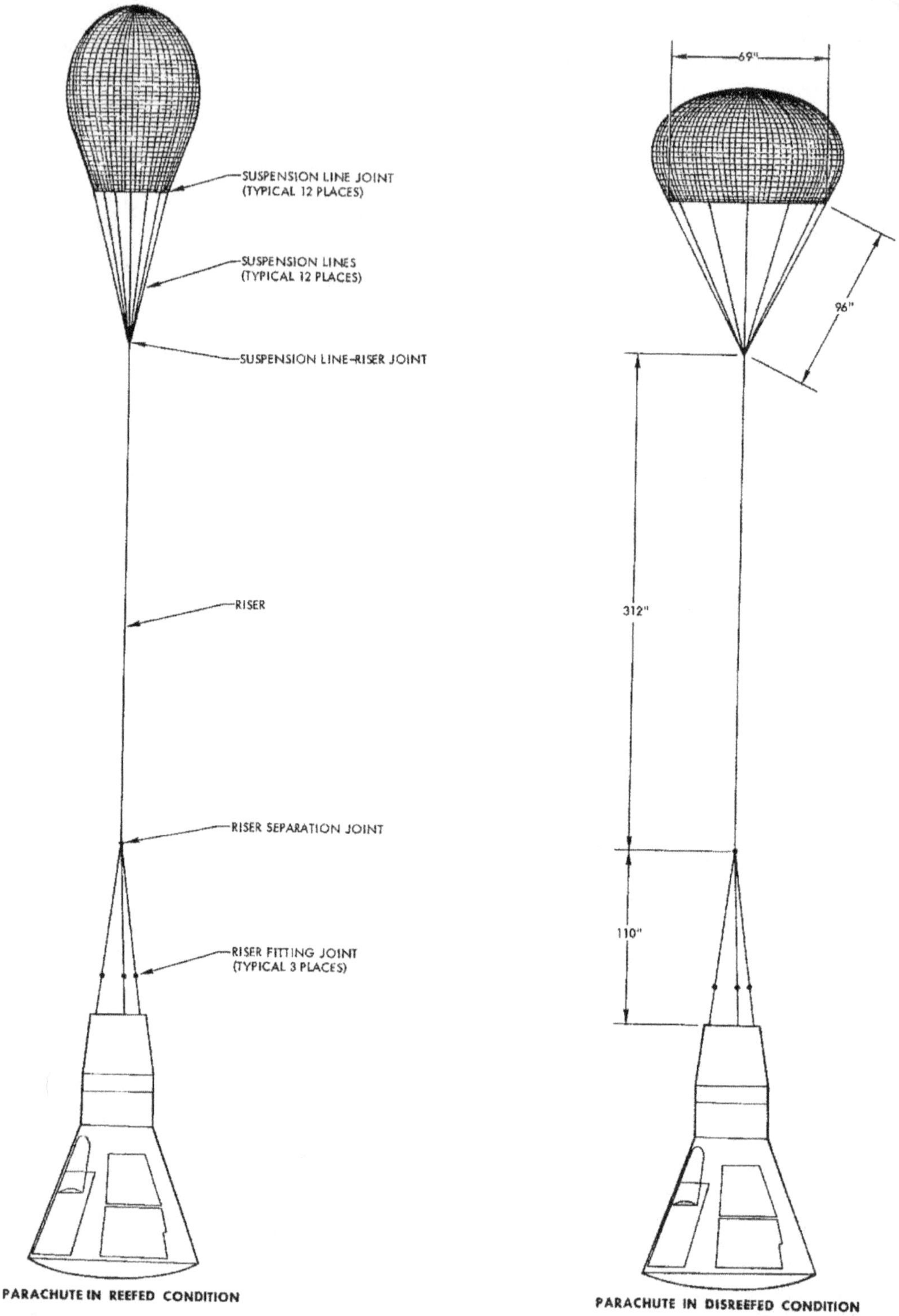

SEDR 300

PROJECT GEMINI

Figure 12-6 Drogue Parachute Assembly

SEDR 300

PROJECT GEMINI

When initially deployed, the drogue chute is reefed to 43% of the parachute diameter in order to reduce the opening shock load. Sixteen seconds after deployment, two pyrotechnic reefing cutters disreef the drogue chute. Initiation of the PARA switch fires three cable guillotines located at the base of the three riser legs. As the drogue chute pulls away from the rendezvous and recovery section, an apex line, which is attached to one of the riser legs, extracts the pilot parachute from the pilot mortar tube. The drogue parachute remains attached to the pilot parachute during the entire descent of the rendezvous and recovery section of the re-entry module.

Drogue Parachute Mortar Assembly

The drogue parachute mortar assembly stores and protects the drogue parachute during flight and deploys the drogue parachute when activated by the HI-ALT DROGUE switch. An insulated metal pan retains the parachute in the mortar tube. The mortar tube has a diameter of 7.15 inches and a length of 9.12 inches. The breech assembly, which is located at the base of the mortar tube, contains two electrically actuated pyrotechnic cartridges and the orifice. The cartridges generate gases that enter the mortar tube through the orifice causing the ejection of the drogue parachute and sabot.

Drogue Mortar Sabot

The drogue mortar sabot is an aluminum cup located in the base of the mortar tube and functions to eject the drogue parachute from the mortar tube with a piston like action. In order to insure the most effective ejection, the sabot is fastened to the base of the mortar tube by a frangible bolt. An "O" ring, located near the base of the sabot, contacts the inner wall of the mortar tube to prevent any escape of gases generated by the two pyrotechnic cartridges. When enough pressure to break the frangible bolt has built up, the sabot and parachute are expelled from the mortar tube. After ejection, the sabot remains attached to the parachute bag and aids in stripping the bag from the parachute.

Drogue Parachute Deployment Bag

The drogue parachute deployment bag protects the drogue parachute during ejection and allows for an orderly deployment of the parachute. The bag is fabricated from cotton sateen and nylon. A 0.35 pound aluminum plate, sewn into the top of the bag, aids in stripping the bag from the canopy during deployment.

PILOT PARACHUTE ASSEMBLY

The pilot parachute assembly (Figure 12-7) decelerates the re-entry module and removes the rendezvous and recovery section from the landing module which results in the deployment of the main parachute. During flight, the pilot parachute assembly is stowed in the pilot mortar tube. The 18.3 foot diameter canopy is of the ringsail type having 16 gores and fabricated from 1.1 and 2.25 ounce per square yard nylon. Sixteen nylon cord suspension lines, which are 17 foot long and have a tensile strength of 550 pounds each, attach the canopy to the riser assembly. A 10.75 foot long split riser, constructed of four layers of 2600 pound tensile strength dacron webbing, holds the pilot parachute assembly to the rendezvous and recovery section of the spacecraft. When initially deployed, the pilot parachute is reefed to 11.5% in order to limit the opening shock load to 3000 pounds. Two pyrotechnic reefing cutters disreef the parachute 6 seconds after deployment. The pilot parachute remains attached to the rendezvous

SEDR 300

PROJECT GEMINI

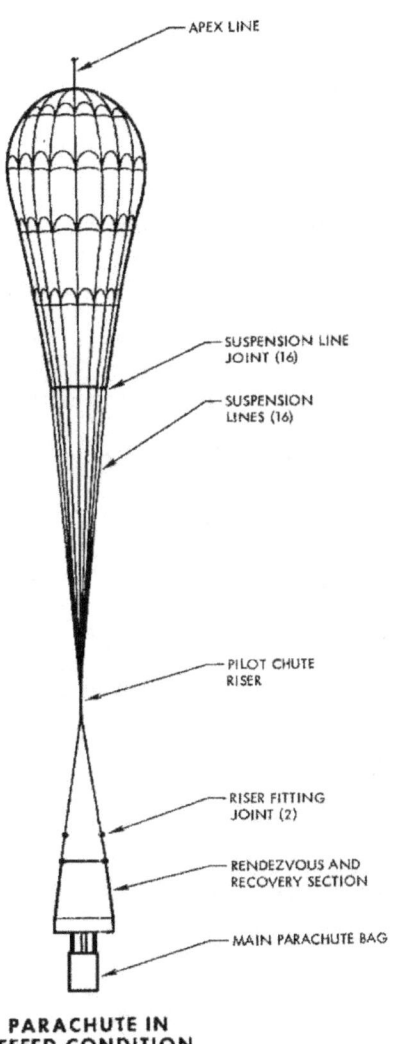
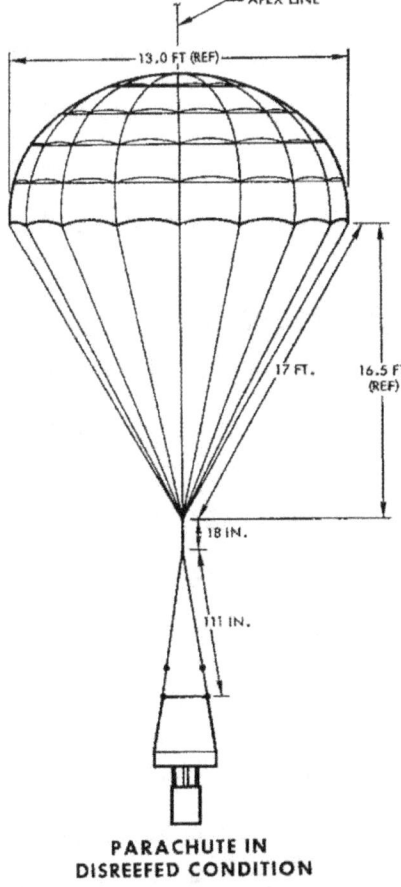

Figure 12-7 Pilot Parachute Assembly

and recovery section throughout the entire descent.

Pilot Parachute Mortar Assembly

The pilot parachute mortar assembly is similar in design and operation to the drogue parachute mortar assembly. During normal operation of the landing system, this assembly serves only to store and protect the pilot parachute. In the event of a failure in the deployment of the drogue parachute, the pilot parachute mortar can be activated to deploy the pilot parachute by initiation of the DROGUE EMERG 10.6 K switch. Actuation of the emergency drogue switch fires the three drogue cable guillotines, the apex line guillotine, and the pilot parachute mortar. After the pilot parachute has been deployed, the landing is completed through the normal sequence of events. Figure 12-5 illustrates the pilot parachute deployment.

Pilot Mortar Sabot

The pilot mortar sabot functions are the same as those of the drogue mortar sabot. Refer to the description of the Drogue Mortar Sabot.

Pilot Parachute Deployment Bag

The pilot parachute deployment bag is similar to the drogue parachute deployment bag in design and use, except for the bag handles attached to the apex line for extraction by the drogue parachute.

MAIN PARACHUTE AND RISER ASSEMBLY

The main parachute (Figure 12-8 and 12-9) is of the ringsail type with a diameter of 84.2 feet. The nylon canopy has seventy-two gores alternating in

SEDR 300

PROJECT GEMINI

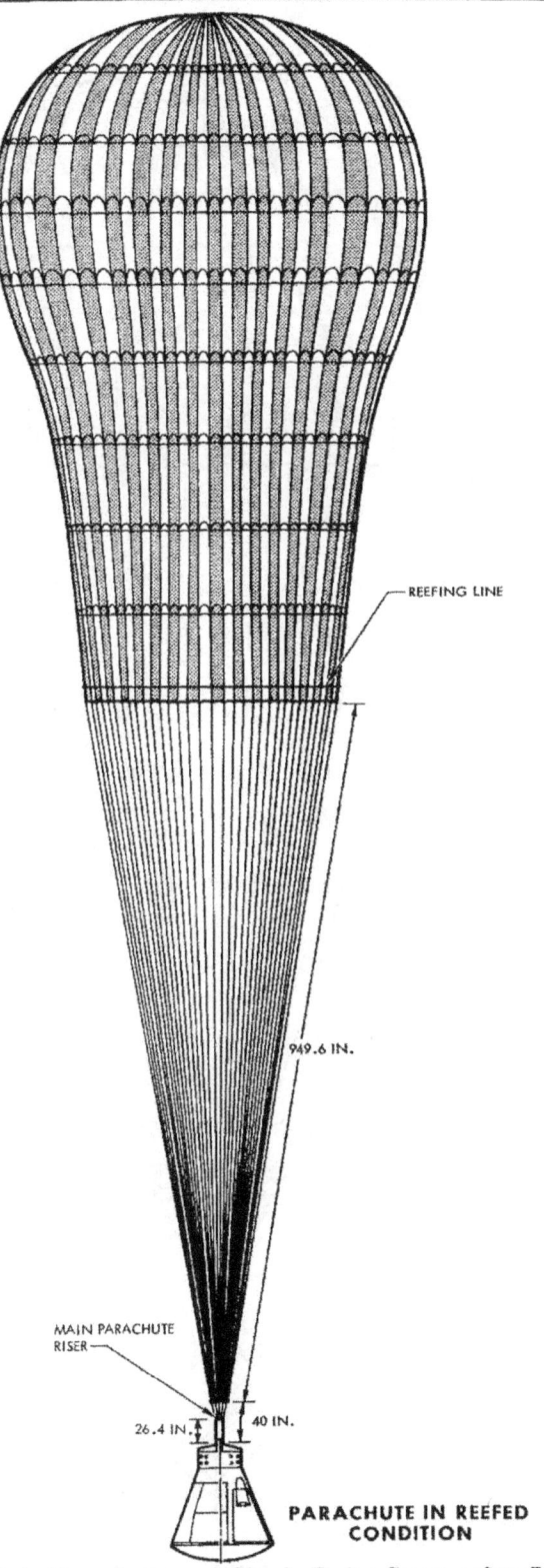

Figure 12-8 Main Parachute and Single Point Suspension System

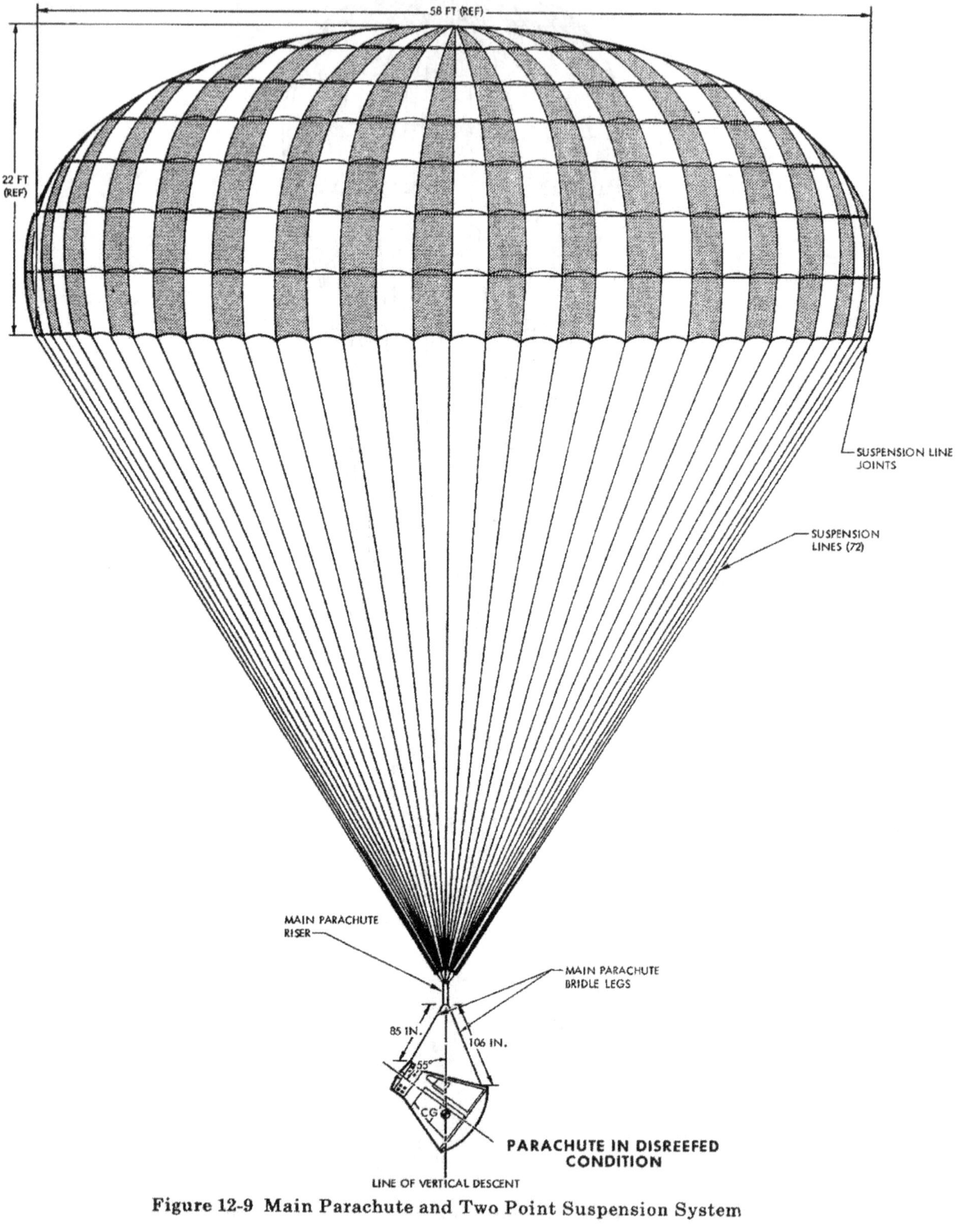

SEDR 300
PROJECT GEMINI

Figure 12-9 Main Parachute and Two Point Suspension System

12-16

PROJECT GEMINI

colors of international orange and white. Seventy-two suspension lines are attached to eight legs of a single integral riser. Each suspension line has a tensile strength of 550 pounds. The 3.25 foot integral riser consists of eight layers of 5,500 pound tensile strength nylon webbing. The canopy is fabricated from 1.1 and 2.25 ounce per square yard nylon and can withstand a dynamic pressure of 120 pounds per foot. However, by reefing the main parachute, a maximum load of 16,000 pounds is experienced at deployment. When initially deployed, the parachute is reefed to 10.5%. The disreefed main parachute allows a maximum average rate of descent of 31.6 feet per second for a module weight of 4,400 pounds.

Main Parachute Deployment Bag and Container Assembly

The main parachute deployment bag and container assembly (Figure 12-1) stows the main parachute. This assembly is located in the aft end of the rendezvous and recovery section of the spacecraft. The deployment bag is fabricated from a cotton sateen material reinforced with nylon webbing. In order to insure a full and orderly deployment of the main parachute, the suspension lines must be stretched out prior to the release of the canopy. Therefore, transverse locking flaps are incorporated in the bag to separate the canopy from the suspension lines. Four restraining straps hold the deployment bag in the container until deployment.

The main parachute container is 22.25 inches in diameter and 21.32 inches long. The container is closed on the forward end and is secured to the rendezvous and recovery section by four vertical reinforcing brackets. At deployment, the restraining straps of the deployment bag are unlocked, the risers and suspension

SEDR 300
PROJECT GEMINI

lines are extended, and the canopy is pulled from the deployment bag. The deployment bag remains attached to the container by four bag handles.

Main Parachute Bridle Assembly

The main parachute bridle assembly (Figure 12-9) provides a two point suspension system in order to achieve the optimum attitude for a water landing. Two separate bridle straps constitute the main parachute bridle assembly. The forward bridle strap is an 85 inch long nylon strap with a looped end connected to the forward bridle disconnect. Prior to single point release, the forward bridle is stowed in the bridle tray (Figure 12-10). The aft bridle is 106 inches long and connects to the aft disconnect which is located immediately forward of the single point hoist loop (Figure 12-10). Constructed of high temperature resistant nylon, the aft bridle is stowed in a trough that extends from the front of the RCS section to the aft disconnect during flight. An insulating cover shields the aft strap in the cable trough until the single point suspension is released, at which time the bridle leg tears through the insulation.

Main Parachute Release

Upon landing in the water, the main parachute is released from the landing module by activation of the PARA JETT switch. This initiates the forward and aft disconnect pyrotechnics and allows the chute to pull away from the landing module.

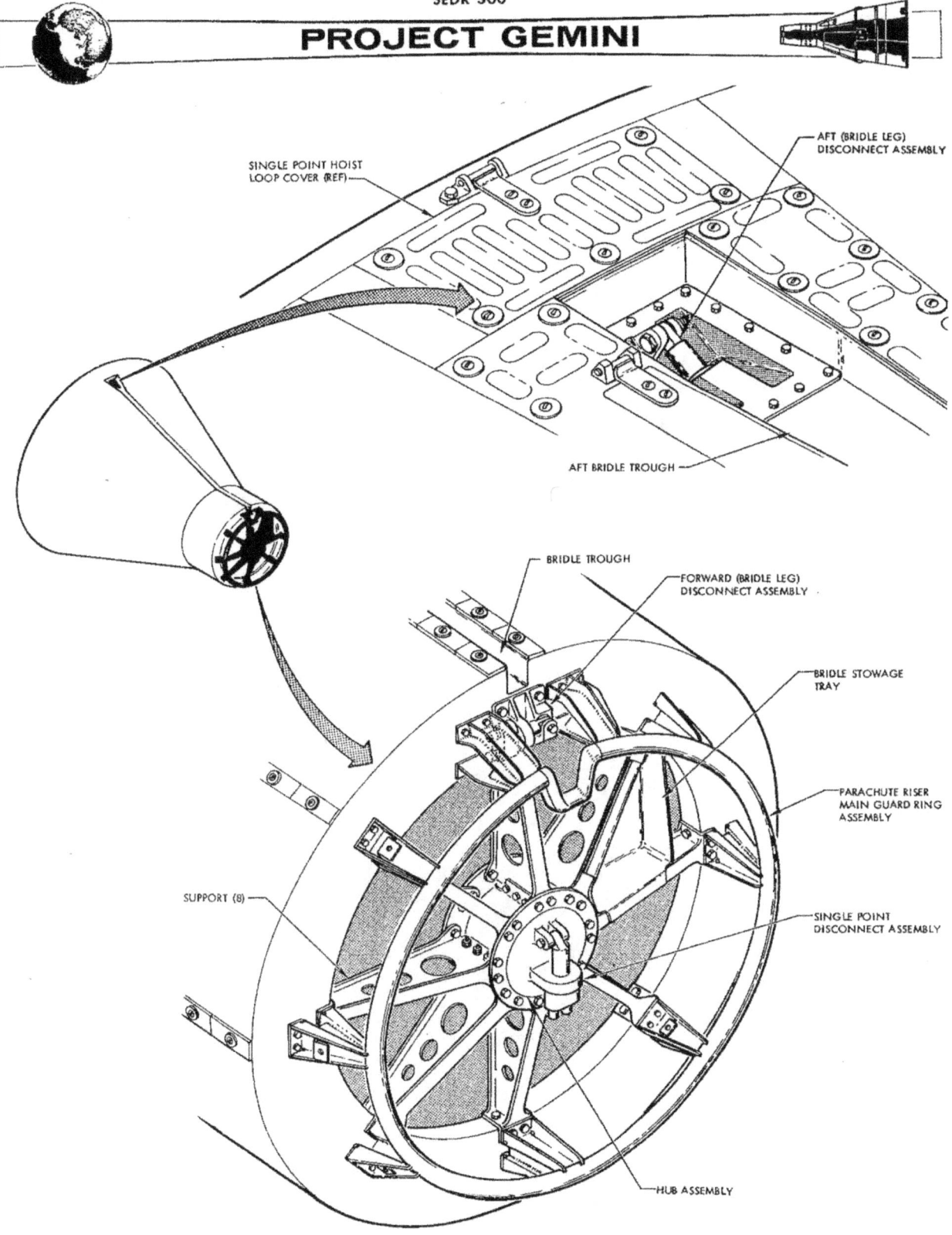

Figure 12-10 Main Parachute Support Assembly

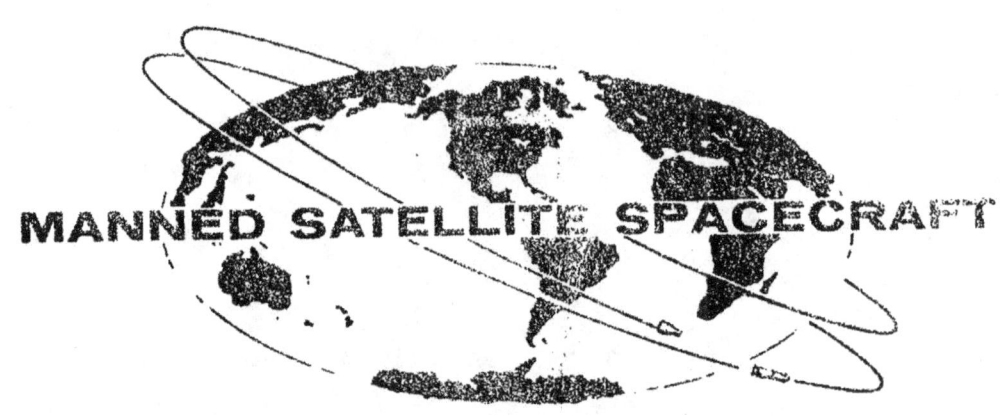

©2011 Periscope Film LLC All Rights Reserved   ISBN #978-1-935700-69-2

www.ingramcontent.com/pod-product-compliance
Lightning Source LLC
Chambersburg PA
CBHW081753300426
44116CB00014B/2107